ASTRONOMY AND
ASTROPHYSICS LIBRARY

T0210981

For further volumes:
http://www.springer.com/series/848

S.K. Saha

Aperture Synthesis

Methods and Applications to Optical
Astronomy

 Springer

Swapan Kumar Saha
Indian Institute of Astrophysics
Sarjapur Road
560034 Bangalore
IInd Block, Koramangala
India
sks@iiap.res.in

ISSN 0941-7834
ISBN 978-1-4614-2702-5   ISBN 978-1-4419-5710-8 (eBook)
DOI 10.1007/978-1-4419-5710-8
Springer New York Dordrecht Heidelberg London

Springer is part of Springer Science+Business Media (www.springer.com)

To my children, Snigdha and Saurabh

Preface

The angular resolution of a single aperture (telescope) is inadequate to measure the brightness distribution across most stellar sources and many other objects of astrophysical importance. A major advance involves the transition from observations with a single telescope to a diluted array of two or more telescopes separated by more than their own sizes, mimicking a wide aperture, having a diameter about the size of the largest separation. Such a technique, called aperture synthesis, provides greater resolution of images than is possible with a single member of the array.

Implementation of interferometry in optical astronomy began more than a century ago with the work of Fizeau (1868). Michelson and Pease (1921) measured successfully the angular diameter of Betelgeuse (α Orionis), using an interferometer based on two flat mirrors, which allowed them to measure the fringe visibility in the interference pattern formed by starlight at the detector plane. Later, Hanbury Brown and Twiss (1954) developed the intensity interferometry (see Sect. 3.3). Unlike Michelson (amplitude) interferometry, this does not rely on actual light interference. Instead, the mutual degree of coherence is obtained from the measurement of the degree of correlation between the intensity fluctuation of the signals recorded with a quadratic detector at two different telescopes. It measures the second-order spatial coherance, where the phase of the signals in separate telescopes was not required to be maintained. However, it ended with the Narrabri intensity interferometer (Hanbury Brown 1974) that was used to measure the diameter of bright stars and the orbit of binaries and was the first to measure the limb-darkening of a star other than the Sun. The survey of stellar diameters by means of this instrument serves as a resource for the effective temperature scale of main-sequence stars. Important results were obtained for the spectroscopic and eclipsing binaries as well.

Obtaining a diffraction-limited image of celestial bodies was one of the major problems faced by the optical astronomers in the past. This is mainly due to the image degradation at optical wavelengths produced by the atmospheric turbulence. Labeyrie (1970) developed speckle interferometry as one way to overcome the degradation due to atmospheric turbulence. Then technological advances overcame many of the problems encountered by Michelson and Pease (1921) allowing further development of phase-preserving optical interferometry, more nearly analogous to radio interferometry. Labeyrie (1975) developed a long baseline interferometer with two small optical telescopes and resolved several stars. This technique depends on the visibility of fringes produced by the amplitude interferences formed by the light collected by two telescopes allowing the measurement of stars much fainter than was possible with intensity interferometry using the same size telescopes.

Following the publication of the article entitled, 'Modern Optical Astronomy: Technology and Impact of Interferometry – Swapan K Saha, 2002, Reviews of Modern Physics, **74**, 551–600,' several astronomers, particularly M. K. Das Gupta, who along with R. C. Jennison and R. Hanbury Brown developed intensity interferometry in radio wavelengths, had requested me to write a monograph, for which I am indebted to. In fact, I had the opportunity to be associated with him during graduate school days and discussed at length on this topic. This monograph, a sequel to my earlier book entitled, 'Diffraction-limited Imaging with Large and Moderate Telescopes', 2007, World-Scientific, is a dossier of knowledge for every graduate student and researcher, who intend to embark on a field dedicated to the long baseline aperture synthesis. I have attempted to make this book self-contained by incorporating more than one hundred and fifty illustrations and tens of footnotes. This monograph addresses the basic principles of interferometric techniques, the current trend, motivation, methods, and path to future promise of true interferometry at optical and infrared wavelengths. Since the basic principle of aperture synthesis imaging in optical astronomy using interferometry is Fourier Optics, this topic along with several fundamental equations is also highlighted in the appendices.

The progress in the field of radio interferometry is exemplary. The success is primarily because of the possibility to preserve phase information for widely separated dishes by using very accurate clocks and time markers in the data streams. Though the principles of optical interferometry are essentially identical to those at radio wavelengths, accurate measurements are more difficult to make: (i) the irregularities in the Earth's atmosphere introduce variations in the path length that are large compared to the wavelength; (ii) it is difficult to achieve the required mechanical stability of the telescopes to obtain interference fringes at a wavelength of the order of 500 nm. The calibration of the instrumental phase is a formidable task; and (iii) the division of the photons incident on each telescope in an array of optical telescopes to estimate the mutual coherence function or the complex visibility over the different possible baselines in the array leads to serious signal-to-noise problems. Despite the differences in technology between radio and optical interferometers, a common characterization of source properties, such as source visibility is adequate to provide a qualitative and quantitative description of the response of a long baseline interferometer.

Optical interferometry is generally performed within the standard atmospheric spectral windows. It requires several optical functions such as spatial filtering, which allows determination of the Fourier transform of the brightness distribution at the spatial frequencies, photometric calibration, polarization control etc., but the practical limitations imposed on these measurements are severe. An instrument of this nature needs extreme accuracies to meet the demands of maintaining the optical pathlengths within the interferometer, constant to a fraction of a wavelength of light, which constrained Long Baseline Optical Interferometers (LBOI) to smaller baselines ($\sim$100 m); mostly they operate at longer wavelengths (in the near- and mid-IR bands). The practical considerations regarding extraction of the Fourier components became important to look at. The first chapter lays the foundation of the mathematical framework that is required to understand the theoretical basis for Fourier Optics, imaging systems, while the second and third chapters address the fundamentals of optical interferometry and its applications.

Speckle interferometry (see Sect. 4.2), a post-processing technique, has successfully uncovered details in the morphology of a range of astronomical objects, including the Sun, planets, asteroids, cool giants and supergiants. Fueled by the rapid advancement of technology such as computational, fabrication, and characterization, development on real time corrections of the atmospheric turbulence, called 'Adaptive Optics' (AO), has given a new dimension in this field (see Sect. 4.3). Combining with LBOI, it offers the best of both approaches and shows great promise for applications such as the search for exoplanets. At this point, it seems clear that interferometry and AO are complementary, and neither can reach its full potential without the other. The fourth chapter introduces the origin and problem of imaging through atmospheric turbulence, and the limitations imposed by the atmosphere on the performance of speckle imaging. Further, it deals with the AO system including discussions of wavefront compensation devices, wavefront sensors, control system etc.

Interferometric technique bloomed during the last few decades. The new generation interferometry with phased arrays of multiple large sub-apertures would provide large collecting areas and high spatial resolution simultaneously. Over the next decades or so, one may envisage the development of hypertelescope (see Sect. 7.5.2). With forthcoming many-aperture systems, interferometry is indeed expected to approach the snapshot imaging performance of putative giant telescopes, the size of which may in principle reach hundreds of kilometers in space. However, daunting technological hurdles may come in the way for implementing these projects. Chapters 5–7 elucidate the current state-of-the art of such arrays. The various types of interferometric applications, for example, astrometry, nulling (see Sect. 5.1.3), and imaging are also described. These applications entail specific problems concerning the type of telescopes that are to be used, beam transportation and recombination, delay-lines, atmospheric dispersion, polarization, coherencing and cophasing, calibration, and detecting fringes using modern sensors (Chap. 6). Proposed ground and space-based interferometry projects (see Sects. 7.5–7.7) are also discussed.

Image-processing is an art and an important subject as well. A power spectrum (second-order moment) analysis provides only the modulus of the Fourier transform of the object, whereas a bispectrum (third-order moment) analysis (see Sect. 8.2.2) yields the phase reconstruction. The latter method is useful for simulations involving a diluted aperture interferometry. Indeed, it is difficult to incorporate adaptive optics system in a hypertelescope. Observations may be carried out by speckle interferometry, using either a redundant or non-redundant many-element aperture. Deconvolution method can also be applied to imaging covering the methods spanning from simple linear deconvolution algorithms to complex non-linear algorithms.

Chapter 8 discusses the methodology of recovering visibility functions of stellar diameter, ratio of brightness of binary components etc., from the raw data obtained by means of interferometry. Various image restoration techniques are also presented with emphasis on the deconvolution methods used in aperture-synthesis mapping.

Many astrophysical problems, such as measuring the diameters and asymmetries of single stars, observing stars as extended and irregular objects with magnetic or thermal spots, flattened or distorted by rapid rotation, determining the orbits of multiple stars, and monitoring mass ejections in various spectral features as they flow towards their binary companions, resolving star-formation regions, distant galaxies, AGNs, need high angular resolution information. Although a relatively new field, the steady progress of interferometry has enabled scientists to obtain results from the area of stellar angular diameters with implications for emergent fluxes, effective temperatures, luminosities and structure of the stellar atmosphere, dust and gas envelopes, binary star orbits with impact on cluster distances and stellar masses, relative sizes of emission-line stars and emission region, stellar rotation, limb-darkening, and astrometry. With the recent interferometers, Very Large Telescope Interferometer (VLTI) in particular, disks around several Young Stellar Objects (YSO), a few debris disks, core of a Luminous Blue Variable (LBV) object and a nova, several Active Galactic Nuclei (AGN) have been resolved. Some of these results obtained by means of optical/IR interferometry are enumerated in chapter nine. Also, it contains discussions on the ability of these instruments to obtain information about the accretion disks, winds and jets, and luminosities of components in binary systems.

I am grateful to A. Labeyrie and V. Trimble for their encouragement and indebted to G. Weigelt, O. Absil, D. Mourard, R. Millan-Gabet, Luc Damé, J. D. Monnier, A. Domiciano de Souza, F. Malbet, P. Lawson, P. M. Hinz, J. P. Lancelot, P. Nisenson, V. Chinnappan, V. Coudé du Foresto, T. R. Bedding, O. Lardière, P. Stee, Ishwara Chandra, P. Hoeflich, D. Soltau, S. LeBohec, A. Subramaniam, S. Golden, D. Braun, D. Bonneau, K. E. Rangarajan, and J. Buckley for providing the images, plots, figures etc., and granting permission for their reproduction. Special thanks are due to R. Ramesh, S. Morel, F. Sutaria, V. Valsan, T. Berkefeld, K. R. Subramaniam, T. P. Prabhu, C. S. Stalin, G. C. Anupama, A. Satya Narayanan, S. P. Bagare, and P. R. Vishwanath for going through selected chapters. I express gratitude for the services rendered by B. A. Varghese, S. Arun, V. K. Subramaniam, R. K. Chaudhuri, and D. Takir as well.

Swapan K. Saha

Principal Symbols

$a(\mathbf{r})$	Complex amplitude of the wave
A_e	Effective area of an antenna
$\mathbf{B}$	Baseline vector
$B(\mathbf{u})$	Atmosphere transfer function
$B_v(T)$	Spectral radiancy
$\mathcal{B}_n(\mathbf{r})$	Covariance function
$\mathcal{C}_n^2$	Refractive index structure constant
$\mathcal{C}_T^2$	Temperature structure constant
$\mathcal{C}_v^2$	Velocity structure constant
D	Diameter of the aperture
$\mathcal{D}_n(\mathbf{r})$	Refractive index structure function
$\mathcal{D}_T(\mathbf{r})$	Temperature structure function
$\mathcal{D}_v(\mathbf{r})$	Velocity structure function
$G(\theta, \phi)$	Antenna gain
H_0	Hubble constant
I	Intensity of light
$\widehat{I}(\mathbf{u})$	Image spectrum
I_v	Specific intensity
j	$= 1, 2, 3$
$J(\mathbf{r}_1, \mathbf{r}_2)$	Mutual intensity function
$\mathcal{J}_{12}$	Interference term
l	Characteristic size of viscous fluid
l_c	Coherence length
l_0	Inner scale length
$\mathcal{L}_\star$	Stellar luminosity
m_v	Apparent visual magnitude
M_v	Absolute visual magnitude

$M_\star$	Stellar mass
$n(\mathbf{r}, t)$	Refractive index of the atmosphere
$\widehat{N}(\mathbf{u})$	Noise spectrum
$O(\mathbf{x})$	Object illumination
$\widehat{O}(\mathbf{u})$	Object spectrum
P	Pressure
$P(\theta, \phi)$	Antenna power pattern
$P(\mathbf{x})$	Pupil transmission function
$\widehat{P}(\mathbf{u})$	Pupil transfer function
$\mathcal{R}$	Resolving power of an optical system
$\mathbf{r}(= x, y, z)$	Position vector of a point in space
R_e	Reynolds number
$\Re$ and $\Im$	Real and imaginary parts of the quantities in brackets
r_0	Fried's parameter
$R_\star$	Stellar radius
$\hat{\mathbf{s}}$	Unit vector
$S(\mathbf{x})$	Point Spread Function
$\langle \widehat{S}(\mathbf{u}) \rangle$	Transfer function for long-exposure image
S_r	Strehl ratio
$\widehat{S}(\mathbf{u})$	Optical Transfer Function
t	Time
T	Period
$T_a(\theta, \phi)$	Antenna temperature
$T_b(\theta, \phi)$	Brightness temperature
$\mathbf{u}$	Spatial frequency vector
$U(\mathbf{r}, t)$	Complex representation of the analytical signal
$V(\mathbf{r}, t)$	Monochromatic optical wave
v_a	Average velocity of a viscous fluid
$\mathcal{V}$	Visibility
$\mathbf{x} = (x, y)$	Two-dimensional space vector
$\gamma(\mathbf{r}_1, \mathbf{r}_2, \tau)$	Complex degree of (mutual) coherence
$\Gamma(\mathbf{r}_1, \mathbf{r}_2, \tau)$	Mutual coherence
$\Gamma(\mathbf{r}, \tau)$	Self coherence
δ	Phase difference
ε	Energy dissipation
(θ, ϕ)	Polar coordinates
κ	Wave number
λ	Wavelength
λ_0	Wavelength in vacuum

$\mu(\mathbf{r}_1, \mathbf{r}_2)$	Complex coherence factor
ν	Frequency
$\Delta\nu$	Spectral width
$\langle\sigma\rangle$	Standard deviation
$\langle\sigma\rangle^2$	Variance
τ_0	Atmospheric coherence time
τ_c	Coherence time
$\Phi_n(\mathbf{k})$	Power spectral density
$\Delta\varphi$	Optical path difference
Ψ	Time-dependent wave-function
ω	Angular frequency
$*$	Complex operator
$\star$	Convolution operator
$\otimes$	Correlation
$\langle\ \rangle$	Ensemble average
$\widehat{}$	Fourier transform operator
∇	Linear vector differential operator
∇^2	Laplacian operator

Some Numerical values of Physical and Astronomical Constants

c	Speed of light	3×10^8 m/s
G	Gravitational constant	6.674×10^{-11} N.m^2/kg^2
h	Planck's constant	6.626196×10^{-34} J.s
k_B	Boltzmann's constant	1.380662×10^{-23} J/K
$\mathcal{L}_\odot$	Solar luminosity	3.839×10^{26} W
$M_\odot$	Solar mass	1.9889×10^{30} kg
$R_\odot$	Solar radius	6.96×10^8 m
$T_\odot$	Solar effective temperature	$5780°$ K
ϵ_0	Permittivity constant	8.8541×10^{-12} F/m
μ_0	Permeability constant	1.26×10^{-6} H/m
σ	Stefan–Boltzmann's constant	5.67×10^{-8} W m^{-2} K^{-4}

List of Acronyms

ACT	Atmospheric Cerenkov Telescope
AGB	Asymptotic Giant Branch
AGN	Active Galactic Nuclei
AMBER	Astronomical Multiple BEam Recombiner
BID	Blind Iterative Deconvolution
BLR	Broad-Line Region
CHARA	Center for High Angular Resolution Astronomy
CMBR	Cosmic Microwave Background Radiation
COAST	Cambridge Optical Aperture Synthesis Telescope
ESA	European Space Agency
ESO	European Southern Observatory
FLUOR	Fiber-Linked Unit for Optical Recombination
FINITO	Fringe-tracking Instrument of Nice and Torino
FSU	Fringe Sensor Unit
GI2T	Grand Interféromètre à deux Télescopes
GMRT	Giant Meterwave Radio Telescope
HR	Hertzsprung–Russell
HST	Hubble Space Telescope
IAU	International Astronomical Union
IMF	Initial Mass Function
IO	Integrated Optics
IOTA	Infrared Optical Telescope Array
IRAS	InfraRed Astronomical Satellite
ISI	Infrared Spatial Interferometer
ISM	InterStellar Medium
I2T	Interféromètre à deux Télescopes
IUE	International Ultraviolet Explorer
KT	Knox–Thomson
laser	Light Amplification by Stimulated Emission of Radiation
LBOI	Long Baseline Optical Interferometry

LBT	Large Binocular Telescope
IUE	International Ultraviolet Explorer
KT	Knox–Thomson
laser	Light Amplification by Stimulated Emission of Radiation
LBOI	Long Baseline Optical Interferometry
LBT	Large Binocular Telescope
LBV	Luminous Blue Variable
LD	Limb-Darkened
LIGO	Laser Interferometer Gravitational-Wave Observatory
LISA	Laser Interferometer Space Antenna
LPV	Long-Period Variables
mas	milliarcseconds
MCAO	Multi-Conjugate Adaptive Optics
MEM	Maximum Entropy Method
MIDI	MID-Infrared Interferometric Instrument
MMT	Multi Mirror Telescope
MROI	Magdalene Ridge Observatory Interferometer
MTF	Modulation Transfer Function
NASA	National Aeronautics and Space Administration
NLR	Narrow-Line Region
NPOI	Navy Prototype Optical Interferometer
NRAO	National Radio Astronomy Observatory
OPD	Optical Path Difference
OTF	Optical Transfer Function
OVLA	Optical Very Large Array
pc	Parsec
PMS	Pre-Main Sequence
PN	Planetary Nebula
PRIMA	Phase-Referenced Imaging & Microarcsecond Astrometry
PSF	Point Spread Function
PTI	Palomar Testbed Interferometer
PTF	Pupil Transmission Function
QUASAR	QUASi-stellAR radio source
RAFT	Real time Active Fringe Tracking
REGAIN	REcombineur pour GrAnd INterféromètre
SAO	Special Astrophysical Observatory
SIM	Space Interferometry Mission
SKA	Square Kilometer Array
SMBH	Super Massive Black Holes
SN	Supernova
SoHO	Solar and Heliospheric Observatory

SOIRDÉTÉ	Synthèse d'Ouverture en Infra Rouge avec DEux TElescopes
SUSI	Sydney University Stellar Interferometer
TC	Triple-Correlation
TPF	Terrestrial Planet Finder
UD	Uniform Disk
VBO	Vainu Bappu Observatory
VEGA	Visible spEctroGraph and polArimeter
VINCI	VLT INterferometer Commissioning Instrument
VLA	Very Large Array
VLBI	Very Long Baseline Interferometry
VLTI	Very Large Telescope Interferometer
VSI	VLTI Spectro-Imager
WR	Wolf–Rayet
YSO	Young Stellar Object

Contents

Chapter 1
Introduction to Wave Optics

1.1 Preamble

Light is an electromagnetic wave propagating as a disturbance in the electric and
magnetic fields. These fields continually generate each other, as the wave propagates
through space and oscillates in time. The Maxwell equations give rise to the wave
equation that enumerates the propagation of electromagnetic waves. In free space,
the propagation of electromagnetic waves is expressed as (Jackson 1999),

$$\nabla^2 \mathbf{E}(\mathbf{r}, t) - \frac{1}{c^2} \frac{\partial^2 \mathbf{E}(\mathbf{r}, t)}{\partial t^2} = 0, \tag{1.1}$$

with

$$\nabla^2 = \frac{\partial^2}{\partial x^2} + \frac{\partial^2}{\partial y^2} + \frac{\partial^2}{\partial z^2}, \tag{1.2}$$

as the Laplacian operator with respect to the Cartesian rectangular coordinates,
$c = 1/\sqrt{\mu_0 \epsilon_0} \approx 3 \times 10^8$ meter (m)/second (s) the speed of light, $\mu_0 (= 1.26 \times 10^{-6}$ H/m) the permeability in free space or in vacuum, $\epsilon_0 (= 8.8541 \times 10^{-12}$ F/m)
the permittivity in vacuum, and $\mathbf{r} = x\mathbf{i} + y\mathbf{j} + z\mathbf{k}$ the position vector.

The solution of Maxwell's equation in free space yields plane-wave solutions:

$$\mathbf{E}(\mathbf{r}, t) = \mathbf{E}_0(\mathbf{r}, \omega) e^{i(\kappa \cdot \mathbf{r} - \omega t)}, \tag{1.3}$$

where $\mathbf{E}_0(\mathbf{r}, \omega)$ is the amplitude (maximum displacement of the wave in either direction from the mean value) of the electric field vectors, $\omega(= 2\pi\nu)$ is the angular
frequency, $\nu = 1/T$ represents the number of oscillations cycles in a unit time, and
T the period of motion and

$$\kappa \cdot \mathbf{r} = \kappa_x x + \kappa_y y + \kappa_z z, \tag{1.4}$$

represents planes in space of constant phase. The Cartesian components of the wave
travel with the same propagation vector, $\kappa = \kappa_x \mathbf{i} + \kappa_y \mathbf{j} + \kappa_z \mathbf{k}$, which provide the
direction of propagation.

S.K. Saha, *Aperture Synthesis*, Astronomy and Astrophysics Library,
DOI 10.1007/978-1-4419-5710-8_1, © Springer Science+Business Media, LLC 2011

The optical field is described in terms of energy, called irradiance (intensity). By taking time average[1] of the quadratic field components over an interval, which is much greater than the time period, $T = 2\pi/\omega$, the irradiance is derived as,

$$I \propto \langle E^2 \rangle = \lim_{T \to \infty} \frac{1}{2T} \int_{-T}^{T} E^2 dt, \tag{1.5}$$

where the angular brackets $\langle .. \rangle$ stands for the time average of the quantity, or in the phasor picture, by

$$I \propto |E|^2$$
$$= E E^* = E^* E. \tag{1.6}$$

The quantity within the sharp brackets is due to the assumed ergodicity[2] of the field. The unit of intensity is expressed as the joule per square meter per second, $(\mathrm{J\,m^{-2}s^{-1}})$ or watt per square meter, $(\mathrm{W\,m^{-2}})$.

The advancement of the theory of light progressed rapidly after the initiation of quantum theory (Planck 1901), and particularly with the statistical interpretation[3] of quantum mechanics introduced by Born (1926). It is the study of the properties of random light. The randomness of photons in fluctuating light fields, which may have different frequencies, arises because of unpredictable fluctuations of the light source or of the medium through which light propagates. Randomness in light may also be generated by scattering from a rough surface, or turbulent fluids imparting random variations to the optical wavefront. In what follows, the fundamentals of wave optics and polarization, diffraction, and image formation are elucidated in brief.

---

[1] The time average over a time that is large compared with the inverse frequency of the product of the two harmonic time-independent functions **a** and **b**, of the same frequency is given by,

$$\langle \mathbf{a}(t) \cdot \mathbf{b}(t) \rangle = \frac{1}{T} \int_0^T \frac{1}{4} \left[\mathbf{a} e^{i\omega t} + \mathbf{a}^* e^{-i\omega t} \right] \cdot \left[\mathbf{b} e^{i\omega t} + \mathbf{b}^* e^{-i\omega t} \right] dt = \frac{1}{2} \Re \left(\mathbf{a} \cdot \mathbf{b}^* \right).$$

[2] Ergodicity implies that each ensemble average is equal to the corresponding time average involving a typical member of the ensemble, while the stationary field implies that all the ensemble averages are independent of the origin of time.

[3] Born (1926) formulated the now-standard interpretation of the probability density function for $\Psi \Psi^*$ in the Schrödinger equation of quantum mechanics (Schrödinger 1926). In quantum mechanics, a probability amplitude is, in general, a complex number whose modulus squared represents a probability or probability density. For example, the values taken by a normalized wave function Ψ are amplitudes, since $\Psi \Psi^* = |\Psi(x)|^2$ provides the probability density at position x. Probability amplitudes, defined as complex-number-valued function of position, may also correspond to probabilities of discrete outcomes. It is a quantity whose value is a definite complex number at any point in space. The probability of finding the particle described by the wave function (e.g., an electron in an atom) at that point is proportional to square of the absolute value of the probability amplitude. That the physical meaning of the wave function is probabilistic was also proposed by Born.

1.2 Complex Representation of Harmonic Waves

A harmonic plane wave represents a wave field spread out periodically in space and time. The notable features are:

- The harmonic variations of the electric and magnetic fields are always perpendicular to each other and to the direction of propagation, κ.
- The field always vary sinusoidally, and vary with the same frequency and are in phase with each other.
- The plane waves described by the wave equation (1.1) in vacuum, are transverse, which implies that the vectors, $\mathbf{E}$ and $\mathbf{B}$ oscillate in a plane perpendicular to the wave number vector, κ. The cross product $\mathbf{E} \times \mathbf{B}$ provides the direction of travel (see Fig. 1.1), in which the unit of the electric field intensity, $\mathbf{E}$, is volt $(V)m^{-1}$, and that for the magnetic flux density $|\mathbf{B}|$, tesla $(T = Wbm^{-2})$.

In a region of an homogeneous medium free of currents and charges, each rectangular component $V(\mathbf{r}, t)$ of the field vectors, obeys the homogeneous wave equation (1.1), In case of solutions representing spherical waves, with the assumption that the function $V(\mathbf{r}, t)$ has spherical symmetry about the origin, i.e., $V(\mathbf{r}, t) = V(r, t)$, where $r = |\mathbf{r}| = \sqrt{x^2 + y^2 + z^2}$ and

$$x = r \sin \theta \cos \phi, \qquad y = r \sin \theta \sin \phi, \qquad z = r \cos \theta. \qquad (1.7)$$

The Laplacian operator for spherical coordinates reads,

$$\nabla^2 = \frac{1}{r^2} \frac{\partial}{\partial r} \left(r^2 \frac{\partial}{\partial r} \right) + \frac{1}{r^2 \sin \theta} \frac{\partial}{\partial \theta} \left(\sin \theta \frac{\partial}{\partial \theta} \right) + \frac{1}{r^2 \sin^2 \theta} \frac{\partial^2}{\partial \phi^2}. \qquad (1.8)$$

Since the spherical wave is spherically symmetric with no dependence on θ and ϕ, the Laplacian operator satisfies the first term of the right hand side of (1.8). An outgoing spherical wave is obtained as,

$$V(r, t) = \frac{a}{r} \cos \left[\omega \left(t - \frac{r}{v} \right) + \psi \right]. \qquad (1.9)$$

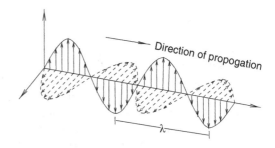

Fig. 1.1 Propagation of a plane wave; the *solid and dashed lines* represent, respectively, the electric and magnetic fields

The (1.9) represents a monochromatic wave, where the amplitude of vibration at any point is constant and the phase varies linearly with time. The amplitude of the spherical wave at a distance $\mathbf{r}$ falls off as $1/\mathbf{r}$, while the irradiance is proportional to the square of the amplitude.

The manipulation of trigonometric function is a difficult task, it is useful to express a wave in complex form. A general time harmonic wave of frequency ω may be defined from the real solution of the wave equation, $V(r,t)$ at a point $\mathbf{r}$ by,

$$V(r,t) = \Re\left\{a(\mathbf{r},v)e^{-i[2\pi v_0 t - \psi(\mathbf{r})]}\right\}, \tag{1.10}$$

$a(\mathbf{r})$ is the amplitude of the wave, t the time, v_0 the frequency of the wave, and $\psi(\mathbf{r})$ the phase functions.

The oscillations of V in (1.10) are bounded by $0 \le |V| \le a$ or, $a \le V \le a$. The physically relevant information is embodied in the relative phase differences among superimposed waves and the relative amplitude ratios. Therefore, one may drop the $\Re$ symbol. This information is encoded in the complex exponential representation by,

$$A(\mathbf{r}) = a(\mathbf{r})e^{i\psi(\mathbf{r})}, \tag{1.11}$$

the complex representation of the analytic signal of a plane wave, $U(r,t)$ becomes,

$$U(r,t) = a(\mathbf{r})e^{-i[2\pi v_0 t - \psi(\mathbf{r})]} \tag{1.12a}$$
$$= A(\mathbf{r})e^{-i2\pi v_0 t}. \tag{1.12b}$$

This complex representation is preferred for linear time invariant systems, because the eigenfunctions of such systems are of the form $e^{-i\omega t}$. The complex representation of the analytic signal of a spherical wave is represented by,

$$U(r,t) = \left[\frac{a(\mathbf{r})}{r}\right]e^{-i[2\pi v_0 t - \psi(\mathbf{r})]}. \tag{1.13}$$

The complex amplitude is a constant phasor in the monochromatic (radiation of single precise energy) case. Therefore, the Fourier transform (FT) of the complex representation of the signal, $U(r,t)$, is given by,

$$\widehat{U}(r,v) = a(\mathbf{r})e^{i\psi}\delta(v - v_0). \tag{1.14}$$

It is equal to twice the positive part of the instantaneous spectrum, $\widehat{V}(r,v)$; here, $\delta(v - v_0)$ is the Dirac delta function.

Unlike monochromatic wave, the amplitude and phase in the case of a quasi-monochromatic wave field undergo irregular fluctuations (Born and Wolf 1984). The fluctuations arise since the real valued wave field, $U^{(r)}$, consists of a large number of contributions independent of each other. Their superposition yields to a fluctuating field, which can be described in statistical terms. Considering $V(\mathbf{r},t)$

as a superposition of monochromatic waves of different frequencies, which may be expressed in the form of a Fourier integral,

$$V(\mathbf{r}, t) = \int_0^\infty a(\mathbf{r}, \nu) \cos[\psi(\mathbf{r}, \nu) - 2\pi \nu t] d\nu$$

$$= \Re \int_0^\infty a(\mathbf{r}, \nu) e^{-i[2\pi \nu t - \psi(\mathbf{r}, \nu)]} d\nu, \tag{1.15}$$

where $a(\mathbf{r}, \nu)$ and $\psi(\mathbf{r}, \nu)$ modulo (2π) are real functions of each monochromatic component of frequency ν.

The (1.15) is the Fourier cosine integral representation of the real valued signal $U^{(r)}(\mathbf{r}, t)$. Invoking Euler's formula, one derives the complex analytic signal $U(\mathbf{r}, t)$ associated with the real function, $U^{(r)}(\mathbf{r}, t)$ as,

$$U(\mathbf{r}, t) = U^{(r)}(\mathbf{r}, t) + i U^{(i)}(\mathbf{r}, t), \tag{1.16}$$

in which the superscript $^{(i)}$ denotes the imaginary function.

For quasi-monochromatic waves, the wavelength range $\Delta\lambda$ is small compared to the mean wavelength, $\bar{\lambda}$, i.e., $\Delta\lambda/\bar{\lambda} \ll 1$. In most applications, the spectral amplitudes have appreciable values in a frequency interval of width $\Delta\nu = \nu - \bar{\nu}$ which is small compared to the mean frequency $\bar{\nu}$. The analytic signal can be expressed in the form,

$$A(\mathbf{r}, t) e^{i\psi(\mathbf{r}, t)} = 2 \int_0^\infty \hat{U}(\mathbf{r}, \nu) e^{-i 2\pi(\nu - \bar{\nu})t} d\nu. \tag{1.17}$$

This phasor is time dependent, although it varies slowly with respect to the variations of exponential frequency term $e^{-i2\pi\bar{\nu}t}$. In terms of A and ψ, one may write,

$$U^{(r)}(\mathbf{r}, t) = A(\mathbf{r}, t) \cos[\psi(\mathbf{r}, t) - 2\pi\bar{\nu}t]. \tag{1.18}$$

While dealing with stationary random processes, it is convenient to define $U(\mathbf{r}, t)$ for all values of t, but in reality, observations are carried out over some finite time $-T \le t \le T$. Following (1.5), the time average of the intensity yields a finite value as the averaging interval is increased indefinitely, i.e.,

$$\left\langle \left| U^{(r)}(\mathbf{r}, t) \right|^2 \right\rangle = \lim_{T \to \infty} \frac{1}{2T} \int_{-T}^{T} \left| U^{(r)}(\mathbf{r}, t) \right|^2 dt. \tag{1.19}$$

If the integral (1.19) approaches a finite limit when $T \to \infty$, the integral $\int_{-\infty}^\infty |U^{(r)}(t)|^2 dt$ becomes divergent. The truncated functions can be analyzed by using Fourier method; hence,

$$U_T^{(r)}(\mathbf{r}, t) = \begin{cases} U^{(r)}(\mathbf{r}, t) & \text{when } |t| \le T, \\ 0 & \text{when } |t| > T, \end{cases} \tag{1.20}$$

with T as some long time interval.

The conjugate functions, $U_T^{(r)}(t)$ and $U_T^{(i)}(t)$ can be defined by,

$$U_T^{(i)}(t) = -\frac{1}{\pi} \int_{-\infty}^{\infty} \frac{U^{(r)}(\tau)d\tau}{\tau - t},$$

where the principal value of the integral is to be considered.

Each truncated function is assumed to be square integrable in the form of a Fourier integral (see Appendix B). The Fourier integral pair are given as,

$$U_T^{(r)}(\mathbf{r}, t) = \int_{-\infty}^{\infty} \widehat{U}_T^{(r)}(\mathbf{r}, v)e^{-i2\pi vt}dv, \tag{1.21a}$$

$$\widehat{U}_T^{(r)}(\mathbf{r}, v) = \int_{-\infty}^{\infty} U_T^{(r)}(\mathbf{r}, t)e^{i2\pi vt}dt. \tag{1.21b}$$

Let $U_T^{(i)}$ be the associated function and U_T, the corresponding analytic signal, i.e.,

$$U_T(\mathbf{r}, t) = U_T^{(r)}(\mathbf{r}, t) + iU_T^{(i)}(\mathbf{r}, t)$$

$$= 2\int_0^{\infty} \widehat{U}_T^{(r)}(\mathbf{r}, v)e^{-i2\pi vt}dv. \tag{1.22}$$

The time average of the intensity is expressed as,

$$\left\langle \left|U_T^{(r)}(\mathbf{r}, t)\right|^2 \right\rangle = \left\langle \left|U_T^{(i)}(\mathbf{r}, t)\right|^2 \right\rangle = \frac{1}{2}\langle U_T(\mathbf{r}, t)U_T^*(\mathbf{r}, t)\rangle$$

$$= 2\int_0^{\infty} \widehat{\Gamma}(\mathbf{r}, v)dv. \tag{1.23}$$

The function $\widehat{\Gamma}(\mathbf{r}, v)$ is the contribution to the light intensity made by all the frequency components in range $(v, v + dv)$, called the 'power spectrum' of the random process characterized by the ensemble of the function $U^{(r)}(t)$ and is also referred as the spectral density of the light vibrations.

1.3 Polarized Waves

Polarization is a state in which rays of light exhibit different properties in different directions. Polarimetry is a key technique in stellar physics, although instrumental polarization may limit its performance. The polarization properties carry interesting physical information. Polarimetry has yielded important results; for example, characterizing the atmospheres and shells of red giants/supergiants (Beiging et al. 2006), modeling the envelopes of Asymptotic Giant Branch (AGB) stars (Gledhill 2005), studying the morphology of Be stars (Wisniewski et al. 2007), and monitoring the short- and long-term behavior of Active Galactic Nuclei (AGN; Moran 2007).

Light can be polarized under natural conditions if the incident light strikes a surface at an angle equal to the polarizing angle of that media. The modified incident polarization caused by the reflection of a mirror is characterized by two parameters:

1. the ratio between the reflection coefficients of the electric vector components which are perpendicular and parallel to the plane of incidence, known as s and p components, respectively, and
2. the relative phase-shift between these electric vibrations.

The polarization of the wave characterizes how the direction of the electric field vector varies at a given point in space as a function of time. If the direction of vibration remains the same with time, the wave is linearly polarized or plane polarized in that direction. If the direction of vibration rotates at the same frequency as the wave, the wave is said to be circularly polarized. Intermediate states are called partially polarized. The amount of order is specified by the degree of polarization.

The general form of polarization is elliptical in which the end points of the instantaneous electric vectors lie on an ellipse (see Fig. 1.2). As the monochromatic wave propagates through space in a (x, y)-plane perpendicular to the propagation of light in z-direction, the end point of the electric vector at a fixed point traces out an ellipse. The shape of the ellipse changes continuously. When the ellipse maintains a constant orientation, ellipticity, and sense in the ellipse, the wave is said to be completely polarized at that point. Let $E_x^{(r)}$, $E_y^{(r)}$ denote the real orthogonal components of the complex electric field vector,

$$E_x^{(r)} = a_x \cos(\psi_1 - \omega t), \tag{1.24}$$

$$E_y^{(r)} = a_y \cos(\psi_2 - \omega t), \tag{1.25}$$

where a_x and a_y are the instantaneous amplitudes along the x and y axes, respectively, ψ_1 and ψ_2 the respective instantaneous phases at a fixed point in space as a function of time.

These signals (1.24, 1.25) fluctuate slowly in comparison with the cosine term at optical frequencies. Figure 1.3 depicts the polarization ellipse circumscribed in a rectangle. The sides of Fig. 1.3a are parallel to the x and y axes in which the angle between the diagonal and the x-axis is γ. The propagation is in the z-direction. Since

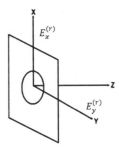

Fig. 1.2 Concept of polarization ellipse

Fig. 1.3 Description
of polarization ellipse (a)
in terms of x, y and (b)
in terms of x', y' coordinates

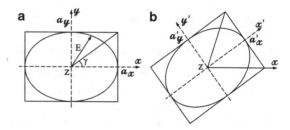

the field is transverse, the x and y components of this electric field are different from
zero. The equation of the trajectory parametrized by (1.24 and 1.25) is obtained by
eliminating ωt,

$$\left(\frac{E_x^{(r)}}{a_x}\right)^2 + \left(\frac{E_y^{(r)}}{a_y}\right)^2 - 2\frac{E_x^{(r)}}{a_x}\frac{E_y^{(r)}}{a_y}\cos\delta = \sin^2\delta, \qquad (1.26)$$

where $\delta = \psi_2 - \psi_1$, $-\pi < \delta \le \pi$, is the phase difference between the orthogonal
components, E_x and E_y.

The (1.26) is an expression of the polarization ellipse of the electric field for a
monochromatic light in which the amplitudes and phases are constant. The cross-
term, $E_x^{(r)}E_y^{(r)}$, implies that the polarization ellipse of the electric field rotates
through an angle θ. The effect of the angle δ on the polarization ellipse is shown
in Fig. 1.4. For $\delta = 0$, the field components E_x and E_y are in phase and the polarization ellipse is reduced to a segment of straight line, known as linearly polarized
(see Fig. 1.4); with $\delta = \pi$, one gets again linear polarization. For $0 < \delta < \pi$, the polarization ellipse is traced with a left hand sense, while for $-\pi < \delta < 0$, is traced
with right hand sense. If the magnitudes of a_x and a_y are equal, but exhibit a phase
difference of $\delta = \pm\pi/2$, the major and minor axes of the ellipse traced by the instantaneous electric vectors coincide with the x- and y-axes. Such a state is said to
be circularly polarized (see Fig. 1.4). When δ has any value other than the afore-
mentioned values, the resultant electric vector traces an ellipse in $x - y$ plane with
the major-axis arbitrarily inclined to the x-axis. This state is called an elliptical polarization.

The relative amplitude of the components E_x and E_y of the field is described by
the angle γ, which is given by,

$$\tan\gamma = \frac{a_y}{a_x} \qquad 0 < \gamma < \frac{\pi}{2}. \qquad (1.27)$$

The shape and the orientation of the polarization ellipse and the senses in which
the said ellipse is traced out are determined by the angles γ and δ. The size of the
ellipse depends on the amplitude of the electric field.

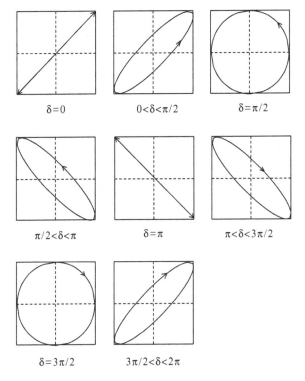

Fig. 1.4 Different Lissajous representations of polarization ellipses for various values of the phase difference δ

1.3.1 Stokes Parameters

The state of polarization of a beam of light is described by four measurable quantities that specify the intensity of the field, the degree of polarization, the plane of polarization and the ellipticity of the radiation at each point and in any given direction. These quantities, known as the Stokes parameters, were introduced by Stokes (1852). These parameters are suitable to deal with partial polarization.

At a particular point in space, the four instantaneous Stokes parameters, are defined by the realizations at time t of the following combinations of the complex analytical signals, $E_x(t)$ and $E_y(t)$. Let the incoming plane wave $\mathbf{E}$ be split into its two orthogonal components E_x and E_y and in the case of monochromatic wave, by using the Stokes parameters (Van de Hulst 1957), the state of polarization of a beam of light is described as,

$$I = E_x E_x^* + E_y E_y^* = a_x^2 + a_y^2, \tag{1.28a}$$

$$Q = E_x E_x^* - E_y E_y^* = a_x^2 - a_y^2, \tag{1.28b}$$

$$U = E_x E_y^* + E_x^* E_y = 2a_x a_y \cos \delta, \tag{1.28c}$$

$$V = i(E_x E_y^* - E_x^* E_y) = 2a_x a_y \sin \delta, \tag{1.28d}$$

where δ the differential phase shift between the two orthogonally linearly polarized components of the optical field, the asterisk denotes the complex conjugate, and $i = \sqrt{-1}$.

These four quantities are used to describe the polarization properties of an electromagnetic wave. The intensity of the field is described by I. The parameters, Q, U, and V together provide the degree of polarization and the characteristics of the polarization ellipse. If the ellipse is imagined to be a combination of a linear and a circular component of polarization, Q and U together provide the magnitude and orientation of the linear component and V gives the sense and the magnitude of the circular component. On inverting (1.28), one finds,

$$E_x E_x^* = \frac{I + Q}{2}, \tag{1.29a}$$

$$E_y E_y^* = \frac{I - Q}{2}, \tag{1.29b}$$

$$E_x E_y^* = \frac{U + iV}{2}, \tag{1.29c}$$

$$E_x^* E_y = \frac{U - iV}{2}, \tag{1.29d}$$

or the polarized brightness,

$$\mathbf{B} = \begin{bmatrix} E_x E_x^* & E_x E_y^* \\ E_x^* E_y & E_y E_y^* \end{bmatrix}$$

$$= \frac{1}{2} \begin{bmatrix} I + Q & U + iV \\ U - iV & I - Q \end{bmatrix}. \tag{1.30}$$

The Stokes parameters for the quasi-monochromatic wave fields are,

$$I = \langle a_x^2 \rangle + \langle a_y^2 \rangle, \tag{1.31a}$$

$$Q = \langle a_x^2 \rangle - \langle a_y^2 \rangle, \tag{1.31b}$$

$$U = 2\langle a_x a_y \rangle \cos \delta, \tag{1.31c}$$

$$V = 2\langle a_x a_y \rangle \sin \delta, \tag{1.31d}$$

a_0, is a vector with amplitude, $[a_x^2 + a_y^2]^{1/2}$; it makes an angle θ with the positive x-axis such that $a_x = a_0 \cos \theta$ and $a_y = a_0 \sin \theta$. In general, θ is called the position angle of polarization.

Both a_x and a_y may take a positive or a negative sign defining the quadrant in which θ lies. In terms of the position angle of polarization and the phase difference δ, the Stokes parameters are rewritten as,

$$I = a_0^2, \tag{1.32a}$$
$$Q = I \cos 2\theta, \tag{1.32b}$$
$$U = I \sin 2\theta \cos \delta, \tag{1.32c}$$
$$V = I \sin 2\theta \sin \delta. \tag{1.32d}$$

On denoting $\mathcal{B}/\mathcal{A} = \tan \chi$, in which $\mathcal{A}$ and $\mathcal{B}$ are the semi-major and semi-minor axes of the ellipse, one finds,

$$\tan 2\phi = \tan 2\theta \cos \delta \tag{1.33a}$$
$$\cos 2\theta = \cos 2\chi \cos 2\phi, \tag{1.33b}$$

where the longitude, 2ϕ and latitude, 2χ, of point P (see Fig. 1.5) are respectively related to the azimuth and the ellipticity angles of the polarization ellipse of the wave; points on the equator, i.e., $2\chi = 0$ represent linearly polarized light, while elliptical polarization states lie between the poles and equator.

The state of polarization is indicated by the location of a point P on the surface of a sphere of radius I in the subspace of Q, U, and V, known as Poincaré's sphere (see Fig. 1.5). Such a sphere provides a convenient graphical tool for utilizing changes in the state of polarization. The three co-ordinates (Q, U, V) of a point (I, χ, ϕ) on Poincaré's sphere are related to the angles ϕ $(0 \le \phi < \pi)$ specifying the orientation of the ellipse and χ $(-\pi/4 \le \chi \le \pi/4)$ that characterizes the ellipticity and sense in which the ellipse is being described. By eliminating θ and δ from the expressions for the Stokes parameter, one obtains,

$$I = a_0^2, \tag{1.34a}$$
$$Q = I \cos 2\chi \cos 2\phi, \tag{1.34b}$$
$$U = I \cos 2\chi \sin 2\phi, \tag{1.34c}$$
$$V = I \sin 2\chi. \tag{1.34d}$$

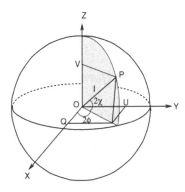

Fig. 1.5 Poincaré's sphere that represents the state of polarization of a monochromatic wave

The state of partially elliptically polarized light is described in terms of combination of two independent components, such as:

1. natural unpolarized light of intensity $I(1 - p_e)$ and
2. fully elliptically polarized light of intensity Ip_e, with $p_e = Ip/I$ as the degree of polarization, Ip the polarized part of the intensity.

Therefore, the Stokes parameters are modified to,

$$Q = Ip_e \cos 2\chi \cos 2\phi, \tag{1.35a}$$
$$U = Ip_e \cos 2\chi \sin 2\phi, \tag{1.35b}$$
$$V = Ip_e \sin 2\chi, \tag{1.35c}$$

where

$$p_e = \frac{[Q^2 + U^2 + V^2]^{1/2}}{I}, \qquad 0 \le p_e \le 1, \tag{1.36}$$

and the factor $p_e \cos 2\chi = p$ is known as the degree of polarization.

Thus, (1.35) translate into,

$$Q = Ip \cos 2\phi, \tag{1.37a}$$
$$U = Ip \sin 2\phi, \tag{1.37b}$$
$$V = Ip_v, \tag{1.37c}$$

where $p_v = p_e \sin 2\chi$ is the degree of ellipticity, which is positive for right-handed elliptical polarization and negative for left-handed polarization.

For a beam of light, the quantities $I, (Q^2 + U^2)$ and V are invariant under a rotation of the reference coordinate system. The degree of linear polarization and the angle which makes axis of the resultant ellipse leads to the following expressions,

$$p = \frac{[Q^2 + U^2]^{1/2}}{I}, \tag{1.38}$$

$$\phi = \frac{1}{2} \tan^{-1} \frac{U}{Q}. \tag{1.39}$$

Since $\delta = 0$ for plane polarized light, (1.33a) implies $\phi = \theta$, a position angle of polarization.

1.3.2 Transformation of Stokes Parameters

The underlying principle in the measurement of polarization of a beam of light is that its state of polarization can be altered in a desired way by introducing appropriate optical elements in its path. The Stokes parameters I', Q', U', and V' of light transmitted through a perfect analyzer with the principal plane containing the optical axis and the incident beam, at position angle ϕ are related to the Stokes parameters I, Q, U, and V of the incident light by the Mueller matrix[4] of transformation equation (Serbowski 1947),

$$
\begin{bmatrix} I' \\ Q' \\ U' \\ V' \end{bmatrix} = \frac{1}{2} \begin{bmatrix} 1 & \cos 2\phi & \sin 2\phi & 0 \\ \cos 2\phi & \cos^2 2\phi & \frac{1}{2}\sin 4\phi & 0 \\ \sin 2\phi & \frac{1}{2}\sin 4\phi & \sin^2 2\phi & 0 \\ 0 & 0 & 0 & 0 \end{bmatrix} \begin{bmatrix} I \\ Q \\ U \\ V \end{bmatrix}. \tag{1.40}
$$

The (1.40) reveals that the Stokes parameters I, Q, U, and V of the incident light can be determined by measuring the intensities of light transmitted by an analyzer with its optical axis at several position angles ϕ. If the analyzer is rotated at a frequency ν, the transmitted light modulates at a frequency 2ν.

In order to create circularly polarized light and to rotate or reverse the polarization ellipse, the quarter-wave plates and half-wave plates are most commonly used, respectively. In the case of combination of half-wave plates and the analyzer, the Stokes parameters can be determined by measuring the intensity of light transmitted by the above combination. The Stokes parameters I', Q', U', and V' of the light transmitted through a perfect retarder is transformed according to the following matrix,

$$
\begin{bmatrix} I' \\ Q' \\ U' \\ V' \end{bmatrix} = \begin{bmatrix} 1 & 0 & 0 & 0 \\ 0 & G + H\cos 4\alpha & H\sin 4\alpha & -\sin\tau\sin 2\alpha \\ 0 & H\sin 4\alpha & G - H\cos 4\alpha & \sin\tau\cos 2\alpha \\ 0 & \sin\tau\sin 2\alpha & -\sin\tau\cos 2\alpha & \cos\tau \end{bmatrix} \begin{bmatrix} I \\ Q \\ U \\ V \end{bmatrix}, \tag{1.41}
$$

in which

$$
G = \frac{1}{2}(1 + \cos\tau), \qquad H = \frac{1}{2}(1 - \cos\tau), \tag{1.42}
$$

---

[4] A Mueller matrix is a generalization of the Jones matrix (Jones 1941) that can be defined for polarizing optical components which allows the nature of the polarization to be easily propagated. Jones vector, $\mathbf{J}$, useful for treating polarized waves is defined by,

$$
\mathbf{J} \equiv \begin{bmatrix} E_x(t) \\ E_y(t) \end{bmatrix} = \begin{bmatrix} E_{0x}e^{i\psi_x} \\ E_{0y}e^{i\psi_y} \end{bmatrix},
$$

where $E_x(t)$ and $E_y(t)$ are the x and y components of the electric field of the light wave, and linear optical elements are represented by Jones matrices. The Mueller matrix is a 4×4 matrix which can be used to reproduce the effect of a given optical element when applied to a Stokes vector, $\mathbf{S}$. The effect of an optical device that transforms the aberrations into the light intensity variations, on the polarization of light is characterized by a Mueller matrix. The description of optical systems in terms of such matrices is applicable to more general situation than the description in terms of Jones matrices. Light which is unpolarized or partially polarized must be treated using Mueller calculus, while fully polarized light may be treated with Jones calculus since the latter works with amplitude rather than intensity of light.

α is the angle between the effective optical axes of the two half-wave plates, and τ the retardance, the phase difference introduced by the retarder between the vibrations in the principal plane and those in the plane perpendicular to it.

If two retarders are kept in series, the square matrix in (1.41) is replaced by the product of two such matrices. The determination of the Stokes parameters thus reduces to the measurement of the intensity of light transmitted by the retarders and the analyzer in the path, for the various position angles of the optical axis of the modulating retarder.

1.4 Diffraction Fundamentals

Diffraction refers to the bending, spreading, and interference of waves around small obstacles and the spreading out of waves past small openings. The amount of bending or spreading depends on the relative size of the slit and the wavelength. Diffraction at an aperture sets a fundamental limit to the resolution of any optical system. It played a major role in elaborating the theory of light, since its introduction in 17th century.

1.4.1 Derivation of the Diffracted Field

In order to describe the evolution of waves through time and space in the case of obstructions, it is useful to utilize the principle of Christian Huygens (1629–1695), which states that every point on a known wavefront in a given medium can be treated as a point source of secondary wavelets, which spread out in all directions with a wave speed characteristic of that medium. The new wavefront at any subsequent time is the envelope of these wavelets. These secondary wavelets have the same period and travel at the same speed as the original propagating wavefront. However, this principle has some shortcomings since it does not take wavelength and relative phases into account. By adding interferences, Augustine Fresnel (1788–1829) proposed the Huygens–Fresnel theory, which states that each point on a wavefront is a source of secondary disturbance and the secondary wavelets emanating from different points mutually interfere. The amplitude of the optical field at any point beyond is the superposition of all these wavelets, considering their amplitudes and relative phases.

Let a monochromatic wave be emitted from a point source, $P_0(\mathbf{r}_0)$, fall on an aperture of an optical element, say a telescope, W, and S be the instantaneous position of a spherical monochromatic wavefront of radius r_0 (see Fig. 1.6). Following (1.13), the disturbance for the elementary contribution, $dU(\mathbf{r})$, of an area, dS, at the point, $Q(\mathbf{r}')$, is given by,

$$dU(\mathbf{r}, t) = K(\chi) \frac{A e^{iKr_0}}{r_0} \frac{e^{iKs}}{s} dS, \qquad (1.43)$$

Fig. 1.6 Fresnel zone
construction

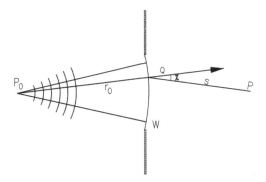

in which Ae^{iKr_0}/r_0 represents the disturbance at $Q(\mathbf{r}')$, s the distance between the
points $Q(\mathbf{r}')$ and $P(\mathbf{r})$, $K(\chi)$ the scaling factor which accounts for the properties of
the secondary wavelet, κ the wave number, χ the angle of diffraction between the
normal at $Q(\mathbf{r}')$, and the direction $Q(\mathbf{r})\, P(\mathbf{r})$.

The scaling factor depends on the angle, χ, being maximum for $\chi = 0$ and zero
for $\chi = \pi/2$. The total disturbance at the point, $P(\mathbf{r})$, is deduced as,

$$U(\mathbf{r}, t) = \frac{Ae^{iKr_0}}{r_0} \int_W \frac{e^{iKs}}{s} K(\chi)d\,s. \tag{1.44}$$

Fresnel evaluated the integral by considering the diffraction apertures of successive zones of constant phase, i.e., the distance s to the point P is constant within $\lambda/2$,

$$K(\chi) = -\frac{i}{2\lambda}(1 + \cos \chi), \tag{1.45}$$

with $K(\chi)$ as scaling factor.

Kirchhoff placed the Fresnel integral (1.44) on a more rigorous mathematical
basis, and showed that the Huygens–Fresnel principle can be regarded as an approximate form of an integral theorem. The solution can be obtained using Green's
functions, by making an assumption that the distances from the aperture are much
greater than the wavelength. Kirchhoff and Sommerfeld derived the resulting solution using Helmholtz's equation and Green's theorem. Scalar wave theory is initially
used to derive the so-called Rayleigh–Sommerfeld diffraction relation, which is then
approximated to the Kirchhoff and finally the Fresnel diffraction approximations.
The Rayleigh–Sommerfeld expression is given by,

$$U(\mathbf{r}) = \frac{Ae^{iKr_0}}{i\lambda r_0} \int_W \frac{e^{iKs}}{s} \cos \chi ds. \tag{1.46}$$

This can be recast in

$$U(\mathbf{r}) = \frac{1}{i\lambda} \int_{-\infty}^{\infty} U(\xi, \eta) \frac{e^{iKs}}{s} \cos \chi ds. \tag{1.47}$$

in which r_0 is the radius of the wavefront W, (ξ, η) the distance coordinates in the aperture plane, and

$$U(\xi, \eta) = \begin{cases} A\dfrac{e^{iKr_0}}{r_0} & \text{within the aperture} \\ 0 & \text{otherwise.} \end{cases} \tag{1.48}$$

The Fresnel–Kirchhoff's diffraction formula, is derived as,

$$U(\mathbf{r}) = \frac{Ae^{iKr_0}}{2i\lambda r_0} \int_{W} \frac{e^{iKs}}{s}(1 + \cos\chi)d\mathsf{S}. \tag{1.49}$$

1.4.2 Near and Far-Field Diffractions

The near-field (or Fresnel) diffraction occurs when a wave passes through an aperture and diffracts in the near field, i.e., the $F \geq 1$ regime produces Fresnel diffraction. Here, $F (\equiv a^2/\lambda s')$ is the Fresnel number (a dimensionless parameter), a the radius of the aperture, λ the wavelength, and s' the distance from the aperture to the projection plane. This can be obtained by approximating the spherical wavefronts as being parabolic. The first order development of s is valid if $x^2 + y^2 \ll s'^2$ and $\xi^2 + \eta^2 \ll s'^2$, in which (x, y) is the distance coordinates in the projection plane. This approximation is equivalent to changing the emitted spherical wave into a quadratic wave. In the case of non-varying optical field, for example in astronomical imaging, near-field approximation is valid for s' greater than 400λ (Papoulis 1968).

Let O be any point in the aperture and assume that the angles which the lines P_0O and OP make with P_0P are not too large (see Fig. 1.7). The factor $1/rs$ is also replaced by $1/r's'$, where r' and s' are the respective distances of P_0 and P from the origin. Thus, (1.49) takes the form,

$$U(\mathbf{r}) \sim \frac{A}{i\lambda}\frac{\cos\delta}{r's'} \int_{A} e^{iK(r+s)}d\mathsf{S}, \tag{1.50}$$

where δ is the angle between the line P_0P and the normal to the screen.

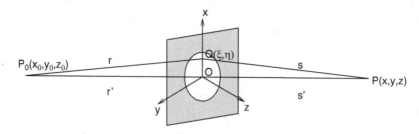

Fig. 1.7 Calculating the optical field at P from the aperture plane

Let (x_0, y_0, z_0) and (x, y, z), respectively, be the co-ordinates of $P_0(\mathbf{r}_0)$ and $P(\mathbf{r})$, and (ξ, η) the co-ordinates of a point Q in the aperture, one gets,

$$r = [(\xi - x_0)^2 + (\eta - y_0)^2 + z_0^2]^{1/2}, \quad r' = [x_0^2 + y_0^2 + z_0^2]^{1/2}, \tag{1.51}$$
$$s = [(x - \xi)^2 + (y - \eta)^2 + z^2]^{1/2}, \quad s' = [x^2 + y^2 + z^2]^{1/2}.$$

Thus, for the incident distance, one finds,

$$s = s' \left[1 + \frac{(x - \xi)^2}{s'^2} + \frac{(y - \eta)^2}{s'^2} \right]^{1/2}. \tag{1.52}$$

The respective near-field (Fresnel) and far-field approximations are:

$$s \approx s' + \frac{1}{2s'} \left[(x - \xi)^2 + (y - \eta)^2 \right], \tag{1.53}$$

$$s \approx s' + \frac{1}{2s'} \left[(x^2 + y^2) \right] - \frac{1}{s'} [x\xi + y\eta]. \tag{1.54}$$

The diffractive field is derived and expressed as a convolution equation:

$$U(x, y) = \frac{A e^{iKs'}}{i\lambda s'} \int\!\!\int_{-\infty}^{\infty} P(\xi, \eta) e^{i(\kappa/2s')[(x-\xi)^2 + (y-\eta)^2]} d\xi d\eta$$

$$= \int\!\!\int_{-\infty}^{\infty} P(\xi, \eta) U_{s'}(x - \xi, y - \eta) d\xi d\eta$$

$$= P(\xi, \eta) \star U_{s'}(x, y), \tag{1.55}$$

where

$$U_{s'}(x, y) = \frac{A e^{iKs'}}{i\lambda s'} e^{i(\kappa/2s')[x^2 + y^2]}, \tag{1.56}$$

the symbol, $\star$, implies convolution parameter and $P(\xi, \eta)$ the pupil function.

In the case of a transparent pupil with aberrations, $P(\xi, \eta)$ represents the aberration function. In $U_{s'}$, two phase factors appear (Mariotti 1988):

1. the first one corresponds to the general phase retardation as the wave travels from one plane to the other and
2. the second one is a quadratic phase term that depends on the positions, O and P.

In the case of Far-field (or Fraunhofer) approximation (see Fig. 1.8a), the diffraction occurs in the limit of small Fresnel number $F \ll 1$, where the diffraction pattern is independent of the distance to the screen, depending on the angles to the screen from the aperture. The conditions of validity:

$$x^2 + y^2 \ll s'^2 \quad \text{and} \quad \xi^2 + \eta^2 \ll 2s'/\kappa = \lambda s'/\pi,$$

are more restrictive.

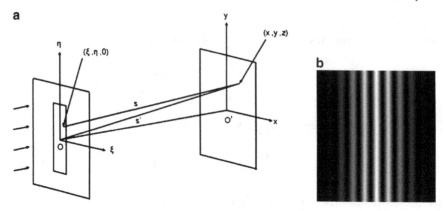

Fig. 1.8 (a) Schematic of Fraunhofer approximation and (b) Double-slit Fraunhofer diffraction pattern on a plane

The distance separating the Fresnel and the Fraunhofer regions is called the Rayleigh distance $s'_R = D^2/\lambda$, in which D is the size of the diffracting aperture. In the Fraunhofer case,

$$U(x, y, s') = \frac{Ae^{iKs'}}{i\lambda s'}e^{i(\kappa/2s')(x^2+y^2)}$$

$$\times \int\!\!\!\int_{-\infty}^{\infty} P(\xi, \eta)e^{-i(\kappa/s')[x\xi+y\eta]}d\xi d\eta. \qquad (1.57)$$

In the far-field, the relative phase fluctuations do not vary that rapidly and the pattern that would be observed can be described by the Fraunhofer diffraction pattern. Since the phase terms outside the integral have no influence, it is discarded. This integral represents the Fourier transform (FT) of the aperture function. The expression for s is written as,

$$(x^2 - \xi^2) + (y^2 - \eta^2) = (x^2 + y^2) - 2(x\xi + y\eta) + \xi^2 + \eta^2. \qquad (1.58)$$

The (1.57) then yields,

$$U(x, y, s') \propto \int\!\!\!\int_{-\infty}^{\infty} P(\xi, \eta)e^{i(\kappa/2s')[\xi^2+\eta^2]}e^{-i(\kappa/s')[x\xi+y\eta]}d\xi d\eta. \qquad (1.59)$$

The Fraunhofer diffraction pattern for the field is proportional to the FT of the transmission function. Here, $s' \gg (\xi^2+\eta^2)/\lambda$, the phase factor is much smaller, and is therefore ignored. Figure 1.8b depicts computer simulated double-slit Fraunhofer diffraction pattern on a plane screen.

1.4.3 Diffraction by a Circular Aperture

Circular apertures, such as telescope mirrors or lenses, produce radially symmetric diffractions, thereby having the same resolutions in all directions. The Fraunhofer diffraction integral (1.59) can be expressed in the form,

$$\widehat{U}(u, v) = C \iint\limits_{-\infty}^{\infty} P(\xi, \eta) \, e^{-i2\pi(u\xi+v\eta)} d\xi d\eta, \tag{1.60}$$

where C is the constant and is defined in terms of quantities depending on the position of the source and the point of observation.

It is pertinent to note that the astronomical objects are two-dimensional (2-D). The 2-D space of spatial frequency is called the Fourier plane or the (u, v) coordinates (a plane perpendicular to the source direction) measured in wavelength, i.e.,

$$u = \frac{x}{\lambda s'}, \qquad v = \frac{y}{\lambda s'}, \tag{1.61}$$

with units of inverse distance. For a circularly symmetric aperture function, it is expressed in terms of polar coordinates, (ρ, θ),

$$P(\rho, \theta) = \begin{cases} 1, & 0 < \rho < a \\ 0, & \text{otherwise} \end{cases} \tag{1.62}$$

where a is the radius of the aperture, and

$$\xi = \rho \cos \theta, \qquad \eta = \rho \sin \theta; \tag{1.63a}$$
$$u = w \cos \phi, \qquad v = w \sin \phi, \tag{1.63b}$$

in which the circular polar coordinates are expressed in terms of u and v.

Wavefronts from a star encountering an aperture of a telescope objective, diffract in the form of the Fraunhofer diffraction pattern, which is seen as the image of the star. The plane wave illuminates the aperture coherently. The intensity pattern in the far-field is described with the Fraunhofer diffraction integral in a general two-dimension. Such an integral at the output point $P(\mathbf{r})$,

$$\widehat{U}(u, v) = C \int_{W} e^{-i2\pi(u\xi+v\eta)} d\xi d\eta,$$

takes the form of the surface integral over the circular aperture, the disturbance per unit area,

$$\widehat{U}(w) = C \int_0^a \rho d\rho \int_0^{2\pi} e^{-i2\pi\rho w \cos(\theta-\phi)} d\theta$$
$$= 2\pi C \int_0^a \rho J_0(2\pi\rho w) d\rho, \tag{1.64}$$

in which $J_0(x)$ and $J_1(x)$ are the Bessel functions of the order zero and one respectively, (w, ϕ) are the coordinates of a point $\mathbf{P(r)}$ in the diffraction pattern, referred to the geometrical image of the source, and $w = \sqrt{u^2 + v^2}$ the sine of the angle which the direction (u, v) makes with the central direction $u = v = 0$.

Thus, the disturbance of a circular aperture is expressed as,

$$\widehat{U}(w) = C\pi a^2 \left[\frac{2J_1(2\pi aw)}{2\pi aw} \right]. \tag{1.65}$$

It is worthwhile to mention that the diffraction pattern of a slit of width a along x-direction is given by the sinc function,

$$\widehat{U}(u) = \frac{\sin \pi ua}{\pi ua}. \tag{1.66}$$

The diffraction pattern of a circular aperture is a series of concentric rings with a bright central spot, and hence taking the property of the first order Bessel function, i.e.,

$$\lim_{x=0} \left[\frac{J_1(x)}{x} \right] = \frac{1}{2}, \tag{1.67}$$

the intensity distribution is given by,

$$\widehat{I}(w) = \widehat{I}_0 \left[\frac{2J_1(2\pi xw)}{2\pi xw} \right]^2, \tag{1.68}$$

where $\widehat{I}_0 [= C^2(\pi x^2)^2]$ is the intensity on-axis ($w = 0$).

When normalized, the intensity distribution in the diffraction pattern through a circular aperture of a point source illustrates the Airy disk having a bright central spot (see Fig. 1.9a) followed by dark and bright rings of considerably lower intensity (see Fig. 1.9b) surrounding it. The central spot has a defined edge surrounded by a

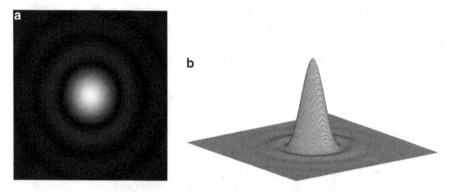

Fig. 1.9 Computer generated intensity distribution of a circular aperture: (**a**) 2-D pattern with a saturated central spot and (**b**) 3-D pattern with faint outer rings

dark ring, referred to as a region of destructive interference, occurring at a radius corresponding to the first zero of the Bessel function $J_1(x)$. Such a diffraction pattern for pupils with or without aberration is referred to as the impulse response or point spread function (PSF).

1.5 Image Formation

An image is the optical appearance of an object, produced by an optical system that redirects the rays to a real image space. A perfect image of a small object lying at a large distance from an imaging system is an exact replica of the object except for its magnification. Such an image is known as the Gaussian image. Let the point $M_1(x_1)$ be the geometrical image of $M_0(x_0)$, in which the position vectors are $x_1 = (x_1, y_1)$ and $x_0 = (x_0, y_0)$, the distribution of the complex disturbances in the case of monochromatic point source of wavelength λ at x_1 is given by,

$$U(x_1) = \int_{-\infty}^{\infty} U_0(x_0) K(x_0; x_1) dx_0, \tag{1.69}$$

where $K(x_0; x_1)$ is the instrument (e.g., telescope) transmission function that characterizes the imaging properties of the system due to a disturbance of unit amplitude and zero phase at the object point x_0.

By applying the Fraunhofer diffraction relation (see Sect. 1.4.2) to the pupil, the transmission function can be determined,

$$K(x_0; x_1) \propto \int_{-\infty}^{\infty} P(x) e^{-i2\pi(x_1 - x_0) \cdot x/\lambda s} dx, \tag{1.70}$$

with

$$P(x) = \begin{cases} U(x) & \text{inside the pupil,} \\ 0, & \text{otherwise,} \end{cases} \tag{1.71}$$

as the pupil function and x the two-dimensional (2-D) space vector.

The (1.70) provides the relationship between the telescope transmission function K and the pupil function $P(x)$. For an iso-planatism[5] (shift-invariance) condition, the telescope transmission function is given by,

$$K(x_0; x_1) = K(x_1 - x_0). \tag{1.72}$$

---

[5] In optics, spatial invariance, known as iso-planatism condition, is an important requirement since the field-dependent geometrical aberrations change the impulse response with respect to the distance to the optical axis. The aberrations in geometrical optics influences causing different rays to converge to different points. These aberrations generate from subsets of a larger centered bundle of spherically aberrated rays. Typically, an optical system suffers from spherical aberration, coma, astigmatism, and chromatic aberration.

In the shift-invariant case, each amplitude spread function depends on only two independent variables. Thus, (1.70) for the transfer function (see Appendix A) can be recast as,

$$K(\mathbf{x}_1 - \mathbf{x}_0) = \int_{-\infty}^{\infty} P(\lambda \mathbf{u}) e^{-i2\pi(\mathbf{x}_1 - \mathbf{x}_0)\cdot\mathbf{u}} d\mathbf{u}, \qquad (1.73)$$

in which the dimensionless variable $\mathbf{u} = \mathbf{x}/\lambda$ and $\mathbf{u}(= u, v)$ the 2-D spatial frequency (see 1.61) vector with magnitude $|\mathbf{u}|$.

The general expression for the intensity of the formed image is given as,

$$I(\mathbf{x}_1) = \iint\limits_{-\infty}^{\infty} K(\mathbf{x}_1 - \mathbf{x}_0')K^*(\mathbf{x}_1 - \mathbf{x}_0'')\langle U_0(\mathbf{x}_0', t)U_0^*(\mathbf{x}_0'', t)\rangle d\mathbf{x}_0' d\mathbf{x}_0'', \qquad (1.74)$$

where $\mathbf{x}_0'(= x_0', y_0')$ and $\mathbf{x}_0''(= x_0'', y_0'')$ are the position vectors of the fields.

The quantity $\langle U_{0'}, U_{0''}^* \rangle$ measures the complex disturbances at two points of the object. If the fields at $\mathbf{x}_0'$ and $\mathbf{x}_0''$ have the same instantaneous amplitudes and a constant phase delay, the situation is said to be coherent (their phase difference at the point of detection remains the same). The term $\langle U_{0'}, U_{0''}^* \rangle$ becomes equal to $U_0(\mathbf{x}_0') \cdot U_0^*(\mathbf{x}_0'')$. The (1.74) translates into,

$$I(\mathbf{x}_1) = \int_{-\infty}^{\infty} K(\mathbf{x}_1 - \mathbf{x}_0')U_0(\mathbf{x}_0')d\mathbf{x}_0' \int_{-\infty}^{\infty} K^*(\mathbf{x}_1 - \mathbf{x}_0'')U_0(\mathbf{x}_0'')d\mathbf{x}_0''. \qquad (1.75)$$

The complex disturbance $U(\mathbf{x}_1)$ is expressed as,

$$U(\mathbf{x}_1) = \int_{-\infty}^{\infty} K(\mathbf{x}_1; \mathbf{x}_0)U_0(\mathbf{x}_0)d\mathbf{x}_0. \qquad (1.76)$$

According to (1.76), U is a convolution (see Appendix B.1.1) of U_0 and K.

If the fields at two different points are fully uncorrelated, the term, $\langle U_0', U_0''^* \rangle$ $[= O(\mathbf{x}_0')\delta(\mathbf{x}_0' - \mathbf{x}_0'')]$, in which $O(\mathbf{x}_0')$ represents the spatial intensity distribution in the object plane, averages out. In this case, the light is assumed to be incoherent; hence, the intensity, $I(\mathbf{x}_1)$ is given by,

$$I(\mathbf{x}_1) = \int_{-\infty}^{\infty} O(\mathbf{x}_0')K(\mathbf{x}_1 - \mathbf{x}_0')d\mathbf{x}_0'$$
$$\times \left[\int_{-\infty}^{\infty} K^*(\mathbf{x}_1 - \mathbf{x}_0'')\delta(\mathbf{x}_0' - \mathbf{x}_0'')d\mathbf{x}_0''\right] d\mathbf{x}_0'$$
$$= \int_{-\infty}^{\infty} O(\mathbf{x}_0')|K(\mathbf{x}_1 - \mathbf{x}_0')|^2 d\mathbf{x}_0'. \qquad (1.77)$$

The (1.77) states that the system is linear with respect to the intensities for the incoherent case.

1.5.1 Optical Transfer Function

The optical transfer function (OTF) describes the imaging quality, which represents the complex factor applied by an optical imaging system to the frequency components of object intensity distribution relative to the factor applied to the zero frequency component. For an object point at the origin, the complex amplitude distribution function in its image may be written from (1.73) as,

$$K(\mathbf{x}_1) = \int_{-\infty}^{\infty} P(\mathbf{x}) e^{-i2\pi \mathbf{x}_1 \cdot \mathbf{x}/\lambda s} d\mathbf{x}. \qquad (1.78)$$

Using the definition of spatial frequency (see 1.61), one may write, $\mathbf{q} = \mathbf{x}/\lambda s'$, with $q(u, v)$ as the spatial frequency vector in the pupil plane, changing $\mathbf{x}$ to $\mathbf{q}$, one writes (1.78) as,

$$K(\mathbf{x}_1) = \int_{-\infty}^{\infty} P(\mathbf{q}) e^{-i2\pi \mathbf{q} \cdot \mathbf{x}_1} d\mathbf{q}. \qquad (1.79)$$

Therefore, the PSF is,

$$|K(\mathbf{x}_1)|^2 = \int\!\!\int_{-\infty}^{\infty} P(\mathbf{q}) P^*(\mathbf{q}') e^{-i2\pi(\mathbf{q}-\mathbf{q}') \cdot \mathbf{x}_1} d\mathbf{q} d\mathbf{q}'. \qquad (1.80)$$

The PSF describes how the radiation from a point source is distributed in the image and is determined by the amplitude and phase of the spherical wavefront as it converges on the focal point, where the photons are detected. The amplitude, measures the intensity of the wavefront at each point, $\mathbf{u}$, on the sphere and is generally uniform across the entire pupil, except in the case of a telescope, where it is obscured by objects in the light path such as the secondary mirror and its support system. The phase, $\psi(\mathbf{u})$ measures the deviation of the wavefront from the perfect focus having zero phase aberration that is measured in units of the observed wavelength of light. To note, a vignetting effect occurs with the increase of the incident angle of the beam, where the entrance pupil gets partially occulted.

Figure 1.10 depicts the normalized intensity PSF of the 2.34 m Vainu Bappu Telescope, Kavalur at the Cassegrain[6] ($f/13$) focus, while Fig. 1.11 displays the fraction of the encircled energy of the said telescope conserving the energy. Putting $\mathbf{q} - \mathbf{q}' = \mathbf{f}$ (a frequency variable), one gets,

---

[6] A Cassegrain telescope is a combination of a parabolic primary mirror that reflects the light rays towards the prime focus and a convex hyperbolic secondary bouncing the rays back through a small hole bored through the center of the main mirror in the shadow of the secondary; a Ritchey–Chrétien Cassegrain system has two hyperbolic mirrors. However, the light gathering power of a telescope is proportional to the area of its main mirror.

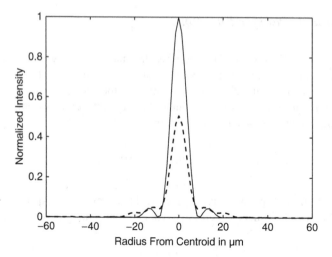

Fig. 1.10 Schematic representation of the normalized intensity PSF of the 2.34 m Vainu Bappu Telescope (VBT) at the Cassegrain ($f/13$) focus. The *solid line* represents PSF for the ideal case and the *dashed line* PSF for the real aberrated case

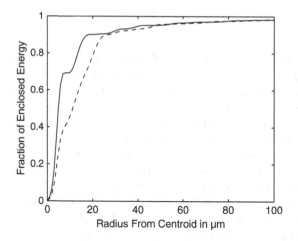

Fig. 1.11 Fraction of encircled energy of the 2.34 m VBT, Kavalur at the Cassegrain focus ($f/13$). The *solid line* represents for an ideal diffraction-limited telescope, while the *dashed line* for a non-ideal case

$$|K(\mathbf{x}_1)|^2 = \iint\limits_{-\infty}^{\infty} P(\mathbf{q}' + \mathbf{f}) P^*(\mathbf{q}') e^{-i2\pi \mathbf{f} \cdot \mathbf{x}_1} d\mathbf{q}' d\mathbf{f}$$

$$= \int_{-\infty}^{\infty} \left[\int_{-\infty}^{\infty} P(\mathbf{q}' + \mathbf{f}) P^*(\mathbf{q}') d\mathbf{q}' \right] e^{-i2\pi \mathbf{f} \cdot \mathbf{x}_1} d\mathbf{f}. \qquad (1.81)$$

Using Fourier inversion transform theorem, one finds

$$\int_{-\infty}^{\infty} P(\mathbf{q'} + \mathbf{f}) P^*(\mathbf{q'}) d\mathbf{q'} = \int_{-\infty}^{\infty} |K(\mathbf{x}_1)|^2 e^{i 2\pi \mathbf{f} \cdot \mathbf{x}_1} d\mathbf{x}_1, \qquad (1.82)$$

with $\mathbf{f}$, as the spatial frequency.

The (1.82) shows that the autocorrelation of the pupil function in terms of frequency variables is the OTF of the optical system. Changing $\mathbf{q'}$ to $\mathbf{q}$, one may write OTF,

$$\widehat{T}(\mathbf{f}) = \int_{-\infty}^{\infty} P(\mathbf{q} + \mathbf{f}) P^*(\mathbf{q}) d\mathbf{q}. \qquad (1.83)$$

The spatial frequency spectrum of the diffracted image of an iso-planatic incoherent object is equal to the product of the spectrum of its Gaussian image and the OTF of the system. The magnitude of the OTF, $|\widehat{T}(\mathbf{f})|$, known as the modulation transfer function (MTF), is a real quantity, and is equivalent to the modulus of the Fourier transform of the PSF:

$$\left|\widehat{T}(\mathbf{f})\right| = \left|\int_{-\infty}^{\infty} |K(\mathbf{x}_1)|^2 e^{i 2\pi \mathbf{f} \cdot \mathbf{x}_1} d\mathbf{x}_1\right|. \qquad (1.84)$$

or taking inverse transform,

$$|K(\mathbf{x}_1)|^2 = \int_{-\infty}^{\infty} \widehat{T}(\mathbf{f}) e^{-i 2\pi \mathbf{f} \cdot \mathbf{x}_1} d\mathbf{f}. \qquad (1.85)$$

The MTF contains no phase information, but is a real quantity. For incoherent imaging, the MTF, $|\widehat{T}(\mathbf{f})| \leq 1$. Typically, it decreases with increasing frequencies, hence the high frequency details in the image are weakened and eventually lost. From (1.85), one obtains,

$$|K(0)|^2 = \int_{-\infty}^{\infty} \widehat{T}(\mathbf{f}) d\mathbf{f}. \qquad (1.86)$$

The normalized form of $\widehat{T}(\mathbf{f})$ is,

$$\widehat{T}_n(\mathbf{f}) = \frac{\widehat{T}(\mathbf{f})}{\widehat{T}(0)}. \qquad (1.87)$$

For a perfect optical imaging system with a uniformly illuminated circular pupil of diameter D, the pupil function may be expressed by,

$$P(\mathbf{x}) = \begin{cases} 1 & \text{inside the pupil} \\ 0 & \text{otherwise,} \end{cases} \qquad (1.88)$$

and the normalized OTF is given by,

$$\widehat{T}_n(\mathbf{f}) = \frac{1}{A_p} \int_{-\infty}^{\infty} P(\boldsymbol{\xi}) P(\boldsymbol{\xi} + \mathbf{f}) d\boldsymbol{\xi} = \frac{1}{A_p} \int_{A_o} d\boldsymbol{\xi}, \qquad (1.89)$$

where $\boldsymbol{\xi} (= \xi, \eta)$ is the position vector of a point in the aperture, A_p the total area of the pupil, and A_o the overlap area of two replicas of the pupil displaced by an amount $\mathbf{f}$.

1.5.2 Influence of Aberrations

The image quality can be characterized by a number called Strehl's ratio, S_r, which is defined as the ratio of the peak intensity of the system's point spread function (PSF) to the peak intensity of the corresponding diffraction-limited PSF (Strehl 1902), i.e.,

$$S_r = \frac{I(\mathbf{r})}{I_G} = \frac{1}{\pi^2} \left| \int_0^1 \int_0^{2\pi} e^{i\left[\kappa\psi - v\rho\cos(\theta - \phi) - \frac{1}{2}u\rho^2\right]} \rho d\rho d\theta \right|^2, \qquad (1.90)$$

where I_G is the Gaussian image (in the absence of aberration), ρ, θ the polar coordinates, r, ϕ the polar coordinates at the image plane, $\kappa = 2\pi/\lambda$ the wave number, and $\kappa\psi$ the deviation due to aberration in phase from a Gaussian sphere about the origin of the focal plane, and u, v the optical coordinates.

In the case of small aberrations, when tilt (a component of ψ linear in the coordinates $\rho\cos\theta$ or $\rho\sin\theta$ in the pupil plane) is removed and the focal plane is displaced to its Gaussian focus, the linear and quadratic terms in the exponential of (1.90) smear. Strehl's ratio simplifies to,

$$S_r = \frac{I(0)}{I_G} = \frac{1}{\pi^2} \left| \int_0^1 \int_0^{2\pi} e^{i\kappa\psi_{ab}(\rho,\theta)} \rho d\rho d\theta \right|^2, \qquad (1.91)$$

in which the subscript $_{ab}$ represents the aberrated quantities, and $\psi_{ab}(\rho, \theta)$ is the optical path error introduced by the aberrations.

It is clear from (1.91) that Strehl's ratio, S_r, is bounded by $0 \le S_r \le 1$. For strongly varying ψ_{ab}, $S_r \ll 1$. This ratio tends to become larger for smaller wavenumber, κ, in the case of any given varying ψ_{ab}. For an unaberrated beam at the pupil, $\psi_{ab}(\rho, \theta) = 0$, S_r turns out to be unity, thus the intensity at the focus becomes diffraction-limited. In general, the aberrations are small, therefore third and higher order of the exponential term are ignored. The (1.91) becomes,

$$S_r = \frac{I_{ab}(0)}{I(0)}$$

$$= 1 + \kappa^2 \left[\int_0^1 \int_0^{2\pi} \psi_{ab}(\rho, \theta) \rho d\rho d\theta \right]^2 - \kappa^2 \int_0^1 \int_0^{2\pi} \psi_{ab}^2(\rho, \theta) \rho d\rho d\theta.$$

$$(1.92)$$

When $\psi_{ab} = 0$ and $S_r \sim 1$, the quality of the image beam is directly related to the root-mean-square phase error. The Strehl ratio for such an error less than $\lambda/2\pi$ is expressed as,

$$S_r = 1 - \left(\frac{2\pi}{\lambda} \right)^2 \langle \sigma \rangle^2 \simeq e^{-(2\pi/\lambda)^2 \langle \sigma \rangle^2}, \qquad (1.93)$$

where σ is the root-mean-square phase error or root-mean-square wavefront error and $\langle \sigma \rangle^2$ the variance[7] of the aberrated wavefront with respect to a reference perfect wavefront,

$$\langle \sigma \rangle^2 = \frac{\int_0^1 \int_0^{2\pi} \left(\psi_{ab}(\rho, \theta) - \bar{\psi}_{ab}(\rho, \theta) \right)^2 \rho d\rho d\theta}{\int_0^1 \int_0^{2\pi} \rho d\rho d\theta}, \qquad (1.94)$$

where $\bar{\psi}_{ab}$ represents the average value of ψ_{ab},

$$\bar{\psi}_{ab} = \frac{1}{\pi} \int_0^1 \int_0^{2\pi} \psi_{ab} \rho d\rho d\theta. \qquad (1.95)$$

1.5.3 Resolving Power of a Telescope

The resolving power of a telescope refers to its ability to form images of two closely separated stars that can be distinguished. The smaller the diffraction disk, the higher

---

[7] The variance measures of statistical dispersion; the higher the variance, the larger the spread of values. It is computed as the average squared deviation of each number from its mean. Mathematically, the variance in a population is given by,

$$\langle \sigma_{st} \rangle^2 = \frac{\sum (X - m)^2}{N},$$

where m is the arithmetic mean, X the variable, and N the number of scores, while the positive square root of the variance is known as the standard deviation that has the same units as the original variable.

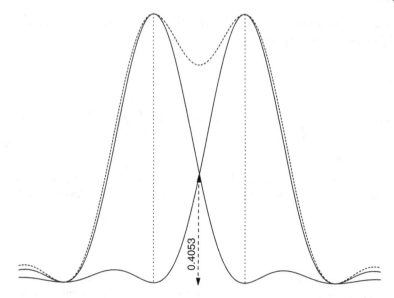

Fig. 1.12 An example of imaging two point sources at the Rayleigh limit of separation

the resolving power. Its resolving power can be determined by a criterion, known as Rayleigh criterion, which states that two points sources can be resolved when the center maximum of the diffraction pattern of one source falls on the first minimum of the diffraction pattern of the other. At this separation, the separate contributions of the two sources to the intensity at the central minima is given by, $4/\pi^2 = 0.4053$ for a rectangular aperture (see Fig. 1.12).

Simplifying (1.12a) to $e^{i\psi}$, the disturbance at a given point, θ, in the focal point of a single telescope is expressed as (Traub 1999),

$$U_{tel}(\theta) = \sum(\text{wavelets}) = \int_{\text{pupil}} e^{i\psi(x)} dx$$

$$= \int_{-D/2}^{D/2} e^{i(2\pi x\theta/\lambda)} dx = D\frac{\sin(\pi\theta D/\lambda)}{\pi\theta D/\lambda}. \qquad (1.96)$$

in which $\psi(x) = 2\pi x \sin\theta/\lambda \simeq 2\pi x\theta/\lambda$ is the phase of a wavelet and D the diameter of the telescope aperture, $x \sin\theta$ is the distance between the incoming wavefront from direction $\theta = 0$ and the outgoing angle wavefront at angle θ.

The observed intensity, I, is the squared magnitude of the disturbance, i.e.,

$$I_{tel}(\theta) = D^2 \left[\frac{\sin(\pi\theta D/\lambda)}{\pi\theta D/\lambda}\right]^2. \qquad (1.97)$$

The intensity pattern is a sinc function. The first zero is the solution of I_{tel} $(\theta) = 0$ and is given by, $\theta_{tel} = \lambda/D$. Similarly, for a circular aperture, the measured intensity, $I_{tel}(\theta)$, is written as,

$$I_{tel}(\theta) = D^2 \left[\frac{2J_1(\pi\theta D/\lambda)}{\pi\theta D/\lambda} \right]^2, \tag{1.98}$$

where J_1 is the first order Bessel function.

In absence of any aberrations such as defects in the optics or free of atmospheric turbulence, the image is diffraction-limited, which is generally described by the angle in radians,

$$\theta_{tel} \sim 1.22 \frac{\lambda}{D}, \tag{1.99}$$

where λ is a wavelength of light and D the diameter of the telescope.

Due to the scattering of the light by inhomogeneities in transparent media, opto-mechanical aberrations of the optical system and diffraction, non-ideal (blur) images are formed. The blurring suffered by such images is modeled as convolution with PSF. According to Strehl's criterion, the resolving power, $\mathcal{R}$, of any telescope is given by the integral of its transfer function,

$$\mathcal{R} = \int_{-\infty}^{\infty} \widehat{S}(\mathbf{u})d\mathbf{u} = \int_{-\infty}^{\infty} \widehat{B}(\mathbf{u})\widehat{T}(\mathbf{u})d\mathbf{u}. \tag{1.100}$$

where $\widehat{B}(\mathbf{u})$ is the wave transfer function, also known as atmosphere transfer function and $\widehat{T}(\mathbf{u})$ the telescope transfer function.

The right hand side of (1.100) implies that the transfer function, $\widehat{S}(\mathbf{u})$, is the product of both the atmosphere and the telescope transfer functions (see Sect. 4.1.6).

Chapter 2
Principles of Interference

2.1 Coherence of Optical Waves

The word 'coherence' describes the ability of radiation to produce interference phenomena and the notion of coherence is defined by the correlation properties between the various quantities of an optical field. The optical coherence is related to the various forms of the correlations of the random processes (Born and Wolf 1984; Mandel and Wolf and Wolf 1995). The interference phenomena stems from the principle of superposition, which states that the resultant displacement (at a particular point) produced by two or more waves is the vector sum of the displacements produced by each one of the disturbances. It reveals the correlations between light waves. The degree of correlation that exists between the fluctuations in two light waves determines the interference effects arising when the beams are superposed. The correlated fluctuation can be partially or completely coherent. A polychromatic point source on the sky produces a fringe packet as a function of an applied path length difference. This fringe packet has an extent referred to as the coherence length,

$$l_c = \frac{c}{\Delta \nu} = \frac{\lambda^2}{\Delta \lambda} = c\tau_c, \qquad (2.1)$$

where $\Delta \nu$ is the effective spectral width, c the speed of light, and

$$\tau_c = \frac{\lambda^2}{c\Delta \lambda} \sim \frac{1}{\Delta \nu}, \qquad (2.2)$$

the coherence time.

The coherence time is defined as the maximum transit time difference for good visibility, $\mathcal{V}$ of the fringes. It is pertinent to note that the visibility, a dimensionless number lying between zero and one, is the attribute of interference fringes describing the contrast between the bright and dark interference regions. Mathematically, it is a complex (phasor) quantity whose magnitude quantifies the fringe contrast, and phase describes the position of the fringes with respect to a phase center. The coherence time is the evolution time of the complex amplitude; a factor less than unity affects the degree of coherence. In order to keep the time correlation close to

S.K. Saha, *Aperture Synthesis*, Astronomy and Astrophysics Library,
DOI 10.1007/978-1-4419-5710-8_2, © Springer Science+Business Media, LLC 2011

unity, the delay τ must be limited to a small fraction of the temporal width or coherence time. The visibility falls to approximately zero when the pathlength difference exceeds the coherence length, l_c, which measures the maximum path difference for which the fringes are still observable. Or equivalently, the visibility approaches zero when the relative time delay exceeds the coherence time. An interferometer based on amplitude division, is generally used to measure the temporal coherence of a source, where it compares a light wave with itself at different moments in time, τ_c; this type of coherence properties encode the spectral content of the intensity distribution. The classical Michelson interferometer is an example of such a method, while the spatial coherence describes the correlation between two light waves at different points in space, for example the Young's double slit experiment. An astronomical interferometer measures spatio-temporal coherence properties of the radiation emerging from a celestial source. The measured interferometric signal depends on both structural and spectral contents of the celestial body concerned.

2.1.1 Interference of Partially Coherent Beams

The Fig. 2.1 is a sketch of a Young's set up where the wave field is produced by an extended polychromatic source. In this, a screen (A) with pin-holes at positions $P_1(\mathbf{r}_1)$ and $P_2(\mathbf{r}_2)$ separated by, h, is placed at a distance, d, from the source, S. The complex disturbance produced at a receiving point, P($\mathbf{r}$), on a second screen (B) situated at distances s_1 and s_2, respectively, from the pin-holes $P_1(\mathbf{r}_1)$ and $P_2(\mathbf{r}_2)$, is expressed as,

$$U(\mathbf{r},t) = K_1 U(\mathbf{r}_1, t - t_1) + K_2 U(\mathbf{r}_2, t - t_2), \qquad (2.3)$$

where $U(\mathbf{r},t)$ represents the associated analytical signal, $t_1 = s_1/c$ and $t_2 = s_2/c$ the transit time of the light from $P_1(\mathbf{r}_1)$ to P($\mathbf{r}$), and from $P_2(\mathbf{r}_2)$ to P($\mathbf{r}$), respectively, and $K_{j=1,2}$ the constant factors that depend on the size of the openings and on

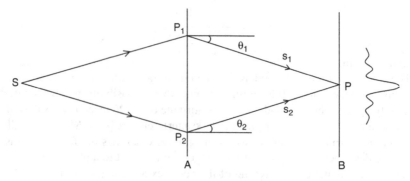

Fig. 2.1 Coherence of the two holes $P_1(\mathbf{r}_1)$ and $P_2(\mathbf{r}_2)$ illuminated by an extended source σ

the geometry of the arrangement, i.e., the angle of incident and diffraction at $P_1(\mathbf{r}_1)$ and $P_2(\mathbf{r}_2)$; the refractive index of the medium between the screens are assumed to be unity.

The complex valued constants are given by,

$$K_1 \simeq \int_{P_1} \frac{\chi(\theta_1)}{i\bar{\lambda}s_1} d S_1, \tag{2.4a}$$

$$K_2 \simeq \int_{P_2} \frac{\chi(\theta_2)}{i\bar{\lambda}s_2} d S_2, \tag{2.4b}$$

in which $\theta_{(j=1,2)}$ are angles indicated in Fig. 2.1 and χ the inclination factor.

These factors K_1 and K_2 are inversely proportional to s_1 and s_2. Since the secondary wavelets from $P_1(\mathbf{r}_1)$ and $P_2(\mathbf{r}_2)$ are out of phase with the primary wave by a quarter of a period, K_1 and K_2 are imaginary numbers. If the pin-holes are small and the diffracted fields are considered to be uniform, the values $|K_j|$ satisfy $K_1^* K_2 = K_1 K_2^* = K_1 K_2$. The diffracted fields are approximately uniform, that is, K_1 and K_2 do not depend on θ_1 and θ_2. In order to derive the intensity of the light at $P(\mathbf{r})$, by neglecting the polarization effects, one may assume that the averaging time is effectively infinite which is valid assumption for true thermal light. The observable intensity (or power), $I(\mathbf{r}, t)$ at $P(\mathbf{r})$ is defined by the formula,

$$I(\mathbf{r}, t) = \langle U(\mathbf{r}, t) U^*(\mathbf{r}, t) \rangle. \tag{2.5}$$

It follows from (2.3, 2.5),

$$\begin{aligned} I(\mathbf{r}, t) = |K_1|^2 \langle |U(\mathbf{r}_1, t - t_1)|^2 \rangle + |K_2|^2 \langle |U(\mathbf{r}_2, t - t_2)|^2 \rangle \\ + K_1 K_2^* \langle U(\mathbf{r}_1, t - t_1) U^*(\mathbf{r}_2, t - t_2) \rangle \\ + K_1^* K_2 \langle U(\mathbf{r}_2, t - t_2) U^*(\mathbf{r}_1, t - t_1) \rangle. \end{aligned} \tag{2.6}$$

The field was assumed to be stationary. One may shift the origin of time in all these expressions. Therefore,

$$\langle U(\mathbf{r}_1, t - t_1) U^*(\mathbf{r}_1, t - t_1) \rangle = \langle U(\mathbf{r}_1, t) U^*(\mathbf{r}_1, t) \rangle = I(\mathbf{r}_1, t), \tag{2.7}$$

$$\langle U(\mathbf{r}_2, t - t_2) U^*(\mathbf{r}_2, t - t_2) \rangle = \langle U(\mathbf{r}_2, t) U^*(\mathbf{r}_2, t) \rangle = I(\mathbf{r}_2, t), \tag{2.8}$$

and if one sets $\tau = t_2 - t_1$, then

$$\langle U(\mathbf{r}_1, t - t_1) U^*(\mathbf{r}_2, t - t_2) \rangle = \langle U(\mathbf{r}_1, t + \tau) U^*(\mathbf{r}_2, t) \rangle, \tag{2.9a}$$

$$\langle U^*(\mathbf{r}_1, t - t_1) U(\mathbf{r}_2, t - t_2) \rangle = \langle U^*(\mathbf{r}_1, t + \tau) U(\mathbf{r}_2, t) \rangle. \tag{2.9b}$$

The coherence function (Zernike 1938) relates the wave fields received at two neighboring (which is considered to as relative to the distance from the source) points in space. Equation (2.9) is called the mutual coherence function and represents a temporal complex cross-correlation between the functions $U(r_1, t - t_1)$ and

$U(r_2, t - t_2)$ during the time interval T. Such a function depends on the time delay, $\tau[=(s_2 - s_1)/c]$, as well as on the separation between P_1 and P_2. In a stationary field, Γ depends on t_1 and t_2 only through the difference $t_1 - t_2 = \tau$; therefore, by defining the mutual coherence function, $\Gamma(\mathbf{r}_1, \mathbf{r}_2, \tau)$, as,

$$\Gamma(\mathbf{r}_1, \mathbf{r}_2; t_1, t_2) = \Gamma(\mathbf{r}_1, \mathbf{r}_2, \tau)$$
$$= \langle U(\mathbf{r}_1, t + \tau)U^*(\mathbf{r}_2, t)\rangle. \tag{2.10}$$

At a point where both the points coincide, the self-coherence reduces to ordinary intensity. When $\tau = 0$,

$$\Gamma(\mathbf{r}_1, \mathbf{r}_1, 0) = I(\mathbf{r}_1, t), \tag{2.11a}$$
$$\Gamma(\mathbf{r}_2, \mathbf{r}_2, 0) = I(\mathbf{r}_2, t). \tag{2.11b}$$

To note, the self-coherence function characterizes the interference effects given by the following equation,

$$\Gamma(\mathbf{r}, \tau) = \langle U^*(\mathbf{r}, t)U(\mathbf{r}, t + \tau)\rangle = |A(\mathbf{r}, t)|^2 e^{-i2\pi\nu\tau}$$
$$= 4\int_0^\infty \widehat{\Gamma}(\mathbf{r}, \mathbf{r}, \nu)e^{-i2\pi\nu\tau}d\nu, \tag{2.12}$$

where $\widehat{\Gamma}(\mathbf{r}, \nu)$ is the power spectrum of the complex light field, $\omega = 2\pi\nu$, and ν the frequency. By denoting $I_j(\mathbf{r}, t) = |K_j|^2\langle|U(\mathbf{r}_j, t - t_j)|^2\rangle$, in which $j = 1, 2$, one derives the intensity at $P(\mathbf{r})$,

$$I(\mathbf{r}, t) = I_1(\mathbf{r}, t) + I_2(\mathbf{r}, t) + 2|K_1 K_2|\Re[\Gamma(\mathbf{r}_1, \mathbf{r}_2, \tau)]. \tag{2.13}$$

The term $|K_1|^2 I(\mathbf{r}_1, t)$ is the intensity observed at $P(\mathbf{r})$ when the pin-hole at $P_1(\mathbf{r}_1)$ alone is opened ($K_2 = 0$) and the term $|K_2|^2 I(\mathbf{r}_2, t)$ has similar interpretation. These intensities may be denoted as, $I_1(\mathbf{r}, t)$ and $I_2(\mathbf{r}, t)$ respectively, i.e.,

$$I_1(\mathbf{r}, t) = |K_1|^2 I_1(\mathbf{r}, t) = |K_1|^2\Gamma(\mathbf{r}_1, \mathbf{r}_1, 0), \tag{2.14a}$$
$$I_2(\mathbf{r}, t) = |K_2|^2 I_2(\mathbf{r}, t) = |K_2|^2\Gamma(\mathbf{r}_2, \mathbf{r}_2, 0). \tag{2.14b}$$

The complex degree of (mutual) coherence of the light vibrations, $\gamma(\mathbf{r}_1, \mathbf{r}_2, \tau)$, of an observed source is a normalized correlation function between the complex fields at the points $P_1(\mathbf{r}_1)$ and $P_2(\mathbf{r}_2)$,

$$\gamma(\mathbf{r}_1, \mathbf{r}_2, \tau) = \frac{\Gamma(\mathbf{r}_1, \mathbf{r}_2, \tau)}{\sqrt{\Gamma(\mathbf{r}_1, \mathbf{r}_1, 0)}\sqrt{\Gamma(\mathbf{r}_2, \mathbf{r}_2, 0)}}$$
$$= \frac{\Gamma(\mathbf{r}_1, \mathbf{r}_2, \tau)}{\sqrt{I_1(\mathbf{r}, t)}\sqrt{I_2(\mathbf{r}, t)}}. \tag{2.15}$$

To note, The normalized form of $\Gamma(\mathbf{r}, \tau)$, known as the complex degree of self-coherence describing the correlation of vibrations at a fixed and a variable point of the light is given by,

$$\gamma(\mathbf{r}, \tau) = \frac{\Gamma(\mathbf{r}, \tau)}{\Gamma(\mathbf{r}, 0)} = \frac{\int_0^\infty \widehat{\Gamma}(\mathbf{r}, \mathbf{r}, \nu) e^{-i2\pi\nu\tau} d\nu}{\int_0^\infty \widehat{\Gamma}(\mathbf{r}, \mathbf{r}, \nu) d\nu}, \tag{2.16}$$

in which $\widehat{\Gamma}(\mathbf{r}, \nu)$ is referred to as the self-spectral density of the two beams.

The complex degree of coherence, $\gamma(\mathbf{r}_1, \mathbf{r}_2, \tau)$, measures both the spatial and the temporal coherence and is characterized by the following properties:

- it is a function with a maximum value at the origin for $\tau = 0$,
- the degree of coherence of the vibration is given by the Cauchy-Schwarz's inequality,[1] $0 \leq |\gamma(\mathbf{r}_1, \mathbf{r}_2, \tau)| \leq 1$, and
- the modulus of the complex degree of coherence is proportional to the contrast or visibility of the interference fringes, therefore, by measuring this one may obtain information about the quality of the source of the interference system.

In general, two light beams are not correlated but the correlation term, $U(\mathbf{r}_1, t)U^*(\mathbf{r}_2, t)$, takes on significant values for a short period of time and $\langle U(\mathbf{r}_1, t)U^*(\mathbf{r}_2, t) \rangle = 0$. Time variations of $U(\mathbf{r}, t)$ are statistical in nature (Mandel and Wolf 1995). Hence, one seeks a statistical description of the field (correlations) as the field is due to a partially coherent source. Depending upon the correlations between the phasor amplitudes at different object points, one would expect a definite correlation between the two points of the field. The effect of $|\gamma(\mathbf{r}_1, \mathbf{r}_2, \tau)|$ is to reduce the visibility of the fringes. There are three following operating regimes for the interference system as a function of the value of $|\gamma(\mathbf{r}_1, \mathbf{r}_2, \tau)|$:

$$|\gamma(\mathbf{r}_1, \mathbf{r}_2, \tau)| \begin{cases} = 1 & \text{completely coherent,} \\ < 1 & \text{partially coherent,} \\ = 0 & \text{incoherent superposition.} \end{cases} \tag{2.17}$$

In the first case, the system is operating in the coherent limit, and the vibrations at $P_1(\mathbf{r}_1)$ and $P_1(\mathbf{r}_1)$ may said to be coherent. In the second case, the vibrations are said to be partially coherent and the source operates with partial degree of coherence. In the case of interferometry, two different polychromatic self-luminous point sources are often considered to be spatially coherent if the fringe packets produced on a detector fall in the same scanning region of the applied path length difference. This coherence is known as partial coherence, which applies to extended sources.

---

[1] The modulus of the inner product of two vectors is smaller than or equal to the product of the norms of these vectors, i.e., for x, y in an inner space $|\langle x, y \rangle| \leq |x| |y|$ and the equality holds if $\{x, y\}$ is a linearly dependent set.

Such a coherence is a property of two waves whose relative phase undergoes random fluctuations which are not enough to make the wave completely incoherent. The third case describes that the system operates in the incoherent limit and the super-position of the two beams do not give rise to any interference effect.

The (2.13) for $I(\mathbf{r}, t)$ can be recast in,

$$I(\mathbf{r}, t) = I_1(\mathbf{r}, t) + I_2(\mathbf{r}, t) + 2\sqrt{I_1(\mathbf{r}, t)}\sqrt{I_2(\mathbf{r}, t)}\Re\left[\gamma(\mathbf{r}_1, \mathbf{r}_2, \tau)\right]. \qquad (2.18)$$

This (2.18) is the general interference law for stationary optical fields. In order to determine the light intensity at $P(\mathbf{r})$ when the two light waves are superposed, the intensity of each beam and the value of the real term, $\gamma(\mathbf{r}_1, \mathbf{r}_2, \tau)$, of the complex degree of coherence must be available.

2.1.2 Source and Visibility

The intensity at the point of superposition varies between maxima which exceeds the sum of the intensities in the beams and minima, which may be zero. The exact pattern depends on the wavelength of the light, the angular size of the source, and the separation between the apertures or slits. The resolution of an interferometer depends on the separation between the slits and is dictated by the spacing between the maxima, which is known as the fringe angular spacing. If one of the apertures is closer to the light source by a half of a wavelength, a crest in one light beam corresponds to a trough in the other beam; hence, these two light waves cancel each other, making the source disappear. The light source reappears and disappears every time the delay between the apertures is a multiple of the wave period. If the apertures are kept sufficiently wide, the source is resolved. When the light emitted by a source σ is quasi-monochromatic with a mean frequency $\bar{\nu}$, it has a spectral range, $\Delta\nu \ll \bar{\nu}$; hence, the complex degree of coherence turns out to be,

$$\gamma(\mathbf{r}_1, \mathbf{r}_2, \tau) = |\gamma(\mathbf{r}_1, \mathbf{r}_2, \tau)|\, e^{i[\Phi(\mathbf{r}_1, \mathbf{r}_2, \tau) - 2\pi\bar{\nu}\tau]}, \qquad (2.19)$$

with

$$\Phi(\mathbf{r}_1, \mathbf{r}_2, \tau) = 2\pi\bar{\nu}\tau + arg\left[\gamma(\mathbf{r}_1, \mathbf{r}_2, \tau)\right]. \qquad (2.20)$$

To note, the normalized complex degree of self-coherence is given by,

$$\gamma(\mathbf{r}, \tau) = |\gamma(\mathbf{r}, \tau)|\, e^{i[\Phi(\mathbf{r}, \tau) - 2\pi\bar{\nu}\tau]},$$

where $\Phi(\mathbf{r}, \tau) = 2\pi\nu_0\tau + arg[\gamma(\mathbf{r}, \tau)]$.

If $|\gamma(\mathbf{r}_1, \mathbf{r}_2, \tau)| = 0$, (2.18) becomes,

$$I(\mathbf{r}, t) = I_1(\mathbf{r}, t) + I_2(\mathbf{r}, t), \qquad (2.21)$$

in which $I_1(\mathbf{r}, t)$, $I_2(\mathbf{r}, t)$ are the intensities of the light in each arm of the interferometer. When the vibrations add up in intensity at $P(\mathbf{r})$, the vibrations are said to be incoherent. The intensity at a point, $P(\mathbf{r})$, in the case of $|\gamma(\mathbf{r}_1, \mathbf{r}_2, \tau)| = 1$ is,

$$I(\mathbf{r}, t) = I_1(\mathbf{r}, t) + I_2(\mathbf{r}, t) + 2\sqrt{I_1(\mathbf{r}, t)}\sqrt{I_2(\mathbf{r}, t)}$$
$$\times |\gamma(\mathbf{r}_1, \mathbf{r}_2, \tau)| \cos[\Phi(\mathbf{r}_1, \mathbf{r}_2, \tau) - \delta], \tag{2.22}$$

where

$$\delta = 2\pi\bar{\nu}\tau = \frac{2\pi(s_2 - s_1)}{\bar{\lambda}} = \bar{\kappa}(s_2 - s_1),$$

with $\bar{\kappa} = 2\pi\bar{\nu}/c = 2\pi/\bar{\lambda}$, $\bar{\kappa}$ as the mean wave number, $\bar{\nu}$ the mean frequency, $\bar{\nu}$ the mean wavelength, and the argument $\Phi(\mathbf{r}_1, \mathbf{r}_2, \tau) - \delta$, in which δ arises from the path difference, s_1 and s_2 two optical path lengths, and $\Phi(\mathbf{r}_1, \mathbf{r}_2, \tau)$ contains information about the source.

The visibility function, $\mathcal{V}$, is related to brightness morphology that an interferometer measures and indicates the extent to which a source is resolved on the baseline used. It is the modulus of the degree of coherence, $\gamma(\mathbf{r}_1, \mathbf{r}_2, \tau)$, at the spatial frequency vector, $\mathbf{u}(= u, v)$, i.e., (projected) baseline vectors.[2] The spatial frequency coordinates depend on the separation of the telescopes projected in the direction of observation,

$$u_{ij} = \frac{B_{x0}}{\lambda} = \frac{\kappa B_{x0}}{2\pi}; \tag{2.23a}$$

$$v_{ij} = \frac{B_{y0}}{\lambda} = \frac{\kappa B_{y0}}{2\pi}. \tag{2.23b}$$

where $\mathbf{B}$ is the baseline vector and λ the wavelength of observation.

Both $\mathbf{B}$ and λ are linear quantities and must be expressed in the same units. The orientation of fringes is normal to the baseline vector and the phase of the fringe pattern is equal to the Fourier phase of the same spatial frequency component. To note, a spatial fringe pattern is a 2-D signal as a function of the image coordinates with fringes along one coordinate enveloped by an Airy disk, while a temporal fringe pattern is a 1-D signal as a function of OPD enveloped by the Fourier transform of the spectrum. Figure 2.2 depicts the snapshots and the corresponding intensity curves of two beam interference fringes obtained from a circular source with various spacings of pin-holes. As the source becomes small compared to the fringe spacing, the visibility approaches unity, while the latter approaches zero as the former becomes very small compared to the source; source extent is equal to the fringe spacing or multiples thereof.

---

[2] The projected baseline is defined as the projection of the baseline vector onto the plane, which is perpendicular to the line-of-sight of the source.

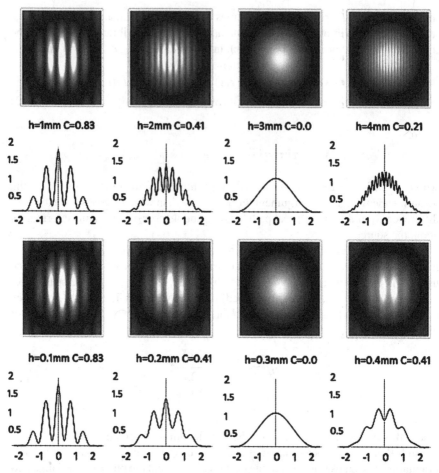

Fig. 2.2 Illustration of the influence on the spatial coherence (on the contrast C) of h, the distance between the holes, and a, the semi-diameter of a circular source, in the Young's two-hole experiment, assuming $d = 1$ m (distance of the source) and a wavelength of 600 nm. For a between 0.1 and 0.4 mm (top simulation), h = 1 mm; for h between 1 and 4 mm (bottom simulation), a = 0.1 mm. The corresponding intensity curves of the two beam interference fringes obtained from the circular source changes with the various spacings of holes and the size of the holes, being zero when 'a' × 'h' = 0.3 (constructive fringe for 'a' × 'h' < 0.3 and destructive fringe for 0.3 < 'a' × 'h' < 0.6). Courtesy: Luc Damé

The function, $\gamma(\mathbf{r}_1, \mathbf{r}_2, \tau)$, can be approximated to $\gamma(\mathbf{r}_1, \mathbf{r}_2, 0)e^{-i2\pi\bar{\nu}\tau}$ for $\tau \ll \tau_c$. The exponential term is nearly constant and $\gamma(\mathbf{r}_1, \mathbf{r}_2, 0)$, measures the spatial coherence. Let $\Phi(\mathbf{r}_1, \mathbf{r}_2)$, be the argument of $\gamma(\mathbf{r}_1, \mathbf{r}_2, \tau)$, thus,

$$
\begin{aligned}
I(\mathbf{r}_1, \mathbf{r}_2, \tau) = {} & I_1(\mathbf{r}, t) + I_2(\mathbf{r}, t) + 2\sqrt{I_1(\mathbf{r}, t)}\sqrt{I_2(\mathbf{r}, t)} \\
& \times \Re\left[|\gamma(\mathbf{r}_1, \mathbf{r}_2, 0)| \, e^{i[\Phi(\mathbf{r}_1, \mathbf{r}_2) - 2\pi\bar{\nu}\tau]}\right].
\end{aligned} \tag{2.24}
$$

The (2.24) illustrates the Young's experiment. For a quasi-monochromatic optical source, $|\gamma(\mathbf{r}_1, \mathbf{r}_2, \tau)|$ is nearly constant (amplitude of fringe envelope) and the phase, $\Phi(\mathbf{r}_1, \mathbf{r}_2, \tau)$, when is considered as functions of τ, varies slowly compared to $\cos 2\pi \bar{\nu}\tau$ and $\sin 2\pi \bar{\nu}\tau$. If the openings at $P_1(\mathbf{r}_1)$ and $P_2(\mathbf{r}_2)$ are sufficiently small, the intensity distributions of the light in the vicinity of P consist of an almost uniform background (no intensity envelope due to diffraction by a finite aperture) $I_1(\mathbf{r}, t) + I_2(\mathbf{r}, t)$ in which a sinusoidal intensity distribution is superimposed with constant amplitude $2\sqrt{I_1(\mathbf{r}, t)}\sqrt{I_2(\mathbf{r}, t)}|\gamma(\mathbf{r}_1, \mathbf{r}_2, \tau)|$. The intensity maxima and minima near $P(\mathbf{r})$ are, therefore given by,

$$I_{max} = I_1(\mathbf{r}, t) + I_2(\mathbf{r}, t) + 2\sqrt{I_1(\mathbf{r}, t)}\sqrt{I_2(\mathbf{r}, t)}\,|\gamma(\mathbf{r}_1, \mathbf{r}_2, \tau)|\,,$$
$$I_{min} = I_1(\mathbf{r}, t) + I_2(\mathbf{r}, t) - 2\sqrt{I_1(\mathbf{r}, t)}\sqrt{I_2(\mathbf{r}, t)}\,|\gamma(\mathbf{r}_1, \mathbf{r}_2, \tau)|\,. \quad (2.25)$$

The modulus of the fringe visibility is estimated as the ratio of high frequency to low frequency energy in the average spectral density. In the vicinity of zero path length difference, the visibility of the fringes $\mathcal{V}(\mathbf{r})$, at a point $P(\mathbf{r})$ in terms of the intensity of the two beams and of their degree of coherence is estimated as,

$$\mathcal{V}(\mathbf{r}) = \frac{I_{max} - I_{min}}{I_{max} + I_{min}}$$
$$= |\gamma(\mathbf{r}_1, \mathbf{r}_2, \tau)|\,\frac{2\sqrt{I_1(\mathbf{r}, t)}\sqrt{I_2(\mathbf{r}, t)}}{I_1(\mathbf{r}, t) + I_2(\mathbf{r}, t)}\,, \quad (2.26)$$

in which τ is a possible delay between the signals transmitted by the two sub-apertures of the interferometer; for a zero time shift τ such a correlation normally yields the maximum value, and if $\tau \ll 1/\Delta\nu$, the visibility turns out to be, $\mathcal{V}(\mathbf{r}) = |\gamma(\mathbf{r}_1, \mathbf{r}_2, 0)|$.

For the special case that both openings transmit an equal power and $I_1(\mathbf{r}, t) = I_2(\mathbf{r}, t)$, the visibility function $\mathcal{V} = |\gamma(\mathbf{r}_1, \mathbf{r}_2, \tau)|$ equals the modulus of the complex degree of coherence, with secondary sources at P_1 and P_2. Although this expression provides a contrast as a function of the location $P(\mathbf{r})$, it should be noted that this contrast is not a function of the size and shape of the collectors. According to the Schwarz's inequality, one finds $|\gamma(\mathbf{r}_1, \mathbf{r}_2, \tau)| \leq 1$, hence from (2.24), it is derived as,

$$I(\mathbf{r}, t) = \begin{cases} [I_1(\mathbf{r}, t) + I_2(\mathbf{r}, t)]^2 & \text{if } \cos[\Phi(\mathbf{r}_1, \mathbf{r}_2, \tau) - \delta] = 1, \\ [I_1(\mathbf{r}, t) - I_2(\mathbf{r}, t)]^2 & \text{if } \cos[\Phi(\mathbf{r}_1, \mathbf{r}_2, \tau) - \delta] = -1. \end{cases} \quad (2.27)$$

The resulting amplitude at $P(\mathbf{r})$ in certain cases is equal to the sum of the amplitudes (vibration in phase). It can also be equal to the difference of the amplitudes (vibrations in anti-phase). The (2.24) can be recast in the form,

$$I(\mathbf{r}, t) = \left[I_1(\mathbf{r}, t) + I_2(\mathbf{r}, t) + 2\sqrt{I_1(\mathbf{r}, t)}\sqrt{I_2(\mathbf{r}, t)}\cos[\Phi(\mathbf{r}_1, \mathbf{r}_2, \tau) - \delta] \right]$$
$$\times |\gamma(\mathbf{r}_1, \mathbf{r}_2, \tau)| + (1 - |\gamma(\mathbf{r}_1, \mathbf{r}_2, \tau)|)[I_1(\mathbf{r}, t) + I_1(\mathbf{r}, t)]\,. \quad (2.28)$$

The first term arises from coherent superposition of the two beams of intensities $|\gamma(\mathbf{r}_1, \mathbf{r}_2, \tau)|I_1(\mathbf{r}, t)$ and $|\gamma(\mathbf{r}_1, \mathbf{r}_2, \tau)|I_2(\mathbf{r}, t)$ and of relative phase difference $\Phi(\mathbf{r}_1, \mathbf{r}_2, \tau) - \delta$ and the second term arises from incoherent superposition of the two beams of intensities, $[1 - |\gamma(\mathbf{r}_1, \mathbf{r}_2, \tau)|]I_1(\mathbf{r}, t)$ and $[1 - |\gamma(\mathbf{r}_1, \mathbf{r}_2, \tau)|]I_2(\mathbf{r}, t)$. Thus, the light which reaches $P(\mathbf{r})$ from all other pin-holes may be regarded as mixture of coherent and incoherent light beams, with intensity in the ratio,

$$\frac{I_{coh}}{I_{incoh}} = \frac{|\gamma(\mathbf{r}_1, \mathbf{r}_2, \tau)|}{1 - |\gamma(\mathbf{r}_1, \mathbf{r}_2, \tau)|}, \tag{2.29a}$$

$$\frac{I_{coh}}{I_{tot}} = |\gamma(\mathbf{r}_1, \mathbf{r}_2, \tau)|, \tag{2.29b}$$

where $I_{tot} = I_{coh} + I_{incoh}$.

The first zero of the intensity pattern for an interferometer is given by,

$$\theta_{int} = \frac{\lambda}{2B} \quad \text{rad.} \tag{2.30}$$

2.1.3 Power-spectral Density of the Light Beam

The power-spectral density refers to the power of a signal or time series as a function of a frequency variable associated with a stationary stochastic process[3] having units of power per frequency (watts per Hz). It is given by the Wiener–Khintchine theorem (see Appendix B). The power-spectral densities are statistical measures, which can be estimated from real data by averaging over the results from many measurements. They can be specified as one-sided functions of positive frequencies, or as two-sided functions of positive and negative frequencies. Optical power densities are, in general, one-sided and can be measured with an optical spectrum analyzer. In the case of ergodicity, temporal coherence is found to be linked to the power spectral density of the source. Following (1.21), one may write,

$$\int_{-\infty}^{\infty} U_T^{(r)}(\mathbf{r}, t + \tau)U_T^{(r)}(\mathbf{r}, t)dt = \int_{-\infty}^{\infty} U_T^{(r)}(\mathbf{r}, t)dt$$
$$\times \left[\int_{-\infty}^{\infty} \widehat{U}_T^{(r)}(\mathbf{r}, \nu)e^{-i2\pi\nu(t+\tau)}d\nu \right]$$
$$= \int_{-\infty}^{\infty} \widehat{U}_T^{(r)}(\mathbf{r}, \nu)e^{-i2\pi\nu\tau}d\nu$$
$$\times \left[\int_{-\infty}^{\infty} U_T^{(r)}(\mathbf{r}, t)e^{-i2\pi\nu t}dt \right]. \tag{2.31}$$

---

[3] A stochastic process, concerns a sequence of events governed by probability theorem. It is a family of interdependent random variables that evolve over time (Papoulis 1984); there is some indeterminacy in its future evolution described by probability distributions. Some basic types of stochastic processes include (1) Markov processes, (2) Poisson processes (for instance, radioactive decay), and (3) time series, with the index variable referring to time.

After dividing both sides by 2T, (2.31) can be recast in,

$$\frac{1}{2T} \int_{-\infty}^{\infty} U_T^{(r)}(\mathbf{r}, t + \tau) U_T^{(r)}(\mathbf{r}, t) dt = \frac{1}{2T} \int_{-\infty}^{\infty} \widehat{U}_T^{(r)}(\mathbf{r}, \nu) \widehat{U}_T^{(r)*}(\mathbf{r}, \nu) e^{-i2\pi\nu\tau} d\nu.$$

(2.32)

The smoothing operation is applied to the right hand side quantity by taking ensemble average over the ensemble of the random function $U^{(r)}$,

$$\left\langle U^{(r)}(\mathbf{r}, t + \tau) U^{(r)}(\mathbf{r}, t) \right\rangle = \int_{-\infty}^{\infty} \widehat{\Gamma}(\mathbf{r}, \nu) e^{-i2\pi\nu\tau} d\nu,$$

(2.33)

in which the function $\widehat{\Gamma}(\mathbf{r}, \nu)$ known as the power-spectral density of the optical disturbance, and is given by,

$$\widehat{\Gamma}(\mathbf{r}, \nu) = \lim_{T \to \infty} \left[\frac{\overline{\widehat{U}_T(\mathbf{r}, \nu) \widehat{U}_T^*(\mathbf{r}, \nu)}}{2T} \right].$$

(2.34)

To note, for an analytical signal $U(t)$ associated with $U^{(r)}(\mathbf{r}, t)$, the self-coherence function, $\Gamma(\mathbf{r}, \tau)$ (see 2.12). A similar procedure may be followed in the case of the spatial coherence,

$$\int_{-\infty}^{\infty} U_T^{(r)}(\mathbf{r}_1, t + \tau) U_T^{(r)}(\mathbf{r}_2, t) dt = \int_{-\infty}^{\infty} \widehat{U}_T^{(r)}(\mathbf{r}_1, \nu) \widehat{U}_T^{(r)*}(\mathbf{r}_2, \nu) e^{-i2\pi\nu\tau} d\nu.$$

(2.35)

After applying the smoothing operation to the right hand side quantity, the ensemble average is given by,

$$\langle U^{(r)}(\mathbf{r}_1, t + \tau) U^{(r)}(\mathbf{r}_2, t) \rangle = \int_{-\infty}^{\infty} \widehat{\Gamma}(\mathbf{r}_1, \mathbf{r}_2, \nu) e^{-i2\pi\nu\tau} d\nu,$$

(2.36)

in which the mutual spectral density or cross-spectral density of the light vibrations at $P_1(\mathbf{r}_1)$ and $P_2(\mathbf{r}_2)$, $\widehat{\Gamma}(\mathbf{r}_1, \mathbf{r}_2, \nu)$ is given by,

$$\widehat{\Gamma}(\mathbf{r}_1, \mathbf{r}_2, \nu) = \lim_{T \to \infty} \left[\frac{\overline{\widehat{U}_T(\mathbf{r}_1, \nu) \widehat{U}_T^*(\mathbf{r}_2, \nu)}}{2T} \right].$$

(2.37)

According to the Hermitian property, $\widehat{\Gamma}(\mathbf{r}_2, \mathbf{r}_1, \nu) = \widehat{\Gamma}^*(\mathbf{r}_1, \mathbf{r}_2, \nu)$; hence, the cross-spectral density function or the cross power spectrum, $\widehat{\Gamma}(\mathbf{r}_1, \mathbf{r}_2, \nu)$ of the light vibrations at points $\mathbf{r}_1$ and $\mathbf{r}_2$ is defined by,

$$\langle \widehat{U}^*(\mathbf{r}_1, \nu) \widehat{U}(\mathbf{r}_2, \nu') \rangle = \widehat{\Gamma}(\mathbf{r}_1, \mathbf{r}_2, \nu) \delta'(\nu - \nu'),$$

(2.38)

in which δ' is the Dirac delta function and the ensemble average $\langle \rangle$ is taken over the different realizations of the field.

The (2.37) implies that the cross-spectral density function, $\widehat{\Gamma}(\mathbf{r}_1, \mathbf{r}_2, \nu)$, is a measure of the correlation between the spectral amplitudes of any specific frequency component of the light disturbances at the points $P_1(\mathbf{r}_1)$ and $P_2(\mathbf{r}_2)$. When these points, coincide, the cross-spectral density function, $\widehat{\Gamma}(\mathbf{r}_1, \mathbf{r}_2, \nu)$ turns out to be a function of one frequency, thus represents the spectral density of the light, $\widehat{\Gamma}_p(\mathbf{r}, \nu)$. For a signal represented by $U(\mathbf{r}, t)$ and associated with $U^{(r)}(\mathbf{r}, t)$, the spectral representation of the mutual coherence function can be derived following (1.21 and 2.36),

$$\Gamma(\mathbf{r}_1, \mathbf{r}_2, \tau) = 4 \int_0^\infty \widehat{\Gamma}(\mathbf{r}_1, \mathbf{r}_2, \nu) e^{-i2\pi\nu\tau} d\nu. \tag{2.39}$$

If $\Gamma^{(r)}(\mathbf{r}_1, \mathbf{r}_2)$ and $\Gamma^{(i)}(\mathbf{r}_1, \mathbf{r}_2)$ denote its real and imaginary parts, i.e.,

$$\Gamma(\mathbf{r}_1, \mathbf{r}_2, \tau) = \Re\left[\Gamma(\mathbf{r}_1, \mathbf{r}_2, \tau)\right] + \Im\left[\Gamma(\mathbf{r}_1, \mathbf{r}_2, \tau)\right], \tag{2.40}$$

the relation between them can be connected by a Hilbert transform (Papoulis 1968). Thus, the correlation between the real and imaginary part of $\Gamma(\mathbf{r}_1, \mathbf{r}_2, \tau)$ is given by,

$$\Im\left[\Gamma(\mathbf{r}_1, \mathbf{r}_2, \tau)\right] = \frac{1}{\pi} P \int_{-\infty}^\infty \frac{\Re\left[\Gamma(\mathbf{r}_1, \mathbf{r}_2, \tau')\right]}{\tau' - \tau} d\tau', \tag{2.41}$$

$$\Re\left[\Gamma(\mathbf{r}_1, \mathbf{r}_2, \tau)\right] = -\frac{1}{\pi} P \int_{-\infty}^\infty \frac{\Im\left[\Gamma(\mathbf{r}_1, \mathbf{r}_2, \tau')\right]}{\tau' - \tau} d\tau', \tag{2.42}$$

where P stands for the Cauchy principal value at $t = \tau$.

The coherence function, $|\Gamma(\mathbf{r}_1, \mathbf{r}_2)|$ considered as a function of τ, is the envelope of $\Re[\Gamma(\mathbf{r}_1, \mathbf{r}_2)]$,

$$\begin{aligned}\Re\left[\Gamma(\mathbf{r}_1, \mathbf{r}_2, \tau)\right] &= 2\langle U^{(r)}(\mathbf{r}_1, t + \tau) U^{(r)}(\mathbf{r}_2, t)\rangle \\ &= 2 \int_{-\infty}^\infty \widehat{\Gamma}(\mathbf{r}_1, \mathbf{r}_2, \nu) e^{-i2\pi\nu\tau} d\nu,\end{aligned} \tag{2.43}$$

and $|\gamma(\mathbf{r}_1, \mathbf{r}_2)|$ is the envelope of the real correlation factor,

$$\begin{aligned}\Re\left[\gamma(\mathbf{r}_1, \mathbf{r}_2, \tau)\right] &= \frac{\Re\left[\Gamma(\mathbf{r}_1, \mathbf{r}_2, \tau)\right]}{\left[\Gamma(\mathbf{r}_1, \mathbf{r}_1, 0)\Gamma(\mathbf{r}_2, \mathbf{r}_2, 0)\right]^{1/2}} \\ &= \frac{\langle U^{(r)}(\mathbf{r}_1, t + \tau) U^{(r)}(\mathbf{r}_2, t)\rangle}{\left[\langle U^{(r)2}(\mathbf{r}_1, t)\rangle \langle U^{(r)2}(\mathbf{r}_2, t)\rangle\right]^{1/2}}.\end{aligned} \tag{2.44}$$

The (2.44) shows that the real part of $\Gamma(\mathbf{r}_1, \mathbf{r}_2, \tau)$ is equal to twice the cross correlation function of the real functions $U^{(r)}(\mathbf{r}_1, t)$ and $U^{(r)}(\mathbf{r}_2, t)$.

2.1.4 Mutual Intensity

Under quasi-monochromatic condition, the time delay, τ that is introduced between the interfering beams must also be small compared to the coherence time τ_c of the light, i.e., $\tau \ll \tau_c$. Thus, by invoking (2.15, 2.19, and 2.39) one obtains,

$$|\Gamma(\mathbf{r}_1, \mathbf{r}_2, \tau)|\, e^{i\Phi(\mathbf{r}_1, \mathbf{r}_2, \tau)} = \sqrt{I_1(\mathbf{r}, t) I_2(\mathbf{r}, t)}\, |\gamma(\mathbf{r}_1, \mathbf{r}_2, \tau)|\, e^{i\Phi(\mathbf{r}_1, \mathbf{r}_2, \tau)}$$

$$= 4\int_0^\infty \widehat{\Gamma}(\mathbf{r}_1, \mathbf{r}_2, \nu) e^{-i2\pi(\nu - \bar{\nu})\tau}\, d\nu. \qquad (2.45)$$

If the time delay is so small that $|(\nu - \bar{\nu})\tau| \ll 1$ for all frequencies for which $|\widehat{\Gamma}(\mathbf{r}_1, \mathbf{r}_2, \nu)|$ is appreciable, i.e., if $|\tau| \ll 1/\Delta\nu$, a small error is introduced if the exponential term of the integrand in (2.45) is replaced by unity. The mutual coherence function, as well as the complex degree of coherence are expressed as,

$$\Gamma(\mathbf{r}_1, \mathbf{r}_2, \tau) \simeq |J(\mathbf{r}_1, \mathbf{r}_2)| e^{i(\Psi(\mathbf{r}_1, \mathbf{r}_2) - 2\pi\bar{\nu}\tau)}$$

$$= J(\mathbf{r}_1, \mathbf{r}_2) e^{-i2\pi\bar{\nu}\tau}, \qquad (2.46a)$$

$$\gamma(\mathbf{r}_1, \mathbf{r}_2, \tau) \simeq |\mu(\mathbf{r}_1, \mathbf{r}_2)| e^{i(\Psi(\mathbf{r}_1, \mathbf{r}_2) - 2\pi\bar{\nu}\tau)}$$

$$= \mu(\mathbf{r}_1, \mathbf{r}_2) e^{-i2\pi\bar{\nu}\tau}, \qquad (2.46b)$$

with

$$J(\mathbf{r}_1, \mathbf{r}_2) = \Gamma(\mathbf{r}_1, \mathbf{r}_2, 0) = \langle U(\mathbf{r}_1, t) U^*(\mathbf{r}_2, t)\rangle$$

$$= \langle A_1(\mathbf{r}_1, t) A_2^*(\mathbf{r}_2, t)\rangle. \qquad (2.47)$$

as the mutual intensity of the light at the apertures $P_1(\mathbf{r}_1)$ and $P_2(\mathbf{r}_2)$,

$$\mu(\mathbf{r}_1, \mathbf{r}_2) = \gamma(\mathbf{r}_1, \mathbf{r}_2, 0) = \frac{\Gamma(\mathbf{r}_1, \mathbf{r}_2, 0)}{\sqrt{\Gamma(\mathbf{r}_1, \mathbf{r}_1, 0)}\sqrt{\Gamma(\mathbf{r}_2, \mathbf{r}_2, 0)}}$$

$$= \frac{J(\mathbf{r}_1, \mathbf{r}_2)}{\sqrt{J(\mathbf{r}_1, \mathbf{r}_1)}\sqrt{J(\mathbf{r}_2, \mathbf{r}_2)}} = \frac{J(\mathbf{r}_1, \mathbf{r}_2)}{\sqrt{I_1(\mathbf{r}, t)}\sqrt{I_2(\mathbf{r}, t)}}, \qquad (2.48)$$

the complex coherence factor of the light, and

$$\Psi(\mathbf{r}_1, \mathbf{r}_2) = \Phi(\mathbf{r}_1, \mathbf{r}_2, 0) = arg\{\gamma(\mathbf{r}_1, \mathbf{r}_2, 0)\} = arg\{\mu(\mathbf{r}_1, \mathbf{r}_2)\}, \qquad (2.49)$$

the phase.

The mutual intensity, $J(\mathbf{r}_1, \mathbf{r}_2)$ is regarded as a phasor amplitude of a spatial sinusoidal fringe, while $\mu(\mathbf{r}_1, \mathbf{r}_2)$ is a normalized version of $J(\mathbf{r}_1, \mathbf{r}_2)$ having the property, $0 \le |\mu(\mathbf{r}_1, \mathbf{r}_2)| \le 1$. If the coherence length of the light is much greater than the maximum pathlength difference (for vacuum) encountered in passage from the source to the interference region of interest, i.e., $\Delta\nu \ll \bar{\nu}$, i.e.,

$$|\Delta\varphi| = |s_2 - s_1| = \frac{\bar{\lambda}}{2\pi}\delta \ll \frac{\lambda^2}{\Delta\lambda}, \tag{2.50}$$

the interference law (2.24) can be recast in,

$$I(\mathbf{r}, t) \sim I_1(\mathbf{r}, t) + I_2(\mathbf{r}, t) + 2\sqrt{I_1(\mathbf{r}, t)}\sqrt{I_2(\mathbf{r}, t)}\,|\mu(\mathbf{r}_1, \mathbf{r}_2)|\cos(\Psi(\mathbf{r}_1, \mathbf{r}_2) - \delta). \tag{2.51}$$

Equation (2.51) represents the basic formula of the quasi-monochromatic theory of partial coherence. The complex coherence factor, $\mu(\mathbf{r}_1, \mathbf{r}_2)$ is often called the complex visibility; it fulfills the condition $|\mu| \leq 1$. Although complex, it can be measured from recorded intensities by varying τ. Given the positions of the secondary point sources, the complex visibility holds information about the light source, which is valid only if the assumptions of quasi-chromaticity and point-like collectors or pin-holes are satisfied. When $I_1(\mathbf{r}, t)$ and $I_2(\mathbf{r}, t)$ are constant in the case of a quasi-monochromatic light, the observed interference pattern has constant visibility and constant phase across the observation region. The visibility in terms of the complex coherence factor is,

$$V = \begin{cases} \dfrac{2\sqrt{I_1(\mathbf{r}, t)}\sqrt{I_2(\mathbf{r}, t)}}{I_1(\mathbf{r}, t) + I_2(\mathbf{r}, t)}\mu(\mathbf{r}_1, \mathbf{r}_2) & \text{when } I_1(\mathbf{r}, t) \neq I_2(\mathbf{r}, t), \\ \mu(\mathbf{r}_1, \mathbf{r}_2) & \text{otherwise.} \end{cases} \tag{2.52}$$

Such a complex coherence factor is usually called visibility by the astronomers. If the complex coherent factor, $\mu(\mathbf{r}_1, \mathbf{r}_2)$, turns out to be zero, the fringes vanish and the two lights are known to be mutually incoherent, and in the case of $\mu(\mathbf{r}_1, \mathbf{r}_2)$ being 1, the two waves are called mutually coherent. For an intermediate value of $\mu(\mathbf{r}_1, \mathbf{r}_2)$, the two waves are partially coherent. The spectral degree of coherence, $\widehat{\mu}(\mathbf{r}_1, \mathbf{r}_2, \nu)$, at frequency ν of the light at the points, $P_1(\mathbf{r}_1)$ and $P_2(\mathbf{r}_2)$, is given by,

$$\widehat{\mu}(\mathbf{r}_1, \mathbf{r}_2, \nu) = \frac{\widehat{\Gamma}(\mathbf{r}_1, \mathbf{r}_2, \nu)}{\sqrt{\widehat{\Gamma}(\mathbf{r}_1, \mathbf{r}_1, \nu)}\sqrt{\widehat{\Gamma}(\mathbf{r}_2, \mathbf{r}_2, \nu)}}. \tag{2.53}$$

The term $\widehat{\mu}(\mathbf{r}_1, \mathbf{r}_2, \nu)$ is also referred as the complex degree of spatial (or spectral) coherence at frequency at ν (Mandel and Wolf and Wolf 1976, 1995).

2.1.5 Propagation of Mutual Coherence

In order to derive the propagation laws for the cross-spectral density function and for the mutual coherence function, the well known Huygens–Fresnel principle for propagation of monochromatic light is elucidated to begin with. The notations relating to the laws for these functions are illustrated in Fig. 2.3. Assume that a wavefront with

Fig. 2.3 Geometry for propagation laws for the cross-spectral density and for the mutual coherence

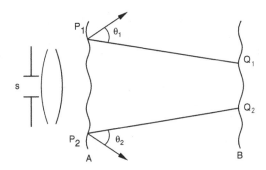

arbitrary coherence properties emerges from an optical system lying on a surface, A, which intercepts the beam from an extended source σ and propagates to a surface, B. Knowing the values for the mutual coherence function, $\Gamma(\mathbf{r}_1, \mathbf{r}_2, \tau)$ (see 2.10), all points $P(\mathbf{r}_1)$ and $P(\mathbf{r}_2)$ on the surface A, one may determine the values of this coherence function,

$$\Gamma(\mathbf{r}'_1, \mathbf{r}'_2, \tau) = \langle U(\mathbf{r}'_1, t + \tau) U^*(\mathbf{r}'_2, t) \rangle, \qquad (2.54)$$

for all points at $Q(\mathbf{r}'_1)$ and $Q(\mathbf{r}'_2)$, on the surface B. Here, $U(\mathbf{r}'_1, t)$ and $U(\mathbf{r}'_2, t)$ are the respective complex light disturbances at $Q_1(\mathbf{r}'_1)$, and $Q_2(\mathbf{r}'_2)$, and $U(\mathbf{r}'_1, \nu)$ and $U(\mathbf{r}'_2, \nu)$ the corresponding spectral amplitudes.

The refractive index of the medium between A and B is considered to be unity. It is found that to a good approximation, the coherence functions depend on time delay τ only through a harmonic term (see 2.46a, b). Following the Huygens–Fresnel principle (see Sect. 1.4.1), the complex amplitudes at the points $Q_1(\mathbf{r}'_1)$ and $Q_2(\mathbf{r}'_2)$, are expressed in terms of the complex amplitudes at all points on the surface B,

$$U(\mathbf{r}'_1, t + \tau) = \int_A \frac{\chi_1(\kappa)}{s_1} U\left(\mathbf{r}_1, t + \tau - \frac{s_1}{c}\right) d s_1, \qquad (2.55a)$$

$$U^*(\mathbf{r}'_2, t) = \int_A \frac{\chi_2^*(\kappa)}{s_2} U^*\left(\mathbf{r}_2, t - \frac{s_2}{c}\right) d s_2, \qquad (2.55b)$$

in which χ_1 and χ_2 are the inclination factors, s_1 and s_2 the distances P_1Q_1 and P_2Q_2 respectively.

If the effective spectral range of the light is sufficiently small, one may replace these factors by $\bar{\chi}_1 = \chi_1(\bar{\kappa})$, $\bar{\chi}_2 = \chi_2(\bar{\kappa})$, where $\bar{\kappa} = 2\pi\bar{\nu}/c$, and $\bar{\nu}$ denotes the mean frequency of the light. By plugging (2.55) into (2.10), the mutual coherence function on the surface B is given by,

$$\Gamma(\vec{r}'_1, \vec{r}'_2, \tau) = \int_A \int_A \frac{\chi_1 \chi_2^*}{s_1 s_2} \Gamma\left(\vec{r}_1, \vec{r}_2, \tau - \frac{s_2 - s_1}{c}\right) d s_1 d s_2. \qquad (2.56)$$

The (2.56) represents the propagation law for the mutual coherence function at points $Q_1(\mathbf{r}_1')$ and $Q_2(\mathbf{r}_2')$ of the surface B in terms of the mutual intensity at all pairs of points on the surface A. Under quasi-monochromatic conditions, one may write,

$$\Gamma\left(\vec{r}_1, \vec{r}_2, \tau - \frac{s_2 - s_1}{c}\right) = J(\vec{r}_1, \vec{r}_2)e^{i\bar{\kappa}(s_2 - s_1)}. \tag{2.57}$$

The (2.56) can be recast as,

$$J(\vec{r}_1', \vec{r}_2') = \int_A \int_A \frac{\chi_1}{s_1} \frac{\chi_2^*}{s_2} J(\vec{r}_1, \vec{r}_2)e^{i\bar{\kappa}(s_2 - s_1)} d S_1 d S_2. \tag{2.58}$$

Equation (2.58) relates the general law for propagation of mutual intensity. When the two points $Q_1(\vec{r}_1')$ and $Q_2(\vec{r}_2')$ coincide at $Q(\vec{r}')$ and $\tau = 0$, the intensity distribution on the surface B is deduced as,

$$I(\mathbf{r}') = \int_A \int_A \frac{\sqrt{I(\mathbf{r}_1)I(\mathbf{r}_2)}}{s_1 s_2} \gamma\left(\mathbf{r}_1, \mathbf{r}_2, \frac{s_2 - s_1}{c}\right) \chi_1 \chi_2^* d S_1 d S_2, \tag{2.59}$$

in which γ is the correlation function (see 2.15).

The (2.59) expresses the intensity at an arbitrary point $P(\mathbf{r}')$ as the sum of surface contributions from all pairs of elements of the arbitrary surface A.

2.2 Van Cittert–Zernike Theorem

The van Cittert–Zernike theorem (van Cittert 1934; Zernike 1938) deals with the field correlations generated by an extended incoherent quasi-monochromatic source. It relates the complex visibility function of the fringes to a unique Fourier component of the impinging brightness distribution. The modulus of the complex degree of coherence in a plane illuminated by an incoherent quasi-monochromatic source is equal to the modulus of the normalized spatial Fourier transform of its brightness distribution.

An extended polychromatic source is considered to be a mosaic of point sources, and hence the response of a telescope or an interferometer can be taken as the summation of the response functions to all these sources. Any two elements of such a source are assumed to be uncorrelated. This source, σ, produces an electromagnetic wave field, represented by the analytic signal, $S(\mathbf{r}', t)$, which is a function of position S and time t. Since it is a scalar quantity, it neglects polarization. This source may be divided into elements $d\sigma_1, d\sigma_2, \ldots$ centered on points $S_1, S_2, \ldots$, which are mutually incoherent, and of linear dimensions small compared to the mean wavelength $\bar{\lambda}$. If $U(\mathbf{r}_{m1}, t)$ and $U(\mathbf{r}_{m2}, t)$ are the complex disturbances at $P_1(\mathbf{r}_1)$ and $P_2(\mathbf{r}_2)$ due to the element $d\sigma$, the total disturbances at these points are,

$$U(\mathbf{r}_1, t) = \sum_m U(\mathbf{r}_{m1}, t), \qquad U(\mathbf{r}_2, t) = \sum_m U(\mathbf{r}_{m2}, t). \tag{2.60}$$

The mutual intensity, $J(\mathbf{r}_1, \mathbf{r}_2)$, at $P_1(\mathbf{r}_1)$ and $P_2(\mathbf{r}_2)$ is given by,

$$
\begin{aligned}
J(\mathbf{r}_1, \mathbf{r}_2) &= \langle U(\mathbf{r}_1, t) U^*(\mathbf{r}_2, t) \rangle \\
&= \sum_m \langle U(\mathbf{r}_{m1}, t) U^*(\mathbf{r}_{m2}, t) \rangle + \sum_{m \neq n} \sum \langle U(\mathbf{r}_{m1}, t) U^*(\mathbf{r}_{n2}, t) \rangle. \quad (2.61)
\end{aligned}
$$

The term $\langle U(\mathbf{r}_1, t) U^*(\mathbf{r}_2, t) \rangle$ is a measure of spatial coherence of the radiation and is proportional to the correlation coefficient between $U(\mathbf{r}_1, t)$ and $U^*(\mathbf{r}_2, t)$.

The light vibrations arising from different elements of the source may be assumed to be statistically independent (mutually incoherent) and of the zero mean value, so that

$$
\begin{aligned}
\langle U(\mathbf{r}_{m1}, t) U^*(\mathbf{r}_{n2}, t) \rangle &= \langle U(\mathbf{r}_{m1}, t) \rangle \langle U^*(\mathbf{r}_{n2}, t) \rangle \\
&= 0; \quad m \neq n. \quad (2.62)
\end{aligned}
$$

The geometrical factors required for derivation of the van Cittert–Zernike theorem are depicted in Fig. 2.4. The field points, $P_1(\mathbf{r}'_1)$ and $P_2(\mathbf{r}'_2)$, on a screen A are illuminated by a self-luminous extended but unresolved quasi-monochromatic source, σ. Here, one assumes that the medium between the source and the screen is homogeneous, and the linear dimensions of the source are small compared to the distance OO' between the source and screen. The complex disturbance due to element $d\sigma_m$ at a point $P_{j=1,2}$ in the screen is,

$$
U_{mj}(t) = A_m \left(t - \frac{s_{mj}}{c} \right) \frac{e^{-i2\pi\bar{\nu}(t - s_{mj}/c)}}{s_{mj}}, \quad (2.63)
$$

where the strength and phase of the radiation coming from element $d\sigma_m$ are characterized by $|A_m|$ and $arg(A_m)$, respectively, and s_{mj} is the distance from the element $d\sigma_m$ to the point P_j.

The mutual coherence function (see 2.10) of P_1 and P_2 turns out to be,

$$
\Gamma(\mathbf{r}_1, \mathbf{r}_2, 0) = \sum_m \left\langle A_m(t) A_m^*(t) \right\rangle \frac{e^{i2\pi\bar{\nu}(s_{m1} - s_{m2})/c}}{s_{m1} s_{m2}}. \quad (2.64)
$$

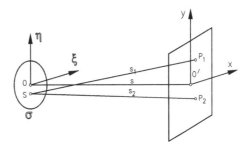

Fig. 2.4 Calculation of degree of coherence of $P_1(\mathbf{r}'_1)$ and $P_2(\mathbf{r}'_2)$ illuminated by σ

Considering a source with a total number of elements so large that it can be regarded as continuous, the sum in (2.64) is replaced by the integral,

$$\Gamma(\mathbf{r}_1, \mathbf{r}_2, 0) = \int_\sigma I(\mathbf{r}) \frac{e^{i\bar{\kappa}(s_1 - s_2)}}{s_1 s_2} d\mathbf{s}. \tag{2.65}$$

in which s_1 and s_2 are the distances of $P_1(\mathbf{r}_1')$ and $P_2(\mathbf{r}_2')$ from a typical point $S(\mathbf{r})$ on the source, respectively, $\bar{\kappa} = 2\pi/\bar{\lambda}$ the wave number, and $I(\mathbf{r})$ the intensity per unit area of the source.

According to (2.48) and (2.65), the complex degree of coherence $\mu(\mathbf{r}_1, \mathbf{r}_2)$ is,

$$\mu(\mathbf{r}_1', \mathbf{r}_2') = \frac{1}{\sqrt{I(\mathbf{r}_1)I(\mathbf{r}_2)}} \left(\frac{\bar{\kappa}}{2\pi}\right)^2 \int_\sigma \frac{1}{s_1 s_2} I(\mathbf{r}) e^{i\bar{\kappa}(s_1 - s_2)} d\mathbf{s}, \tag{2.66}$$

where

$$I(\mathbf{r}_j) = J(\mathbf{r}_j', \mathbf{r}_j') = \left(\frac{\bar{\kappa}}{2\pi}\right)^2 \int_\sigma \frac{I(\mathbf{r})}{s_j^2} d\mathbf{s}, \tag{2.67}$$

in which $j = 1, 2$ and $I(\mathbf{r}_j)$ the corresponding intensities at $P_j(\mathbf{r}_j')$.

The (2.66) is the result of van Cittert–Zernike theorem. It expresses the complex degree of coherence at two fixed points, $P_1(\mathbf{r}_1')$ and $P_2(\mathbf{r}_2')$ in the field illuminated by an extended quasi-monochromatic source in terms of the intensity distribution $I(\mathbf{r})$ across the source and the intensity $I(\mathbf{r}_1)$ and $I(\mathbf{r}_2)$ at the corresponding points, $P_1(\mathbf{r}_1')$ and $P_2(\mathbf{r}_2')$.

Let (ξ, η) be the coordinates of the source plane, $S(\mathbf{r})$, referred to axes at O,

$$s_1 = \sqrt{s^2 + (x_1 - \xi)^2 + (y_1 - \eta)^2} \simeq s + \frac{(x_1 - \xi)^2 + (y_1 - \eta)^2}{2s},$$

$$s_2 = \sqrt{s^2 + (x_2 - \xi)^2 + (y_2 - \eta)^2} \simeq s + \frac{(x_2 - \xi)^2 + (y_2 - \eta)^2}{2s}, \tag{2.68}$$

where (x_j, y_j) are the coordinates in the observation plane, and the term in $x_j/s, y_j/s, \xi/s$, and η/s are retained.

On setting,

$$p = \frac{(x_1 - x_2)}{s}, \qquad q = \frac{(y_1 - y_2)}{s}, \tag{2.69a}$$

$$\Psi(\mathbf{r}_1, \mathbf{r}_2) = \bar{\kappa} \left[\frac{(x_1^2 + y_1^2) - (x_2^2 + y_2^2)}{2s}\right]. \tag{2.69b}$$

The quantity $\Psi(\mathbf{r}_1, \mathbf{r}_2)$ represents the phase difference $2\pi(\mathrm{OP}_1 - \mathrm{OP}_2)/\bar{\lambda}$ and may be neglected if $(\mathrm{OP}_1 - \mathrm{OP}_2) \ll \bar{\lambda}$. By normalizing, the van Cittert–Zernike theorem (2.66) yields,

$$\mu(\mathbf{r}_1', \mathbf{r}_2') = e^{i\Psi(\mathbf{r}_1,\mathbf{r}_2)} \frac{\displaystyle\iint_{-\infty}^{\infty} I(\xi, \eta) e^{-i\bar{\kappa}(p\xi + q\eta)} d\xi d\eta}{\displaystyle\iint_{-\infty}^{\infty} I(\xi, \eta) d\xi d\eta}. \tag{2.70}$$

The (2.70) states that for an incoherent circular source, the complex coherence factor far from the source is equal to the normalized Fourier transform of its brightness distribution.

Chapter 3
Applications of Interferometry

3.1 Early Stellar Interferometry

In astronomy, interferometry can be traced back to 1868 when Fizeau (1867) proposed to the Académie des Sciences that this technique could be used to measure stellar diameters. This scheme consists of installing a mask with two small openings (apertures) separated in distance at the entrance of a telescope. When pointing at a star, the superposition of the two images in the focal plane, where both beams intersect, creates interference fringe (classical Young's fringes) across the combined image, and thereby measuring the fringe visibility as a function of aperture separation. He realized that a relationship exists between the aspect of interference fringes and the angular diameter of the light source, in this case a star.

Five years later, an experimental illustration for this statement was demonstrated by the work of Stéphan (1874) by masking a 28-inch refractor to define two apertures. He could detect visually a series of destructive and constructive patterns, which appeared within the common Airy disk of the sub-apertures. These fringes remain visible in presence of seeing, but unable to notice any significant drop of fringe visibility even when the slits were separated by the full diameter of the telescope. Therefore, he concluded that none of the observed stars approached the diffraction-limit of the telescope. The apparent diameters of stars are less than $0.158''$, the smallest angle measurable with the available aperture.

Later, Michelson (1890) came up with the same idea and discussed the expected results for uniform disk-like star and limb-darkened stars, as well as for binary systems. He had opined that for useful measurements of stellar diameters to be made, apertures separated by up to 10 m would be required. Michelson (1891) measured the diameter of Jupiter's Galilean satellites using Fizeau mask on top of the 40-inch Yerkes refractor. This work was followed up by Schwarzschild and Villiger (1896). Later, Anderson (1920) who placed a similar interferometer (Fizeau mask) on top of the 100 inch telescope at Mt. Wilson Observatory and measured the angular separation (ρ) of the spectroscopic binary star, α Aurigae (Capella).

The first successful measurement of the angular diameter of the red supergiant star, Betelgeuse (α Orionis), was performed by Michelson and Pease (1921). They reported a diameter of 0.047 arcseconds at optical wavelengths with an uncertainty

S.K. Saha, *Aperture Synthesis*, Astronomy and Astrophysics Library,
DOI 10.1007/978-1-4419-5710-8_3, © Springer Science+Business Media, LLC 2011

of 10%. Pease (1931) continued the work, resolving six more stars, namely α Boötis, α Tauri, β Pegasi, α Herculis, α Ceti, and α Scorpii. Although this instrument was not sensitive enough to enlarge their investigation, the value of these results led to the construction of a 50-ft interferometer by Pease, which added measurement of one more star, β Andromeda, to the list. The effects of poor seeing and flexure became more severe with the larger instrument; the 50-ft long girder flexed far too much. Difficulties encountered in the subsequent operation due to the lack of technology brought the development of stellar interferometry to a halt for several decades, until it was revitalized by the development of intensity interferometry (Hanbury Brown and Twiss 1958).

3.1.1 Fizeau–Stéphan Interferometer

Fizeau–Stéphan interferometry, also known as image-plane interferometry, has intrinsic pathlength compensation and a correspondingly wide field. There is no need for modulating the fringes actively since the atmospheric turbulence may introduce relative delays and cause the fringe pattern to slide back and forth, smearing out the fringes.

The starlight is imaged through a mask having a pair of sub-apertures separated by a baseline vector, $\mathbf{B}$, kept over a telescope aperture, which effectively allows two small beams to interfere in the focal plane. The incoming broadband light is made quasi-monochromatic by filtering it. The beams propagate from different directions and are focused on the same spot to make an image of the object. The images are superposed so that interference fringes form across the combined image. As one moves along the image plane, the differential tilt between the beams produces fringes in the PSF.

Interference fringes can be envisaged with amplitude proportional to the modulus of the pupil mutual intensity function, $\mathbf{J}$ and a spatial phase identical with the phase of $\mathbf{J}$. According to the van Cittert–Zernike theorem (see Sect. 2.2), such an intensity function is a scaled version of the 2-D Fourier transform of the object intensity distribution. The measurement of the parameters of this fringe provides the object spectrum at spatial frequency $\mathbf{u}$ (see 1.61). The intensity in the focal plane can be written as follows:

$$I(\mathbf{x}) = a(D.x)\left[I_1 + I_2 + 2\sqrt{I_1 I_2}\mid \gamma_\mathbf{B}(\mathbf{0})\mid \cos\left\{ \frac{2\pi}{\lambda}\left(\frac{\mathbf{x}\cdot\mathbf{B}}{f} + d \right) - \psi_\mathbf{B} \right\} \right],$$

$$(3.1)$$

where $\mathbf{B}$ is the baseline vector, D the diameter of the sub-apertures, f the focal length, the envelope, $a(D.x)$ the image of each sub-apertures (Airy disk in ideal case), $\psi_\mathbf{B}$ the phase of $\gamma_\mathbf{B}(\mathbf{0})$, $2\pi d/\lambda$ the incidental non-zero OPD between the fields, and d the extra optical path in front of one aperture.

Fig. 3.1 Fizeau–Stéphan interferometer using mask over the 1-m telescope at Vainu Bappu Observatory, Kavalur, India (Saha et al. 1987)

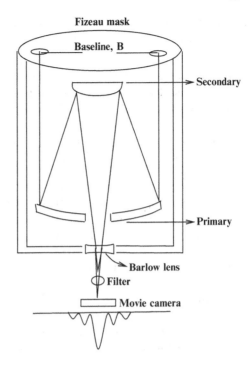

From (2.26), the fringe visibility, $\mathcal{V}$, may be derived. In order to measure the angular diameter of the source (see 1.99), the baseline between the two sub-apertures should be gradually increased until the fringes first vanish. A similar experiment has also been conducted by Saha et al. (1987) at the Vainu Bappu Observatory (VBO), situated at Kavalur, India, using the 1-m telescope and recording the fringes of several bright stars through a broadband filter in the blue region using a 16-millimeter (mm) movie camera giving an exposure of 33 milliseconds (ms) per frame. A Fizeau mask with two holes of 10 cm in diameter each, separated by a distance B = 69.4 cm was placed in front of the telescope. A Barlow lens (double concave of focal length 12 cm) was inserted in the $f/13$ converging beam at the Cassegrain end of the said telescope (see Fig. 3.1). This lens magnifies the image adequately. The image scale of this telescope was calibrated by this experiment. With a peak wavelength of 4,500 Å, the separation of the fringes is $\lambda/B = 0.130$ arcseconds. On the film, these had a linear separation of 0.035 mm, which can yield to an image scale of $3.70''/$mm.

The notable advantage of Fizeau–Stéphan interferometer is that the telescope acts as both collector and correlator. Temporal coherence is, therefore, automatically obtained due to the built-in zero optical path difference (OPD). The spatial modulation frequency, as well as the required sampling of the image change with the separations of sub-apertures. Since the maximum fringe angular spacings that can be explored are limited by the physical diameter of the telescope, the number of stellar sources for measuring diameters is limited.

3.1.2 *Michelson Stellar Interferometer*

Michelson stellar interferometer compares the amplitude of the light beams at two
separated points. It was designed in a periscopic fashion of four siderostat mirrors
(two outer and two inner), M_1, M_2, M_3, and M_4 (see Fig. 3.2), mounted onto the
100-inch telescope at Mt. Wilson, USA. These mirrors, about 152 mm in diameter, inclined 45° to the base, are mounted on slides. Two mirrors, M_1 and M_4, kept
at a distance of 20 feet (6.09 m), ride on an external rail so as to enable one to
adjust the baseline length, B. Two pencil beams from a star, which hit each of the
outer mirrors at the same time, are reflected from the outer mirrors to the inner mirrors, M_2 and M_3, that re-direct the beams through the diaphragm onto the 100 inch
primary mirror, to be eventually reunited in the eyepiece in the form of interference fringes. By first setting M_1 and M_4 symmetrically on the beam, equality of two
paths is maintained. These outer mirrors are adjustable about two horizontal axes
by means of fine screws at the end of 228 mm lever arms. Their distances from M_2
and M_3 must be equal, which are permanently fixed except that M_3 has a motion of
several millimeters along its slide parallel to the beam. The coherence of the star is
measured by the baseline, B, so this value governs the visibility of the source. The
distance between the outer mirrors is varied to effect a change in fringe visibility.
The beam combination used a smaller (internal combination) baseline, B_0, between
these two inner mirrors. This baseline, $B_0 = 1.14$ m provides the fringe pattern a
fixed spacing equal to 0.02 mm. The modulation pattern in the focal plane is set by
this internal combination baseline.

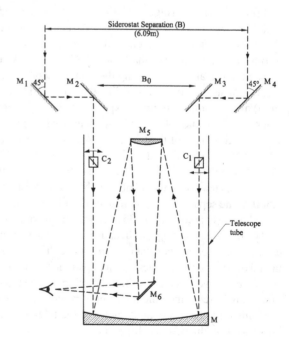

Fig. 3.2 Schematic diagram
of Michelson stellar
interferometer depicting the
collecting baseline, B, as well
as the internal combination
baseline B_0, which determine
the fringe spacing

A plane parallel glass plate, C_1, was inserted in one of the beams to maintain the geometrical paths in coincidence in the focal plane, thereby effectively making the wavefronts parallel at the entrance pupil. This glass plate can be inclined in any direction in order to compensate for small alignment errors in the relay mirrors and for bending of the structure. Michelson also introduced two opposite glass wedges into the other beam, which provides a variable glass thickness, and thereby compensate for inequalities of optical paths at all wavelengths. This compensation is essential since the white light is visible only near zero order, and was controlled by observing the channeled spectrum in a small spectroscope (Born and Wolf 1984). Assume that a photo-detector located at the focal plane of the interferometer collects the light, so the intensity, I_{int}, is given by,

$$I_{int}(\theta) = 2I_{tel}(\theta) \left[1 + \mathcal{V} \cos(2\pi\theta B_0/\lambda)\right], \tag{3.2}$$

in which B_0-dependent cosine term expresses the modulation of the envelope I_{tel} and B-dependent visibility $\mathcal{V}$ term expresses the degree of modulation, i.e.,

$$\mathcal{V} = \frac{2J_1(\pi\theta_s B/\lambda)}{\pi\theta_s B/\lambda}, \tag{3.3}$$

with J_1 as the first order Bessel function, θ_s the apparent diameter of the source, B the baseline length, and λ the operating wavelength.

It is reiterated that according to the van Citter–Zernike theorem, the amplitude function of the Fourier transform of its brightness distribution is derived from the visibility of the fringes. The degree of coherence of the light depends on the angular diameter of a star, the wavelength of light, and the baseline vector, **B**, between the two outermost mirrors. If the baseline, B, is small, the degree of partial coherence is nearly unity, and hence, the fringes are discernible; with the increase of B, it tends to diminish. The smallest angular diameter that can be determined by the maximum separation (6.1 m) of the outer mirrors was about 0.02 arcseconds at $\lambda = 5,500$Å. Figure 3.3 depicts the variation of fringe visibility with the changes of baseline between the outermost mirrors. With an eyepiece, Michelson noticed interference fringes at the focal plane of the telescope and following which the fringe visibility was estimated (Michelson and Pease 1921).

Fig. 3.3 Fringe visibility with the separation of mirrors for a Michelson interferometer; the visibility turns out to be zero when the source is resolved

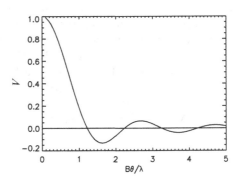

The disadvantage of the Michelson mode is a very narrow field-of-view (FOV) compared to the Fizeau's. The other notable drawbacks of such a system are:

1. The measurement of interference fringes that are displaced by an amount proportional to the phase shift, ψ, in a direction parallel to the line joining the openings. For a quasi-monochromatic light having a narrow bandwidth, $\Delta \nu \ll \nu_0$, the effect is to introduce a phase shift, ψ, into all frequency components of the light. If the relative phase at the two small mirrors is changed by π, the destructive and constructive bands of fringes are interchanged. As such phenomena occur with faster speed than the observing time, the fringe visibility is reduced. In an actual experiment such a measurement is difficult to make with a large interferometer due to the stringent requirement of pointing accuracy; the entire instrument was required to be carefully aligned as well. The pathlength difference in the two arms needs to be maintained equal within a fraction of the coherence length (see 2.1) of the light. In order to collect more light the broadest possible bandwidth of light was used in this setup.

2. The turbulent atmosphere, the physics of which is discussed in Chap. 4, places strict constraints on the optical interferometer, its effect hampers the fringe visibilities. The fluctuations caused by the atmosphere and instrument enter strongly into the cosine term, in (3.2), which limits its utility. Variations of refractive index above the sub-apertures of the interferometer cause the interference pattern to move as a whole. Since the star light passes through the turbulent atmosphere before reaching the sub-apertures, it introduces phase shift $\psi_{j=1,2}$. In such a situation, the complex degree of coherence, $\gamma'(\mathbf{r}_1, \mathbf{r}_2, \tau)$, between the light at the two sub-apertures is given by,

$$\langle \gamma'(\mathbf{r}_1, \mathbf{r}_2, \tau) \rangle = \gamma(\mathbf{r}_1, \mathbf{r}_2, \tau) e^{i(\psi_1 - \psi_2)}, \tag{3.4}$$

in which $\gamma(\mathbf{r}_1, \mathbf{r}_2, \tau)$ is the complex degree of coherence (see 2.15) in the absence of an atmosphere, and $\langle \rangle$ denotes the time average. In a case where the phase shifts are random, uncorrelated, and distributed with even probability over the range $-\pi$ to π, i.e., $\langle \psi_1 \rangle = \langle \psi_2 \rangle = \langle \psi_1 \psi_2 \rangle$, therefore, (3.4) turns out to be,

$$\langle \gamma'(\mathbf{r}_1, \mathbf{r}_2, \tau) \rangle = 0. \tag{3.5}$$

This (3.5) reveals that if the complex degree of coherence, $\gamma'(\mathbf{r}_1, \mathbf{r}_2, \tau)$, is taken over periods that are comparable with the characteristic period of the fluctuations in phase, the time average of such coherence tends to zero; the time average of the fringe visibility, $\mathcal{V}$, is reduced as well.

3. The introduction of path difference between the two beams of light reaching the focal plane of the interferometer, which are significantly larger than a wavelength or period of the light, may reduce the fringe visibility by a factor, $\sin(\pi \Delta \nu \tau)/(\pi \Delta \nu \tau)$, in which $\Delta \nu$ is the optical bandwidth.

3.2 Radio Interferometry

Optical telescopes have been the main tools for the astronomical community for the last few centuries. The discoveries made using these telescopes are numerous, but are usable only in the visible range of the electromagnetic spectrum. However, stellar bodies emit radiation in other frequencies such as in radio and InfraRed (IR) wavebands as well. The resolving power of an optical telescope is limited by the atmospheric turbulence to about $\sim 1''$ with much smaller diffraction limit. The radio waves are about thousand to million times longer in wavelength compared to the light waves. So the phase variations introduced by the intervening Earth's atmosphere are less severe and does not introduce any serious propagation effects and the telescopes are mainly diffraction limited.

Astronomy in radio wavebands began with the detection of cosmic noise by Jansky (1932), which appeared to be coming from the direction of the center of the Milky-way. This discovery was subsequently corroborated by the work of Reber (1940). His radio map from the observations carried out with a radio telescope at the wavelength of 1.87 meters (m) showed significant radio emission from the direction of the constellations of Cygnus and Cassiopeia. Radio emission from the disturbed Sun, known as active Sun[1] was detected a few years later by J. S. Hey in 1942 during the World War II period with Army radar equipments when trying to identify the locations of jamming radio signals (Hey 1946). A few discrete sources of radio emission were also discovered by him. Subsequently, in 1942 Southworth (1945) detected thermal radiation from the undisturbed Sun at wavelength of 3.2 cm. It was

---

[1] Solar activity reaches a maximum every 11 years when the Sun's magnetic field is stronger than average, which gives rise to sunspots (dark spots) on the photosphere. These spots are relatively cooler ($\sim 4300° $K) compared to the rest of the photosphere (where the average temperature is $\sim 5800° $K) and are highly magnetized. The magnetic field in a sunspot may be 1,000 times stronger than in the surrounding area, where it is about 1 G (gauss). The packed magnetic field lines provide a barrier preventing hot gas from being convected into the sunspots. Such spots, in general, occur in pairs connected by a loop. When hot ionized gas travels along the looped field lines, it forms solar prominence rising up to about 100,000 km above the photosphere. Bright areas called faculae near spots and the overlying plages associated with enhanced chromospheric heating are some of the manifestation of magnetic activity. Magnetic field lines near a large group of sunspots can suddenly snap triggering solar flares. Flares are energetic phenomena that occur in the solar atmosphere, wherein a sudden (on a time scale of about 1 h) outburst of energy (10^{29}–10^{32} erg) is released from localized active regions in the form of radiation and fast particles representing a dramatic gust in the solar wind. The solar wind is the flux particles, mainly protons and electrons together with nuclei of heavier elements in smaller numbers, which are accelerated by the high temperatures of solar corona to velocities large enough to escape from the Sun's gravitational field. This wind is responsible for deflecting both the tail of the Earth's magnetosphere and the tails of comets away from the Sun. When the matter (particles) in the solar flare passes into the Earth's upper atmosphere, complicated geomagnetic and ionospheric storms occur due to which radio communication gets disturbed. Bursts of solar radiation at widely different wavelengths sometimes occur during the observation of a solar flare and their individual characteristics differ greatly. The duration of the emission may range from less than a minute to several hours. The growth phase is usually $\sim$minutes and then the flare fades slowly (tens of minutes).

noted that radio emission from the undisturbed Sun decrease in intensity as the wavelength is increased, while the irregular enhanced radiation from solar flares and other manifestations of the active Sun increases in intensity at longer wavelengths.

The strongest emissions reaching the Earth from outer space emanate from the solar outbursts (which are transient phenomena and fairly randomly distributed in time within any particular year). These signals are intense at low radio frequencies ($\leq$150 MHz) and can be detected with a simple dipole antenna connected to a stable receiver. Later the detection of line radiation at a wavelength of 21 cm from a hyperfine transition in neutral hydrogen atom[2] and the family of OH lines at a wavelength around 18 cm in the interstellar space had helped radio astronomers to measure the radial velocity of galaxies by Doppler shift, and determine their structure.

Real progress in astronomical observations at radio wavelengths was during the post war period. A group of radio and electronic engineers, experienced in development of radar directional antenna system resumed scientific investigations, mostly in radio astronomy, at the universities and research institutes. To note, an antenna is a transducer designed to transmit or receive electromagnetic waves. It converts such waves into electric currents and vice versa. Various types of antennas such as dipole, yagi, loop, and dish are in use. Based on the transmitting or receiving frequencies, they can be built. In a directional antenna system, most of the energy is condensed in the main lobe of the radiation/reflection pattern of the antenna.

Many radio astronomical observatories have now sprung up across the globe, where colossal single dish radio telescopes, large array of antennas operating in the interferometric mode with excellent phase stability, dynamic spectrographs are constructed with the aim of collecting as much of radio energy as possible. Developments of long baseline interferometry and very long baseline interferometry (VLBI), as well as usage of sophisticated image processing techniques have brought high dynamic range images with milliarcsecond (mas) resolution.

3.2.1 The Radio Telescope

Unlike an optical telescope where mirrors are used to receive visible light frequencies, a ground based radio telescope can receive waves having frequencies from $\sim$10 MHz to 300 GHz. While the lower limit is due to reflection in the ionosphere, the upper limit is caused by molecular absorption bands of water and oxygen in the atmosphere. Such a telescope can be a dipole antenna or paraboloid (single-dish)

---

[2] The spin-flip transition of neutral hydrogen in its two closely spaced energy levels in the ground state leads to the emission of a photon with a wavelength of 21.1 cm, which corresponds to a frequency of 1420.4 MHz. This line falls within the radio spectrum and was first detected by Ewan and Purcell (1951). Subsequently, it is used in radio astronomy extensively, since it can penetrate dust clouds that are opaque to visible wavelengths. Most of the hydrogen gas in the InterStellar Medium (ISM) is in cold atomic or molecular form. The temperature of atomic hydrogen gas is of the order of 100–3,000° K. The atomic hydrogen gas in the Milky Way (the Galaxy) has been mapped by the observation of the 21 cm line of hydrogen.

equipped with a feed at its focus, depending on the frequency of observation. The first part of the radio receiver, referred to as the front-end, is placed right below the dipole or at the focus in the case of a dish antenna. The energy that falls on the surface of the dipole/dish forms the input to the front end receiver. The output from the front end receiver goes through coaxial optical fiber cables to the main observatory building and processed by a super-heterodyne receiver. In some cases, the latter may be kept near the antenna itself and could be a part of the front end receiver chain.

Heterodyne is the process of down converting high frequency radio frequency (RF) signal to an intermediate one by mixing it with a coherent local oscillator (LO). Such a conversion makes it easy to amplify the signal at low frequency; the expense of the electronic components are also less. The intermediate frequency (IF) encompasses a range of frequencies: those greater than the LO are part of the upper sideband, while those less are in the lower sideband. For instance, if a high frequency radio signal encompasses of frequency, ω_0, is mixed with a local oscillator of frequency, ω_l ($\omega_l > \omega_0$), the upper sideband has a frequency of $\omega_l + \omega_0$ and a lower sideband has a frequency of $\omega_l - \omega_0$. The upper side band is usually rejected by using an appropriate filter. To note, a filter restricts the range in frequencies and define the bandwidth; it also helps to cut out interference from other signals outside the band.

In order to preserve the characteristics of the observed RF signal in the IF signal, the LO signal should, in principle, be monochromatic and the phase relationship of LO signal at each mixer[3] must be coherent (Fomalont and Wright 1974). For single dish observations, one needs to measure the average noise power. This is performed by passing the IF signal corresponding to the RF signal from the celestial radio source through a square-law detector. The primary component of the latter is a diode whose DC output is proportional to the square of the input IF voltage.

3.2.1.1 Antenna Pattern

An antenna picks up radiation from a limited region of the sky, which is characterized by normalized power pattern, $P(\theta, \phi)$, where by convention $P(0, 0) = 1$; (θ, ϕ) are the polar coordinates, $0 \leq \theta \leq \pi$ the azimuthal angle, and $0 \leq \phi \leq 2\pi$ the polar angle. The power pattern quantifies the antenna response as a function of direction, which is essentially the point spread function (PSF) or 'beam' for the

---

[3] A mixer is a device that changes the frequency of the input RF signal by shifting to a lower intermediate frequency (IF). It has two inputs, one for the RF signal whose frequency is to be changed, the other input is usually a sine wave generated by a tunable signal generator, the local oscillator (LO). Mathematically, it can be expressed as,

$$e^{i2\pi\nu t} \times e^{-i2\pi\nu_{LO}t} = e^{i2\pi(\nu-\nu_{LO})t}.$$

Sometimes there could be more than one mixer along the signal path, creating a series of intermediate frequencies, one of which is optimum for signal transport, another which is optimum for amplification.

antenna. The electric field response across the aperture is described by the aperture distribution (grading) function, i.e., the apodizing function in optics. It depends on the illumination by the feed. The extent of the concentration of the 'beam' and its width depend upon the size of the antenna and the wavelength of the incident signal. The ability of an antenna to receive waves (radio) incident upon it is described in terms of its effective aperture, A_e, which is the ratio of the power per unit flux interval (W / Hz) available at the antenna output terminals to the energy flux per unit frequency interval, i.e., the flux density (W/m^2/Hz) of the radio source that is being measured. The units are given by m^2 and is effectively null outside the antenna beam. If an antenna detects single polarization (half of the total power), one defines it as an effective area, A_e. Usually, the effective area, A_e, is less than the geometric cross-sectional area of the main reflector, $A_g (\sim \pi R^2)$, in which R is the radius of the antenna. With uniform field over the aperture, in general A_g is the effective aperture, A_e. Any antenna with a single output is sensitive to radiation of one particular polarization which is matched to the antenna. The available power at the output is proportional to the flux density of the matched polarized component of the the incident radiation. Consequently, the definition of the antenna effective aperture should be expressed in terms of the flux density, S, of the matched polarized component; this is equal to $S/2$ in case of randomly polarized wave. If the antenna delivers all the matched polarized energy falling on its surface then its effective aperture will be equal to its geometric area. The effective area is a function of the angle of arrival of the incident wave, i.e., $A_e = A_e(\theta, \phi)$. Often this directional property is expressed by the power pattern, $P(\theta, \phi)$, which is the effective area normalized to be unity at the maximum, i.e.,

$$P(\theta, \phi) = \frac{A_e(\theta, \phi)}{A_e^{max}}. \tag{3.6}$$

The power pattern has a primary maximum, called main lobe and several other subsidiary maxima, known as sidelobes. The angle across the main lobe of an antenna pattern between the two directions, where the sensitivity/gain of the antenna is half the value at the center of the lobe ($\theta = 0$ direction) is referred to as the half-power beamwidth (the angular separation of the half-power points of the radiated pattern).

The directive gain, $G(\theta, \phi)$, of an antenna is the angular selectivity it has over a lossless isotropic antenna; the beam solid angle for an isotropic antenna is 4π. The antenna gain is the measure in dB (decibel), how much more power an antenna radiates in a certain direction with respect to that which would be radiated by a reference antenna. Such a gain, $G(\theta, \phi)$, depends upon both the directivity and efficiency of the antenna, i.e.

$$G(\theta, \phi) \equiv \eta_a D(\theta, \phi) = \frac{4\pi A_e}{\lambda^2}, \tag{3.7}$$

where λ is the wavelength, $A_e (= \eta_a A_g = \eta_a \pi R^2)$ the effective collecting area, η_a the aperture efficiency defined by the ratio of the effective area, A_e, of an antenna measuring its ability to respond to radiation of a particular polarization, to its geometric area, A_g, and $D(\theta, \phi)$ the directivity that is defined as the ratio of the maximum radiative power per unit solid angle to the radiation intensity of an antenna averaged over a sphere (Krauss 1966),

$$D(\theta, \phi) = \frac{4\pi P(\theta, \phi)}{\int_{4\pi} P(\theta, \phi) d\Omega}, \qquad (3.8)$$

in which $d\Omega (= \sin\theta \, d\theta \, d\phi)$ the differential element of solid angle in spherical polar coordinates.

The beam solid angle[4] of the antenna, Ω_a i.e., the effective region of the sky to which the antenna is sensitive, is related to its effective area, A_e, by

$$A_e(\theta, \phi) \equiv A_e^{max} P(\theta, \phi)$$

$$= \frac{\lambda^2 P(\theta, \phi)}{\int_{4\pi} P(\theta, \phi) d\Omega} = \frac{\lambda^2}{\Omega_a}, \qquad (3.9)$$

where $\int_{4\pi} P(\theta, \phi) d\Omega \sim (\lambda/D)^2$ for a reflecting telescope, so that $A_e^{max} \sim D^2$, and D is the diameter of the telescope; the maximum effective aperture area is assumed to be equal to the geometric area of the reflector.

Hence, from (3.8 and 3.9), it is found,

$$D(\theta, \phi) = \frac{4\pi A_e(\theta, \phi)}{\lambda^2}. \qquad (3.10)$$

The directional gain is defined by,

$$G(\theta, \phi) = p \, A_g \, \eta_a \, P_n(\theta, \phi), \qquad (3.11)$$

where $P_n(\theta, \phi)$ is the normalized power pattern, p the fraction of signal with polarization of antenna; for a thermal (randomly polarized) source $p = 1/2$; the factor of $1/2$ arises because a single antenna will be able to detect only one polarization, and therefore can receive only half the power from an unpolarized source.

The power due to a point source of flux density,[5] S, by an antenna of geometric area, A_g, is given by,

$$w = S \, A_g \, \Delta f, \qquad \text{watts.} \qquad (3.12)$$

---

[4] The beam solid angles of antennas with normalized power patterns (1) a top hat function with $P = 1$ for $\theta < \theta_0$ and $P = 0$ for otherwise is $2\pi(1 - \cos\theta_0)$, (2) a narrow Gaussian function with $P(\theta, \phi) = e^{-\theta^2/\theta_0^2}$ and $\theta_0 \ll \pi$ is $\pi\theta_0^2$ and (3) a doughnut function, with $P(\theta, \phi) = \sin^2\theta$, (the pattern of a dipole),

$$\Omega_a = \int P(\theta, \phi) d\Omega = \int_{\theta=0}^{\pi} \int_{\phi=0}^{2\pi} \sin^2\theta \sin\theta d\theta d\phi = 2\pi \int_0^{\pi} \sin^3\theta \, d\theta = \frac{8\pi}{3},$$

and the effective area, $A_e = \lambda^2/\Omega_a = 3\lambda^2/(8\pi)$ m^2.

[5] In general, the flux density, S is applied to point sources, and the sky brightness, $B(\nu)$ or intensity, $I(\nu)$, is applied to extended sources.

The signal, in this case, is assumed to be broadband noise from the source and Δf the electrical bandwidth (in Hz). The flux density is the power received per unit area per unit frequency ($\mathrm{W\,m^{-2}\,Hz^{-1}}$). In radio astronomy, the flux densities are often expressed in jansky, Jy, where $1\,\mathrm{Jy} = 10^{-26}\,\mathrm{W\,m^{-2}\,Hz^{-1}}$.

If the source is not on the axis of the antenna, but at (θ, ϕ), the power is expressed as,

$$w = \frac{1}{2} S\, A_e\, P(\theta, \phi)\Delta f. \tag{3.13}$$

As far as signal from an extended source is concerned, it can be handled by breaking it down into many tiny angular patches. Such a patch at position (θ, ϕ) has a small solid angle, $\Delta\Omega$ and contributes a flux density ΔS to overall flux density, i.e.,

$$\Delta S = B(\theta, \phi)\Delta\Omega,$$

where $B(\theta, \phi)$ is the surface brightness as a function of position over source, measured in $\mathrm{W\,m^{-2}Hz^{-1}sr^{-1}}$ (sr stands for steradian[6]).

The flux density of an extended source is therefore,

$$S = \int_{\mathrm{sky}} B(\theta, \phi)d\Omega. \tag{3.14}$$

When the source is observed with an antenna of power pattern, $P_n(\theta, \phi)$, the flux density, S, becomes,

$$S = \int_{\mathrm{source}} B(\theta, \phi)P_n(\theta, \phi)d\Omega. \tag{3.15}$$

Owing to the directional pattern of the antenna, the observed flux density will be less than the true value. Antenna may be pointed towards the source as well as away from the source. The difference in the two outputs providing the contributions from the source, is expressed in terms of either brightness temperature, $T_b(\theta, \phi)$, or antenna temperature representing unwanted noise, T_a, which varies from one telescope to the other.

3.2.1.2 Brightness Temperature

Brightness is defined to be the amount of power received per unit area per unit bandwidth per unit solid angle. Often the sky brightness is measured in temperature units, which is defined as the Rayleigh–Jeans temperature of an equivalent black

---

[6] Steradian (sr) is the solid angle subtended at the center of a sphere of radius r by an area on the surface of the sphere having an area r^2. The total solid angle of a sphere is 4π steradian.

body[7] that provides the same power as the source. The brightness temperature, T_b, can be derived from,

$$I_\nu = B_\nu(T_b), \tag{3.16}$$

where I_ν is the specific intensity and B_ν is given by Planck's law,

$$B_\nu(T) = \frac{2h}{c^2} \frac{\nu^3}{e^{(h\nu/k_B T)} - 1}, \tag{3.17}$$

which has units of energy per unit time per unit surface area per unit solid angle per unit frequency ($J\,s^{-1}\,m^{-2}\,Sr^{-1}\,Hz^{-1}$). Here, $B_\nu(T)$ is the spectral radiancy, ν the frequency, $k_B = 1.380662 \times 10^{-23} J\,K^{-1}$ the Boltzmann's constant, and T the temperature in Kelvin (K).

This is approximated using the Rayleigh–Jeans' law[8] in the case of radio astronomy ($h\nu/k_B T \ll 1$),

$$
\begin{aligned}
B_\nu(T_b) &= \frac{2h\nu^3}{c^2} \frac{1}{e^{(h\nu/k_B T)} - 1} \qquad W\,m^{-2}Hz^{-1}sr^{-1} \\
&= \frac{2h\nu^3}{c^2} \frac{1}{1 + (h\nu/k_B T) + \cdots - 1} \\
&\approx \frac{2k_B \nu^2 T}{c^2} = \frac{2k_B T}{\lambda^2},
\end{aligned}
\tag{3.18}
$$

with $h = 6.626196 \times 10^{-34}$ joules (J) seconds (s)) as Planck's constant, ν the frequency, k_B the Boltzmann's constant, T the temperature, and c the speed of light.

This approximation to Planck's spectrum is called the Rayleigh–Jeans approximation and is valid at the radio wavelengths. Therefore the flux density from an extended black body source with a temperature distribution, $T(\theta, \phi)$ across the sky is given by,

$$S = \frac{2k_B \nu^2}{c^2} \int_{-\infty}^{\infty} T(\theta, \phi) d\Omega, \tag{3.19}$$

and solving for the sources that are not black body radiators by assigning them a brightness temperature, T_b, defined by,

$$T_b(\theta, \phi) = \frac{c^2 B(\theta, \phi)}{2k_B \nu^2}. \tag{3.20}$$

---

[7] A black body absorbs all of the electromagnetic radiation incident on its surface without reflection. It emits radiation of all wavelengths and the wavelength distribution of the radiation follows Planck's law.

[8] Rayleigh–Jeans' law is the limit of Planck's law when $h\nu \ll k_B T$.

If the source is a black body and Rayleigh–Jeans law applies, T_b becomes equal to T. Generally, the brightness temperature is a function of frequency.

The radio brightness distributions are the properties of the celestial sources, which vary across them and change according to radiative transfer equations. If the radiation passes through a cloud, which is both emitting and absorbing, the observed brightness temperature is modified. The absorbing properties of such a cloud can be expressed by a dimensionless quantity, τ, known as optical depth. The latter is a measure of transparency, which is defined as the fraction of radiation that is scattered between a point and the observer. An optical depth of unity is that thickness of absorbing gas from which a fraction of $1/e$ photons can escape. This defines the visible edge of a star since it is at an optical depth of unity that the star becomes opaque. The optical depth of a given medium is different for different colors of light. For certain sources, like the quiet Sun and H II region, the emission mechanism is thermal bremsstrahlung.[9] The observed spectrum is the Rayleigh–Jeans tail of the black body spectrum. In this case, the brightness temperature is directly related to the physical temperature of the electrons in the source.

The strength of the signal emitted from a large cloud of neutral hydrogen depends upon its temperature and the total number of atoms in the line of sight. The cloud is opaque if there is a very large number of atoms. The signal strength corresponds to the temperature of the cloud provided that the cloud fills the antenna beam. With a small angular diameter of the cloud, the signal strength falls in the ratio of the solid angle subtended by it to the solid angle of the said beam. The brightness temperature, T_b, from the cloud is given by,

$$T_b = T_c \left(1 - e^{-\Delta}\right),$$ (3.21)

where T_c is the actual temperature of the cloud ($^\circ$K) and Δ the optical thickness[10] of the cloud. As the thickness of the cloud increases, the brightness temperature,

---

[9] Bremsstrahlung is a process whereby an electron is deflected by the electromagnetic field of an oppositely charged ion. It is produced by a high energy particle, such as an electron, deflected in the electric field of an atomic nucleus. It has a continuous spectrum.

[10] Optical thickness is defined as the integrated extinction coefficient (fractional depletion of radiance per unit pathlength) over a vertical column of unit cross section. Adding all the elementary contributions along a ray, the brightness temperature in the direction of the ray is expressed as,

$$T_b = \int_{-\infty}^{\infty} T_c e^{-\tau_c} d\tau_c,$$

where τ_c is the optical depth to each element.

For a thick layer of isothermal homogeneous cloud of temperature, T_c, and optical thickness, Δ, the brightness temperature, T_b is given by,

$$T_b = \int_0^{\Delta} T_c e^{-\tau_c} d\tau_c = T_c \left(1 - e^{-\Delta}\right).$$

T_b, approaches the cloud temperature, T_c. In the case of a source of brightness B_s observed through a cloud which both emits and absorbs. The observed brightness is given by,

$$B = B_s e^{-\Delta} + B_i \left(1 - e^{-\Delta}\right), \qquad (3.22)$$

in which B_s is the brightness of the source (W m^{-2} Hz^{-1} rad^{-2}) and B_i the intrinsic brightness of cloud.

The first term of the right hand side of the above (3.22) represents the loss in brightness of the irradiating source of brightness B_s due to the absorption of the cloud (Krauss 1966), while the second term represents the contribution to the observed brightness due to emission and absorption by the cloud. In terms of temperature, (3.21) is recast as,

$$T_b = T_s e^{-\Delta} + T_i \left(1 - e^{-\Delta}\right), \qquad (3.23)$$

with T_b, T_s, and T_c as the observed brightness, source, and cloud temperatures, respectively.

If the cloud is thin (almost transparent), the optical thickness is small, Δ is much less than unity, hence T_b becomes equal to $T_c \Delta$, while in the case of the thick cloud (opaque), Δ is greater than unity and therefore $T_b = T_c$.

3.2.1.3 Antenna Temperature

The antenna temperature, T_a, is a measure of the power absorbed by the antenna, which is proportional to the power per unit bandwidth received from a source. In an ideal case, this is equal to the brightness temperature if the intensity of the received radiation is constant within the main lobe. If the source's angular dimension is smaller than the main lobe, the antenna temperature turns out to be equal to the brightness temperature multiplied by the ratio of the solid angle subtended by the source to the effective solid angle of the antenna. The equivalent temperature of the power, P, received at an antenna from a source,

$$T_a = \frac{P}{k_B \Delta f} = GS, \qquad (3.24)$$

where G is the antenna gain, Δf the bandwidth, k_B the Boltzmann's constant, and S the measured flux density impinging on the antenna from the source.

However, the equivalent temperature also represents the unwanted noise produced by ground radiation, attenuation in the transmission lines, and other sources (in the absence of a signal from a source). The noise power, P_N is given by (Crane and Napier 1994),

$$P_N = k_B T_{sys} \Delta f,$$

with $T_{sys} = T_{sky} + T_{spill} + T_{loss} + T_r$, T_{sky} as the contribution of the background sky brightness, for instance, temperature arising from the radio sky, which is strong emitter of non-thermal continuum radiation; at all frequencies, the sky contributes at least $3°K$ from the cosmic background,[11] T_{spill} the spillover noise; often the feed antenna picks up stray radiation, e.g., from the ground (which radiates at $\approx 300°K$); T_{loss} the noise due to lossy element in the feed path, and T_r the receiver temperature representing internal noise from the receiving amplifier.

It is pertinent to note that at the low frequencies ≤ 100 MHz, T_{sys} is mainly due to the contribution from the background radiation and at frequencies ≥ 50 GHz, the system temperature is determined by the front-end electronics. In the case of a signal from an extended source over an unpolarized region of the sky of surface brightness, $B(\theta, \phi)$, the power received per unit bandwidth by an antenna is,

$$w = \frac{P}{\Delta f} = \frac{1}{2}A_e \int_{sky} B(\theta, \phi) P_n(\theta, \phi) d\Omega, \qquad (3.25)$$

with $w(= k_B T)$ as the spectral power, P the power received by the antenna from a radio source, $B(\theta, \phi)$ the sky brightness, $P_n(\theta, \phi)$ the normalized power pattern over the solid angle $d\Omega$ at (θ, ϕ).

If the sky is uniformly bright ($B = $ constant), the spectral power, w is defined as,

$$w = \frac{1}{2}A_e B \int_{sky} P_n(\theta, \phi) d\Omega$$
$$= \frac{1}{2}B A_e \Omega_a \quad W\,Hz^{-1}. \qquad (3.26)$$

For a point source of flux density S, the antenna temperature from source alone is given by,

$$T_a = \frac{w}{k_B} = \frac{S A_e}{2 k_B}, \qquad (3.27)$$

in which $w = S A_e/2$, is the received power and T_a a measure of the received power and is called the antenna temperature provided (1) the source fills the beam of the antenna and (2) the source is thermal, with a Planck's spectrum; it is not the physical temperature of the antenna.

Since the beam solid angle, Ω_a, of the antenna is related to its effective area, A_e (3.9), (3.26) for the spectral power from a uniformly bright source simplifies to,

$$w = \frac{1}{2}B\lambda^2 = k_B T_a. \qquad (3.28)$$

---

[11] While calibrating the system temperature of a radio telescope, Penzias and Wilson (1965) discovered the Cosmic Microwave Background Radiation (CMBR), for which they were awarded Nobel prize in 1978. This radiation has a thermal $3°K$ black body spectrum, which peaks in the microwave range. This excess $3°K$, coming from the sky, was identified with the radiation from the Big Bang and was one of the pieces of evidence in favor of Big Bang model.

Therefore, the observed flux density of the source is expressed as,

$$S = \int_{-\infty}^{\infty} B(\theta, \phi) P_n(\theta, \phi) d\Omega$$

$$= \frac{2k_B}{\lambda^2} \int_{-\infty}^{\infty} T_b(\theta, \phi) P_n(\theta, \phi) d\Omega$$

$$= \frac{2k_B}{\lambda^2} T_a \Omega_a = \frac{2k_B T_a}{A_e}, \tag{3.29}$$

or,

$$T_a = \frac{A_e}{\lambda^2} \int_{-\infty}^{\infty} T_b(\theta, \phi) P_n(\theta, \phi) d\Omega$$

$$= \frac{1}{\Omega_a} \int_{-\infty}^{\infty} T_b(\theta, \phi) P_n(\theta, \phi) d\Omega, \tag{3.30}$$

in which $T_b(\theta, \phi)$ is the brightness temperature of a region of sky, T_a the antenna temperature, and $P_n(\theta, \phi)$ the normalized antenna power pattern.

If the temperature is measured at a remote point from the antenna, the attenuation caused by the transmission line should be taken into account; therefore, the observed flux density, S, of the source is expressed as,

$$S = \frac{2k_B T_{tr}}{\eta_{tr}} A_e, \tag{3.31}$$

where A_e is the effective aperture of antenna, $T_{tr} = \eta_{tr} T_a$ the temperature due to source measured at the end of transmission line, and η_{tr} the efficiency factor for the transmission line of same length ($0 \le \eta_{tr} \le 1$); it is a dimensionless quantity.

If the angular size of the source is $\Omega_s \ll \Omega_a$, so that $P_n(\theta, \phi) \simeq 1$, the observed flux becomes,

$$S = \frac{2k_B}{\lambda^2} T_b \Omega_s, \tag{3.32}$$

and therefore, the observed antenna temperature of the source is recast as,

$$T_a = \frac{1}{\Omega_a} \int_{\Omega_s} T_b(\theta, \phi) d\Omega = \frac{\Omega_s}{\Omega_a} T_{av}, \tag{3.33}$$

with T_{av} as the average temperature over source.

It is noted that antenna temperatures vary from telescope to telescope, while brightness distributions are properties of the celestial sources.

3.2.1.4 Sensitivity

Sensitivity provides a measure of the weakest source that can be detected by the system. Generally, narrow bandwidths are used in radio astronomy to minimize the effects of interfering signals from outside the observing band. The adjacent components beat together to give (1) quasi-sinusoidal underlying waveform $\sim \sin 2\pi \nu_0 t$ and (2) envelope fluctuating randomly with a coherence time (see 2.2). A detector extracts this fluctuating envelope, the output of which has a mean level having a noise signal with a standard deviation, σ equaling the mean, M. This noise has an uncertainty in the mean level due to fluctuating nature of the signal. The solution to this is to average the signal, albeit with independent samples. The number of independent measurements is $N = \Delta f \Delta t$, in which Δf is the bandwidth and Δt the integration time. The uncertainty in a measurement drops as the square-root of the number of samples, so that the S/N ratio is,

$$S/N = \sqrt{\Delta f \Delta t}. \tag{3.34}$$

In the case of photon noise (Poisson or shot noise), the S/N ratio is $S/N = \sqrt{n}$, with n as the number of photons received. The sensitivity of an antenna is equal to the root-mean-square (RMS) noise temperature of the system. The antenna temperature may be composed of several contributions, i.e., $T_a = T_s + T_{sky}$, in which T_s arises from the source, T_{sky} as described earlier. The signal-to-noise (S/N) ratio is dictated by,

$$S/N = \frac{T_s \sqrt{\Delta f \Delta t\, N}}{T_{sys}}, \tag{3.35}$$

in which N is the number of records averaged.

The (3.35) is the fundamental expression for the sensitivity of a single-dish telescope. If the S/N ratio is sufficiently large, one can detect the source. This equation may be modified to,

$$S/N = \frac{S A_e \sqrt{\Delta f \Delta t\, N}}{2k_B T_{sys}}, \tag{3.36}$$

where $T_a = S A_e/(2k_B)$.

3.2.1.5 Brightness Distribution

The relation between the geometry of the dish and the size and shape of the beam can precisely be determined by polar diagram[12] (radiation pattern), which is an analog of the optical case of the Fraunhofer diffraction (see Sect. 1.4.2) pattern of an aperture

---

[12] Polar diagram is the graphical representation of the variation of the radiated power (far field) as a function of angle in the case of a radiating antenna.

of similar shape. This implies that the far-field radiation pattern of a telescope is the Fourier transform of the field distribution in its aperture. Let the telescope be described in terms of an aperture plane on which currents are induced by the incoming radiation. The voltage pattern, $V(\alpha)$ of the aperture plane is described by,

$$V(\alpha) = \int_{aperture} g(\mathbf{x})e^{-i2\pi\mathbf{x}\cdot\alpha}d\mathbf{x}, \tag{3.37}$$

in which $g(\mathbf{x})$ is the grading providing relative response over the aperture, $\mathbf{x} = x, y$, α the direction of the response with respect to the direction of the pointing axis.

It is to note that the amplitude pattern of a rectangular aperture telescope along x axis is the same as the amplitude pattern of a slit x wavelengths wide (see Sect. 1.4.3) and is given by the Fourier transform of a rectangular distribution which varies as sinc function $\sin \pi x/\pi x$ (see 1.66), while the field diffraction polar diagram of a circular dish is given by the Fourier transform, $J_1\pi(x)/\pi(x)$ (see 1.68), in which J_1 denotes a Bessel function of the first order and x the diameter of the dish in terms of the wavelengths. Figure 3.4a, b depict the normalized amplitude and intensity distribution patterns, respectively, of a rectangular telescope aperture, as well as a circular aperture. The transforms are related to the field distribution of the electric vector, whereas the quantity that may be measured is usually the intensity of power output of the receiver. In this case, the power polar diagram (beam pattern) is proportional to the squared modulus of the voltage radiation pattern. The antenna power radiation pattern is given by the autocorrelation theorem of Fourier transforms,

$$\mathcal{F}[P(\mathbf{x})] = g \otimes g^* = \int_{-\infty}^{\infty} g(\boldsymbol{\xi})g^*(\boldsymbol{\xi} - \mathbf{u})d\boldsymbol{\xi} = \widehat{P}(\mathbf{u}), \tag{3.38}$$

in which $\widehat{P}(\mathbf{u})$ is the instrumental transfer function (autocorrelation function of the grading), * the complex conjugate, $\otimes$ the autocorrelation, $\mathbf{u} = u, v$ the 2-D spatial

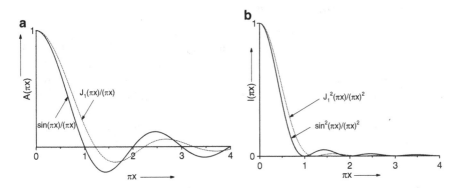

Fig. 3.4 Directivity pattern of a *rectangular telescope* aperture (*solid line*) and a *circular aperture* (*dotted line*); (**a**) amplitude and (**b**) intensity

frequency vector; u, v are spatial frequency components in East–West and North–South directions on the ground, and $\mathcal{F}[]$ stands for the 2-D Fourier transform.

The (3.38) relates that the power pattern is the Fourier transform of the auto-correlation of the aperture current distribution. The far-field directivity pattern or polar diagram of a large telescope is dependent upon the dimensions of the aperture, the way in which the signal is distributed in amplitude and phase across this aperture, and of course the wavelength of the signal. The precise relationship can be obtained by Fourier analysis. Since the radio sources have a finite angular extent, their appearance is characterized by its brightness distribution. When a point source is moved across the aperture, the polar diagram of an antenna may be traced directly by a power recorder on the output of its associated receiver. The narrower the polar diagram of an antenna, the greater is its ability to resolve detail of small angular structure. For an incoherent source with an intensity distribution, I, the output from a receiver connected to an antenna of polar diagram is given by the sum of the products of the contributions from the different parts of the source with their associated ordinates on the beam pattern as it is swept through the source.

As the radiation propagates, it begins to diverge, which gives rise to a wider profile. In two closely spaced radiating points, each of which contributes to the output of the receiver. The output recorder traces the sum of their contributions. When an antenna is employed to scan a radio source its finite beam smooths out any spatial discontinuities in the source. The observed response of a radio telescope to a sky brightness distribution is proportional to the convolution of the true brightness distribution, $B(\mathbf{x})$, with the antenna beam pattern, i.e.,

$$S(\mathbf{x}) = \int_{-\infty}^{\infty} B(\mathbf{x}')P(\mathbf{x} - \mathbf{x}')d\mathbf{x}', \tag{3.39}$$

where $S(\mathbf{x})$ is the observed flux-density distribution and $P(\mathbf{x})$ the mirror image of normalized antenna power pattern.

3.2.2 The Radio Interferometer

Observations at radio frequencies in the early days were severely limited by, (1) poor angular resolution and (2) low sensitivity of the receiving system. In order to achieve angular resolution of 1 arcsecond ($''$) at 5 cm, a radio telescope must have a paraboloid of ≈ 10 km diameter. Developing huge radio telescopes to achieve a resolution comparable to that of their optical counterparts was a major task for radio astronomers before the era of modern interferometers. The simple two-element interferometer constructed by Ryle[13] and his colleagues (Ryle 1952)

---

[13] Martin A. Ryle was awarded the Nobel Prize in 1974 for Physics, along with A. Hewish, for his work on Earth-rotation aperture-synthesis (see Sect. 5.3.2.1) technique in radio interferometry. A. Hewish received it for discovery of pulsars.

at the Cambridge Observatory is the starting point of the large interferometer arrays that were constructed subsequently and started producing high resolution images employing the technique of aperture-synthesis. Bigger reflector dishes were made but the prohibitive cost, engineering difficulties, and the imperfection of the surface made a 300 m aperture about the limit, for example the Arecibo telescope.

An interferometer array, that is a group of antennas or antenna elements arranged to provide the desired directional characteristics, collects the signal from a distant cosmic radio source, and measures the complex visibilities of the source. To note, the output from a single antenna as a function of intensity is positive (always), that from an interferometer oscillates about the mean intensity, following the relative phase angle between the signals in the two antennas as the source passes through the beam. The phase of these fringes depends on the relative length of the ray paths from each point in the source to the combing elements. This is analogous to the phenomena of Young's slit experiment (see Sect. 2.1.1) in optics. The major advantages of such a method in radio wavelengths over optical wavebands are: (1) uniform phase over individual apertures, (2) time integration, (3) phase stability of delay-lines, and (4) electric delay-lines. In this technique, a pair of directional antennas separated by a distance, $\mathbf{B}$ called baseline vector, are tuned to receive radio emissions from a source in a desired radio frequency (RF) band.

In a tracking interferometer, two identical antennas should be made accurately steerable in order to follow the motion of a source in its diurnal motion around the sky for hours simultaneously, so that the maximum sensitivity is obtained and the variation in baseline, which connects the phase centers of these antennas, is exploited in determining the source structure. Both these antennas track the radio source as it moves, one of the antennas receives the signal delayed by the geometric time delay,

$$\tau_g = \frac{\mathbf{B} \cdot \hat{\mathbf{s}}}{c}, \tag{3.40}$$

in which $\mathbf{B}$ the baseline vector defined by the two antennas along the unit vector, $\hat{\mathbf{s}}$, pointing to the source, and c the speed of light.

The path delay, τ_g is compensated using a signal delay component, known as instrumental delay, τ_i, into one arm of the interferometer to approximately equalize the pathlength over both paths to the multiplier (Thompson et al. 2001); pathlength can change due to the troposphere and ionosphere.[14]

A multiplier is a device in which two voltages are correlated or multiplied to give what is called visibility coefficient. In the quasi-monochromatic approximation, the multiplier output of a two element interferometer is given by, $|\mathcal{V}| \cos(2\pi \nu \tau_g + \Psi_\nu)$, in which $|\mathcal{V}|$ and Ψ_ν are the amplitude and phase of the visibility, respectively. In the case of the aperture-synthesis, the quantities required for mapping a source are

---

[14] The refractive index of the atmosphere differs from unity, hence there would be an additional delay in traversing the atmosphere. The ionosphere also causes a frequency-dependent refraction; it is worse at low frequencies. In addition to refraction, the atmosphere absorbs/emits radiation, which in turn, reduces flux and increases noise.

$|\mathcal{V}|$ and $\Psi_\mathcal{V}$ for all pairs of telescopes. These quantities are measured by canceling the term $2\pi\nu\tau_g$ by delay tracking and fringe stopping. Compensation of $2\pi\nu_{LO}\tau_g$ is referred to as fringe stopping, which can be achieved either (1) by changing the phase of the local oscillator signal by an amount ψ_{LO}, so that

$$2\pi\nu_{LO}\tau_g - \psi_{LO} = 0,$$

or (2) digitally by multiplying the sampled time series by $e^{-i\psi_{LO}}$.

3.2.2.1 Simple Radio Interferometer

In general, a radio telescope is configured to receive the RF signal over a finite bandwidth from the radio source of interest and process the same into a form suitable IF signal for transmission to the correlator. The correlator output is proportional to the cross-correlation of the disturbance of the electric field generated by the signal incident at the telescopes. The IF signal is suitably amplified in order to increase the voltage of the signals for subsequent processing. Usually a filter is used to reject the high frequency signal (upper sidebands), as well as to limit the IF frequency range within the specifications of the combiner, delay-line, and IF amplifier bandpass. The output signals are combined and detected by the receiver. Neglecting the effect of finite bandwidth, the response $V(\mathbf{r}_1)$, and $V(\mathbf{r}_2)$, of the two antennas placed at $\mathbf{r}_1$ and $\mathbf{r}_2$ respectively to the emission from a point source may be expressed as,

$$V(\mathbf{r}_1) = E(\mathbf{r}_1)\cos\omega t, \tag{3.41a}$$
$$V(\mathbf{r}_2) = E(\mathbf{r}_2)\cos(\omega t + \psi), \tag{3.41b}$$

where

$$\psi = \frac{2\pi\mathbf{B}\cdot\hat{\mathbf{s}}}{\lambda}, \tag{3.42}$$

is the phase difference between two signals, E the constant, ω the frequency of observation, t the time, $\mathbf{B}$ the baseline vector of the interferometer, $\hat{\mathbf{s}}$ the unit vector in the direction of the source, and λ the observing wavelength.

In the total power interferometer (see Fig. 3.5), these two voltages are added and squared and the output, R, from the receiver is proportional to power,

$$\begin{aligned} R &= E^2(\mathbf{r}_1)\langle\cos^2\omega t\rangle + E^2(\mathbf{r}_2)\langle\cos^2(\omega t + \psi)\rangle \\ &\quad + E(\mathbf{r}_1)E(\mathbf{r}_2)\langle 2\cos\omega t\cos(\omega t + \psi)\rangle \\ &= R_0 + R_1\cos\psi, \end{aligned} \tag{3.43}$$

where R_0, R_1 are the constants, $\langle\cos^2\omega t\rangle = \langle\cos^2(\omega t + \psi)\rangle = 1/2$, while

$$\langle 2\cos\omega t\cos(\omega t + \psi)\rangle = \langle\cos\psi + \cos(2\omega t + \psi)\rangle = \cos\psi.$$

Fig. 3.5 Geometry of adding interferometer receiver

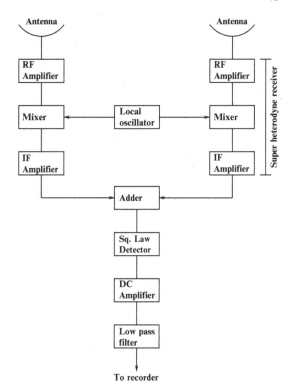

With identical antennas, $R_0 = R_1$ and $E(\mathbf{r}) = E_n(\mathbf{r}) \cos(\psi/2)$, in which $E_n(\mathbf{r})$ is the normalized field pattern of the individual array element, the output may be expressed as,

$$R \propto |E_n(\mathbf{r})|^2 (1 + \cos\psi)/2$$
$$\propto |E_n(\mathbf{r})|^2 \cos^2 \psi/2. \qquad (3.44)$$

The output R is a function of the source intensity and the source-baseline geometry. If the source moves in the plane passing through the baseline, a fringe pattern is envisaged; the separation between two fringes is governed by $\lambda/(\mathbf{B} \times \hat{\mathbf{s}})$. With equal length of the feeder cables, the interferometer pattern follows the voltage pattern of either antenna taken separately, modulated by a system of cosine fringes whose period depends upon the separation of the antennas. However, if the separation between the two antennas is more than a few hundred wavelengths, i.e., $B \gg \lambda$, then the length of the feeder system becomes larger and the attenuation during transmission of the signals becomes severe, resulting in the appreciable loss of signal-to-noise (S/N) ratio. This effect is a serious limiting factor while observing weak sources.

The diurnal motion of the source produces a variation in the angle to the source, θ, that is measured from the baseline direction. As the path difference between the two antennas varies due to the movement of the source in the sky, the output of the receiver is modulated by fringes. The corresponding changes in the phase difference, ψ would vary rapidly. The output of the total power interferometer, therefore exhibits a variation of amplitude imposed on a level varying approximately according to the background sky. It is essential that the gains of both of the antennas and the losses in the feeder cables from the individual antennas should be identical, or else the modulation of the total power from the source would be incomplete, hence an erroneous measure of the visibility may arise; the electrical lengths of the antenna should be identical as well.

Visibility is computed by measuring amplitude of these fringes relative to that of the unmodulated output. In order to compute the distribution of source brightness, one determines the visibility at different separations with an array of antennas. It is to be noted that the amplitude function of the Fourier transform (FT) of the brightness distribution across a source may be determined from the visibility of the fringes in the reception pattern of the interferometer. Figure 3.6 represents the response of an individual antenna, a simple two element adding interferometer, the interference pattern and the combined output for a point source (see Appendix C for the aperture distribution, spatial frequency response of an adding interferometer).

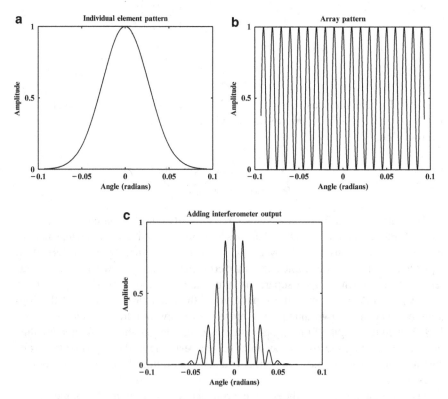

Fig. 3.6 (a) Individual element pattern for the ideal case, (b) array pattern, and (c) the resultant interferometer pattern of a point source

Measuring the position of a source on the sky as a function of time with a long baseline interferometer involves the following:

- The apparent motion of the source is roughly sidereal, and the measurement of source position may be carried out at discrete intervals of angular separation.
- The orientation of the interference pattern on the sky, which can be determined from the antenna surveys and the phase relationships between the signal path from different antennas all the way till the corresponding final IF outputs; the loci of the maxima, minima, and zero crossover(s) of the pattern can be found.
- The phase stability of the equipment should also be very high; even a high quality antenna/feeder system is subject to atmospheric changes, and local heating of a part of the feeder chain may cause appreciable changes in the phase of the signal, and therefore the apparent position or structure of the source.

3.2.2.2 Phase-switched Radio Interferometer

For measurement of accurate positions, the simple two-element adding radio interferometer that is described above is not well suited. A great deal of improvement to the simple interferometer was introduced by Ryle (1952), in which a switching mechanism was incorporated whereby an extra section of line having electrical length of one-half wavelength at the signal frequency is periodically switched in/out in the path of one of the two antennas. An extra half wavelength of transmission cable introduces a phase-shift of $180°$ in the corresponding signal path. The output of the receiver is then arranged to record only the difference of noise level between these two states by inserting a synchronized relay in series with the detector. The notable advantage of such an instrument is that it is not affected by large variations in the background radiation of the sky as well as gain variations in the receiver system. The requirements of stability are reduced considerably and the sensitivity of the receiver can be kept constant by employing an automatic gain control without affecting appreciably the fringe size in the output. A variation of this technique is to rotate the phase of the signal in one of the two arms of the interferometer so as to generate artificial fringes in the presence of radiation from a small source. The periodic change in feeder length provides two alternating interference patterns, one of which is, of course, the same as that shown in Fig. 3.7a and the other can be found by considering the effect of a $180°$ phase shift. Irrespective of the instantaneous value of a sinusoidal wave, a $180°$ phase shift results in the negative of that value. Figure 3.7b shows the output of a phase-switched interferometer (see Appendix C for the spatial frequency response of a phase-switched interferometer).

A voltage, E, proportional to the field produced by the source, is generated at each antenna, but at different time defined by τ. The voltages at the antenna output are given by,

$$V(\mathbf{r}_1) = Ee^{i\omega t}, \tag{3.45a}$$

$$V(\mathbf{r}_2) = Ee^{i\omega(t-\tau)} = Ee^{i(\omega t - 2\pi B \cos\theta/\lambda)}. \tag{3.45b}$$

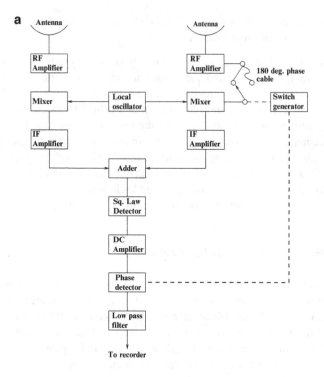

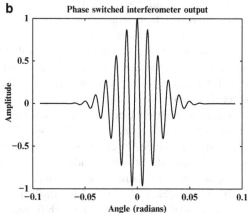

Fig. 3.7 (a) Geometry of a phase-switched interferometer and (b) Fringe pattern of an interferometer in phase

Earth's motion causes the angle between the point source and the line connecting the antenna, θ, to vary with time, τ. The power output $R(\tau)$ of the multiplier, following low pass filtering, is expressed as,

$$R(\tau) = A\cos(2\pi B_\lambda \cos\theta(\tau))$$
$$= A\cos(2\pi \mathbf{B}_\lambda \cdot \hat{\mathbf{s}}(\tau)), \tag{3.46}$$

where A is the fringe amplitude, and the term $\cos\theta B/\lambda$ is written as $\mathbf{B}_\lambda \cdot \hat{\mathbf{s}}$.

Unlike the adding interferometer, in a correlation interferometer, such voltages are multiplied. The signals from the two antennas are fed to a voltage multiplier where the signal voltages are multiplied followed by time-average to filter out high frequencies. The convolution of the signal amplitudes from these antennas yields the cross-correlation function, $\Gamma_{x,y}(t)$ (see Appendix B). It is important to note that an interferometer pair can detect fringes as long as the cross-correlation performed on the radiation arriving at the two antennas is done within coherence length (see 2.1).

In practice, the output voltages, $V(\mathbf{r}_1, t)$ and $V(\mathbf{r}_2, t - \tau)$ of the two antennas are multiplied to obtain the cross-correlated power as function of time, τ, hence,

$$R(\tau) = \frac{1}{2T} \int_{-T}^{T} V(\mathbf{r}_1, t) V(\mathbf{r}_2, t - \tau) dt. \tag{3.47}$$

For the 1-D example, one may write,

$$R(\tau) = Ae^{i2\pi B\cos\theta(\tau)/\lambda} \equiv Ae^{i\alpha'}, \tag{3.48}$$

in which $\alpha'[= 2\pi B\cos\theta(\tau)/\lambda]$, is the fringe phase.

The practical limitations imposed on the measurements of fringe visibility are severe and render conventional techniques untenable if the baseline of the interferometer exceeds a few kilometers. In such a situation, transmission of the RF signal incident on the antenna via cables to a centralized receiver room is not practically feasible.

3.2.2.3 The Shortcomings

Simple radio interferometer faces several disadvantages, viz.,

- Its response to the total power from large resolved features in the sky, which do not give rise to fringes, but may cause large changes in the mean level of the signal (see Appendix C).
- Its inability to obtain accurate measure of the normal visibility at short spacings or of the power from the source, more so for a source of small intensity, e.g., a faint source situated near sidelobes of the instrumental response pattern, where uncertainty in the contribution of the background radiation prevails.
- Its requirements for amplitude stability of the antennas and feeder system should be exceptionally high so that the sensitivity does not vary appreciably during a single observation.

- Its vulnerability to the stability of frequency; the effect of a small change in the frequency of the receiving system is equivalent to a change in the baseline of the apertures and causes a corresponding motion of the fringes.
- Its proneness to serious errors in the amplitude and phase of the signals received at the apertures due to the occurrence of scintillation in the ionosphere[15]; the effect of which is more pronounced when the baseline extends beyond the distance over which the fluctuations are correlated.

The main source of error in the computation of source distributions in a phase-switched interferometer is in the variation of the gain of the movable antenna and its associated pre-amplifier[16] when removed to different sites. This difficulties may be overcome by employing the gains of which are invariant at the either extreme of the baseline. The baseline uncertainty arises due to small unbalances in the phase switch. A certain amount of total power response is often present, which response has the effect of shifting the baseline to one side of the axis. A crossover may be located approximately as precisely as the axis is known, more or less so as the crossover is more nearly parallel or perpendicular to the axis.

The other notable sources are interference and scintillation, which strike directly on the recorder. The phase-switched interferometer cannot reject interference that consists of radio emissions of terrestrial origin having directional properties. Making crossover measurements through interference is either impossible or subject to uncertainty.

A phase-switched interferometer is more sensitive to the amplitude scintillation near maxima and minima, where the pattern is relatively independent of small phase shifts. Strong amplitude scintillation near a crossover may cause the record to drop to the baseline, instrument or the noise level producing uncertainty as to the location of the crossover. Conversely, the high slope of output voltage versus phase near and at crossovers makes the pattern more sensitive to the phase scintillation, since phase scintillation is a change in the direction of the source, and therefore equivalent to a change of phase in the interferometer. Scintillation may affect the precise positions of expected crossovers. The changes of crossover position are of the same type as those produced by the traveling waves in the ionosphere. Fortunately severe scintillations do not occur regularly and their effect is less marked on high frequency observations.

---

[15] The ionosphere is the part of the Earth's upper atmosphere that is ionized by solar radiation. By ionization process, an atom gains or looses one or more electrons due to which the atom becomes an ion instead of being neutral. The atom becomes positively charged after losing an electron, and it possesses negative charge if it has gained one; high temperature is particularly conducive to ionization. The ionosphere plays an important role in atmospheric electricity and forms the inner edge of the magnetosphere. It influences radio propagation to distant places on the Earth. Also it reflects back radio waves of frequency less than ~ 10 MHz from the cosmic radio sources

[16] A pre-amplifier boasts the voltage, $V(t)$, by a large factor turning the μV level input signal from the antenna to mV level, which is strong enough not to be degraded or lost in further processing.

3.2.2.4 Large Antenna Arrays

In spite of the fact that the diffraction-limited resolution of the largest radio telescope is about a minute of arc, the spectacular progress has been witnessed in the case of the radio interferometry with large arrays. These instruments have led to several new discoveries. In what follows, a few large arrays are described in brief.

1. Very Large Array (VLA): Located at Socorro, New Mexico, USA, this array consists of 27 radio antennas, each of which has a dish diameter of 25 m. The antennas are distributed along the three arms of a Y-shaped configuration spreading over three 21 km tracks providing 351 baselines. Located at an altitude of 2,124 m above sea level, this array is a component of the National Radio Astronomy Observatory (NRAO). The data from the antennas is combined electronically. Since this instrument is a 2-D aperture, synthesized from 27 smaller apertures, it can make an image in snapshot mode. The smallest angular resolution that can be reached is about 0.05 arcseconds at Q band (40–50 GHz). The VLA is a multi-purpose instrument is used to investigate various celestial objects such as radio galaxies, quasars, pulsars, protoplanetary disks around young stars, supernova remnants, gamma ray bursts, and radio-emitting stars (Leahy and Perley 1991; Bridle et al. 1994; Barger et al. 2001; Cannon et al. 2005; Gopal-Krishna and Wiita 2005; and references therein).

2. Giant Meterwave Radio Telescope (GMRT): The facility for radio astronomical research using the meterwavelengths range of the radio spectrum, known as the Giant Meterwave Radio Telescope (GMRT), is situated at Narayangaon, Pune, India. It consists of 30 fully steerable parabolic dishes of 45 m diameter, capable of operating at six different frequency bands such as 50 MHz, 153 MHz, 233 MHz, 325 MHz, 610 MHz, and 1,420 MHz (Swarup et al. 1991). All these feeds provide dual polarization outputs; in some configurations, dual-frequency observations are also possible. Fourteen dishes are randomly arranged in the central square, with the remaining sixteen arranged in three arms of a Y-shaped array giving an interferometric baseline of about 25 km. The multiplication (correlation) of radio signals from all the 435 possible pairs of antennas over several hours may enable radio images of celestial objects to be synthesized with a resolution of about 1 arcsecond at the frequency of neutral hydrogen (1420 MHz). This instrument is used to observe different astronomical objects, such as sun and solar winds, galactic center, transient sources, pulsars, supernovae, and galaxies (Freire et al. 2004; Roy and Rao 2004; Cameron et al. 2005; Kantharia et al. 2005; Mercier et al. 2006).

3. Other instruments: The Expanded Very Large Array (EVLA) which is an ongoing project will enhance the instrument's (i.e., the existing VLA) sensitivity, frequency range, resolution, and imaging capability. This instrument will be capable of operating at any frequency between 1.0 and 50 GHz, with up to 8 GHz bandwidth per polarization. Another system, called The Multi-Element Radio Linked Interferometer Network (MERLIN) telescope array, operated by Jodrell Bank Observatory, is an array of seven radio telescopes spread across Great

Britain, with maximum separations of up to 217 km. It can operate at frequencies between 151 MHz and 24 GHz. Such an instrument will provide radio imaging, spectroscopy and polarimetry with 10–150 mas resolution. This instrument uses microwave links to send data, having limited bandwidth (30 MHz). The major upgradation of this system, called e-MERLIN, is to replace the current link by optical fibre links carrying a bandwidth of 4 GHz. Another development is frequency flexibility, which will have the ability to alter the observing band of the entire array in a short time using rotating carousels of receivers. When e-MERLIN becomes operational the telescope may switch rapidly between 1.4, 5, 6, and 22 GHz. Among others, MERLIN is used to observe (1) radio-loud galaxies, Messier 87, (2) quasars, 3C 418, and (3) spectral line observations of Hydroxyl (OH) in interstellar gas clouds (Lovell 1985). The Australia Telescope Compact Array (ATCA) is an array of six 22-m antennas is another facility at Narrabri in Australia. It is a premier instrument of its kind in the southern hemisphere and is operated by the CSIRO, Australia. The ATCA has a range of East–West configurations that are well suited to imaging with complimentary arrays from 214 m to 6 km. The observing frequencies are 1.75, 2.45, 5.5, 17, 33, 43, and 93 GHz. The Westerbork Synthesis Radio Telescope (WSRT) is another major radio astronomy facility that is in operation at the Netherlands. It has 14 dish antennas of 25 m diameter each, along an East–West line. Ten of the antennas have a fixed location, while the other four are movable on rail track. The maximum baseline length is 2.7 km. The observing frequencies range from 300 MHz to 8 GHz. The original design was to operate the instrument at 1,400 MHz to observe Hydrogen line mentioned before, from the cosmic radio sources. The Owens Valley Radio Observatory (OVRO) in California, USA has a millimeter wavelength array of six radio telescopes, each 10.4 m in diameter and movable along a T-shaped railroad track to give the equivalent resolution of a single 300-m dish. The Berkeley-Illinois-Maryland Association (BIMA) array at Hat Creek in northern California is another millimeter wave array. Both the above millimeter arrays are being merged to for a much powerful millimeter wave telescope at a higher elevation site.

Work on two more new radio interferometric array projects such as (1) Low Frequency ARray (LOFAR) and (2) Square Kilometer Array (SKA) are in progress. The former array, a low frequency array distributed across the countries of Europe, is a large radio telescope consisting of phased array of ~10,000 dipole antennas. This mission aims to survey the Universe at radio frequencies from ~10 to 240 MHz with a resolution of better than an arcsecond at 240 MHz (Röttgering 2003). The latter array, the next generation of large array radio telescope, is intended to have a collecting area of approximately one square kilometer, which will be in operation by the end of next decade. It will operate at frequencies of 0.10–25 GHz, with a goal of 0.06–35 GHz. The SKA will create images of distant radio sources using aperture-synthesis technique (Carilli and Rawlings 2004).

3.2.3 Very Long Baseline Interferometry

Very long baseline interferometry (VLBI) is a technique that measures the coherence or correlation between the RF signal from the same cosmic source incident on antennas kept at different locations spaced more than thousands of kilometers apart. It is distinguished from other forms of interferometry by not requiring communication between the individual antennas while observations are being made. Independent, high quality frequency standard replace a distributed local oscillator signal. Development of such a technique was an important breakthrough since it removed the limitation to the size of the effective aperture. Intercontinental links have produced an effective apertures of 10,000 kilometers providing a milliarcsecond angular resolution at cm wavelengths. For instance, the European VLBI Network (EVN) formed in 1980 by Max Planck Institute for Radio Astronomy (MPIfR) – Germany, Institute of Radio astronomy (IRA) – Italy, Netherlands Institute for Radio Astronomy (ASTRON), Onsala Space Observatory – Sweden, and Jodrell Bank Observatory – UK, is an interferometric array of radio telescopes spread throughout Europe and beyond, which conducts high resolution observations of cosmic radio sources. Employing the technique of VLBI, another instrument, called Very Long Baseline Array (VLBA), is a system of 10 antennas, each with a dish 25 m in diameter. The longest baseline in the array is 8,611 km. Such a system is remotely controlled from the the National Radio Astronomy Observatory (NRAO) in Socorro, New Mexico. The ground-based VLBI experiments are limited by the size of the Earth of baselines of about 11,000 km. This limitation was successfully overcome by Space VLBI observations wherein simultaneous observations with the EVN and the Japanese VSOP (VLBI Space Observatory Programme) project's HALCA satellite. The latter has a 8 m diameter radio telescope in Earth orbit. The maximum possible baseline achievable in the Space VLBI observations is about up to three times longer than those achievable on ground based VLBI on Earth. These instruments provide ultra-high angular resolution information of the phenomena, including supernovae, pulsars, flare stars, star-forming regions in molecular clouds, the environment surrounding nearby and distant galaxies, gravitational lenses, starburst galaxies, active galactic nuclei, and black holes. With interferometric and VLBI techniques, methanol (CH_3OH) and water (H_2O) maser associations with of high-mass Young Stellar Objects (YSO) have been detected (Moscadelli 2005). Compact symmetric objects are of particular interest in the study of the physics and evolution of active galaxies (Xiang et al. 2005, 2008) as well. An important point to be noted here is that the source must be very compact. The minimum detectable flux density, S_{min}, is given by,

$$S_{min} = \frac{2k_B T_b}{A_e}, \tag{3.49}$$

where T_b is the source brightness temperature, k_B the Boltzmann's constant, $A_e(\lambda^2/\Omega_a)$ the effective area of the antenna, and $\Omega_a[\simeq \pi(\lambda/2B)^2]$ the beam solid angle; the beam width of the interferometer is equal to λ/B, in which B is the baseline between the antenna.

After rearranging (3.49), the source brightness temperature, T_b is expressed as,

$$T_b = \frac{2B^2 S_{min}}{\pi k_B}. \tag{3.50}$$

The main drawback of VLBI techniques is that the observed complex visibility function is corrupted by effects of propagation and instruments, as well as uncertainties in the geometry; the sampling in the u, v is also incomplete and irregular. However, simultaneous observations of several sources or frequencies may improve the accuracy of VLBI substantially. The relative visibility functions among the sources or frequencies can be obtained with inaccurate baselines, and source positions, as well as with drifts in oscillator. The measurement of positions in the order of milli-arcsecond demands precise calibration, and measurements of effects like continental drift, Earth tides, precision rotation of the Earth, atmospheric and ionospheric pathlength changes, and relativistic bending of radio waves (Fomalont and Wright 1974).

3.2.3.1 Principles of VLBI

Since the baselines are very long for direct (real-time) transmission of the signal from the antennas to the central recovery station, the VLBI dispenses the links between them. Each station is equipped with (1) an antenna, (2) an amplifier capable of preserving fringe phase, (3) a mixing system with phase stable oscillator, (4) a means of synchronizing the clocks at each of the stations, (5) a device for storing the received signal and precise timing information, and (6) an identical playback system at each observatory for correlating offline the recorded data. The radio frequency signals, primarily in the ∼GHz frequency range, from the distant cosmic radio sources are received and amplified at each antenna using normal radio astronomical equipment. They are mixed with a local oscillator signal derived from the VLBI frequency standard, to bring one edge of the observing pass band to 0 Hz, i.e., to the video or the base band. The mixing and amplification generally occur in several intermingled steps. The signals are then filtered and digitized. After digitization, the data are formatted and recorded on a magnetic tape. Extra information, notably the time, is encoded with the data. These data are transmitted from the distant station to the home station for correlation studies. Figure 3.8 depicts the schematic of a VLBI system.

1. Baseline coordinates: The coordinates of the baselines among the VLBI antennas must be derived to a precision of ∼30 cm. This would enable the observer to measure (1) time (UT1) to ±0.0002 s, (2) variations in the rate of the Earth's rotation, and (3) polar motion to an accuracy of 30 cm (Downes 1988).
2. Intermediate frequency response: The IF output of each element is recorded separately on high speed recording device. The two IF responses are multiplied in a computer. These responses should be lined up with proper time delay before multiplication in order to observe the interference fringes on the astronomical

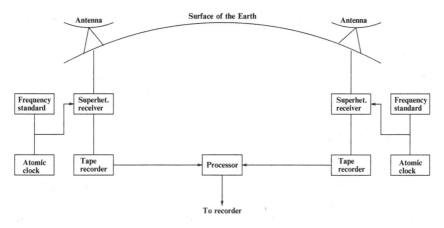

Fig. 3.8 Schematic of a VLBI system

sources. The time-delay occurs due to the position difference between the antennas, which is measured with a precise atomic clock (see Sect. 3.2.3.2) installed at each antenna.

3. Time keeping: If the signal bandwidth is Δf, the coherence time, τ_c, the time over which it is predictable, turns out to be approximately $1/\Delta f$. The time relation between the two recorded signals must also be kept less than the reciprocal of the bandwidth; the accuracy of the timing is required to be better than 1 μs for a bandwidth of 1 MHz. In this case fringes may be located immediately; with a higher signal bandwidth, such as 10 MHz, there is a need to search for fringes over 10 possible delays.

4. Local oscillator stability: One needs to integrate for a time before fringes are strong enough to envisage; signal phase must not wander. The random frequency drift, δf in the frequencies of the local oscillator during the integration time, Δt, should be $\leq 1/2\pi$ cycles or the phase drifts must be ≤ 1 radian during the integration time, Δt (Downes 1988), i.e., $\delta f\ \Delta t\ <\ 1/2\pi$, and therefore the relative frequency stability of the local oscillator should be,

$$\frac{\delta f}{f} < \frac{1}{2\pi \nu_0 \Delta t}. \tag{3.51}$$

Here, f is the frequency.

In order to preserve the fringe phase, the local oscillator must be phase-locked to very accurate atomic frequency standards. This has become feasible with the advent of high performance frequency standards that use the discrete atomic energy transitions (see Sect. 3.2.3.2).

5. Data storage: Data recorded on magnetic tape at various observatories are shipped to the correlator for processing at a convenient time. The essential functions of the playback system (Booth 1985) are:

• Alignment of the tapes containing data with a time offset to compensate the interferometer delay,

- Lobe rotation in order to compensate the Doppler frequency shift[17] between the two observatories, and
- cross-correlation of the data stream.

In general, the deconvolution algorithms based on least-square optimization and maximum entropy method (see Sect. 8.3.3) are used for image reconstruction.

3.2.3.2 Atomic Clocks

Atomic clock is a device regulated by the resonance frequency of atoms or molecules of certain substances. It is used to generate standard frequency. Many long- and medium wave broadcasting stations have installed such a clock to deliver a very precise carrier frequency. Atomic clocks are also used in Observatories, radio interferometry in particular (McCarthy and Seidelmann 2009).

The definition of universal time[18] (UT) proved inadequate when clocks controlled by resonance properties of quartz crystal revealed that the second so defined was constantly changing by small amounts. The direct measurement of the length of the day can be made with a precision of five parts in 10^8 (Essen 1963). The precision of quartz clock is about an order magnitude higher. By using clocks controlled by these oscillators it emerged that the definition of second in terms of the rotation of the Earth undergoes a slow periodic change in addition to annual variation; the length of the day is increasing at the rate of 0.00164 s per century with fluctuations in the rate of rotation of the Earth occurring at irregular intervals (Brouwer 1951). The causes of irregularities affecting the uniformity of the observed time are due to precession and nutation, the forces of tidal friction mostly in shallow seas, winds, earthquakes, magnetic field, polar motion of the Earth, seasonal variation of the speed of rotation of the Earth, and various other factors (Smart 1956; Essen 1963).

Subsequently, it was felt that the ephemeris time[19] (ET) would be more stable and any minor variations can be calculated and corrected. For this, the Earth's orbital position is determined by observing the relative position of the Sun with respect to stars. The unit of ET is expressed in terms of a tropical year which is defined as the interval between the passages of the fictitious mean Sun, through the mean vernal equinox (see Sect. 5.2.1.1). The International Astronomical Union (IAU) in 1955

---

[17] The Doppler-shift is observed as a contraction and expansion in the wavelength of light wave. Wavelength contraction corresponds to increased frequency and bluer light, while wavelength expansion corresponds to decreased frequency and redder light. The Doppler-shift of a spectral line can be transformed into rotational velocities by means of the following formula,

$$\frac{\Delta\lambda}{\lambda} = \frac{v_r}{c},$$

with v_r as the radial velocity that measures variations in the star's relative velocity, c the speed of light, and $\Delta\lambda$ the measured shift in wavelength.

[18] Universal time (UT) is defined as the period of rotation of the Earth round its axis.

[19] Ephemeris time (ET) is measured by the orbital movements of the Earth round the Sun.

recommended that the ET should be determined from lunar observations employing lunar ephemeris, which is directly computed from Brown's theory (Brouwer and Hori 1962). The practical determination of ET depends on the actual observation of the Sun's position, which cannot be made very precisely. On the other hand, the position of the Moon can be determined more accurately by the dual-rate camera (Markowitz 1960). The Moon's position have also been tabulated in ET. The table of lunar ephemeris provides the value of ET corresponding to the values of right ascension (RA) and declination (dec; see Sect. 5.2.1.1) as obtained by Moon camera observations. These observations are capable of determining an interval of 1 year (Essen 1963) with a precision of about 2 parts in 10^9. The frequency stability of a quartz clock is, however, not sufficient to permit a subdivision of the 1 year interval so obtained into equal seconds and for the accurate realization of the unit of ET.

Development of physics and electrical engineering made it possible to control clocks by using as a pendulum one or another of the natural frequencies associated with transitions between energy states in atoms and molecules. In order to develop a device having the desired long term stability, the National Bureau of Standards, USA built the first atomic clock in 1948 based on microwave absorption in ammonia; stability of the order of 1 part in 10^9–10^{10} were predicted at that time. The new time scale was known as atomic time (AT). Later on, good transition for time keeping were found in the elements cesium (Cs), hydrogen (H), oxygen (O), and thallium; the isotope Cs^{133} proved to be the most suitable. In 1967, the International Bureau of Weights and Measures decided to replace the definition of second by the following: 'The second is the duration of 9,192,631,770 periods of radiation corresponding to the transition between the two hyperfine levels (the hyperfine levels are $F = 4, m_F = 0$ and $F = 3, m_F = 0$) of the ground state of the Cs^{133} atom'.

Both the time scales: the International Atomic Time (TAI; Temps Atomique International), a time scale based on the international second and the Coordinated Universal Time (UTC), which is used for civil time-keeping all over the Earth's surface, were unified. The former is generated by the Bureau International de l'Heure (BIH), Paris from the input of numerous time-keeping laboratories around the globe, while the latter is based on atomic second, which is derived from TAI using information on UT obtained by several astronomical observatories. UTC is the official worldwide legal time scale, having replaced the Greenwich mean time (GMT) in 1986, and follows TAI exactly except for an integral number of seconds, presently 34 as of 2009. These leap seconds are inserted to ensure that, on average over the years since 1972, the Sun is overhead within 0.9 s of 12:00:00 UTC on the meridian of Greenwich.

In atomic standards, in general, the absorption phenomenon is used to control the frequency of an auxiliary microwave oscillator through a closed loop feedback control system. The three most commonly used types of atomic clock are the cesium atomic beam, the hydrogen maser, and the rubidium gas cell. These oscillators are based on atomic resonances, and are less susceptible to the environmental factors. However, the temperature sensitive mechanisms required to interrogate the resonance frequency can perturb the resonance frequency (Eidson 2006). The sharpness of the absorption line, which is a measure of Q for the transition, determines the residual error and consequently the accuracy of the frequency standard. Hence, the shorter the line width is, the greater the frequency stability is. The natural half power line width (Levy 1968) and the equivalent Q of the resonance are respectively given by,

$$\Delta\omega = \frac{1}{\tau},\tag{3.52a}$$

$$Q = \frac{2\omega_0}{\Delta\omega} = 2\tau\omega_0,\tag{3.52b}$$

where τ is the mean life time of an atom in a given state and ω_0 the angular frequency of the emitted wave.

The (3.52b) reveals that the greater the mean life time of an atom in a given state, the greater the Q value and sharper the spectral line. Hence, the high Q value is an essential prerequisite for the excellent stability performance of standard oscillators. The slope of the $m \pm 1$ state in the case of hydrogen is relatively large, 1.4×10^5 Hz/gauss. Moreover, hydrogen transition has a frequency of 1.4 GHz as compared to that of cesium's 9.2 GHz. Resonance distortions caused by the cavity imperfection and electronic problems at the high frequency are the predominate factors which limit the accuracy of a Cs standard.

The stability of an atomic clock can be characterized in the frequency domain or in the time domain (Barnes et al. 1971; Hellwig et al. 1975; Howe 1976). The instantaneous fractional frequency deviation $y(t)$ from the nominal frequency ν_0 is related to the instantaneous phase deviation, $\psi(t)$ from the nominal phase, $2\pi\nu_0 t$ by,

$$y(t) \equiv \frac{\psi(t)}{2\pi\nu_0 t}.\tag{3.53}$$

In the frequency domain, frequency stability is defined as the one sided spectral density, $S_y(t)$ of $y(t)$. The function, $S_y(t)$ has the dimension of Hz^{-1}. The frequency stability in time domain is defined as the square root of the two sample Allan variance,

$$\sigma_y^2(\tau) = \langle\sigma_y^2(N = 2, T \approx \tau, \tau)\rangle$$

$$= \frac{\langle(\bar{y}_{k+1} - \bar{y}_k)^2\rangle}{2} = \frac{1}{2(M-1)}\sum_{1}^{M-1}(\bar{y}_{k+1} - \bar{y}_k)^2.\tag{3.54}$$

The Allan variance[20] is a special case of general N sample variance. Here, y_k is the kth individual sample averaged over a sampling time τ, $\langle\,\rangle$ denote the expectation

---

[20] The Allan variance, named after D. W. Allan, is a measurement of frequency stability in clocks and oscillators, to characterize the bias stability of gyroscopes. It is defined as one half of the time average of the squares of the differences between successive readings of the frequency deviation sampled over the sampling period. The advantage of this variance over the classical variance is that it converges for most of the commonly encountered kinds of noise, whereas the classical variance

value; for a finite data set, it is taken as the average value of the quantity enclosed in the brackets, T, the time interval between the beginning of the successive samples, N, the number of sample used in computation of the variance, M, the number of data values available, and $M - 1$, the number of difference averaged. This measure of frequency stability is dimensionless.

The generation of standard output frequency from a practical device, involves several steps of physical and technical processing, giving rise to many undesirable side effects leading to biases in the generated frequency. Both the accuracy and stability of the real devices have been limited by the fluctuations of these biases and additional sources of noise. Major causes of the biases in the various types of primary frequency standards include the second order Doppler shift, cavity phase shift in caesium beam, wall collision shift, cavity pulling in hydrogen maser (Mungall et al. 1974).

The frequency standards are described usually by their relative frequency stability, $\delta v / v_0$, which is determined in the laboratory from measurements of difference between two standards over a time interval, Δt. For instance, with $\Delta t \sim 100$ s and $v = 10^{10}$ Hz, the requirement of frequency stability would be 2×10^{-13}. Such a stability may be achieved with a hydrogen maser frequency standard that has $\delta v / v \sim 10^{-14}$ over 100 s and drifts of less than 1 μs per year.

Clocks based on optical transitions exhibit significantly improved accuracy over the cesium standards at microwave frequency (9.2 GHz). Recently, Chou et al. (2010) have built an optical atomic clock based on quantum logic spectroscopy of an aluminum ions (Al^+) that has a fractional frequency inaccuracy of 8.6×10^{-18}. Such a clock confines aluminum and beryllium (Be^+) ions closely together in an electromagnetic trap and slowed by lasers to near absolute zero temperatures. The aluminum clock is insensitive to background magnetic and electric fields, and also to temperature.

3.3 Intensity Interferometry

A novel technique, called intensity interferometry (originally known as post-detection interferometry), was conceived by R. Hanbury Brown. The philosophy of his idea, as mentioned in his book entitled, 'Boffin' (Hanbury Brown 1991), was that if the radiation received at two places is mutually coherent, the fluctuation in the intensity of the signals received at those two places is also correlated. The formal mathematical basis was put forward by Hanbury Brown and Twiss (1954).

Having succeeded in building an intensity interferometer at radio wavelengths, Hanbury Brown and Twiss (1956a) set out to demonstrate that interference could be detected in the intensity of light (see Sect. 3.3.3.2), despite the fact that light was detected as a stream of photons. This was the first instrument in quantum optics, the

---

does not always converge to a finite value. Flicker noise and random walk noise are two examples which commonly occur in clocks.

Fig. 3.9 Principle of an
intensity interferometer

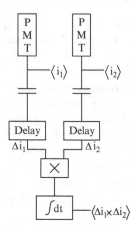

principle of which is demonstrated in Fig. 3.9. In this, both detectors measure fluctuations in the intensity of light from a particular star. The intensities are transformed into electrical currents in the photo-detectors at the foci. The current output of each detector is proportional to the instantaneous intensity of the incident radiation. The fluctuations in the current output by these detectors are partially correlated. This correlation gradually changes as the telescopes are moved apart from one another. The Fourier transform of the size and shape of the source along the projected baseline is obtained as a function of telescope separation. However, the experiment raised some controversy soon after the announcement of this concept, as it appeared to contradict the established beliefs of quantum interference. A question was raised, as it was thought, it violates quantum mechanics. This confusion got resolved by considering the quantum theory of photon detection and correlation (Glauber 1963). The correlation between the two signals suggests that the photons tend to arrive in pairs at the detector, i.e., they are bunched.[21] This observation can be explained by statistical fluctuation of the instantaneous intensity, $I(t)$, having a probability distribution,

$$\mathcal{P}(I)dI = \frac{I}{\bar{I}}e^{-I/\bar{I}}dI. \tag{3.55}$$

For a mean irradiance, $\bar{I}$ recorded in a time interval of δt, the mean number of emitted electrons $\bar{N} = \bar{I}\eta_d \delta t/\hbar\omega$, in which η_d is the quantum efficiency (QE) of the detector. The mean value of $\bar{N}$ depends on the mean irradiance and the time over which the averaging is carried out.

The correlation, $\mathcal{C}(\mathbf{B})$, which is proportional to the square of the degree of coherence of the radiation at the two detectors is given by,

---

[21] Photon bunching refers to the statistical tendency for photons in a seemingly uncorrelated thermal beam to be detected in close-by pairs. For thermal (chaotic) light and zero delay time, the second-order degree of temporal coherence, $\gamma^{(2)}(\mathbf{r}_1, \mathbf{r}_1, \tau) = 2$, so that the detection rate is twice that for long delay time. With zero time delay the photons arrive in pairs, while they arrive independently at long time delays.

$$\frac{C_n(\mathbf{B})}{C_n(0)} = \frac{\langle \Delta i(\mathbf{r}_1, t) \Delta i(\mathbf{r}_2, t) \rangle}{\langle i(\mathbf{r}_1) \rangle \langle i(\mathbf{r}_2) \rangle}$$

$$= |\gamma(\mathbf{r}_1, \mathbf{r}_2, 0)|^2 = \mathcal{V}^2, \tag{3.56}$$

in which $|\gamma(\mathbf{r}_1, \mathbf{r}_2, 0)|^2$ is the complex degree of coherence, $\Delta i(\mathbf{r}_1, t)$ and $\Delta i(\mathbf{r}_2, t)$ are the fluctuations in the two currents, $\mathbf{r}_1$, $\mathbf{r}_2$ the position vectors, $\langle \rangle$ the long-term averaging, $\mathcal{V}$ the corresponding fringe visibility, $C_n(\mathbf{B})$ the normalized correlation, $\mathbf{B}$ the baseline vector, and $C_n(0)$ the constant of the equipment.

The (3.56) indicates that the covariance of the time-integrated product of the current fluctuations, $\langle \Delta i(\mathbf{r}_1, t) \Delta i(\mathbf{r}_2, t) \rangle$, is the squared modulus of the covariance of the complex amplitude. This correlation is proportional to the square of the modulus of the degree of coherence, $|\gamma(\mathbf{r}_1, \mathbf{r}_2, 0)|^2$, of the light at the two detectors. This phenomenon, now known as Hanbury Brown and Twiss effect (HBT effect), represents a key element of quantum optics, which opened up to consider about the behavior of light in terms of quanta. In quantum optics, some devices do take advantage of correlation and anti-correlation effects in beams of photons. With the development of the techniques based on photo-electric detectors and fast electronic circuitry, the field of radiation fluctuations has acquired a new significance. Radiations in an enclosure can be treated by statistical mechanics. The subject 'radiation fluctuations' concerning with deviation from the equilibrium was generalized in applying Bose–Einstein statistics[22] to photons, irrespective of their spectral distribution. These treatments were confined to a closed system rather than to propagating light beam. Following (3.17), the energy fluctuation formula (Einstein gas) for black body radiation in an enclosure at equilibrium temperature T is given by (Fowler 1929),

$$\overline{(\Delta E)^2} = k_B T^2 \frac{d\overline{E}}{dT}$$

$$= h\nu \overline{E}_\nu \left[1 + \frac{1}{e^{(h\nu/k_B T)} - 1} \right], \tag{3.57}$$

where T stands for the source temperature in Kelvin (K), k_B the Boltzmann's constant, E the energy of the individual photons, ν their spectral frequency, and $\overline{E}$ the mean energy averaged over a large ensemble of similar systems, $\Delta E = E - \overline{E}$ the deviation from the mean energy, and $\overline{(\Delta E)^2}$ the variance.

---

[22] Bose–Einstein (B-E) statistics determines the statistical distribution of identical indistinguishable bosons (particles with integer-spin) over the energy states in thermal equilibrium. B-E statistics was introduced initially for photons by Bose and generalized later by Einstein. The expected number of particles in i^{th} energy level for B-E statistics is given by,

$$n_i = \frac{g_i}{e^{(\epsilon_i - \mu)/k_B T} - 1},$$

with $\epsilon_i > \mu$, and n_i as the number of particles in state i, g_i the degeneracy of state i, ϵ_i the energy of the ith state, μ the chemical potential, k_B the Boltzmann's constant, and T the absolute temperature.

Applying (3.57) to the energy dE_ν of the radiation lying within a narrow frequency interval $d\nu$, with $\overline{dE_\nu}$ given by Planck's distribution law for black body radiation in an enclosure of volume, V,

$$\overline{dE_\nu} = \frac{8\pi V \nu^2 d\nu}{c^3} \frac{h\nu}{e^{(h\nu/k_B T)} - 1}, \tag{3.58}$$

one obtains for classical particles and classical waves,

$$\overline{(\Delta dE_\nu)^2} = h\nu \overline{dE_\nu} \left[1 + \frac{1}{e^{(h\nu/k_B T)} - 1} \right]$$

$$\simeq h\nu \overline{dE_\nu} + \frac{c^3 \overline{(dE_\nu)^2}}{8\pi V \nu^2 d\nu}. \tag{3.59}$$

The second term of the right hand side of (3.59) follows from the classical Rayleigh–Jeans formula for radiation at low frequencies. The total energy due to the classical particles of energy $h\nu$ yields (3.59).

Intensity interferometry was one of the inspirations for the foundation of quantum optics (Glauber 1963; Sudarshan 1963). Today, Hanbury Brown–Twiss correlations have many variants in other fields of physics, from condensed matter to high energy. They are applied in elementary particle physics to investigate the size of the source of pion[23] emitters emerging from nuclear collisions, by observing correlations between arrivals at a pair of detectors as a function of their separations (Boal et al. 1990; Baym 1998; Alexander 2003). Other noted applications of such a method are for synchrotron X-ray beam diagnostics (Yang et al. 1994; Tai 2000), correlations in atomic beam (Yasuda and Shimizu 1996), to measure the properties of ultra-short optical pulses.

3.3.1 Derivation of the Separation of Two Points on a Star

Instead of evaluating the electromagnetic fields in two different locations, an intensity interferometer evaluates the statistical average of the correlations between pairs of point sources. These sources radiate white light and are independent of one another. Hanbury Brown (1974) illustrated elegantly this correlation by considering two harmonic emitting points on the surface of the star (see Fig. 3.10), with slightly different frequencies, ω_1 and ω_2, in order that these frequencies may beat together

---

[23] In particle physics, pions are short-lived sub-atomic mesons with a neutral form (π^0 having mass of the order of 135 MeV) and a charged form (π^+, π^- having mass of 139 MeV). The charged pion decays into muons (μ^+, μ^-) and neutrinos, while the neutral form decays into γ-ray photons. They belong to the hadron class of elementary particles, and consist of pairs of quarks. They play an important role in binding forces within the nucleus of an atom. The pions have spin 1, and therefore are bosons.

Fig. 3.10 Illustrating the correlation of two harmonic emitting points in a source with slightly different frequencies. Telescopes A and B are being close together, they receive beats almost at the same time. This corresponds to a high degree of correlation. Telescopes A and C being further apart do not receive beats in phase and the degree of correlation becomes lower (LeBohec et al. 2008; courtesy: S. LeBohec)

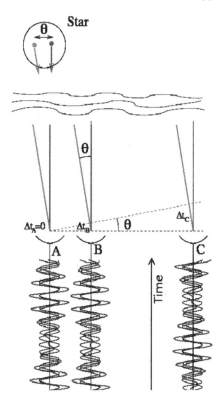

at each detector. These two points in the source give rise to the Fourier components of light, $E_1 \sin(\omega_1 t + \psi_1)$ and $E_2 \sin(\omega_2 t + \psi_2)$, which reach the detectors via the respective telescopes of the interferometer. Assuming simple linear polarization, the output current from the detector A, i_A, which is proportional to the intensity of the light, is given by the following equation,

$$i_A = K_A[E_1 \sin(\omega_1 t + \psi_1) + E_2 \sin(\omega_2 t + \psi_2)]^2, \qquad (3.60)$$

for telescope A. Here, K_A is the constant dependent upon the detector. E_1, E_2 and ψ_1, ψ_2 are the amplitudes and phases of the light from the two points on the surface of the star, separated by angle θ (see Fig. 3.10).

Similarly, the same Fourier components received via second telescope situated at B give rise to a current,

$$i_B = K_B\{E_1 \sin[\omega_1 t + \psi_1] + E_2 \sin[\omega_2 (t + d_B/c) + \psi_2]\}^2, \qquad (3.61)$$

where d_B is the path difference (at telescope B) between the light from the aforementioned two points on the source. The corresponding time difference is $\Delta t_B = d_B/c$. To note, there is no path difference at telescope, A, i.e., $\Delta t_A = 0$.

Expanding (3.60) and (3.61),

$$i_A = \frac{1}{2} K_A \{ (E_1^2 + E_2^2) - E_1^2 \cos 2(\omega_1 t + \psi_1) - E_2^2 \cos 2(\omega_2 t + \psi_2)$$

$$-2E_1 E_2 \cos[(\omega_1 + \omega_2)t + (\psi_1 + \psi_2)]$$

$$+2E_1 E_2 \cos[(\omega_1 - \omega_2)t + (\psi_1 - \psi_2)]\}, \tag{3.62a}$$

$$i_B = \frac{1}{2} K_B \{ (E_1^2 + E_2^2) - E_1^2 \cos 2[\omega_1 t + \psi_1]$$

$$-E_2^2 \cos 2[\omega_2 (t + d_B/c) + \psi_2]$$

$$-2E_1 E_2 \cos[(\omega_1 + \omega_2)t + (\psi_1 + \psi_2) + \omega_2 d_B/c]$$

$$+2E_1 E_2 \cos[(\omega_1 - \omega_2)t + (\psi_1 - \psi_2) - \omega_2 d_B/c]\}. \tag{3.62b}$$

The first term of (3.62) are DC components proportional to the total flux incident on the detector. These signals are applied to filters, which reject all the components except those of differential frequencies of the form $(\omega_1 - \omega_2)$. According to Hanbury Brown (1974), if these difference frequencies fall within the pass-band (1–100 MHz) of the filters, they would reach multiplier, and therefore two components at the multiplier are,

$$i_A = K_A E_1 E_2 \cos[(\omega_1 - \omega_2)t + (\psi_1 - \psi_2)], \tag{3.63a}$$

$$i_B = K_B E_1 E_2 \cos[(\omega_1 - \omega_2)t + (\psi_1 - \psi_2) - \omega_2 d_B/c]. \tag{3.63b}$$

It is evident from (3.63) that the components i_A and i_B are correlated. It is assumed that $\omega_1 \approx \omega_2 = \omega$, therefore i_A and i_B differ in phase by $\omega d_B/c$. Their product or correlation is given by,

$$C_n(\mathbf{B}) = K_A K_B E_1^2 E_2^2 \cos(\omega d_B/c), \tag{3.64}$$

The phase difference of the correlated components (i_A, i_B) is the difference between the relative phases of the two Fourier components at the detectors; the correlation does not depend on the phase difference of the light at the two detectors (see Fig. 3.10). By simple geometry, (3.64) may be rewritten as,

$$C_n(\mathbf{B}) = K_A K_B E_1^2 E_2^2 \cos(2\pi \mathbf{B}_\lambda \theta), \tag{3.65}$$

with $\mathbf{B}_\lambda = \mathbf{B}/\lambda$, $\mathbf{B}$ as the baseline vector between the telescopes A and B, λ the interferometer operating wavelength, and $d_B = \mathbf{B}_\lambda \theta$.

On integrating (3.65) over all possible pairs of points on the stellar disk, over all possible pairs of Fourier components lying within the optical bandpass and over all difference frequencies that lie the band pass of the electrical filter, Δf, Hanbury Brown and Twiss (1957a) obtained the result given in (3.56). The output signal from the correlator is proportional to square of the visibility that would be observed

Fig. 3.11 Fringe visibility
with the separation of two
antennas for an intensity
interferometer

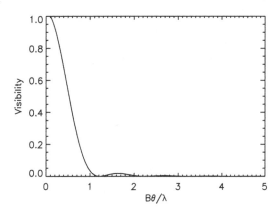

with a Michelson type interferometer with the same baseline length. The square of
visibility is expressed as,

$$V^2 = \left| \frac{2J_1^2(\pi\theta_s B_\lambda)}{(\pi\theta_s B_\lambda)^2} \right|, \qquad (3.66)$$

where J_1 is the first order Bessel function.

On measuring the correlation, $C_n(\mathbf{B})$ as a function of the baseline separation, $\mathbf{B}$,
between the detectors, the angular diameter of a star can be measured. In the case
of a source that can be modeled as a uniform disk with angular diameter, θ_s, the
various baselines used in the observation sample an Airy function. For observations
at wavelength, λ, the degree of coherence, $\gamma(\mathbf{r}_1, \mathbf{r}_2, 0)$, cancels when $B = 1.22/\lambda$.
Figure 3.11 depicts the variations of the fringe visibility (normalized) of an uniform
disk with the changes of baselines between the two arms of an intensity interferometer. However, a two element intensity interferometer is not phase-sensitive, which
is a serious disadvantage if it is applied to the study of source structure.

3.3.2 Intensity Interferometer at Radio Wavelengths

The initial instrument was developed for radio wavelengths where the term, intensity, is used for the radio equivalent of the photometric illuminance. In order to
check the principle of intensity interferometry, a smaller version was developed in
1950 by Hanbury Brown, which was able to measure the apparent angular diameter of the radio Sun at 125 MHz (Jennison 1966 and references therein). This had
demonstrated the correctness of his theoretical ideas.

Subsequently, Hanbury Brown et al. (1952) had developed a radio intensity
interferometer with a long baseline (see Fig. 3.12). Such an instrument had two sets
of dipole arrays that were configured in a fashion so as to improve the resolution,
sensitivity, or field of view of their response to the astronomical sources relative to

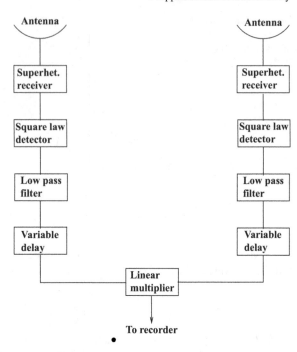

Fig. 3.12 Principle of an intensity interferometer; two heterodyne receivers were tuned to 125 MHz with bandwidth, Δf, of 200 KHz

that provided by a single antenna. One of them was kept fixed at the home station, and the other could be dismantled and transported for the erection at various locations, several kilometers away. A separate square-law detector in each arm of the system followed by two identical low-pass filters with bandwidth, Δf, from 1 to 2 KHz, feeding a linear multiplier. In order to compensate for the time taken for difference in the time of arrival of the signals at the two antennas when the direction of the source is not normal to the baseline, a variable delay is introduced in the individual signal paths from the two antennas. Four baselines of different lengths (of the order of a few kilometers) and orientations were used to determine the angular dimensions of the two strong following radio sources:

1. Cygnus A (3C 405.0): Cygnus A is one of the brightest radio galaxies, lying in the constellation Cygnus. It was once thought to be two galaxies of comparable size in collision, but recent ideas suggest that it is a giant elliptical galaxy with a large black hole at its core. Emanating from this central region are two opposing jets of hot gas that end in radio- and X-ray-emitting lobes called hot spots (Harris et al. 1994). Infrared spectroscopic observations (Ward et al. 1991) of the central regions of this radio galaxy showed neutral gas (molecular hydrogen). Jennison and Das Gupta (1953, 1956) have measured the apparent angular structure of Cygnus A, which is highly elongated in a roughly East–West direction (see Fig. 3.13). Its brightness profile showed two bumps of increased intensity of radio emission, coming from two distinct regions of space. They found the first evidence of a radio galaxy having two almost symmetrical blobs on either side,

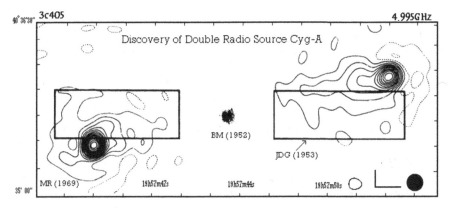

Fig. 3.13 First identification of Cygnus A (3C405) as a double radio source (*rectangular boxes*). Also superposed on the figure are the observations of Baade and Minkowski (1952) and Mitton and Ryle (1969). The figure is adapted from Thorne (1994)

each extending about one minute of arc long in the direction of the major axis and about thirty seconds of arc wide in the perpendicular direction. The centers of emission are spaced apart by about 1.5 arcminutes near position angle 100° at 127 MHz. The results for Cygnus A were found to be similar with the observations made with conventional radio interferometers (Mills 1952; Smith 1952). Subsequent observations in a wavelength of 11 cm by Hogg et al. (1969) confirmed the dimensions of Cygnus A measured by Jennison and Das Gupta (1956), which reveal the source as having two components separated by 123 arcseconds along a position angle of 111.0° with each component having dimensions of 30 arcseconds along the major axis and about 15–20 arcseconds across the axis.

2. Cassiopeia A (3C 461): Cassiopeia A is the remnant of a galactic supernova (type IIb) explosion that occurred over 300 years ago in the constellation Cassiopeia. It is bright at all energies of the electromagnetic spectrum, particularly at radio wavelengths having a flux of 2,720 Jy at 1 GHz. Cassiopeia A was found to be a circularly symmetric radio source (Hanbury Brown et al. 1952). Since intensity interferometry measures the square of the visibility function so that it is inherently incapable of measuring the phase of any object. Later, Jennison and Latham (1959) measured amplitude and phase (hence its structure) of the visibility function at 127 MHz along an East–West line with the phase sensitive interferometer, and concluded that its brightness distribution consists of a limb-brightened disk about 4.1 arcminute across, with an extension to the East about 4 arcminute in length.

In an instrument of this kind, matching paths in the arms of the interferometer was easier than for a conventional amplitude interferometer. The tolerance was set by the maximum frequency of the filtered low-frequency signals whose correlation was being measured and not set by the radio frequency signal. The DC component is proportional to the total flux falling on the square-law detectors and is rejected by the filters. The output of the detectors contain the cross beats between all the elementary radiators in the source that are selected prior to detection in the radio frequency bands of the two channels. Separate local oscillators were used and the output of one channel may be transmitted to the mixer via an incoherent radio link. The fluctuations of the electrical signals from the two receivers are compared by a linear multiplier where the interference between these fluctuations takes place; the bandwidth of the correlated signal was kept at 2 kHz. An integrator that integrator averages the fluctuating output of the detector to determine its mean level precisely, measures the correlation of the currents from which the square of the modulus of the complex degree of coherence is derived.

The intensity interferometry at radio wavelengths depends essentially on the square-law detector, and was therefore, relatively insensitive to weak signals and is of little practical use unless the S/N ratio exceeded unity. Although the major advantage of such a system at radio wavelengths was that of the insignificant effect of scintillation due to inhomogeneous ionosphere on the measurements of correlation of radio sources (Hanbury Brown 1974), its use as a radio interferometer had to cease.

3.3.3 Optical Intensity Interferometry

An optical analog of the radio intensity interferometer can also be developed by employing a pair of optical telescopes equipped with photomultiplier tubes (PMT), kept at a distance. The correlation is made after the detection of photons at the two photo-cathodes for coherent incident light. The requirement of opto-mechanical accuracy in such an instrument is less stringent, since it depends on the electrical bandwidth of the detectors, but not the wavelength of light. The alignment tolerances are extremely relaxed since the pathlengths need to be maintained to a fraction of $c/\Delta f$. It requires control of the light paths to an accuracy determined by the light coherence time ($\sim$100 ns); therefore, it is much less sensitive to atmospheric phase fluctuations; the signals input to the correlator follow roughly the same path through the atmosphere, for at least small separation between the two telescopes. Such an instrument eliminates the need for optical delay-lines as well. Being insensitive to seeing conditions, interferometer of this kind relaxes the site requirements. However, certain amount of accuracies are needed for the delay tracking and local oscillator stability. Another important point to be noted is that the necessity of high optical astronomy quality light collectors are not needed as they are required only to isochronally concentrate the light on a PMT.

3.3.3.1 Laboratory Experiment

The methodology in intensity correlation involves measuring intensity observed by each detector to be sent to the central correlator facility. The shot noise from both channels is responsible for most of the multiplier output fluctuations (Hanbury Brown 1974). Another component, called wave noise, can also be envisaged as the beating between the different Fourier components of the light reaching the detectors. The wave noise correlation providing a measurement of the square of the degree of coherence of the light at the two detectors is compared to the sensitivity of a specific experiment to be estimated. Such a correlation does not depend on the phase difference of the light at the two detectors. The root-mean-square (RMS) signal-to-noise (S/N) ratio, according to Hanbury Brown and Twiss (1957b), is given by,

$$(S/N)_{RMS} = A_e \eta_d n \, |\gamma(\mathbf{r}_1, \mathbf{r}_2, 0)|^2 \, (\Delta f \, T_0/2)^{1/2}, \tag{3.67}$$

where A_e is the telescope collection area, η_d the quantum efficiency (QE) of the photo-detector, Δf, the electronics signal bandwidth, T_0, the time interval over which the multiplier output is averaged, n, the source spectral density in photons per unit optical bandwidth, per unit area and unit time, and $|\gamma(\mathbf{r}_1, \mathbf{r}_2, 0)|^2$ the complex degree of coherence.

Note that the signal-to-noise (S/N) ratio is independent of the optical bandwidth, $\Delta \nu$. Let $I(t)$ be the intensity of the light at the photo-cathode, averaged over a few cycles, therefore the probability of an emitted electron would be $\eta_d I(t)dt$, in which the detector quantum efficiency, η_d, is assumed to be constant over the optical bandwidth, $\Delta \nu$. The mean number of photon, $\bar{n}_p$ that is counted in an interval, T, turns out to be,

$$\bar{n} = \eta_d \bar{I} T, \tag{3.68}$$

and from Mandel (1963), the variance, $\overline{(\Delta n_p)^2}$, in this count would become,

$$\overline{(\Delta n_p)^2} = \begin{cases} \bar{n}_p(1 + \bar{n}_p) & \text{when } T \ll 1/\Delta \nu, \\[2mm] \bar{n}_p(1 + \bar{n}_p \tau_c / T) & \text{otherwise,} \end{cases} \tag{3.69}$$

with

$$\tau_c = \int_{-\infty}^{\infty} |\gamma(\mathbf{r}_1, \mathbf{r}_2, \tau)|^2 \, dt, \tag{3.70}$$

as the coherence time.

From (3.69), the variance $\overline{(\Delta n_p)^2}$ in photon number, n_p, in a given volume and its relationship to the mean value $\bar{n}_p$,

$$\overline{(\Delta n_p)^2} = n_p \left[1 + \frac{\bar{n}_p}{N_{cell}} \right], \tag{3.71}$$

in which $N_{cell} = 2 A \tau \Omega \nu_0^2 \Delta \nu c^{-2}$, i.e., the number of phase cells in the volume occupied by the photons, A the telescope aperture, τ the time interval over which a single measurement in the ensemble of measurements of n_p is made, Ω the solid angle, and ν_0 the central frequency.

The first term in the bracket of (3.71) corresponds to the fluctuations in a classical assembly of particles, and the second term to the wave noise. With the introduction of the emissivity of the body, ϵ, the analysis provides,

$$\overline{(\Delta n_p)^2} = n_p \left[1 + \frac{\epsilon \bar{n}_p}{N_{cell}} \right]. \tag{3.72}$$

This (3.72) agrees with formula given in the case of a black body ($\epsilon = 1$), but provides a lower wave noise for a gray body (Hanbury Brown 1974).

Let two photo-multipliers be kept at points $\mathbf{r}_1$, $\mathbf{r}_2$, separated by a distance, B, that are illuminated by unpolarized quasi-monochromatic light, $\Delta v / \bar{v} \ll 1$, in which $\bar{v}$ is the mean frequency. The two counters, N_1 and N_2 register the number of photons in each channel and the counter, N_c registers a coincidence when two pulses arrive within the coherence time, τ_c. The average counting rates are given by,

$$\bar{N}_1 = \eta_{d1} \bar{I}_1, \quad \bar{N}_2 = \eta_{d2} \bar{I}_2. \tag{3.73}$$

Following Mandel (1963), the coincidence rate is expressed as,

$$\bar{N}_c = \bar{N}_1 \bar{N}_2 \left[2\tau_c + \frac{1}{2} |\gamma(\mathbf{r}_1, \mathbf{r}_2, 0)|^2 \int_{-\infty}^{\infty} |\gamma(\mathbf{r}_1, \mathbf{r}_1, \tau)|^2 d\tau \right], \tag{3.74}$$

where $\gamma(\mathbf{r}_1, \mathbf{r}_1, \tau)$ is the normalized autocorrelation function of the incident light given by (2.15).

It is assumed here that the spectral distribution of the light at the two detectors is identical and their quantum efficiencies are constant over the optical bandwidth. The (3.74) can be simplified as,

$$\bar{N}_c = \begin{cases} \bar{N}_1 \bar{N}_2 2\tau_c [1 + \frac{1}{2} |\gamma(\mathbf{r}_1, \mathbf{r}_2, 0)|^2 \tau_0 / 2\tau_c] & \text{when } 2\tau_c \gg 1/\Delta v, \\ \\ \bar{N}_1 \bar{N}_2 2\tau_c [1 + \frac{1}{2} |\gamma(\mathbf{r}_1, \mathbf{r}_2, 0)|^2] & \text{otherwise.} \end{cases} \tag{3.75}$$

In practical case for measuring stars, where the resolving time is long compared with coherence time ($2\tau_c \gg 1/\delta v$). From (3.75), the S/N ratio is written as,

$$(S/N)_{RMS} = \frac{\tau_0}{2} |\gamma(\mathbf{r}_1, \mathbf{r}_2, 0)|^2 \sqrt{N_1 N_2} \sqrt{T_0 / 2\tau_c}. \tag{3.76}$$

Hanbury Brown and Twiss (1956a) demonstrated the potential of intensity interferometry at optical wavelengths by investigating the effect experimentally in the laboratory (see Fig. 3.14). In this, they used the beam from a mercury vapor lamp and a half-silvered mirror to split the beam into two. By measuring the intensity correlations between the two separated beams, the intensities at two different points in unseparated beam were compared. The output from two photo-tubes would correlate if the light falling on them is mutually coherent; they were measured in

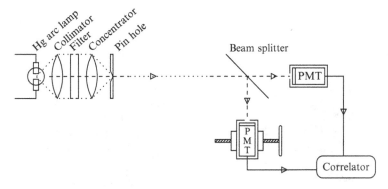

Fig. 3.14 Schematic diagram of the experimental apparatus used for demonstrating the correlation between the intensity fluctuations of coherent light beam by Hanbury Brown and Twiss (1956a)

a different recorder as the total flux arrived from a source. By varying the relative pathlengths between the mirror and the detectors they could vary the time delay, τ, of the points. They observed no correlation of intensities at large τ, while the correlations increased with decreasing τ; the characteristic time scale is the coherence time. This experiment was the important demonstration of photon bunching. The correlation was observed in close agreement with the theoretical prediction. This observation demonstrated that intensity correlations could be measured by detecting individual photons and could be used to determine coherence area or time for chaotic light.

3.3.3.2 A Model of an Optical Intensity Interferometer

Hanbury Brown and Twiss (1958) applied intensity interferometry to the astronomical problem of determining angular diameters of visible stars. In this arrangement, two large anti-aircraft searchlights with 5-foot mirrors were used as telescopes. The starlight arriving at the two searchlights was focused onto two photo-multiplier tubes and the correlation of fluctuations in the photo-currents was measured as a function of mirror separation. The magnitude of this correlation agreed reasonably well with what was expected. A device placed in one of the branches of the interferometer introduced a variable delay that permits the compensation of difference in the time of arrival of the wavetrains on the photomultipliers.

Figure 3.15 shows the experiment conducted at Jodrell Bank, UK by which the observations of a main sequence star, α CMa (Sirius), at a distance of 2.7 parsecs, were carried out. However, the star reaches here a maximum of 20° elevation restricting the observations to within 2 h of transit. Another difficulty came from the excessive extinction (see Sect. 9.1.1). The significant effect also came from scintillation induced by the atmosphere (see Sect. 4.1.6.2). Nevertheless, the star resolved at a baseline of 30 feet where the magnitude of the correlation was at the minimum, thus demonstrating the ability of the instrument. The measurements showed

Fig. 3.15 The optical intensity interferometer at Jodrell Bank in 1956

the angular diameter of the said star to $0.0068'' \pm 0.0005'' = 3.1 \times 10^{-8}$ rad (Hanbury Brown and Twiss 1956b). A theoretical estimate of the star's diameter according to parameters such as effective temperature, bolometric magnitude etc., gave a value of 6.9 milliarcseconds. Due to the factors stated above, the total observing time over a period of 5 months (between November 1955 and March 1956) was limited to 18 h.

3.3.3.3   Narrabri Intensity Interferometer

Success of the afore-mentioned experiment at Jodrell Bank prompted raising of necessary funds for the development of stellar intensity interferometer at Narrabri Observatory, Australia (Hanbury Brown et al. 1967). This instrument used analog electronics and a pair of light collectors (telescopes) having diameter of 6.5 m (larger than any other optical telescope at that time), with 11 m (with a tolerance of ± 0.15 m), focal length. These telescopes, equipped with photo-multiplier tubes, could be moved along a circular track to preserve the relative timing while tracking stars across the sky. These telescopes were regular 12 sided polygons of which the usable reflecting area in each was 30 m^2 (see Fig. 3.16).

Each mirror was paraboloid in shape and its surface was formed by a mosaic of 252 hexagonal glass mirrors with spherical curvature, each approximately 38 cm between the opposite sides and 2 cm thick; 251 aligned on the signal detector and 1 on a separate star guidance detector. These telescopes were mounted on turntables carried by the mobile trucks that are connected to the control building, by catenary cable and were capable of their independent motions. Having movable telescopes removes the need for additional delay-lines to bring the signals in time. The separation of the mirrors could be varied from 10 m up to 188 m.

Fig. 3.16 Stellar intensity interferometer at Narrabri, Australia

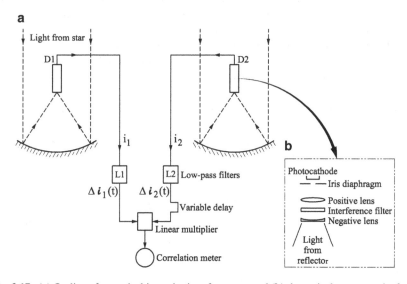

Fig. 3.17 (a) Outline of an optical intensity interferometer and (b) the optical system at the focus

Starlight collected by the reflectors was sent to an optical system at the focus (see Fig. 3.17a). This optical system was mounted on a steel tube about 11 m long which projected from the center of the reflector. The converging beam from the reflector was collimated by a negative lens and was passed through an interference filter centered on 443 nm, with a passing bandwidth of 10 nm. The filtered light was then focused by a positive lens through an adjustable iris diaphragm (see Fig. 3.17b), on to the cathode of a PMT, which converted the light intensity to a current with 25%

quantum efficiency at 440 nm and a 60-MHz effective bandwidth. The outputs of their detectors are limited in frequency by two identical low pass filters with unity gain. The signals were sent to the control building where the correlator was located.

The DC component of the signals was measured and recorded while the AC components were sent to the input of the correlator, a four transistor based linear multiplier. The correlator measured the correlation between the fluctuations in the anode currents of the photomultiplier detectors at the foci of the two telescopes. It multiplied the fluctuations in the two channels together and the correlation was a unidirectional output superimposed on random noise. The DC component to be measured on the output of the multiplier is very small in comparison to the random fluctuations ($\sim$100 dB). This difficulty was resolved by integrating the output of the multiplier over 100-s time intervals. The drifting effects of small offsets were eliminated, by applying the technique of phase switching consisting of inverting the phase of the signals on the input of the multiplier at rates of 10 kHz and 10 Hz, respectively. On the output of the multiplier, these switching frequencies were amplified and demodulated before the signal was integrated, thus removing any significant contamination due to offsets.

The measurement of correlation for different baselines, B, of the two photomultipliers were taken. As the photomultiplier moved, the degree of coherence $|\gamma(\mathbf{r}_1, \mathbf{r}_2)|$ at the two photo-cathodes varied in a manner that was calculable from the geometry of the arrangement. The experimental curve obtained in this way approaches the theoretical curve that provides $|\gamma(\mathbf{r}_1, \mathbf{r}_2, 0)|^2$ as a function of B, from which one derives $|\gamma(\mathbf{r}_1, \mathbf{r}_2, 0)|$, and hence the angular diameter of the star.

Intensity correlation technique has the drawback of requiring a large number of photons hence a stronger limitation in sensitivity, which was limited by the narrow bandwidth[24] filters that are used to increase the speckle life time. The correlated fluctuations can be obtained if the detectors are spaced by less than a speckle width. For an unresolved ($|\gamma(\mathbf{r}_1, \mathbf{r}_2, 0)| = 0$), the star possessing magnitude (see Sect. 9.1.1) $m_v = 0$ observed during 1 h, the predicted S/N ratio amounts to $\sim$130. However, the actual S/N ratio was degraded by a factor of 5 because of optical losses and excess noise of various origins. This limited observations to stars brighter than 2.5 m_v. Theoretical calculations (Roddier 1988) show that the limiting magnitude that can be observed with such a system is of the order of 2.

In order to improve the sensitivity and resolving power to measure stars with a limiting magnitude of $m_v = 7$, Hanbury Brown (1974) suggested to design a new interferometer with a possibility of increasing (1) the light collecting area, (2) the electrical bandwidth (Δf), the exposure time remains same as compared with those at Narrabri system, and (3) multiply the number of independent optical channels used in the measurements. The sensitivity does not depend on the optical bandwidth, so this can be easily achieved by simultaneously using several narrow optical bands.

---

[24] The electrical bandwidth (100 MHz) implies that the paths from the photo-multipliers to the correlator must be equal to about 10 nanoseconds (ns) to avoid loss of correlation due to temporal coherence.

3.3.4 *Intensity Correlations in Partially Coherent Fields*

It is to reiterate that an intensity interferometer measures the correlation between the fluctuations of intensity at two separate points in a partially coherent field. Following Mandel (1963), the instantaneous intensities $I(\mathbf{r}_1, t)$ and $I(\mathbf{r}_2, t)$ at $\mathbf{r}_1$ and $\mathbf{r}_2$, where the two photomultipliers are located, are given by,

$$I(\mathbf{r}_1, t) = V(\mathbf{r}_1, t)V^*(\mathbf{r}_1, t), \tag{3.77a}$$
$$I(\mathbf{r}_2, t) = V(\mathbf{r}_2, t)V^*(\mathbf{r}_2, t), \tag{3.77b}$$

in which $V(\mathbf{r}_1, t)$ and $V(\mathbf{r}_2, t)$ are the analytic signals associated with real vibrations.

The cross-correlation function of the intensities is given by,

$$\langle I(\mathbf{r}_1, t + \tau)I(\mathbf{r}_2, t)\rangle = \langle V(\mathbf{r}_1, t + \tau)V^*(\mathbf{r}_1, t + \tau)V(\mathbf{r}_2, t)V^*(\mathbf{r}_2, t)\rangle. \tag{3.78}$$

The term, $\langle I(\mathbf{r}_1, t + \tau)I(\mathbf{r}_2, t)\rangle$, is said to be the correlation of the second order. Following (1.16), the terms, $V(\mathbf{r}_1, t)$ and $V(\mathbf{r}_2, t)$ may be decomposed into real and imaginary parts,

$$V(\mathbf{r}_1, t) = V^{(r)}(\mathbf{r}_1, t) + iV^{(i)}(\mathbf{r}_1, t), \tag{3.79a}$$
$$V(\mathbf{r}_2, t) = V^{(r)}(\mathbf{r}_2, t) + iV^{(i)}(\mathbf{r}_2, t), \tag{3.79b}$$

so that the cross-correlation of the intensities becomes,

$$\begin{aligned}
\langle I(\mathbf{r}_1, t + \tau)I(\mathbf{r}_2, t)\rangle = {} & \langle |V^{(r)}(\mathbf{r}_1, t + \tau)|^2 \, |V^{(r)}(\mathbf{r}_2, t)|^2\rangle \\
& + \langle |V^{(r)}(\mathbf{r}_1, t + \tau)|^2 \, |V^{(i)}(\mathbf{r}_2, t)|^2\rangle \\
& + \langle |V^{(i)}(\mathbf{r}_1, t + \tau)|^2 \, |V^{(r)}(\mathbf{r}_2, t)|^2\rangle \\
& + \langle |V^{(i)}(\mathbf{r}_1, t + \tau)|^2 \, |V^{(i)}(\mathbf{r}_2, t)|^2\rangle. \tag{3.80}
\end{aligned}$$

The four real correlation functions on the right hand side of (3.80), can be evaluated separately from the known probability distributions of $V^{(r)}(\mathbf{r}_1, t)$, $V^{(r)}(\mathbf{r}_2, t)$, etc. By using the following equations,

$$\begin{aligned}
\int_{-\infty}^{\infty} V^{(r)}(\mathbf{r}_1, t + \tau)V^{(r)}(\mathbf{r}_2, t)dt &= \int_{-\infty}^{\infty} V^{(i)}(\mathbf{r}_1, t + \tau)V^{(i)}(\mathbf{r}_2, t)dt, \\
\int_{-\infty}^{\infty} V^{(r)}(\mathbf{r}_1, t + \tau)V^{(i)}(\mathbf{r}_2, t)dt &= -\int_{-\infty}^{\infty} V^{(i)}(\mathbf{r}_1, t + \tau)V^{(r)}(\mathbf{r}_2, t)dt,
\end{aligned}$$

$$\tag{3.81}$$

one obtains the expression,

$$\langle V(\mathbf{r}_1, t + \tau)V(\mathbf{r}_2, t)\rangle = 0, \tag{3.82}$$

from which, one obtains,

$$\langle V^{(r)}(\mathbf{r}_1, t + \tau)V^{(r)}(\mathbf{r}_2, t)\rangle = \langle V^{(i)}(\mathbf{r}_1, t + \tau)V^{(i)}(\mathbf{r}_2, t)\rangle, \qquad (3.83a)$$

$$\langle V^{(i)}(\mathbf{r}_1, t + \tau)V^{(r)}(\mathbf{r}_2, t)\rangle = -\langle V^{(r)}(\mathbf{r}_1, t + \tau)V^{(i)}(\mathbf{r}_2, t)\rangle, \qquad (3.83b)$$

Of course with $\tau = 0$, and $V(\mathbf{r}_1, t) = V(\mathbf{r}_2, t)$, one obtains,

$$\langle V^{(r)^2}(\mathbf{r}_1, t)\rangle = \langle V^{(i)^2}(\mathbf{r}_1, t)\rangle = \frac{1}{2}|V(\mathbf{r}_1, t)|^2, \qquad (3.84)$$

and

$$\langle V^{(r)}(\mathbf{r}_1, t)V^{(i)}(\mathbf{r}_1, t)\rangle = 0. \qquad (3.85)$$

This (3.85) states that the values of $V^{(r)}(\mathbf{r}_1, t)$ and $V^{(i)}(\mathbf{r}_1, t)$ have no correlation at time t. From (3.77) and (3.83), the following relationship emerges,

$$\langle V^{(r)^2}(\mathbf{r}_1, t)\rangle = \langle V^{(i)^2}(\mathbf{r}_1, t)\rangle = \frac{\bar{I}(\mathbf{r}_1)}{2}. \qquad (3.86)$$

Following (3.79 and 3.83), the mutual coherence function, $\Gamma(\mathbf{r}_1, \mathbf{r}_2, \tau)$, may be expressed as,

$$\begin{aligned}
\Gamma(\mathbf{r}_1, \mathbf{r}_2, \tau) &= \langle V^{(r)}(\mathbf{r}_1, t + \tau)V^{(r)}(\mathbf{r}_2, t)\rangle + \langle V^{(i)}(\mathbf{r}_1, t + \tau)V^{(i)}(\mathbf{r}_2, t)\rangle \\
&\quad + i\langle V^{(i)}(\mathbf{r}_1, t + \tau)V^{(r)}(\mathbf{r}_2, t)\rangle - i\langle V^{(r)}(\mathbf{r}_1, t + \tau)V^{(i)}(\mathbf{r}_2, t)\rangle \\
&= 2\langle V^{(r)}(\mathbf{r}_1, t + \tau)V^{(r)}(\mathbf{r}_2, t)\rangle - 2i\langle V^{(r)}(\mathbf{r}_1, t + \tau)V^{(i)}(\mathbf{r}_2, t)\rangle
\end{aligned}$$

$$(3.87)$$

which, together with (3.83a and 3.83b), leads to,

$$\langle V^{(r)}(\mathbf{r}_1, t + \tau)V^{(r)}(\mathbf{r}_2, t)\rangle = \frac{1}{2}\Re\left[\Gamma(\mathbf{r}_1, \mathbf{r}_2, \tau)\right] \qquad (3.88a)$$

$$\langle V^{(i)}(\mathbf{r}_1, t + \tau)V^{(r)}(\mathbf{r}_2, t)\rangle = \frac{1}{2}\Im\left[\Gamma(\mathbf{r}_1, \mathbf{r}_2, \tau)\right], \qquad (3.88b)$$

where $\Re$ and $\Im$ are the real and imaginary parts of the quantities in brackets.

The terms $V^{(r)}(\mathbf{r}_1, t)$, $V^{(r)}(\mathbf{r}_2, t)$, $V^{(i)}(\mathbf{r}_1, t)$, and $V^{(i)}(\mathbf{r}_2, t)$ are all Gaussian variables. On applying the relation, $\langle a^2 b^2\rangle = \langle a^2\rangle \langle b^2\rangle + 2\langle ab\rangle^2$, in which $a(t)$ and $b(t)$ are two real functions of time, in (3.80), its first term (right hand side) is recast as,

$$\begin{aligned}
\langle V^{(r)^2}(\mathbf{r}_1, t + \tau)V^{(r)^2}(\mathbf{r}_2, t)\rangle &= \frac{1}{4}\bar{I}(\mathbf{r}_1)\bar{I}(\mathbf{r}_2) \\
&\quad + 2[\langle V^{(r)}(\mathbf{r}_1, t + \tau)V^{(r)}(\mathbf{r}_2, t)\rangle]^2 \\
&= \frac{1}{4}\bar{I}(\mathbf{r}_1)\bar{I}(\mathbf{r}_2) + \frac{1}{2}\{\Re\left[\Gamma(\mathbf{r}_1, \mathbf{r}_2, \tau)\right]\}^2.
\end{aligned}$$

$$(3.89)$$

Similarly, for the remaining three terms of the right hand side of (3.80), one may write,

$$\langle V^{(r)^2}(\mathbf{r}_1, t + \tau) V^{(i)^2}(\mathbf{r}_2, t) \rangle = \frac{1}{4} \bar{I}(\mathbf{r}_1) \bar{I}(\mathbf{r}_2) + \frac{1}{2} \{\Im [\Gamma(\mathbf{r}_1, \mathbf{r}_2, \tau)]\}^2,$$

$$\langle V^{(i)^2}(\mathbf{r}_1, t + \tau) V^{(r)^2}(\mathbf{r}_2, t) \rangle = \frac{1}{4} \bar{I}(\mathbf{r}_1) \bar{I}(\mathbf{r}_2) + \frac{1}{2} \{\Im [\Gamma(\mathbf{r}_1, \mathbf{r}_2, \tau)]\}^2,$$

$$\langle V^{(i)^2}(\mathbf{r}_1, t + \tau) V^{(i)^2}(\mathbf{r}_2, t) \rangle = \frac{1}{4} \bar{I}(\mathbf{r}_1) \bar{I}(\mathbf{r}_2) + \frac{1}{2} \{\Re [\Gamma(\mathbf{r}_1, \mathbf{r}_2, \tau)]\}^2.$$

$$(3.90)$$

By summing up these results,

$$\langle I(\mathbf{r}_1, t + \tau) I(\mathbf{r}_2, t) \rangle = \bar{I}(\mathbf{r}_1) \bar{I}(\mathbf{r}_2) + |\Gamma(\mathbf{r}_1, \mathbf{r}_2, \tau)|^2 \qquad (3.91a)$$

$$= \bar{I}(\mathbf{r}_1) \bar{I}(\mathbf{r}_2) \left[1 + |\gamma(\mathbf{r}_1, \mathbf{r}_2, \tau)|^2 \right]. \qquad (3.91b)$$

This result may be derived in the form of a correlation between the fluctuations of intensity, $I(\mathbf{r}_1)$ and $I(\mathbf{r}_2)$, above the respective mean values, $\bar{I}(\mathbf{r}_1)$ and $\bar{I}(\mathbf{r}_2)$. If $\Delta I(\mathbf{r}_1, t)$ and $\Delta I(\mathbf{r}_2, t)$ denote the instantaneous deviations between the intensities, $I(\mathbf{r}_1, t)$ and $I(\mathbf{r}_2, t)$, with respect to $\bar{I}(\mathbf{r}_1)$ and $\bar{I}(\mathbf{r}_2)$, one gets,

$$\Delta I(\mathbf{r}_1, t) = I(\mathbf{r}_1, t) - \bar{I}(\mathbf{r}_1), \qquad (3.92a)$$

$$\Delta I(\mathbf{r}_2, t) = I(\mathbf{r}_2, t) - \bar{I}(\mathbf{r}_2), \qquad (3.92b)$$

with $\langle \Delta I(\mathbf{r}_1, t) \rangle = 0$ and $\langle \Delta I(\mathbf{r}_2, t) \rangle = 0$, and from which

$$\langle I(\mathbf{r}_1, t + \tau) I(\mathbf{r}_2, t) \rangle = \bar{I}(\mathbf{r}_1) \bar{I}(\mathbf{r}_2) + \langle \Delta I(\mathbf{r}_1, t + \tau) \Delta I(\mathbf{r}_2, t) \rangle. \qquad (3.93)$$

By comparing (3.91) and (3.93), one derives,

$$\langle \Delta I(\mathbf{r}_1, t + \tau) \Delta I(\mathbf{r}_2, t) \rangle = |\Gamma(\mathbf{r}_1, \mathbf{r}_2, \tau)|^2 \qquad (3.94a)$$

$$= \bar{I}(\mathbf{r}_1) \bar{I}(\mathbf{r}_2) |\gamma(\mathbf{r}_1, \mathbf{r}_2, \tau)|^2. \qquad (3.94b)$$

Thus, it appears that, as long as there is coherence between the light at $\mathbf{r}_1$ and $\mathbf{r}_2$ (where the photomultipliers are located), there is correlation between the intensity fluctuations. Moreover, the degree of coherence between points in an optical field can be explored by measuring intensity correlations, as well as by examining interference fringes. If the points $\mathbf{r}_1$ and $\mathbf{r}_2$ coincide, (3.91) reduces to the auto-correlation function of the intensity:

$$\langle I(\mathbf{r}_1, t + \tau) I(\mathbf{r}_1, t) \rangle = \bar{I}^2(\mathbf{r}_1) \left[1 + |\gamma(\mathbf{r}_1, \mathbf{r}_1, \tau)|^2 \right], \qquad (3.95)$$

while, (3.94) turns out to be,

$$\langle \Delta I(\mathbf{r}_1, t + \tau) \Delta I(\mathbf{r}_2, t) \rangle = \bar{I}^2(\mathbf{r}_1) |\gamma(\mathbf{r}_1, \mathbf{r}_1, \tau)|^2. \tag{3.96}$$

In particular, for $\tau = 0$, one can find the variance $\langle (\Delta I(\mathbf{r}_1))^2 \rangle = \bar{I}^2(\mathbf{r}_1)$.

In the form of normalized second-order degree of temporal coherence is defined as,

$$\gamma^{(2)}(\mathbf{r}_1, \mathbf{r}_1, \tau) = \frac{\langle I(\mathbf{r}_1, t + \tau) I(\mathbf{r}_1, t) \rangle}{\bar{I}^2(\mathbf{r}_1)}, \tag{3.97a}$$

$$= 1 + |\gamma(\mathbf{r}_1, \mathbf{r}_1, \tau)|^2. \tag{3.97b}$$

For chaotic light the range of $\gamma^{(2)}(\mathbf{r}_1, \mathbf{r}_1, \tau)$ is $1 \leq \gamma^{(2)}(\mathbf{r}_1, \mathbf{r}_1, \tau) \leq 2$. This (3.97) provides a connection between the second-order and the first-order correlation function for chaotic light. For a practical instrument, it is necessary to deal with unpolarized light in which the orthogonal components of the fields are uncorrelated; therefore, the correlation is half that expected for the case of the linearly polarized light, i.e., the (3.94) translates into:

$$\langle \Delta I(\mathbf{r}_1, t + \tau) \Delta I(\mathbf{r}_2, t) \rangle = \frac{1}{2} \bar{I}(\mathbf{r}_1) \bar{I}(\mathbf{r}_2) |\gamma(\mathbf{r}_1, \mathbf{r}_2, \tau)|^2. \tag{3.98}$$

In a partially polarized beam, the instantaneous intensity, $I(\mathbf{r}, t)$, at a point, $\mathbf{r}$, may be defined as,

$$I(\mathbf{r}, t) = |V_x(\mathbf{r}, t)|^2 + |V_y(\mathbf{r}, t)|^2 \tag{3.99a}$$

$$= I_x(\mathbf{r}, t) + I_y(\mathbf{r}, t), \tag{3.99b}$$

where $V_x(\mathbf{r}, t)$ and $V_y(\mathbf{r}, t)$ represent the disturbances in two perpendicular directions normal to the direction of propagation, and $I_x(\mathbf{r}, t)$ and $I_y(\mathbf{r}, t)$ the partial intensities associated with x and y components of the electromagnetic wave.

By splitting the beam of light with a beam-splitter[25] and correlating the intensity fluctuations emerging from mutually orthogonal polarizers placed in each beam, the degree of coherence, $|\gamma_{x,y}(\mathbf{r}, \mathbf{r}, 0)|$ between the x and y components can be measured; the maximum value of $|\gamma_{x,y}(\mathbf{r}, \mathbf{r}, 0)|$ as the polarizers are rotated, while the remaining orthogonal, is the degree of polarization, p (Mandel 1963). On choosing partial intensities, $I_x(\mathbf{r}, t)$, $I_y(\mathbf{r}, t)$ in a fashion so that,

$$\bar{I}_x(\mathbf{r}_1) = \bar{I}_y(\mathbf{r}_1) = \frac{1}{2} \bar{I}(\mathbf{r}_1) \tag{3.100a}$$

$$\bar{I}_x(\mathbf{r}_2) = \bar{I}_y(\mathbf{r}_2) = \frac{1}{2} \bar{I}(\mathbf{r}_2). \tag{3.100b}$$

---

[25] The function of the beam-splitter (half-silvered mirror) is to superimpose (image) one mirror onto the other; usually 50:50 splitter is employed.

Wolf (1960) deduced a formula on the particle picture of a light beam as,

$$\langle \Delta I(\mathbf{r}_1, t + \tau') \Delta I(\mathbf{r}_2, t) \rangle = \frac{1}{2} \bar{I}(\mathbf{r}_1) \bar{I}(\mathbf{r}_2)(1 + p^2), \qquad (3.101)$$

with τ' as a possible time shift introduced between the coherent beams at $\mathbf{r}_1$ and $\mathbf{r}_2$. If $\mathbf{r}_1$ and $\mathbf{r}_2$ coincide, the intensity auto-correlation may be expressed as,

$$\langle (\Delta I(\mathbf{r}, t))^2 \rangle = \frac{1}{2} \bar{I}^2(\mathbf{r})(1 + p^2), \qquad (3.102)$$

and for unpolarized light,

$$\langle \Delta I(\mathbf{r}, t + \tau) \Delta I(\mathbf{r}, t) \rangle = \frac{1}{2} \bar{I}^2(\mathbf{r}) |\gamma(\mathbf{r}, \mathbf{r}, \tau)|^2. \qquad (3.103)$$

3.3.5 Correlation Between the Signals of the Photo-detectors

A photo-detector is a device capable of sensing an optical signal and converting into an electrical signal containing the same information. It uses the principle of photo-electric effect, in which light falling on a photosensitive surface releases bound electrons which flow as a current to a charge-measuring device. These processes are referred to as quantum processes, which is statistical. Given a certain electric field, the number of charges N released is statistically related to the exposure W (wave power multiplied by time) by Einstein's relation, $W = \bar{N}\hbar\omega$, with N as an integer having mean value $\bar{N}$, and obeying Poissonian statistical distribution.[26] A photomultiplier tube (PMT) is a highly sensitive detector of light. It is a combination of an electron-emitter followed by a current amplifier in one structural unit, which makes possible a very large amplification by as much as 10^8 of the electric current (a photo-current) by the photosensitive layer from a faint light source. In such a tube, photons impinging on the photo-cathode liberate electrons, which are directed to an electrode, called dynode. The electrons leave the photo-cathode, having the energy of the incoming photon minus the work function of the

---

[26] In the photo-electric detection, a constant irradiance onto the detector provides photo-electron emission pulses having Poisson statistics. The probability of N electrons being emitted in a certain time interval and with a mean number $\bar{N}$ is,

$$\mathcal{P}(N) = \frac{\bar{N}^N}{N!} e^{-\bar{N}}.$$

The variance or mean square fluctuation for the Poisson distribution is equal to its mean,

$$\langle N^2 \rangle - (\bar{N})^2 = \bar{N}.$$

photo-cathode. The secondary electron emitted by the first dynode can be directed onto the second dynode which functions in the same manner as the first. Each dynode is held at a more positive voltage than the previous one. This process may be repeated many times. If a multiplier has n such dynodes, each with same amplification factor, δ, the total gain or amplification factor for the PMT is δ^n.

Let a plane wave of linearly polarized light be impinging on the surface of a photoelectric detector placed at a point $P_1(\mathbf{r}_1)$, which gives rise to an output current, $i(\mathbf{r}_1, t)$. The probability of emission of a photo-electron, $\mathcal{P}(\mathbf{r}_1, t)$, is proportional to the light intensity, $I(\mathbf{r}_1, t)$, i.e.,

$$\mathcal{P}(\mathbf{r}_1, t) = \eta_d I(\mathbf{r}_1, t) \Delta t, \tag{3.104}$$

in which η_d is the quantum efficiency of the detector.

The output current, $i(\mathbf{r}_1, t)$ is given by,

$$i(\mathbf{r}_1, t) = \eta_d e I(\mathbf{r}_1, t), \tag{3.105}$$

where e is the electric charge.

The fluctuations in the output of a photo-detector may also be treated as the fluctuations in the output of a square-law detector where the intensity is proportional to the square of the electric vector of the incident wave. For simplicity, let the sensitivity of these multipliers be constant over the spectral range $\Delta \nu$ of the incoming light. In practice, the photomultipliers do not provide currents proportional to the luminous intensities, $I(\mathbf{r}_1, t)$ and $I(\mathbf{r}_2, t)$ (Mandel 1963), but are proportional to the mean values of these currents over a short period, T, that is the time of resolution of the photomultipliers,

$$i(\mathbf{r}_1, t) = \frac{e\eta_d}{T} \int_{-T/2}^{T/2} I(\mathbf{r}_1, t + t') dt', \tag{3.106a}$$

$$i(\mathbf{r}_2, t) = \frac{e\eta_d}{T} \int_{-T/2}^{T/2} I(\mathbf{r}_2, t + t') dt'. \tag{3.106b}$$

The multiplier measures the correlation, $\langle i(\mathbf{r}_1, t) i(\mathbf{r}_2, t) \rangle$, and with zero time shift, τ, which yields maximum value. Since luminous vibration is stationary, therefore one derives,

$$\langle i(\mathbf{r}_1, t) i(\mathbf{r}_2, t) \rangle = \frac{e^2 \eta_d^2}{T^2} \int\!\!\int_{-T/2}^{T/2} \langle I(\mathbf{r}_1, t + t') I(\mathbf{r}_2, t + t'') \rangle dt' dt''$$

$$= \frac{e^2 \eta_d^2}{T^2} \int\!\!\int_{-T/2}^{T/2} \langle I(\mathbf{r}_1, t + t' - t'') I(\mathbf{r}_2, t) \rangle dt' dt''.$$

$$\tag{3.107}$$

Following (3.91) and (3.107), the following relationship for unpolarized light due to Mandel (1963) emerges,

$$\langle \Delta i(\mathbf{r}_1, t) \Delta i(\mathbf{r}_2, t) \rangle = \frac{e^2 \eta_d^2 \bar{I}(\mathbf{r}_1) \bar{I}(\mathbf{r}_2)}{2T^2} \int\limits_{-T/2}^{T/2}\!\!\int |\gamma(\mathbf{r}_1, \mathbf{r}_2, t' - t'')|^2 \, dt' dt''.$$

(3.108)

Since the spectrum of the light is identical in the two photo-detectors, one may write,

$$\gamma(\mathbf{r}_1, \mathbf{r}_2, \tau) = \gamma(\mathbf{r}_1, \mathbf{r}_2, 0)\gamma(\mathbf{r}_1, \mathbf{r}_1, \tau),$$

(3.109)

where $\gamma(\mathbf{r}_1, \mathbf{r}_1, \tau)$ represents the autocorrelation function.

By substitution, one obtains,

$$\langle \Delta i(\mathbf{r}_1, t) \Delta i(\mathbf{r}_2, t) \rangle = \frac{e^2 \eta_d^2 \bar{I}(\mathbf{r}_1) \bar{I}(\mathbf{r}_2)}{2T^2} |\gamma(\mathbf{r}_1, \mathbf{r}_2, 0)|^2 \left(\frac{\tau_c}{T}\right),$$

(3.110)

with as the coherence time (see 3.70).

The time of resolution T of the photomultipliers is larger in comparison with the coherence time, τ_c (see 2.2). Let the resolving time, T, be represented by a low-pass filter of bandwidth, Δf, thus,

$$\langle \Delta i(\mathbf{r}_1, t) \Delta i(\mathbf{r}_2, t) \rangle = e^2 \eta_d^2 \bar{I}(\mathbf{r}_1) \bar{I}(\mathbf{r}_2) |\gamma(\mathbf{r}_1, \mathbf{r}_2, 0)|^2 \frac{\Delta f}{\Delta \nu},$$

(3.111)

in which $\tau_c/T = 2\Delta f/\Delta \nu$.

The term, $\bar{I}(\mathbf{r}_1)$ in (3.111) is measured by the multiplier followed by an integrator. Therefore, one obtains the degree of partial coherence, $|\gamma(\mathbf{r}_1, \mathbf{r}_2, 0)|$ (see 2.16), and the angular profile of the stars.

3.4 Interferometer for Cosmic Probe

In a classic paper, Einstein (1905a) introduced the concept of special theory of relativity by completing Maxwell's electrodynamics. He modified the foundations of classical mechanics in order to remove the apparent contradictions between mechanics and electrodynamics. He showed that space and time are closely linked and affect each other. Space, time, simultaneity, mass are all relative, but speed of light is absolute being independent of the notion of the source and the observer. His subsequent work of the general theory on relativity (Einstein 1916) led to the search for gravity waves.

In general relativity, a weak gravitational wave is described by a metric perturbation $h_{\mu\nu}$. Typically, the detectable $h_{\mu\nu}$ for the astrophysical gravitational wave sources is of the order of $\sim 10^{-22}$. Gravitational radiation caused by the astrophysical events like coalescing binaries, or exploding stars such as supernovae, gamma-ray bursts, causes a time-dependent strain in space-time distances acting as a dielectric. For example, when stellar objects like the binary pulsars undergo accelerations in their orbits, gravity waves of measurable strengths are emitted. Such an exotic system can be a test-bed for general relativity, because relativistic effects can be seen in the timing of the pulsar pulses.

Gravitational waves travel with the speed of light but with very small amplitudes so they need very sensitive detector for their detection. Key to such detection is the accurate measurement of small changes in distance. The effect of a gravitational wave is an apparent strain in space, transverse to the direction of propagation causing a differential change of the pathlength between the arms of the interferometer, thereby introducing a phase-shift. This shift is due to the changes in the distances between pairs of mirrors that are kept mutually perpendicular in vacuum chambers. Such wave passing through the instrument shortens one arm, while lengthening the other. This relative change in length of the two arms can be measured. One may experience a tiny amplitude-relative oscillations due to gravitational waves. In spite of the smallness of the gravitational coupling constant,[27] experimental validations of the general theory of relativity are being carried out. The scalar potential, $\varphi(\mathbf{r})$ in a region empty of charge (free space) obeys the Laplace equation,

$$\nabla^2 \varphi = 0, \qquad (3.112)$$

where ∇^2 is the Laplacian operator (see 1.2).

The (3.112) is valid for the time-independent Newtonian gravity, while for the time-dependent Einsteinian metric gravity, the tensor field, $\varphi_{\mu\nu}(\mathbf{r}, t)$, in which μ and ν run from 1 to 4 (Scully and Zubairy 1997) satisfies a wave equation of the form,

$$\nabla^2 \left[\varphi_{\mu\nu}(\mathbf{r}, t) \right] - \frac{1}{c^2} \frac{\partial^2}{\partial t^2} \left[\varphi_{\mu\nu}(\mathbf{r}, t) \right] = 0. \qquad (3.113)$$

Figure 3.18 depicts a schematic diagram of a Michelson interferometer. A laser[28] beam drives this interferometer which is influenced by an incident gravity wave

---

[27] The gravitational coupling constant is the coupling constant characterizing the gravitation attraction between electrons and protons. It is a dimensionless quantity, and is given by,

$$\alpha_G = \frac{Gm_e^2}{\hbar c} = \left(\frac{m_e}{m_P} \right)^2 \approx 1.752 \times 10^{-45},$$

where $G (= 6.674 \times 10^{-11} \text{N.m}^2/\text{kg}^2)$ is the gravitational constant, m_e the electron mass, c the speed of light in vacuum, $\hbar$ reduced Planck's constant, and m_P Planck's mass.

[28] A laser (acronym for Light Amplification by Stimulated Emission of Radiation) is a quantum optical device that produces a nearly monochromatic, and coherent beam of light by exciting atoms to a higher energy level and causing them to radiate their energy in phase. The laser output can be continuous or pulsed and is contributing significantly to the science, communications, and technology.

Fig. 3.18 Schematic diagram
of a Michelson interferometer

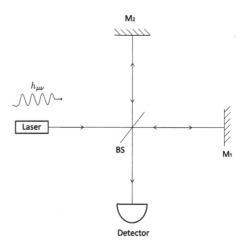

denoted by $h_{\mu\nu}$. Like in a classical Michelson interferometer, the laser beacon is divided into two beams by a beam-splitter, oriented at $45°$ with the direction of the beam. These are reflected at mirrors M_1 and M_2 and return to beam splitter, where they are recombined before entering the detector via a focusing device. Another passive system can be used by replacing the mirrors by Fabry–Perot cavities in both arms in order to fold the light several times, which lengthens the effective optical pathlengths in each arm.

Consider that the gravitational wave of frequency, ν_g, causes the length, l_0, to vary between the mirror, M_2 and the beam-splitter, as,

$$l_0 = l\left[1 + A_0 \cos(\nu_g t)\right], \tag{3.114}$$

in which l is the length of the interferometer arm in the absence of a gravitational wave and A_0 the amplitude of the gravitational wave and the sensitivity of such an amplitude is of the order of $\leq 10^{-21}$ for envisioned sources.

A phase-shift, δ is expected between the light traversing the two arms of the interferometer by an amount,

$$\delta = \kappa(l_0 - l_0') = \kappa l A_0 \cos(\nu_g t). \tag{3.115}$$

The recorded intensity I_0 by the detector is given by,

$$I_0 = \frac{1}{2}I(1 + \cos\delta), \tag{3.116}$$

in which I is the intensity of an incident wave.

In practice, cavities are employed in the two arms of the Michelson interferometer. Due to a gravity wave the signal translates into a time-dependent phase-shift, hence the effective pathlength, $\bar{l}$, that is essentially the number of bounces times the length of the arm, l, is considered. The gravity induced phase-shift for times $t \ll 1/\nu_g$ is expressed as,

$$\Delta\theta_p = \frac{\nu}{c} A_0 \bar{l}, \tag{3.117}$$

in which ν is the frequency of the laser beam.

The fundamental quantum limit for such an experiment arises from the photon shot noise. Such a noise comes from the statistical fluctuations in the number of photons detected at the output of the interferometer. From the Heisenberg uncertainty principle (Heisenberg 1927), $\Delta n_p \Delta\psi \sim 1$, in which n_p is the number of photons detected at the output and $\Delta\psi$ the phase difference of the light in the two arms of the interferometer, one derives the expression for the limiting sensitivity of the amplitude for a Michelson interferometer of arm length l. Assuming unit quantum efficiency (QE) for a detector, the phase uncertainty due to such noise for a measurement of duration, t_m is,

$$\Delta\theta_n \simeq \frac{1}{\sqrt{\langle n_p \rangle}} = \sqrt{\frac{\hbar\nu}{P t_m}}, \tag{3.118}$$

where $\langle n_p \rangle$ is the average number of laser photons; the ensemble mean is determined by $\langle \rangle$, P the power at the detector, and $\hbar = h/2\pi$.

In addition to the photon shot noise, the thermal noise is also expected to be the limiting factor of the sensitivity of the interferometric detectors. In the case of the former, the effects occur at the high frequency end of the operating regions of these detectors, while in the case of the latter these effects occur at low to mid frequency region (Robertson 2000). From (3.117 and 3.118), one may derive the minimum detectable amplitude of the gravity wave for this system,

$$A_{0'} \simeq \frac{c}{\nu\bar{l}} \sqrt{\frac{\hbar\nu}{P t_m}} \tag{3.119}$$

$$= \frac{\varphi}{\nu} \sqrt{\frac{\hbar\nu}{P t_m}}, \tag{3.120}$$

with $\varphi(= c/\bar{l})$ as the cavity decay rate.

Advances in modern laser optics technology have made it possible to develop extremely sensitive, kilometer-scale gravity wave detectors. Several ground-based gravity-wave detectors are in operation. They are described in brief:

1. LIGO: This instrument, called LIGO (Laser Interferometer Gravitational-Wave Observatory) comprises two facilities at widely separated sites, viz., the LIGO Livingston Observatory in Livingston, Louisiana and the LIGO Hanford Observatory, on the Hanford Nuclear Reservation in USA. Both house a 4-km L-shaped interferometer (Sigg 2006). The interferometer at each site consists of mirrors suspended at each of the corners of the 'L'. A pre-stabilized laser emits a beam of up to 35 W that passes through an optical mode cleaner before reaching a

beam splitter at the vertex of the 'L'. The beam splits into two paths, one for each arm, which contains Fabry–Perot cavities that store the beams and increase the effective pathlength.

2. VIRGO: This interferometer with a baseline of 3 km is located near Pisa, Italy. It consists mainly a Michelson laser interferometer consisting of two orthogonal arms. The frequency range of VIRGO extends from 10 to 10,000 Hz. An elaborate seismic isolation system, with six-stage pendulums, would measure gravity waves down to frequencies of 10 Hz (Acernese et al. 2006).

3. GEO: This detector, with an arm length of 600 m, is located near Hannover, Germany (Lück et al. 2006). It is a second generation instrument which aims at the direct detection of gravitational waves. However, the necessary high detection sensitivity requires ultra-stable high-power lasers with good beam quality; the power stabilization of a laser down to frequencies of 100 Hz has been achieved (Seifert et al. 2006).

4. TAMA: With a baseline of 300 m, this instrument adopts a Fabry-Perot Michelson interferometer with power recycling. It is situated at the National Astronomical Observatory near Tokyo in Japan (Ando et al. 2005).

5. AIGO: A small interferometer with 80 m baseline is being built near Perth, Australia (McChelland et al. 2006).

The measurement of gravitational wave signals is facing a daunting task against many types of noise entering the detectors. Among the noise sources, the laser noise and shot noise are prominent. In addition, the seismic and 'gravity-gradient noise (so called Newtonian noise) are also considered to be the limiting factors in the sensitivity of such ground-based interferometers. The seismic noise is due to vibration of Earth's crust and is present all over its surface. A very high degree of seismic isolation is essential to reduce the effects of seismic motion for the ground-based gravitational wave detectors. To note, the resonant mass detectors consisting of the bar type detectors and spheres can also be used for ground-based detection. But because of their scalability and their broad band-width the laser interferometers are favored. Such detectors may be useful at high frequencies, because the shot-noise in the interferometers is large.

Such ground-based instruments are able to cover the frequency range from a few Hz to a few kHz due to the afore-mentioned factors. The solution is to go into space, where the detection of signals around 1 Hz is possible. The laser interferometer space antenna (LISA) is a jointly sponsored mission (Bender et al. 1998) by the European Space Agency (ESA) and the National Aeronautical and Space Administration (NASA). One of the goals of this mission is to detect the gravitational waves generated by binaries within the Galaxy and by massive black holes in distant galaxies (Shaddock 2009 and references therein).

LISA consists of three identical spacecrafts flying in an equilateral triangle separated by 5×10^9 m located 20° (50 million km) behind the Earth in its orbit. Lasers will be used to measure relative movements of each spacecraft generated by passing gravitational waves. This instrument is based on heterodyne interferometry, in which two lasers with different frequencies would interfere to produce a beat note at their frequency difference. Displacement is detected by measuring the phase of this beat note, where one optical wavelength of pathlength change would produce a phase-shift of one cycle in the beat note. The displacement readout is, therefore self-calibrating down to the laser wavelength. LISA may be able to measure the amplitude, direction, and polarization of gravitational waves simultaneously. As the interferometer orbits the Sun in the course of a year, the observed gravitational waves are Doppler-shifted by the orbital motion. The changing inclination of the detector provides amplitude modulation of gravitational waves, while a slight frequency modulation results from Doppler shifts. This allows the determination of the direction of the source and assessment of some of its characteristics (Hughes 2006).

New generation gravity wave detectors, when in operation by the end of the next decade or so, are expected to (1) make direct detection of gravitational waves, (2) measure the Hubble constant, and (3) test of general relativity in the strong field regime. Plans for building third generation gravitational wave detectors more sensitive than the current ones are now underway an example of which is the Einstein telescope.

Chapter 4
Single-dish Diffraction-limited Imaging

4.1 Turbulence

Turbulence is a state of the flow of a fluid in which apparently random irregularities occur in the instantaneous velocities, often producing major deformations of the flow. In the terrestrial atmosphere, large-scale temperature inhomogeneities caused by non-uniform heating of different portions of the Earth's surface produce random micro-structures in the spatial distribution of temperature causing the fluctuations in the refractive index of air. The large-scale refractive index inhomogeneities are broken up by non-uniform winds and convection, spreading the scale of the inhomogeneities to smaller sizes, until thermal molecular dissipation dominates. The disturbance takes the form of distortion of shape of the incoming wavefront and affects the intensity distribution, thereby introducing phase fluctuations. This leads to blurring of the image.

The diffraction-limited resolution of celestial objects viewed through such turbulence could be achieved by employing post-detection processing of a large data set of short exposure images. Labeyrie (1970) suggested a method called speckle interferometry, where post-detection data processing algorithms are used to decipher diffraction-limited spatial image features of stellar objects. Significant improvements in the techniques during the past few decades have led to achievement of real time compensation of wavefront distortions (Hardy 1998; Roddier 1999). This active is known as adaptive optics (AO) system. In what follows, a brief introduction on the behavior of the atmosphere and its effect on the plane wavefront from a stellar source, the formation of speckles in the case of non-coherent quasi-monochromatic source and of ways to detect them, and the salient features of an adaptive optics system are discussed.

4.1.1 Spectral Description of Turbulence

The atmosphere is difficult to study due to the high Reynolds number, R_e, a dimensionless quantity, characterizing the turbulence. This parameter developed by Reynolds (1883) quantifies the relative importance of two types of forces for given flow conditions. When the average velocity, v (m/s), of a viscous fluid of characteristic size, l (m), is gradually increased, two distinct states of fluid motion are observed (Tatarski 1961), namely, (1) laminar (regular and smooth in space and time), at very low v and (2) unstable and random fluid motion at v greater than some critical value. Between these two extreme conditions, the flow passes through a series of unstable states. The Reynolds number is defined as the ratio of inertial forces to viscous forces, i.e.,

$$R_e = \frac{v\,l}{v_v}, \tag{4.1}$$

where v_v is the kinematic viscosity of the fluid expressed in m^2/s.

For a pipe flow, if the value of R_e is below 2,100, it indicates laminar flow, while values nearing 4,000 suggest turbulence. For $R_e \sim 10^6$, the medium becomes chaotic in both space and time, exhibiting considerable spatial structure. The turbulent motions often take the form of eddies (or vortices), which in a Eulerian reference frame[1] manifests itself as series of waves.

Since turbulence is a non-linear process, Kolmogorov (1941a,b) had used a statistical approach based on the assumption that wind velocity fluctuations are approximately homogeneous locally and isotropic random fields for scales less than the largest wind eddies. The velocity fluctuations occur on a wide range of space and time-scales. The spectral energy cascades proceed at a constant rate governed by the eddy turn over time. The energy enters the flow at scale length, L_0, called outer-scale, and spatial frequency, $\kappa_{L_0} = 2\pi/L_0$, as a direct result of the non-linearity of the Navier–Stokes equation governing fluid motion,

$$\frac{\partial \mathbf{v}}{\partial t} + \mathbf{v}(\nabla \cdot \mathbf{v}) = -\frac{\nabla p}{\rho} + v_v \nabla^2 \mathbf{v}, \tag{4.2}$$

$$\nabla \cdot \mathbf{v}(\mathbf{r}, t) = 0, \tag{4.3}$$

with

$$\nabla = \mathbf{i}\frac{\partial}{\partial x} + \mathbf{j}\frac{\partial}{\partial y} + \mathbf{k}\frac{\partial}{\partial z},$$

as the linear vector differential operator, $p(\mathbf{r}, t)$ the dynamic pressure, and ρ the constant density, and $\mathbf{v}(\mathbf{r}, t)$ the velocity field, $\mathbf{r}(= x, y, z)$ the position vector and t the time.

The large-scale fluctuations, referred to as large eddies, have the size of the geometrically imposed outer-scale length L_0. The outer-scale reflects the size of the largest eddies that can be supported by the energy of the turbulent motion. These eddies are not universal with respect to flow geometry; they vary according to the local conditions. Several attempts have been made at measuring the size of this outer-scale length using a variety of different methods (Colavita et al. 1987; Coulman et al. 1988; Buscher et al. 1995; Linfield et al. 2001; Ziad et al. 2004). But there has been substantial variation in the measured values ranging from a few meters to the order of kilometers. Also, Conan et al. (2000) derived a mean value

---

[1] In a Eulerian reference frame, velocities are related to the fixed reference frame.

$L_0 = 24$ m for a von Kármán spectrum (see (4.23)) from the data obtained at Cerro Paranal, Chile. Observations made with the Sydney University Stellar Interferometer (SUSI) at different baselines (Davis et al. 1995) showed a significant departure from the 5/6th power law even for the 5-m baseline.

At small-scales (≤ 1 cm), the turbulent energy is dissipated through the viscosity of the air (Roddier 1981). The length-scale at which the mechanism of kinetic energy dissipation changes from turbulent breakup to frictional heating is called the inner-scale of turbulence, l_0. The turbulent energy is cascading into smaller and smaller loss-less eddies until at a small enough Reynolds number, the kinetic energy of the flow is converted into heat by viscous dissipation. This results in a rapid drop in power-spectral density (for definition see Sect. 2.1.3), $\Phi_n(\kappa)$ for $\kappa > \kappa_0$, in which κ_0 is critical wave number and κ the wave vector. The changes in such spectral density are characterized by the inner-scale length, l_0, and spatial frequency, $\kappa_{l_0} = 2\pi/l_0$, where l_0 varies from a few millimeter near the ground to a centimeter (cm) high in the atmosphere.

In the smallest perturbations with sizes, l_0, the rate of dissipation of energy into heat is determined by the local wind velocity gradients in these smallest perturbations. By keeping the viscosity term which is dominant at l_0, the energy dissipated as heat, ε, is given by,

$$\varepsilon \sim \frac{\nu_v v^2}{l^2} \sim \frac{\nu_v v_0^2}{l_0^2}, \tag{4.4}$$

in which v_0 the velocity and l_0 the local spatial scale.

The unit of ε is expressed as energy per unit mass of the fluid per unit time joules s^{-1}kg^{-1} = m^{2}s^{-3}. Thus the energy is given by,

$$v_0^2 \sim \varepsilon^{2/3} l_0^{2/3} \tag{4.5}$$

The (4.5) gives rise to the scaling law, popularly known as the two-third law. The mean square difference of the velocities at two points in each fully developed turbulent flow is proportional to the two-third power of their distance. From the relationships, $v_0 \sim (\varepsilon l_0)^{1/3}$ and (4.4), one may obtain,

$$l_0 \sim \left(\frac{v_v^3}{\varepsilon}\right)^{1/4} \sim L_0 R_e^{-3/4}. \tag{4.6}$$

The 3-dimensional (3-D) power-spectral density, $\Phi(\kappa)$, is proportional to κ,

$$\Phi(\kappa) \propto \kappa^{-11/3}, \tag{4.7}$$

while for one-dimensional (1-D) energy spectrum is,

$$\mathcal{E}(\kappa) \propto \kappa^{-5/3}. \tag{4.8}$$

The energy spectrum holds good for any conserved passive additive factors such as refractive index etc. Small-scale fluctuations with sizes $l_0 < r < L_0$, known as the inertial subrange, where r is the radial distance, have universal statistics (scale-invariant behavior) independent of the flow geometry. To note, the range over which viscous effects are not important can be called inertial range. The inertial sub-range is the range of length-scales in between the large (slow) scale of turbulence production and the small (fast) scale of molecular dissipation. It is of fundamental importance to derive the useful predictions for turbulence within it. The value of inertial subrange would be different at various locations on the site.

4.1.2 Structure Function for Deriving Kolmogorov Turbulence

The variance of the velocity of turbulent motion is infinite, therefore in order to circumvent the problem of infinite covariances, a function called the structure function, which characterizes the random process, is used in turbulence theory. Let $f(t)$ be a non-stationary random variable and $F_t(\tau) = f(t + \tau) - f(t)$ is a difference function that is stationary for small τ. The structure function is defined as,

$$\mathcal{D}_f(\tau) = \langle |F_t(\tau)|^2 \rangle = \langle |f(t + \tau) - f(t)|^2 \rangle, \qquad (4.9)$$

is a measure of fluctuations of $f(t)$ over a time-scale $\leq \tau$, in which the angular brackets indicate an average over time, t.

According to Kolmogorov (1941a,b), the structure function in the inertial range (homogeneous and isotropic random field) depends on $r = |\mathbf{r}|$, in which $\mathbf{r} = \boldsymbol{\rho}' - \boldsymbol{\rho}$ and $\mathbf{r}$ is the 3-D position vector, as well as on the values of the rate of produc-tion or dissipation of turbulent energy ε and the rate of production or dissipation of temperature inhomogeneities η.

The velocity structure function, $\mathcal{D}_v(\mathbf{r})$, due to the eddies of sizes r, i.e., $\mathcal{D}(\mathbf{r}) \sim v^2$, is usually assumed to have a Gaussian random distribution with the following second order structure function,

$$\begin{aligned}\mathcal{D}_v(\mathbf{r}) &= \langle |v(\mathbf{r}) - v(\boldsymbol{\rho} + \mathbf{r})|^2 \rangle, \\ &= 2\left[\langle v(\mathbf{r})^2 \rangle - \langle v(\mathbf{r})v(\boldsymbol{\rho} + \mathbf{r}) \rangle\right] \qquad l_0 \ll r \ll L_0, \qquad (4.10)\end{aligned}$$

The (4.10) expresses the variance of the velocities at two points of distance $\mathbf{r}$ apart which varies between the inner-scale, l_0 and outer-scale, L_0. The structure function is related to the covariance function, $\mathcal{B}_v(\mathbf{r})$, through

$$\mathcal{D}_v(\mathbf{r}) = 2[\mathcal{B}_v(0) - \mathcal{B}_v(\mathbf{r})], \qquad (4.11)$$

in which $\mathcal{B}_v(\mathbf{r}) = \langle v(\boldsymbol{\rho}) v(\boldsymbol{\rho} + \mathbf{r}) \rangle$ and the covariance is the 3-D Fourier transform (FT) of the spectrum, $\Phi_v(\kappa)$.

Above the inner-scale of turbulence, l_0, the kinematic viscosity, v_v, plays no role in the value of the structure function. For f to be dimensionless, one should have $f(x) = x^{2/3}$. Asserting $\mathcal{D}_v(\mathbf{r})$ is a function of $\mathbf{r}$ and ε, one writes,

$$\mathcal{D}_v(\mathbf{r}) \propto \mathcal{C}_v^2 r^{2/3} \qquad l_0 \ll r \ll L_0, \qquad (4.12)$$

in which $\mathcal{C}_v^2$ is the velocity structure constant.

Turbulent flow produces temperature inhomogeneities by mixing adiabatically atmospheric layers at different temperatures. In this case, buoyancy becomes a source of atmospheric instability. The atmospheric stability may be measured by another dimensionless quantity, called Richardson number that is expressed as,

$$R_i = \frac{g}{T} \frac{\partial \bar{\theta}/\partial z}{(\partial \bar{u}/\partial z)^2}, \tag{4.13}$$

where g is the acceleration due to gravity, T the mean absolute temperature, $\partial \bar{\theta}/\partial z$ the gradient of the mean temperature, and $\partial \bar{u}/\partial z$ the gradient of the flow velocity.

When the term, $\partial \bar{\theta}/\partial z$, becomes negative, a parcel of air brought upward becomes warmer than surrounding air so that its upward motion would be maintained by buoyancy producing an instability, while in the reverse condition, i.e., if $\partial \bar{\theta}/\partial z$ is positive buoyancy produces stability. It is important to note that flows with $R_i > 0.25$ are stable, while flows with $R_i < 0.25$ are unstable. The temperature structure function is described as,

$$\mathcal{D}_T(\mathbf{r}) = \langle |T(\rho + \mathbf{r}) - T(\rho)|^2 \rangle, \tag{4.14}$$

which expresses its variance at two points. The temperature structure function is the mean squared variation of the difference between the temperature at two points separated by the horizontal vector, ρ. From the simple dimension considerations one gets,

$$\mathcal{D}_T(\mathbf{r}) \propto \eta \varepsilon^{-1/3} r^{2/3} \qquad \text{or} \tag{4.15}$$
$$\mathcal{D}_T(\mathbf{r}) \propto C_T^2 r^{2/3}, \tag{4.16}$$

where C_T^2 is known as the temperature structure constant, which is a measure of the local intensity of the temperature fluctuations.

4.1.3 Refractive Index Power-spectral Density

The variations of temperature, density, as well as water vapor are needed to calculate the refractive index structure function; the idea of a conserved passive additive[2] is also required (Tatarski 1961). Light traveling through the atmosphere is affected by fluctuations of the refractive index. The refractive index at a point r and at a time t is given by,

$$n(\mathbf{r}, t) = n_0 + n_1(\mathbf{r}, t), \tag{4.17}$$

---

[2] A passive additive is a quantity, which does not affect the dynamics of the flow. It does not disappear through some chemical reaction in the flow.

where $n_0 \approx 1$ is the mean refractive index of air, $n_1(\mathbf{r}, t)$ the randomly fluctuating term, and t the time.

The dependence of the refractive index of air upon pressure, P (millibar) and temperature, T (Kelvin), at optical wavelengths is given by (Ishimaru 1978),

$$n_1 \cong n - 1 = 77.6 \times 10^{-6} \frac{P}{T}. \tag{4.18}$$

Following (4.10), the structure function for the range of values $l_0 \ll r \ll L_0$ is also found to be related to the covariance function, $\mathcal{B}_n(\mathbf{r})$, i.e.,

$$\mathcal{D}_n(\mathbf{r}) = 2\left[\mathcal{B}_n(\mathbf{0}) - \mathcal{B}_n(\mathbf{r})\right], \tag{4.19}$$

where $\mathcal{B}_n(\mathbf{r}) = \langle n(\mathbf{r}')n(\mathbf{r}' + \mathbf{r})\rangle$, $n(\mathbf{r})$ is the refractive index at position $\mathbf{r}$, $\mathcal{D}_n(\mathbf{r})$ the statistical variance in refractive index between two parts of the wavefront in an atmospheric layer, and the covariance is the 3-D FT of the spectrum, $\Phi_n(\kappa)$ (Roddier 1981).

Following Kolmogorov's model, the structure function, $\mathcal{D}_n(\mathbf{r})$, for the case of an isotropic turbulent layer depends on the strength of the turbulence. The scale of the phase perturbations falls exactly in the range where a function represented in the form,

$$\mathcal{D}_n(r) = C_n^2 r^{2/3}, \tag{4.20}$$

in which C_n^2, known as the structure constant of the refractive index fluctuations, is valid.

C_n^2 is a constant of proportionality describing the strength of the turbulence. Knowledge of this parameter is essential for the development and optimal operation of an adaptive optics (see Sect. 4.3) system in optical astronomy, Multi-Conjugate Adaptive Optics (MCAO) system (see Sect. 4.4.6) in particular. For the case of an atmosphere stratified into a series of horizontal layers, C_n^2 is taken as a function of the height h above the ground (see Fig. 4.1). In terms of spectral density function, $\Phi_n(\kappa)$ of structure function $\mathcal{D}_n(\mathbf{r})$, one may write,

$$\mathcal{D}_n(\mathbf{r}) = 2\int_{-\infty}^{\infty} (1 - \cos \kappa \cdot \mathbf{r})\Phi_n(\kappa)d\kappa$$
$$= 8\pi \int_0^{\infty} \Phi_n(\kappa)\kappa^2 \left[1 - \frac{\sin \kappa r}{\kappa r}\right]d\kappa, \tag{4.21}$$

where $\Phi_n(\kappa)$ is the spectral density in 3-D space wave numbers of the distribution of the amount of inhomogeneity in a unit volume.

The form of this function corresponding to two-third law for the concentration $\mathbf{n}$. By noting,

$$\int_0^{\infty} x^a \left[1 - \frac{\sin bx}{bx}\right]dx = -\frac{\Gamma(a)\sin(\pi a/2)}{b^{a+1}}, \qquad -3 < a < -1.$$

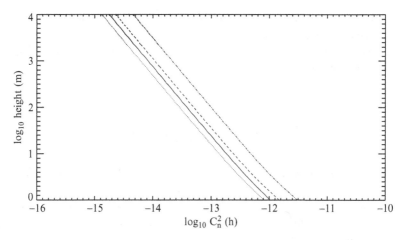

Fig. 4.1 Variation of C_n^2 as a function of height at different times, 8 AM (*solid line*), 10 AM (*dotted line*), 12 noon (*dashed line*), and 2 PM (*dashed-dot line*) IST, observed on 10th June, 2007, at Hanle, India, using the SHABAR equipment (courtesy: K. E. Rangarajan and S. P. Bagare)

The power-spectral density, $\Phi_n(\kappa)$, for the wave number, $\kappa > \kappa_0$, in the inertial sub-range, can be equated as,

$$\Phi_n(\kappa) = \frac{\Gamma(8/3)\sin\pi/3}{4\pi^2}C_n^2\,\kappa^{-11/3} = 0.033\,C_n^2\,\kappa^{-11/3}. \tag{4.22}$$

The (4.22) states that the power-law spectrum of the fluctuation scales as the $-11/3$ power of the wave vector, κ. This model is valid within the inertial subrange and is widely used for astronomical purposes (Tatarski 1993). The important property of Kolmogorov's law is that the refractive index fluctuations are largest for the largest turbulent elements up to the outer-scale of the turbulence. Owing to the non-integrable pole at $\kappa = 0$, mathematical problems arise to use this equation for modeling the spectrum of the refractive index fluctuations when, $\kappa \to 0$. The integral over $\Phi_n(\kappa) \propto \kappa^{-11/3}$ is infinite, i.e., the variance of the turbulent phase is infinite. This is a well known property of Kolmogorov turbulence of the atmosphere. Robbe et al. (1997) reported from observations using the Interféromètre à deux Télescopes (I2T) that most of the measured temporal spectra of the angle of arrival exhibit a behavior compatible with the 2/3 power law.

Since Kolmogorov's spectrum is not defined outside the inertial range, for a finite outer-scale, the von Kármán spectrum can be used to perform the finite variance (Ishimaru 1978). In order to accommodate the finite inner- and outer-scales, the power spectrum can be written as,

$$\Phi_n(\kappa) = 0.033C_n^2\left(\kappa^2 + \kappa_0^2\right)^{-11/6}e^{-\kappa^2/\kappa_i^2}, \tag{4.23}$$

with $\kappa_0 = 2\pi/L_0$ and $\kappa_i = 5.9/l_0$.

The refractive index variations lead directly to phase fluctuations influencing the performance of the imaging system. By producing a profile of C_n^2 as a function of altitude, one may decide on the type of adaptive optics system, which is needed at a particular telescope. Following Roddier (1981), the two structure constants C_n^2 and C_T^2 are related by,

$$C_n^2(h) = \left(80 \times 10^{-6} \frac{P(h)}{T^2(h)}\right)^2 C_T^2(h), \tag{4.24}$$

where $P(h)$ and $T(h)$ are the pressure in millibar and the absolute temperature T in K, respectively, at height h in meter (m); a pressure equilibrium is assumed.

As the temperature, T, and humidity, $\mathcal{H}$, are both functions of height in the atmosphere, turbulent mixing creates inhomogeneities of temperature and humidity at scales comparable to eddy size.

4.1.4 Turbulence and Boundary Layer

The spatial and temporal characteristics of wavefront distortions convey essential information. However, the numerical evaluation of the critical parameters requires the knowledge of the refractive index structure constant, C_n^2 and wind profiles as a function of altitude. The former varies with the geographic location and also with altitude and can be perturbed very easily by artificial means such as a flying aircraft. Its unit is $m^{-2/3}$. Since most of the above parameters are directly or indirectly related to C_n^2, a particular optical path needs to be modeled. The most popular for scintillation (see Sect. 4.1.6.2) effects is Hufnagel–Valley model (Hufnagel 1974; Valley 1980) due to its simplicity. The refractive index structure coefficient is described as,

$$C_n^2(h) = 5.94 \times 10^{-53}(v_w/27)^2 h^{10} e^{-h/1000}$$
$$+2.7 \times 10^{-16} e^{-h/1500} + A e^{-h/100}, \tag{4.25}$$

where h is the height (m), v_w the wind speed, and A the $C_n^2(0)$ at the Earth's surface, which is normally set to $1.7 \times 10^{-14} m^{-2/3}$.

A and v_w can be adjusted to achieve the desired low-altitude shape. The most common value for v_w is 21 m/sec; the ground wind speed parameter is usually 5 m per second. The significant scale-lengths, in the case of the local terrain, depend on the local objects which primarily introduce changes in the inertial subrange and temperature differentials. Such scale-lengths in this zone depend on the nearby objects while the refractive index structure constant, C_n^2 is proportional to $h^{-2/3}$. In real turbulent flows, turbulence is usually generated at solid boundaries. Near the boundaries, shear is the dominant source, where scale-lengths are roughly constant. This can be attributed to:

- surface boundary layer due to the ground convection, extending up to a few km height of the atmosphere, ($C_T^2 \propto z^{-2/3}$); the surface layer turbulence varies the most, and generally, contributes dominantly to seeing at low elevations, the resulting image blur decreasing with a scale-height of 3.5 m (Racine 2005),
- the free convection layer associated with orographic disturbances, where the scale-lengths are height dependent, ($C_T^2 \propto z^{-4/3}$), and
- in the tropopause and above, where the turbulence is due to the wind shear as the temperature gradient vanishes slowly.

The turbulence which reaches a minimum just after the sunrise and steeply increases until afternoon, is primarily due to the solar heating of the ground. It decreases to a secondary minimum after sunset and slightly increases during night (Roddier 1981 and references therein).

4.1.5 Statistics of the Amplitude and Phase Perturbations

The effect of turbulence on the wavefront can be perceived as phase screens (phase distortion of the wavefront introduced by the atmosphere), which are driven by the wind in front of the telescope. The spatial correlational properties of the turbulence-induced field perturbations are evaluated by combining the basic turbulence theory with the stratification and phase screen approximations. The variance of the ray can be translated into a variance of the phase fluctuations. For calculating the same, Roddier (1981) used the correlation properties for propagation through a single (thin) turbulence layer and then extended the procedure to account for many such layers. Several investigators (Goodman 1985; Troxel et al. 1994) have argued that individual layers can be treated as independent provided the separation of the layer centers is chosen large enough so that the fluctuations of the log amplitude and phase introduced by different layers are uncorrelated.

4.1.5.1 Computation of Phase Structure Function

Turbulence is thought to exist in thin layers at varying heights within the atmosphere. Following Roddier (1981), consider a monochromatic horizontal plane wave of wavelength, λ, from a distant star at the zenith, propagating through the atmosphere towards the ground-based observer. The radiation field at a distance, z, from the detector can be represented by the complex amplitude $(\mathbf{x}, z)$, where $\mathbf{x}$ represents two-dimensional (2-D) coordinates in the plane perpendicular to the line of sight. The complex field described by a vertical height, h above the ground and a horizontal position denoted by a co-ordinate vector, $\mathbf{x}$, in the atmosphere is given by,

$$U_h(\mathbf{x}) = |U_h(\mathbf{x})| \, e^{i\psi(\mathbf{x})}. \tag{4.26}$$

At the top of the atmosphere, the unperturbed complex field is normalized to unity $[U_\infty(\mathbf{x}) = 1]$. As the radiation propagates down through the atmosphere, the field evolves as phase fluctuations are imprinted by atmospheric turbulence and as the phase and amplitude of field are modified by diffraction from these fluctuations during propagation. In order to quantify these effects, an uniform atmosphere containing a single thin turbulent layer of thickness, dh, located at a height h above the detector, is considered. At each height, the phase, $\psi(\mathbf{x})$, is taken with respect to its average value, so that for any height $\langle\psi(\mathbf{x})\rangle = 0$. This thickness is considered to be large compared to the scale of turbulent eddies but small enough for the phase screen approximation (diffraction effects is negligible over the distance, δh). The atmosphere is assumed to be non-absorbing and horizontally stratified so that its statistical properties depend on h. If the wave propagates vertically downward at the zenith through such a thin horizontal layer, the complex field after passing through the layer is expressed as,

$$U_h(\mathbf{x}) = e^{i\,\psi(\mathbf{x})}. \tag{4.27}$$

At the top of the layer, $U_{h+\delta h}(\mathbf{x}) = 1$. The refractive index fluctuations, $n(x, z)$, give rise to phase shift, $\psi(\mathbf{x})$, inside the layer, i.e.,

$$\psi(\mathbf{x}) = \kappa \int_h^{h+\delta h} n(x, z)dz, \tag{4.28}$$

where $\kappa = 2\pi/\lambda$ is the wave number.

In this case, the rest of the atmosphere is thought to be calm and homogeneous. From the bottom of the layer to the ground at $h = 0$ (more precisely, in the plane of the detector), the propagation of the wave can be described by the Fresnel approximation (which corresponds to keeping only the leading term in the expansion of the optical path length in powers of $x = |\mathbf{x}|$). Hence, the effect of the turbulent layer on the radiation field at the detector is,

$$U_0(\mathbf{x}) = U_h(\mathbf{x}) \star \frac{e^{i\pi x^2/\lambda h}}{i\lambda h}, \tag{4.29}$$

with respect to the variable $\mathbf{x}$. Here, the symbol $\star$ the denotes convolution parameter.

The coherence function, $\mathcal{B}_h(\boldsymbol{\xi})$, of the complex amplitude after transmitting through the layer at height h is expressed as,

$$\mathcal{B}_h(\boldsymbol{\xi}) = \langle U_h(\mathbf{x})U_h^*(\mathbf{x} + \boldsymbol{\xi})\rangle$$
$$= \langle e^{i[\psi(\mathbf{x})-\psi(\mathbf{x}+\boldsymbol{\xi})]}\rangle = e^{-\frac{1}{2}\mathcal{D}_\psi(\boldsymbol{\xi})}, \tag{4.30}$$

in which the term $\mathcal{D}_\psi(\xi)(= \langle |\psi(\xi) - \psi(\rho + \xi)|^2\rangle)$ is the 2-D structure function of the phase, $\psi(\mathbf{x})$ (Fried 1966), which is given by,

$$\mathcal{D}_\psi(\xi) = \kappa^2 \delta h \int_{-\infty}^{\infty} [\mathcal{D}_n(\xi, \zeta) - \mathcal{D}_n(0, \zeta)] \, d\zeta, \tag{4.31}$$

with

$$\mathcal{D}_n(\xi, \zeta) = 2 [\mathcal{B}_n(0, 0) - \mathcal{B}_n(\xi, \zeta)], \tag{4.32}$$

as the refractive index structure function and $\mathcal{B}_n(\xi, \zeta) = \langle n(\mathbf{x}, z)n(\mathbf{x} + \xi, z')\rangle$ the 3-D refractive index covariance.

The refractive index structure function defined in (4.20) is evaluated as,

$$\mathcal{D}_n(\xi, \zeta) = C_n^2 (\xi^2 + \zeta^2)^{1/3}, \tag{4.33}$$

and using (4.20), (4.31) can be integrated to yield,

$$\mathcal{D}_\psi(\xi) = 2.91 \kappa^2 C_n^2 \xi^{5/3} \delta h. \tag{4.34}$$

The covariance of the phase is deduced by substituting (4.34), in (4.30),

$$\mathcal{B}_h(\xi) = e^{-\frac{1}{2}[2.91\kappa^2 C_n^2 \xi^{5/3} \delta h]}. \tag{4.35}$$

Using the Fresnel approximation, Roddier (1981) finds the covariance of the phase at the ground level due to a thin layer of turbulence at some height off the ground is,

$$\mathcal{B}_0(\xi) = \mathcal{B}_h(\xi). \tag{4.36}$$

For high altitude layers the complex field fluctuates both in phase and in amplitude (scintillation), and therefore, the wave structure function, $\mathcal{D}_\psi(\xi)$, is not strictly true as at the ground level. The turbulent layer acts like a diffracting screen; however, correction in the case of astronomical observation remains small (Roddier 1981).

The wave structure function after passing through N layers can be expressed as the sum of the N wave structure functions associated with the individual layer. For each layer, the coherence function should be multiplied by the right hand side of (4.35), and therefore the coherence function at ground level translates into,

$$\mathcal{B}_0(\xi) = \prod_{j=1}^{N} e^{-\frac{1}{2}[2.91\kappa^2 C_n^2(h_j)\xi^{5/3}\delta h_j]}$$
$$= e^{-\frac{1}{2}\left[2.91\kappa^2 \xi^{5/3} \sum_{j=1}^{N} C_n^2(h_j)\delta h_j\right]}. \tag{4.37}$$

This expression (4.37) may be generalized for a star at an angular distance γ away from the zenith viewed through all over the turbulent atmosphere,

$$\mathcal{B}_0\left(\xi\right) = e^{-\frac{1}{2}\left[2.91\kappa^2\xi^{5/3}\sec\gamma\int C_n^2(h)dh\right]}. \tag{4.38}$$

4.1.5.2 Fried's Parameter, r_0

Fried's parameter, r_0, is described by the size of the atmospheric correlation region (Fried 1966). It is a measure of the strength of turbulence in terms of the diameter of a telescope that gives the same resolution as an image taken with a long-exposure time[3] through the atmosphere. It represents in a single number the integrated effect of the refractive-index fluctuations for the entire atmosphere along the path of interest.

Fried's parameter is a function of the turbulence strength constant, C_n^2, wavelength of the light λ, altitude, slant angle of beam direction, and of course path length. Figure 4.2 depicts C_n^2 and the value of r_0 measured at Hanle, using SHABAR. According to (4.38), the covariance, $\mathcal{B}(\mathbf{u})$, in which $\mathbf{u}(= u, v)$ is the 2-D spatial frequency vector (see 1.61), is expressed as,

$$\mathcal{B}(\mathbf{u}) = \mathcal{B}_0(\lambda\mathbf{u}) = e^{-\frac{1}{2}\left[2.91\kappa^2(\lambda u)^{5/3}\sec\gamma\int C_n^2(h)dh\right]}. \tag{4.39}$$

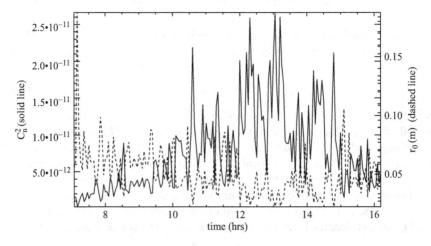

Fig. 4.2 Relation between r_0 and C_n^2 factor observed on 10 June 2007 at Hanle, India, using the SHABAR equipment (courtesy: K. E. Rangarajan and S. P. Bagare)

---

[3] When a detector is exposed to the incident light for more than the freezing time of the atmosphere generally called the long-exposure time.

Introducing r_0, (4.39) takes the form,

$$\mathcal{B}(\mathbf{u}) = e^{-3.44(\lambda u/r_0)^{5/3}}. \tag{4.40a}$$

$$\mathcal{B}_0(\xi) = e^{-3.44(\xi/r_0)^{5/3}}. \tag{4.40b}$$

The phase structure function, $\mathcal{D}_\psi(\xi)$, can be described in terms of the parameter, r_0,

$$\mathcal{D}_\psi(\xi) = 6.88\left(\frac{\xi}{r_0}\right)^{5/3}. \tag{4.41}$$

By comparing (4.41) with (4.39), yields an expression for r_0 in terms of the angle away from the zenith and an integral over the refractive index structure constant of the atmospheric turbulence,

$$r_0 = \left[0.423\kappa^2 \sec\gamma \int \mathcal{C}_n^2(h)dh\right]^{-3/5}. \tag{4.42}$$

This relation (4.42) points out the link between the value of r_0 at the telescope and the varying value of $\mathcal{C}_n^2$ with height, which features strong contributions at different altitudes. The dependence of r_0 as a function of height is depicted in Fig. 4.3. Fried's parameter also depends on the zenith angle, γ, as well as the wavelength. Its value is an essential parameter in designing an adaptive optics (AO) system for an astronomical telescope.

Fried (1965) and Noll (1976) noted that r_0 corresponds to the aperture diameter for which the variance $\langle\sigma\rangle^2$ of the wavefront phase averaged over the aperture comes approximately to unity:

$$\langle\sigma\rangle^2 = 1.03\left(\frac{D}{r_0}\right)^{5/3}. \tag{4.43}$$

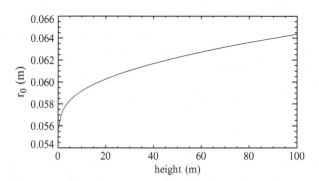

Fig. 4.3 Relation between r_0 and height observed on 10 June 2007 at Hanle, India, using SHABAR equipment (courtesy: K. E. Rangarajan and S. P. Bagare)

This (4.43) represents a commonly used definition for r_0. The relation between the structure function, $\mathcal{D}_\psi(\boldsymbol{\xi})$, and the power-spectral density, $\Phi_\psi(\kappa)$ is given by (Noll 1976),

$$\mathcal{D}_\psi(\boldsymbol{\xi}) = 2 \int_{-\infty}^{\infty} \Phi_\psi(\kappa) \left[1 - \cos\left(2\pi\kappa \cdot \boldsymbol{\xi}\right)\right] d\kappa. \tag{4.44}$$

This relationship along with (4.41), one may derive the spatial power spectrum of the phase fluctuations for Kolmogorov's spectrum,

$$\Phi_\psi(\kappa) = 0.023 r_0^{-5/3} \kappa^{-11/3}. \tag{4.45}$$

The Wiener spectrum of phase gradient after averaging with the telescope aperture, D, due to Kolmogorov turbulence is deduced as,

$$\Phi_\psi(\kappa) = 0.023 r_0^{-5/3} \kappa^{-11/3} \left| \frac{2J_1(\pi D\kappa)}{\pi D\kappa} \right|^2, \tag{4.46}$$

where J_1 is the first order Bessel function describing Airy disk, the diffraction-limited PSF, which is the FT of the circular aperture.

4.1.5.3 Atmospheric Coherence Time, τ_0

Atmospheric coherence time, τ_0, is determined by the stability of the atmosphere along the line of sight to the source. For a Long Baseline Optical Interferometer (LBOI), it is the interval during which the fringe phase remains stable, changing by ≤ 1 radian (rad). τ_0 is a function of Fried's parameter, r_0 and the transverse component of the wind velocity. With the knowledge of the spatial properties of the structure function and the wind velocity, the temporal behavior of the perturbations can be derived. The temporal fluctuation arises from the wind-driven motion of a frozen layer of turbulence across the telescope or an optical interferometer, for example LBOI. By employing Taylor's frozen-turbulence approximation (Taylor 1921), the variations of the turbulence caused by a single layer may be modeled by a frozen screen that is moved across the telescope aperture by the wind in that layer. In this case, the wavefront evolution time is assumed to be much greater than the time for the screen to blow across a telescope aperture. Assuming that a static layer of turbulence moves with constant speed $\mathbf{v}$ in front of the telescope aperture, the temporal phase structure function is,

$$\mathcal{D}_\psi(\mathbf{v}\tau) = \langle |\psi(\mathbf{x}, t) - \psi(\mathbf{x} - \mathbf{v}\tau, t)|^2 \rangle. \tag{4.47}$$

This provides the mean square phase error associated with a time delay, τ. Such a structure function depends individually on the two coordinates parallel and perpendicular to the wind direction. Though time evolution is complicated in the case of multiple layers contributing to the total turbulence, the temporal behavior is characterized by atmospheric coherence time τ_0. In the direction of the wind speed, an estimate of the correlation time yields the temporal coherence time, τ_0

$$\tau_0 = 0.314 \frac{r_0}{v_w}, \tag{4.48}$$

in which v_w is a characteristic wind velocity.

To reiterate, the fluctuations due to the atmospheric turbulence are induced regarding both modulus and phase of the complex amplitude at the ground introducing spatial (r_0) and temporal (τ_0) losses. Measurements of both these parameters are important for the design of any LBOI; the former limits the usable area of the collecting aperture, while the latter limits the coherent integration time.

4.1.5.4   Measurements of r_0 and τ_0

Fried's parameter characterizes the effect of seeing at a particular wavelength. Its value is dependent on the integrated magnitude of refractive index variations in the atmosphere. Small values of r_0 correspond to strong turbulence and poor seeing, while large values mean weak turbulence and good seeing (Hardy 1998). Typically, the size of r_0 ranges between 3 and 30 cm at 500 nm. For instance, a site like Mauna Kea[4] in Hawaii, has unique seeing conditions, which are mainly due to upper atmosphere variations.

An instrument, called Differential Image Motion[5] Monitor (DIMM), can be used to measure Fried's parameter. It measures the differential centroid motion of two images resulting from a single bright star between two separate optical paths through the atmosphere. This can be accomplished with a modest telescope having a two-aperture (diameter a few centimeters) mask on the front and a prism displacing the light from one aperture to produce two well-separated subimages of one and the same star onto a detector. These image motions are due to a wavefront disturbance spectrum resulting from the turbulent medium, which can be interpreted in terms of r_0. It is prudent to note that the common mode motion, namely tracking

---

[4] Mauna Kea observatory is situated at an elevation of 4,200 m above the Pacific ocean, atop a dormant volcano on the big island of Hawaii. The summit site, surrounded by thousands of miles of relatively thermally-stable ocean, is suitable for observation due to the stabilizing effect of large water body on the atmosphere. It has no nearby mountain ranges to spoil the upper atmosphere or throw light reflecting dust into the air. The coherence scale is long due to high laminar flow of the Pacific ocean winds over the peak of the mountain. However, the fast winds of the overhead jet stream result in very short coherence time that lies between 1.5 and 10 ms.

[5] A similar instrument called, Solar Differential Image Motion Monitor (SDIMM), can also be used to measure day time seeing. It uses the edge (limb) of the solar disk for the differential motion measurement. A narrow slit is placed at right angles across the solar limb and imaged into two separate images for each of the two apertures on the detector as is done in the stellar DIMMs. The variable position differences of the two limbs is the SDIMM signal which is used to calculate r_0.

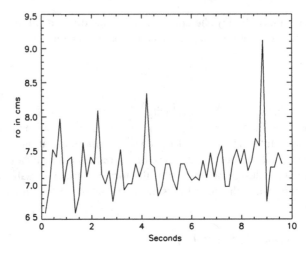

Fig. 4.4 Fluctuations of r_0 as observed at the 2.34 m VBT, Kavalur, India on 21 February, 1997 (Saha and Yeswanth 2004)

errors and wind shakes, affect each image in the same manner and do not introduce relative image motion. Hence, the variation of the subimage separation provides r_0 integrated over the entire atmosphere.

Another method is to use speckle interferometric technique for obtaining Fried's parameter. Figure 4.4 depicts the microfluctuations of r_0 at a step of $\sim$150 ms observed at the 2.34 m Vainu Bappu telescope (VBT), Kavalur, India. The transfer function $\langle|\widehat{S}(\mathbf{r})|^2\rangle$ is obtained by calculating Weiner spectrum of the instantaneous intensity distribution from each of the afore-mentioned stars. Here, $\mathbf{r} = (x, y)$ is a 2-D position vector, and $|\ |$ the modulus.

The atmospheric coherence time, τ_0, is typically between 1 and 20 ms at 500 nm and scale with wavelength in the same manner as r_0. For adaptive optics, a pre-detection compensation technique, the reciprocal of the coherence time indicates the required bandwidth of the closed loop correction system. Davis and Tango (1996) have devised a method for measuring the atmospheric coherence time that can be carried out in parallel with the determination of the fringe visibility in an amplitude interferometer. The measured coherence time was found to be in the range of $\sim$1 $-\sim$7 ms, with the Sydney University Stellar Interferometer (SUSI).

4.1.6 Imaging Through Atmospheric Turbulence

The observed illumination at the focal plane of a telescope in presence of the turbulent atmosphere as a function of the direction $\boldsymbol{\alpha}$,

$$S(\boldsymbol{\alpha}) = \langle U(\boldsymbol{\alpha})U^*(\boldsymbol{\alpha})\rangle = \frac{1}{A_p}\left|\mathcal{F}[\widehat{U}(\mathbf{u})\widehat{P}(\mathbf{u})]\right|^2. \tag{4.49}$$

The term $S(\boldsymbol{\alpha})$ is known as the instant point spread function (PSF; see Sect. 1.5.1) produced by the telescope, the atmosphere, and $\widehat{P}(\mathbf{u})$ the pupil transfer function. The instant transfer function of the telescope and the atmosphere takes the form,

$$\widehat{S}(\mathbf{f}) = \int_{-\infty}^{\infty} U(\boldsymbol{\alpha})U^*(\boldsymbol{\alpha})e^{-i2\pi\boldsymbol{\alpha}\cdot\mathbf{f}}d\boldsymbol{\alpha}$$

$$= \int_{-\infty}^{\infty} S(\boldsymbol{\alpha})e^{-i2\pi\boldsymbol{\alpha}\cdot\mathbf{f}}d\boldsymbol{\alpha}, \qquad (4.50)$$

where $\mathbf{f}$ is the spatial frequency vector expressed in rad^{-1} and $|\mathbf{f}|$ is its magnitude.

According to the autocorrelation theorem (see Appendix B), the Fourier transform of the squared modulus (4.49) is the autocorrelation of $\widehat{U}(\mathbf{u})\widehat{P}(\mathbf{u})$, hence,

$$\widehat{S}(\mathbf{f}) = \frac{1}{A_p} \int_{-\infty}^{\infty} \widehat{U}(\mathbf{u})\widehat{U}^*(\mathbf{u}+\mathbf{f})\widehat{P}(\mathbf{u})\widehat{P}^*(\mathbf{u}+\mathbf{f})d\mathbf{f}. \qquad (4.51)$$

The (4.51) describes the spatial frequency content $\widehat{S}(\mathbf{f})$ of images taken through the turbulent atmosphere. For a perfect non-turbulent atmosphere, $\widehat{U}(\mathbf{u}) = 1$, (4.51) shrinks to the telescope transfer function.

4.1.6.1 Seeing-limited Images

Astronomical seeing describes atmospheric effects on the quality of a stellar image in a telescope. It is the total effect of distortion in the path of starlight via different contributing layers of the atmosphere to the detector placed at the focus of the telescope. Degradation in image can occur due to: (1) variations of airmass $(X \sim 1/\cos Z)$ or of its time average between the target object and the reference star, (2) seeing differences between the Modulation Transfer Function (MTF) for the object and its estimation from the reference, (3) deformation of mirrors or to a slight misalignment while changing its pointing direction, (4) bad focusing, and (5) thermal effect from the telescope. Further, it degrades due to the temperature differentials in and around the telescope dome and difference in temperature of the primary mirror surface with its surroundings.

The conventional method of measuring seeing from the star image is to measure the full width half maximum (FWHM) of a long-exposure stellar image (expressed in arcseconds) at zenith. The spread in the image due to seeing varies as $\lambda^{-1/5}$, the seeing errors being less at longer wavelengths. A ten-fold increase of λ reduces the seeing by a factor of 0.63. Seeing affects the measurements of fringe visibility with a Long Baseline Optical Interferometer (LBOI; Labeyrie 1975), by introducing phase aberrations across the wavefronts incident on the interferometer, therefore, the relative phase of the wavefronts at the apertures changes with time and also varies with the optical paths through the two arms.

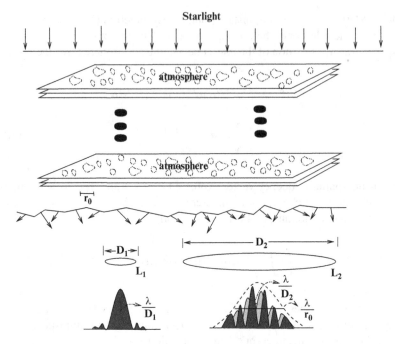

Fig. 4.5 Plane-wave propagation through the multiple turbulent layers. L_1 and L_2 represent the small and large telescopes with respective diameters D_1 and D_2

Let the modulation transfer function of the atmosphere and a simple lens based telescope in which the PSF is invariant to spatial shifts be described as in Fig. 4.5. If a point-like object is observed through such a telescope, the light reaches the entrance pupil of a telescope with patches of random excursions in phase. When such aberrated wavefronts are focused onto the focal plane of a telescope, all details in the image smaller than λ/r_0 get obliterated in long-exposure images. The PSF of such images is defined by the ensemble average, $\langle S(\mathbf{x}) \rangle$, where $S(\mathbf{x})$ is the instantaneous illumination of a point source, which is independent of the direction. In such a case, the distribution of intensities becomes uncorrelated in time and the statistics of irradiance turn out to be Gaussian over the seeing-disk. If the object emits incoherently, the relation between the irradiance $\langle I(\mathbf{x}) \rangle$ in the image space and the radiance $O(\mathbf{x})$ of an object obeys convolution relationship, i.e.,

$$\langle I(\mathbf{x}) \rangle = \int_{-\infty}^{\infty} O(\mathbf{x}') \langle S(\mathbf{x} - \mathbf{x}') \rangle d\mathbf{x}'$$
$$= O(\mathbf{x}) \star \langle S(\mathbf{x}) \rangle, \tag{4.52}$$

with $\mathbf{x}(= x, y)$ as the 2-D position vector and $\mathbf{x}'$ the deviation of a stellar image from its mean position,

Using 2-D FT, (4.52) translates into,

$$\langle \widehat{I}(\mathbf{u}) \rangle = \widehat{O}(\mathbf{u}) \cdot \langle \widehat{S}(\mathbf{u}) \rangle, \tag{4.53}$$

where, $\widehat{O}(\mathbf{u})$ denotes the object spectrum and $\langle \widehat{S}(\mathbf{u}) \rangle$ the transfer function for long-exposure images.

The transfer function is the product of the atmosphere transfer function, $\widehat{B}(\mathbf{u})$, and the optical transfer function (OTF; see Sect. 1.5.1) of the telescope, $\widehat{T}(\mathbf{u})$,

$$\langle \widehat{S}(\mathbf{u}) \rangle = \int_{-\infty}^{\infty} \langle \widehat{U}(\mathbf{u})\widehat{U}^*(\mathbf{u} + \mathbf{u}') \rangle \widehat{P}(\mathbf{u})\widehat{P}^*(\mathbf{u} + \mathbf{u}') d\mathbf{u}$$
$$= \widehat{B}(\mathbf{u}) \cdot \widehat{T}(\mathbf{u}). \tag{4.54}$$

with

$$\widehat{B}(\mathbf{u}) = \langle \widehat{U}(\mathbf{u})\widehat{U}^*(\mathbf{u} + \mathbf{u}') \rangle.$$

Following (4.40), the OTF of the atmospheric turbulence is expressed as,

$$OTF_{turb}(\boldsymbol{\xi}) = e^{-3.44(u|\xi|/r_0)^{5/3}}, \tag{4.55}$$

while the expression for the short-exposure[6] OTF of the turbulence (Fried 1966) is given by,

$$OTF_{turb}(\boldsymbol{\xi}) = e^{-3.44(u|\xi|/r_0)^{5/3}[1-\alpha(u|\xi|/D)^{1/3}]}. \tag{4.56}$$

The term α is a constant and takes the value of 1 if the telescope is in the near field of the turbulence. Following (1.100), the resolving power of a telescope is given by,

$$\mathcal{R} = \int_{-\infty}^{\infty} \widehat{T}(\mathbf{u}) d\mathbf{u} = \frac{1}{A_p} \int_{-\infty}^{\infty} \widehat{P}(\mathbf{u})\widehat{P}^*(\mathbf{u} + \mathbf{u}') d\mathbf{u}$$
$$= \frac{1}{A_p} \left| \int_{-\infty}^{\infty} \widehat{P}(\mathbf{u}) d\mathbf{u} \right|^2, \tag{4.57}$$

where A_p is the pupil area in wavelength squared units.

The resolution is limited either by the telescope aperture or by the atmosphere, depending on the relative width of the two functions, $\widehat{B}(\mathbf{u})$ and $\widehat{T}(\mathbf{u})$, i.e.,

$$\mathcal{R} = \int_{-\infty}^{\infty} \widehat{T}(\mathbf{u}) d\mathbf{u} = \frac{\pi}{4} \left(\frac{D}{\lambda} \right)^2 \quad D \ll r_0. \tag{4.58a}$$
$$= \int_{-\infty}^{\infty} \widehat{B}(\mathbf{u}) d\mathbf{u} = \frac{\pi}{4} \left(\frac{r_0}{\lambda} \right)^2 \quad D \gg r_0. \tag{4.58b}$$

---

[6] Short-exposure is the frame integration time required to freeze the atmosphere, which is generally less than 20 ms in optical domain and in infrared wavebands, it may go up to the order of 100 m.

The long-exposure resolution through an optical telescope with an infinite diameter in presence of atmospheric turbulence is characterized by Fried's parameter, r_0, which is the same as the resolution through a telescope of diameter r_0 in the absence of turbulence.

4.1.6.2 Scintillation

The rapid variations in apparent brightness (or color) of a distant star viewed through the atmosphere is known as scintillation. It occurs due to spatial differences in refractive index in the atmosphere. Scintillation is caused by small temperature variations (of the order of 0.1–$1°C$) in the atmosphere. The temporal variation of higher order aberrations due to the movement of small cell causes dynamic intensity fluctuations. Two aspects of poor seeing can be seen, amplitude scintillation, which is a rapid quasi-random fluctuation in the observed intensity of the source and phase scintillation, which is a similar fluctuation in the apparent direction of the source.

Scintillation depends on the height of the turbulent layer. As in Sect. 4.1.5.1, assuming that the turbulence is concentrated in a thin horizontal layer between altitude h and $h + \delta h$, the complex disturbance $U_h(\mathbf{x})$ at the layer output is written as,

$$U_h(\mathbf{x}) = e^{i\psi(\mathbf{x})} \simeq 1 + i\psi(\mathbf{x}). \tag{4.59}$$

Here, the phase delay $\psi(\mathbf{x})$ is introduced by the layer, which is assumed to be $\ll 1$. At the ground level, the disturbance, $U_0(\mathbf{x})$ is obtained by placing (4.59) in (4.29),

$$U_0(\mathbf{x}) = [1 + i\psi(\mathbf{x})] \star \frac{e^{i\pi x^2/\lambda h}}{i\lambda h} = 1 + \psi(\mathbf{x}) \star \frac{e^{i\pi x^2/\lambda h}}{\lambda h}. \tag{4.60}$$

The second term of right hand side of (4.60) describes the relative fluctuations of the disturbance at the ground level, whose real and imaginary parts are:

$$\chi(\mathbf{x}) = \psi(\mathbf{x}) \star \frac{1}{\lambda h} \cos \pi \frac{x^2}{\lambda h} \tag{4.61a}$$

$$\psi_0(\mathbf{x}) = \psi(\mathbf{x}) \star \frac{1}{\lambda h} \sin \pi \frac{x^2}{\lambda h}. \tag{4.61b}$$

For simplicity, let the argument of these functions at the ground level be dropped with the understanding that they denote values in the plane of the detector unless otherwise indicated. The statistics of these fluctuations are embodied in the spatial covariance functions, defined by,

$$B_\chi(\boldsymbol{\xi}) = \langle \chi(\mathbf{x})\chi(\mathbf{x} + \boldsymbol{\xi}) \rangle, \tag{4.62a}$$

$$B_\psi(\boldsymbol{\xi}) = \langle \psi(\mathbf{x})\psi(\mathbf{x} + \boldsymbol{\xi}) \rangle, \tag{4.62b}$$

where the brackets denote a spatial average over the $\mathbf{x}$ coordinate.

The Fourier transforms of these functions are the power spectra of the respective fluctuations:

$$W_\chi(\mathbf{f}) = \int B_\chi(\boldsymbol{\xi})e^{-i2\pi\boldsymbol{\xi}\cdot\mathbf{f}}d\boldsymbol{\xi}, \tag{4.63a}$$

$$W_\psi(\mathbf{f}) = \int B_\psi(\boldsymbol{\xi})e^{-i2\pi\boldsymbol{\xi}\cdot\mathbf{f}}d\boldsymbol{\xi}. \tag{4.63b}$$

Here, f is the 2-D spatial frequency, with units of inverse length. Since the phase fluctuations, $\psi(\mathbf{x})$ has Gaussian statistics, $\chi(\mathbf{x})$ and $\psi_0(\mathbf{x})$ are random function with Gaussian statistics as well. Their power spectra are related to the local 2-D power spectrum, $W_\chi(\mathbf{f})$, of the phase fluctuation, $\psi(\mathbf{x})$, induced by the turbulent layer, as:

$$W_\chi(\mathbf{f}) = W_\psi(\mathbf{f})\sin^2(\pi\lambda h f^2), \tag{4.64a}$$
$$W_{\psi_0}(\mathbf{f}) = W_\psi(\mathbf{f})\cos^2(\pi\lambda h f^2). \tag{4.64b}$$

The 2-D power spectra of $W_\chi(\mathbf{f})$ and $W_{\psi_0}(\mathbf{f})$, arising from a single layer at h and thickness δh are given by (Roddier 1981),

$$W_\chi(\mathbf{f}) = 0.38\lambda^{-2}f^{-11/3}\sin^2(\pi\lambda h f^2)C_n^2(h)\delta h, \tag{4.65a}$$
$$W_{\psi_0}(\mathbf{f}) = 0.38\lambda^{-2}f^{-11/3}\cos^2(\pi\lambda h f^2)C_n^2(h)\delta h, \tag{4.65b}$$

where $C_n^2(h)$ is the structure constant of index-of-refraction fluctuations within the turbulent layer.

Since the layer properties are statistically independent, their power spectra add linearly as well, any continuous distribution of turbulence may produce power spectra of the form,

$$W_\chi(\mathbf{f}) = 0.38\lambda^{-2}f^{-11/3}\int_0^\infty \sin^2(\pi\lambda h f^2)C_n^2(h)\delta h \tag{4.66a}$$

$$W_{\psi_0}(\mathbf{f}) = 0.38\lambda^{-2}f^{-11/3}\int_0^\infty \cos^2(\pi\lambda h f^2)C_n^2(h)\delta h. \tag{4.66b}$$

The atmospheric $C_n^2(h)$ profile determines the power spectra of both disturbance and phase fluctuations at ground level, and thereby, the contribution of the atmosphere to seeing, over scales for which Kolmogorov's spectrum holds. Its measurement is the primary goal of the scintillometer array technique. The scintillation index, $\langle\sigma_I\rangle^2$, is defined as the variance of the relative intensity fluctuations which is given by,

$$\langle\sigma_I\rangle^2 = 19.2\lambda^{-7/6}(\sec\gamma)^{11/6}\int_0^\infty C_n^2(h)h^{5/6}dh. \tag{4.67}$$

This (4.67) is valid for small apertures. Scintillation is reduced for larger apertures since it averages over multiple independent sub-apertures. This changes the amplitude of the intensity fluctuations; it changes the functional dependence on zenith angle, wavelength, and turbulence height as well. Considering the telescope filtering function, $|\widehat{P}_0(\mathbf{f})|^2$, and with aperture frequency cut-off $f_c \sim D^{-1}$ sufficiently small, so that, $\pi\lambda h f_c^2 \ll 1$, (4.67) translates into,

$$\langle\sigma_I\rangle^2 \propto D^{-7/3} (\sec\gamma)^3 \int_0^\infty \mathcal{C}_n^2(h)h^2 dh. \tag{4.68}$$

A few following equipment can be employed to measure the strength of the optical turbulence, $\mathcal{C}_n^2(h)$, for example:

1. SCIDAR: The SCIntillations Detection And Ranging (SCIDAR) measures the strength of the turbulence, $\mathcal{C}_n^2(h)$ and its dependence on altitude. This concept, proposed by Vernin and Roddier (1973), is based on an analysis of stellar scintillation images of a star produced by the turbulent layers. In this, the autocorrelation of scintillation pattern, obtained from a large number of short-exposure telescope pupil images of a star, is used to probe the strength of the atmospheric turbulence. From the autocorrelation of the pupil images obtained by a single star SCIDAR (Garnier et al. 2005), one retrieves $\mathcal{C}_n^2(h)$ from a lone peak. A binary star SCIDAR is also used for the said purpose although it suffers from locating a binary pair in the direction of interest. In this case, the autocorrelation image has three peaks. The position of side peaks in correlation profile indicates height of layer and height of peaks indicates of turbulence.

2. SHABAR: SHAdow BAnd Ranger (SHABAR) measures scintillation. Light from an extended source, for instance the Sun, is allowed to fall on the array of sensors as shown in Fig. 4.6. Each photo-diode, fitted with a diffuser and color filter, responds to turbulence within a cone whose diameter increases with height.

Fig. 4.6 SHABAR uses six photo-diodes attached to a single aluminum I-Beam providing a maximum baseline of 46.87 cm, which was mounted over the telescope with the Solar Differential Image Motion Monitor (SDIMM). The position of the individual sensors in the array are placed at a different locations, which act as a miniature telescopes with $7°$ field-of-view (FOV). They are separated by 15 non-redundant baselines. The cross-correlations of the measurements over several possible detector separations can be inverted to estimate r_0 as a function of height. This instrument was procured from National Solar Observatory (NSO), USA and installed at the Indian Astronomical Observatory, Hanle (courtesy: S. P. Bagare)

Intensity fluctuations occur at each sensor. The cones of any two detectors begin to overlap at an altitude (h) that is proportional to the separation (d) between the detectors. Turbulence in the region of overlap produces correlated intensity fluctuations at the two detectors. With enough independent baselines spanning a sufficiently large range of separation, the vertical structure of the turbulence can be determined. These observations allow the determination of seeing in the atmosphere upto a height of 100 m and in the ground layer. From the scintillometer array observations, $C_n^2(h)$ is determined for heights along the line-of-sight (Beckers 2001).

3. Micro-thermal sensor unit: This equipment consists of (1) sensors that sample the temperature with rapid response time, (2) amplifiers for the measured signal, and (3) a data-acquisition system. It can be used for site characterization, through the measurement of differential temperatures for a given separation, at chosen heights. Since the turbulence varies on the time-scale of tenths (or hundredths) of a second, the temperature variation on these time-scales is of the order of a few millidegrees, which is termed as micro-thermal variation. As stated earlier that the refractive index structure constant, C_n^2 is connected with the temperature structure constant, C_T^2 of the microthermal field variations (see (4.24)) producing fluctuations in the refractive at visible wavelengths. For a Kolmogorov's spectrum, the magnitude of the variation is dependent on the separation of the sensors (this separation should be between the inner-scale and the outer-scale) and characterized by a single value of temperature structure constant, C_T^2. Thus, one may infer the power in the Kolmogorov's spectrum by measuring the mean-squared variation at only one separation between two sensors. A simple set-up, capable of resolving a temperature change of about 3.7 millidegrees, is developed for site characterization at Hanle (Srinivasan 2008). The principle used in this is to excite a balanced Wheat-stone bridge by a precision reference voltage to provide a null output. The bridge accommodates two differential temperature probes that are made from micro-thin copper resistance wire temperature detectors. The probes sense any turbulence/change in temperature through a resistance change. The bridge provides a proportional voltage, corresponding to the change in the differential temperature of the two probes. This voltage is converted to its temperature equivalent.

4.2 Speckle Interferometry

If a point source is imaged through the telescope, with a short-exposure, each patch of the wavefront with diameter, r_0, would act independently of the rest of the wavefront yielding in breaking up into speckles (see Fig. 4.7) spread over the area defined by the long-exposure image (see Sect. 4.1.6.1). These speckles occur randomly along any direction within an angular patch of diameter, r_0.

To note, an image of a double star, with spacing smaller than the turbulence angle, consists of a superposition of two identical speckle patterns that are shifted by an

Fig. 4.7 Instantaneous
specklegram of a close binary
star HR 4689 taken with a
speckle interferometer (Saha
et al. 1999a) at the Cassegrain
focus of 2.34-m Vainu Bappu
Telescope (VBT), Kavalur,
India on 21 February, 1997

amount smaller than the image size. This image is difficult to analyze visually. However, its diffraction pattern, generated by a laser beam, shows a set of parallel equispaced fringes whose spatial frequency is proportional to the double star separation.

The speckle size of a point source is of the same order of magnitude as the Airy disk given by the telescope in the absence of seeing and aberrations. The lifetime of speckles is of the order of 0.1–0.01 s, and is determined by $\tau_s \approx r_0/\Delta v$, in which Δv is the velocity dispersion in the turbulent seeing layers across the line of sight. Integration time of each exposure varies from a few milliseconds to twenty milliseconds, depending on the condition of seeing. The number of speckles, n_s, per image is defined by the ratio of the area occupied by the seeing-disk, $1.22\lambda/r_0$ to the area of a single speckle, which is of the same order of magnitude as the Airy disk of the aperture (see 1.99),

$$n_s \approx \left(\frac{D}{r_0}\right)^2. \tag{4.69}$$

As the seeing improves, the number of speckles decreases. However, the number of photons, n_p, per speckle is independent of its diameter. Speckle interferometry, introduced by Labeyrie (1970), estimates a power spectrum (see Appendix B) from a set of specklegrams of the object of interest. However, the major limitation is that it cannot be used to take images of very faint objects. For such objects light captured by short-exposure is not sufficient enough for analysis.

4.2.1 Deciphering Information from Specklegrams

A specklegram represents the resultant of diffraction-limited incoherent imaging of the object irradiance, $O(\mathbf{x})$, convolved with the function representing the combined effects of the turbulent atmosphere and the telescope, $S(\mathbf{x})$. An ensemble of such specklegrams, $I_k(\mathbf{x}), k = t_1, t_2, t_3, \ldots, t_M$, constitute an astronomical speckle observation. By integrating the autocorrelation function of these successive narrow bandpass exposures, the spatial resolution of the objects at low light levels can be obtained. The transfer function of $S(\mathbf{x})$, is estimated by calculating Wiener spectrum of the instantaneous intensity from an unresolved star. The size of the data sets is constrained by the consideration of the signal-to-noise (S/N) ratio. Usually specklegrams of the brightest possible reference star are recorded to ensure that the S/N ratio of reference star is much higher than the S/N ratio of the programme star.

The variability of the corrugated wavefront yields 'speckle boiling' and is the source of speckle noise that arises from difference in registration between the evolving speckle pattern and the boundary of the PSF area in the focal-plane. In general, the specklegrams have additive noise contamination, $N_j(\mathbf{x})$, which includes all additive measurement of uncertainties. This may be in the form of

- photon statistics noise and
- all distortions from the idealized iso-planatic model represented by the convolution of $O(\mathbf{x})$ with $S(\mathbf{x})$ that includes non-linear geometrical distortions.

For each of the short-exposure instantaneous record, a quasi-monochromatic incoherent image system with additive noise is modeled mathematically as,

$$I(\mathbf{x}) = O(\mathbf{x}) \star S(\mathbf{x}) + N(\mathbf{x}), \tag{4.70}$$

with

$$\widehat{I}(\mathbf{u}) = \int_{-\infty}^{\infty} I(\mathbf{x}) e^{-i 2\pi \mathbf{u} \cdot \mathbf{x}} d\mathbf{x}, \tag{4.71a}$$

$$\widehat{O}(\mathbf{u}) = \int_{-\infty}^{\infty} O(\mathbf{x}) e^{-i 2\pi \mathbf{u} \cdot \mathbf{x}} d\mathbf{x}, \tag{4.71b}$$

$$\widehat{S}(\mathbf{u}) = \int_{-\infty}^{\infty} S(\mathbf{x}) e^{-i 2\pi \mathbf{u} \cdot \mathbf{x}} d\mathbf{x}, \tag{4.71c}$$

where $N(\mathbf{x})$ the noise contamination (photon noise and detector noise), $\mathbf{x}(= x, y)$ the 2-D spatial coordinate, $\mathbf{u}$ the 2-D spatial frequency, $\widehat{I}(\mathbf{u})$ the FT of the degraded image, $\widehat{O}(\mathbf{u})$ the visibility of the object, and $\widehat{S}(\mathbf{u})$ the transfer function of the telescope-atmospheric combination.

Denoting the noise spectrum as $\widehat{N}(\mathbf{u})$, the Fourier space relationship between the object and the image translates to,

$$\widehat{I}(\mathbf{u}) = \widehat{O}(\mathbf{u}) . \widehat{S}(\mathbf{u}) + \widehat{N}(\mathbf{u}). \tag{4.72}$$

In the spatial domain, by means of direct inversion, i.e., dividing the two sides of (4.72) by $\widehat{S}(\mathbf{u})$, yielding,

$$\widehat{O}(\mathbf{u}) = \frac{\widehat{I}(\mathbf{u})}{\widehat{S}(\mathbf{u})} - \frac{\widehat{N}(\mathbf{u})}{\widehat{S}(\mathbf{u})}. \tag{4.73}$$

The estimated visibility is,

$$\widehat{O}'(\mathbf{u}) = \frac{\widehat{I}(\mathbf{u})}{\widehat{S}(\mathbf{u})}. \tag{4.74}$$

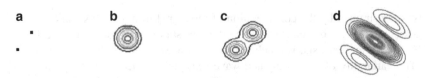

Fig. 4.8 (a) Simulated binary system, (b) Gaussian PSF, (c) convolved binary with the PSF, and (d) power spectrum of (c)

The estimated image $O'(\mathbf{x})$ is the inverse FT of $\widehat{O}'(\mathbf{u})$. However, the direct inversion method has a few drawbacks such as: (1) the transfer function, $\widehat{S}(\mathbf{u})$, has values of zero or near zero; at these points the division operation is not defined and (2) for the points having very small $|\widehat{S}(\mathbf{u})|$, the ignored noise would be amplify to an intolerable extent.

The speckle facility, after the end of each run, provides the accumulated power spectrum of the collected speckle frames. Figure 4.8 shows a simulated binary system convolved with a Gaussian PSF and its power spectrum. Since $|\widehat{S}(\mathbf{u})|^2$ is a random function in which the detail is continuously changing, the ensemble average of this term becomes smoother. The smooth function can be performed on a point source, obtained from observations of a nearby star, yields $\langle|\widehat{S}(\mathbf{u})|^2\rangle$. The averaged power spectrum of the object star is divided by this averaged power spectrum reference star canceling out in this way the contribution of the atmospheric turbulence affecting the observation. This image power spectrum is also needed both for removing some features caused by the possible repetitive noise induced on the camera signal and to eliminate a typical cross-shaped disturbance occurring when the speckle image of the object is not entirely contained in the camera field-of-view (FOV). The ensemble average of the Wiener spectrum is found by writing,

$$\langle|\widehat{I}(\mathbf{u})|^2\rangle = |\widehat{O}(\mathbf{u})|^2.\langle|\widehat{S}(\mathbf{u})|^2\rangle + \langle|\widehat{N}(\mathbf{u})|^2\rangle. \tag{4.75}$$

The power spectrum of this ratio is inverted via fast Fourier Transform (FFT), and the autocorrelation function of the brightness distribution of the astronomical target is obtained. By the Wiener–Khintchine theorem (see Appendix B.1.1), the inverse FT of (4.75) provides the autocorrelation of the object, $\mathcal{A}[O(\mathbf{x})] = \mathcal{F}^-[|\widehat{O}(\mathbf{u})|^2]$.

The transfer function of $S(\mathbf{x})$, is generally estimated by calculating the power spectrum of the instantaneous intensity from a nearby point source star. The autocorrelation function of a typical image generally decreases away from the origin. Since the power spectrum of an image is the Fourier transform of its autocorrelation function, the power spectrum generally decreases with frequency. Typical noise sources have a flat power spectrum or one that decreases with frequency more sluggishly than typical image power spectra. Thus the expected situation is for the signal to dominate the spectrum at low frequencies and the noise to dominate at high frequencies. Since the magnitude of the deconvolution filter increases with frequency, in general, it enhances high frequency noise. In order to get rid of such

Fig. 4.9 Autocorrelation of specklegrams of HR 4689 depicting signature of binary component (Saha and Maitra 2001). The axes of the figure are the pixel values; each pixel value is 0.015″. The central contours represents the primary component. One of the two contours on either side of the central one displays the secondary component. The contours at the corners are the artifacts

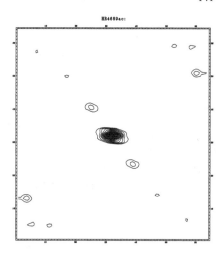

a high frequency noise as much as possible, Saha and Maitra (2001) developed an algorithm, where a Wiener parameter, w_1, is added to PSF power spectrum i.e.,

$$|\widehat{O}(\mathbf{u})|^2 = \frac{\langle|\widehat{I}(\mathbf{u})|^2\rangle}{\left[\langle|\widehat{S}(\mathbf{u})|^2\rangle + w_1\right]}. \tag{4.76}$$

The notable advantage of such an algorithm is that the object can be reconstructed with a few frames. In this process, a Wiener filter is employed in the frequency domain. The technique of Wiener filtering damps the high frequencies and minimizes the mean square error between each estimate and the true spectrum.

In the case of a binary system, the autocorrelation function shows the characteristic behavior of a central peak with two opposite and symmetric secondary peaks (see Fig. 4.9). The distance between the central peak and one of the secondary ones is the separation between the two components while the position angle is given by the orientation of the secondary peak with uncertainty. However, the speckle interferometric technique in such a case retrieves the separation, position angle with 180° ambiguity, and the relative magnitude difference at low light levels.

4.2.2 Benefit of Short-exposure Images

In the long-exposure images, the image is spread during the exposure by its random variations of the tilt. The image sharpness and the MTF are affected by the wave-front tilt, as well as by the more complex shapes, while in the case of a short-exposure image, the image sharpness and MTF are insensitive to the tilt. The random factor associated with the tilt is extracted from the MTF before taking the average, where in the long-exposure case, no such factor is removed (see Sect. 4.1.6.1).

Consider a pair of seeing cells separated by a vector $\lambda \mathbf{u}$, in the telescope pupil. When a point source is imaged through the telescope consisting of two apertures (θ_1, θ_2), which can be considered as imaging by two pupil functions corresponding to two seeing cells, a fringe pattern is produced with narrow spatial frequency bandwidth. The major component $\widehat{I}(\mathbf{u})$, at the frequency, $\mathbf{u}$, is produced by contributions from all pairs of points with separations $\lambda \mathbf{u}$, with one point in each aperture. If the major component is averaged over many frames, the resultant for frequencies greater than r_o/λ, tends to zero since the phase-difference, $\theta_1 - \theta_2$; mod 2π, between the two apertures is distributed uniformly between $\pm\pi$, with two mean. Averaging the component, $\widehat{I}(\mathbf{u})$, over many frames as in the case of long-exposure image, the Fourier component performs a random walk in the complex plane and average to zero, i.e.,

$$\langle \widehat{I}(\mathbf{u}) \rangle = 0, \qquad u > r_0/\lambda. \tag{4.77}$$

The argument of (4.53) is expressed as,

$$arg\,|\widehat{I}(u)| = \psi(u) + \theta_1 - \theta_2, \tag{4.78}$$

where $\psi(\mathbf{u})$ is the Fourier phase at $\mathbf{u}$ and $arg|\ |$ stands for, 'the phase of'.

While in the case of autocorrelation technique, the autocorrelation of $\mathcal{I}(\mathbf{x})$ is the correlation of $\mathcal{I}(\mathbf{x})$ and $\mathcal{I}(\mathbf{x})$ multiplied by the complex exponential factor with zero spatial frequency. The major Fourier component of the fringe pattern is averaged as a product with its complex conjugate and so the atmospheric phase contribution is eliminated and the averaged signal is non-zero, i.e.,

$$\langle |\widehat{I}(\mathbf{u})|^2 \rangle \neq 0. \tag{4.79}$$

The argument of (4.75) is given by the expression,

$$\begin{aligned} arg\,|\widehat{I}(\mathbf{u})|^2 &= \psi(\mathbf{u}) + \theta_1 - \theta_2 + \psi(-\mathbf{u}) - \theta_1 + \theta_2 \\ &= 0. \end{aligned} \tag{4.80}$$

where $\psi(\mathbf{u})$ is the Fourier phase at $\mathbf{u}$ and $\theta_{j=1,2}$ the apertures, corresponding to the seeing cells.

4.3 Adaptive Optics

Yet another method of obtaining diffraction-limited imaging is that of using adaptive optics (AO) system. The technology has evolved over years due to contribution of numerous scientist and engineers. Babcock (1953) proposed a scheme to correct the rapidly changing atmospheric seeing effects, however, it could not be implemented until recent years due to the lack of technological advancement. An adaptive optics system adapts to compensate for optical effects introduced by the medium between the object and its image. This system removes temporally and spatially varying optical wavefront aberrations by introducing an optical imaging system in the light path, which performs two main functions: (1) sense the wavefront perturbations and (2) compensate for them in real time (Roggemann et al. 1997). For an unresolved source, adaptive optics attempts to put as many photons in as small an image area as possible, thus enhancing the image contrast against the sky background thereby improving the resolution, and allowing better interferometric imaging with telescope array. For a resolved source, the improved resolution extends imaging to fainter and more complex objects. However, most of the results obtained thus far are in the near-IR band; results at visible wavelengths continue to be sparse. In what follows, a succinct description of AO system is given.

4.3.1 Atmospheric Compensation

Apart from the defined atmospheric parameters (see Sect. 4.1), the need of a few more parameters is required and characterized accurately (Tyson 1991). Strehl's ratio, S_r, (see Sect. 1.5.2) quantifies the performance of an AO system. It is an exponential of the wavefront variance, $\langle\sigma\rangle^2$, (see 1.93), which is the sum of all the wavefront error variances, i.e.,

$$\langle\sigma\rangle^2 = \langle\sigma_F\rangle^2 + \langle\sigma_D\rangle^2 + \langle\sigma_P\rangle^2 + \langle\sigma_{\theta_0}\rangle^2, \tag{4.81}$$

where $\langle\sigma_F\rangle^2$ is the deformable mirror fitting error variance, $\langle\sigma_D\rangle^2$ the detection error variance, $\langle\sigma_P\rangle^2$ the prediction error variance, and $\langle\sigma_{\theta_0}\rangle^2$ the variance that occurs from aniso-planatism.

The main sources for errors in an AO systems are wavefront fitting errors, $\langle\sigma_F\rangle$, which depends on how closely the wavefront corrector matches the detection error, $\langle\sigma_D\rangle$, and the prediction error, $\langle\sigma_P\rangle$. The detection error is the reciprocal to the S/N ratio of the wavefront sensor (see Sect. 4.4.2) and the prediction error is due to the time delay between the measurement of the wavefront disturbances and their correction.

The wavefront fitting error is described by,

$$\langle\sigma_F\rangle^2 = k\left(\frac{d}{r_0}\right)^{5/3}, \tag{4.82}$$

where k is a coefficient that depends on influence functions of deformable mirror and on the geometry of the actuators.

If a plane wave is fitted to the wavefront over a circular area of diameter, d, and its phase is subtracted from the wavefront phase (tip-tilt removal), the mean square phase distortion reduces to,

$$\langle\sigma_{\psi'}\rangle^2 = 0.134\left(\frac{d}{r_0}\right)^{5/3}. \tag{4.83}$$

The prediction error is due to the time delay between the measurement of the wavefront disturbances and their correction. The variance due to the time delay depends on the mean propagation velocity, v_w, i.e.,

$$\langle \sigma_\tau \rangle^2 = 6.88 \left(\frac{v_w \tau}{r_0} \right)^{5/3}. \tag{4.84}$$

This (4.84) shows that the time delay error, $\langle \sigma_\tau \rangle^2$ depends on two parameters, viz., v_w and r_0 which vary with time independent of each other. The detector noise that may be in the form of photon noise as well as read noise can also deteriorate the performance of correction system for low light level. In a system that consists of a segmented mirror controlled by a Shack–Hartmann sensor (see Sect. 4.4.2.2), let θ'' be the width of a subimage. Assume that each sub-aperture is larger than Fried's parameter, r_0, each sub-image is blurred with angular size, $\theta'' \simeq \lambda / r_0$, hence the variance, $\langle \sigma_\delta \rangle^2$, can be derived as,

$$\langle \sigma_\delta \rangle^2 = \frac{\lambda^2}{n_p r_0^2}, \tag{4.85}$$

where n_p is the number of photons.

From (4.83), the fitting error for a segmented mirror in terms of optical path fluctuations, one may express,

$$\langle \sigma_F \rangle^2 = 0.134 \frac{1}{\kappa^2} \left(\frac{d}{r_0} \right)^{5/3}, \tag{4.86}$$

in which $\kappa = 2\pi/\lambda$ is the wavenumber and d the spot size.

4.3.1.1 Iso-planatic Angle

The angular distance (from a reference star) over which atmospheric turbulence is unchanged is known as iso-planatic angle, θ_0. Aniso-planaticity occurs when the volume of turbulence experienced by the reference object differs from that experienced by the target object of interest, and therefore experience different phase variations. This yields in non-linear degradations causing a severe problem in compensating seeing, both for post-processing imaging technique like speckle interferometry (see Sect. 4.2), as well as for adaptive optics system. There are four main classifications of aniso-planatism, such as angular, temporal, displacement, and chromatic aniso-planatism. For a given distance, θ, between the target of interest and the guide star, the wavefront variance is deduced as,

$$\langle \sigma_\theta \rangle^2 = 2.914 \kappa^2 \sec \gamma \int_{-\infty}^{\infty} C_n^2(h)(\theta h \sec \gamma)^{5/3} \delta h$$

$$= \left(\frac{\theta}{\theta_0} \right)^{5/3} \quad \text{rad}^2, \tag{4.87}$$

where the iso-planatic angle, θ_0 is defined as the angular separation between the object and the reference object that results in a mean squared phase error of 1 rad^2,

$$\theta_0 \simeq \left[2.914\kappa^2(\sec\gamma)^{8/3} \int_{-\infty}^{\infty} C_n^2(h)h^{5/3}\delta h \right]^{-3/5}, \qquad (4.88)$$

and $h \sec \gamma$ the distance at zenith angle γ with respect to the height h.

This (4.88) is analogous to Fried's parameter, r_0 (see 4.42), which provides,

$$\theta_0 = (6.88)^{-3/5} \frac{r_0}{h \sec \gamma} = 0.314 \frac{r_0}{h \sec \gamma}. \qquad (4.89)$$

This parameter limits the distance between guide star and the celestial object of interest. The variance error due to aniso-planaticity, $\langle\sigma_{\theta_0}\rangle^2$, is expressed as,

$$\langle\sigma_{\theta_0}\rangle^2 = 6.88 \left(\frac{\theta \bar{h} \sec \gamma}{r_0} \right)^{5/3}. \qquad (4.90)$$

4.3.1.2 Greenwood Frequency

Greenwood frequency, f_G, characterizes the temporal dynamics of phase fluctuations resulting from the pupil-plane phase distorting layer moving at a certain (wind) velocity. Since the wind velocity dictates the speed, the dynamic behavior of atmospheric turbulence is necessary to be corrected. Two parameters such as the gain and the bandwidth are adjusted according to the number of Zernike modes (see Appendix D). Greenwood (1977) determined the required bandwidth, f_G (the Greenwood frequency), for full correction by assuming a system in which the static case corrects the wavefront and the remaining aberrations were due to finite bandwidth of the control system. He derived the mean square residual wavefront error as a function of servo-loop bandwidth for a first order controller, which is given by,

$$\langle\sigma_{cl}\rangle^2 = \left(\frac{f_G}{f_c} \right)^{5/3} \quad \text{rad}^2, \qquad (4.91)$$

where f_c is the frequency at which the variance of the residual wavefront error is half the variance of the input wavefront, known as 3 db closed-loop bandwidth of the wavefront compensator, and f_G the required bandwidth.

For a single turbulent layer, f_G is defined by the relation,

$$f_G = \frac{0.426v_w}{r_0}, \qquad (4.92)$$

with v_w as the velocity of the wind in meters/sec (m/s). It must be noted that the required bandwidth for adaptive optics does not depend on height, but instead is proportional to v_w/r_0, which is proportional to $\lambda^{-6/5}$. If the turbulent layer moves at a speed of 10 m/s, the closed loop bandwidth for $r_0 = 11\ cm$, in the optical band (550 nm) is around 39 Hz.

4.3.1.3 Image Jitter

In comparison with the characteristic time of small-scale perturbations, the movement of large-scale perturbations in the turbulent atmosphere is slower. It is reiterated that the tilt of the corresponding atmospheric layers produces a deviation of all incident rays. Due to refractive transmission of the beam by eddies of sizes larger than the beam diameter, the wavefront tilt takes place. The dynamic tilt or jitter is generally expressed as an angular variance or as its square root, i.e., the root-mean-square deviation of an angle; the dynamics are expressed in terms of a power-spectral density. The first order tilt aberration of a beam is expressed as,

$$\langle \sigma_j \rangle^2 = 1.83 C_n^2 \lambda^{-1/6} \bar{h}^{17/6}. \tag{4.93}$$

The introduction of such a jitter into an imaging system reduces its performance. Frequency of wander is approximately $1/10^{th}$ of f_G (see 4.92). Without affecting the quality, tilt displaces position of the image. In a large mirror telescope, small local deviations from perfect curvature are manifested as high spatial frequency aberrations. The variance of the wavefront tilt over an aperture of diameter, D, is given by,

$$\langle \sigma_{tilt} \rangle^2 = 0.364 \left(\frac{D}{r_0} \right)^{5/3} \left(\frac{\lambda}{D} \right)^2 \qquad \text{rad}^2. \tag{4.94}$$

By defining seeing-disk, θ_s, as the FWHM of a Gaussian function fitted to a histogram of image position in arcsec one finds,

$$r_0 = 1.009 D \left(\frac{\lambda}{\theta_s D} \right)^{6/5}. \tag{4.95}$$

By comparing the expression for r_0 (4.42) with this expression, one may find that the seeing-disk is independent of wavelength. For an aperture of the order of 1 m (4.95) is approximately,

$$r_0 \sim \left(\frac{\lambda}{\theta_s} \right)^{6/5}. \tag{4.96}$$

The (4.96) states that Fried's parameter, r_0 increases with wavelength as the six-fifth power, implying that the width of seeing-limited images, $1.22\lambda/r_0 \propto \lambda^{-1/5}$ varies with λ. With the increase of wavelength, r_0 increases; therefore, the effect of atmospheric turbulence is more severe for ultra-violet light and less so in the infra-red (IR). It can derived that a r_0 value of 10 cm at 550 nm corresponds to 53 cm at 2.2 μm, while at centimetric radio wavelengths, it is of the order of 30 km.

4.3.1.4 Thermal Blooming

The thermal blooming effect, represented by the blooming distortion number N_B, is caused by the resonant absorption of the high energy laser beam with atmospheric molecules. This is due to the non-linear response of the atmosphere. There is a critical power, which can be transmitted through the atmosphere for which this effect does not occur. This effect essentially needs to be compensated during the propagation of the high power laser beams through the atmosphere. An adaptive optics (AO) system partially corrects this effect. For a zero wind velocity case, thermal blooming appears, since the lowest index of refraction occurs near the center of the beam. This atmospheric negative lens causes the beam to defocus. An artificial disturbance in the wind due to the wind slewing causes the beam to take on a characteristic crescent shaped pattern. A variance contribution from the non-linear effects of thermal blooming, $\langle \sigma_{bl} \rangle^2$, is a function of the blooming strength, N_b, given by,

$$N_b = \frac{-2.94 P_b \alpha h^2}{\pi a^3 n \rho v C_P} \left(\frac{dn}{dT} \right),$$ (4.97)

in which α is the linear absorption coefficient, h the propagation path, a the radius of the beam, P_b the beam power, C_P the specific heat at constant pressure, ρ the density of air, dn/dT the change of refractive index with temperature, and n the refractive index of the medium.

The variance contribution from the non-linear effects of thermal blooming, $\langle \sigma_{bl} \rangle^2$, that is a function of the blooming strength, N_b, and the number of modes corrected, N_{mod}, should be taken into account as well. The approximation is given by,

$$\langle \sigma_{bl} \rangle^2 = \frac{\sqrt{2}}{5\pi^4} \frac{N_b^2}{N_{mod}^{2.5}}.$$ (4.98)

4.4 Required Components for an AO System

An adaptive optics system is based on the hardware-oriented approach, where a Cassegrain type telescope is normally used. The other required components are: (1) image stabilization devices that are the combination of deformable reflecting surfaces, (2) wavefront sensor, (3) wavefront phase error computation, and (4) post-detection image restoration. In addition, a laser guide star may also be needed to improve the signal-to-noise (S/N) ratio for the wavefront signal since the natural guide stars are not always available within iso-planatic patch.

Figure 4.10 illustrates a schematic of an adaptive optics system. In order to remove the low frequency tilt error, generally the incoming collimated beam is fed by a tip-tilt mirror. After traveling further, it reflects off of a deformable mirror that eliminates high frequency wavefront errors. A beam-splitter divides the beam into two parts: one is directed to the wavefront sensor to measure the residual error in the wavefront and to provide information to the actuator control computer to compute the deformable mirror voltages; the other is focused to form an image.

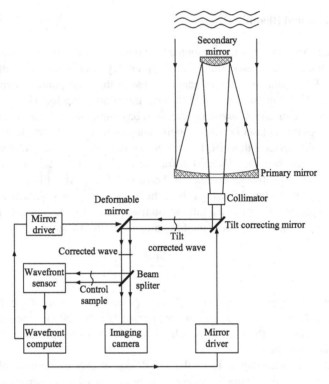

Fig. 4.10 Schematic of an adaptive optics system

4.4.1 Wavefront Correcting Systems

Wavefront corrector is used to compensate for any aberrations due to atmospheric or due to imperfections in the telescope optics. It is an electro-opto-mechanical device, which is driven by a voltage that depends on the wavefront and the characteristics of the corrector. The physical movement of an actuator does not need to be proportional to the applied voltage.

4.4.1.1 Tip-tilt Mirror

Tip-tilt mirrors are effectively used in active and adaptive optics for various dynamic applications including precision scanning, tracking, pointing, and laser beam or image stabilization. They are used in solar and stellar imaging, and are useful for long baseline optical interferometry. A tip-tilt mirror may be tilted fast about its axis of the spring/mass system, in order to direct a image in x, y-plane. It corrects the mean tilt of the wavefront across the telescope aperture, which allows operation of a corrector element with a smaller stroke at the expense of loss of light in the additional reflection (Roddier 1999). This is equivalent to taking out the mean of image motions.

Implementation of the dynamically controlled active optical components consisting of a tip-tilt mirror system in conjunction with closed-loop control electronics has several advantages such as (1) conceptually the system is simple, (2) fainter guide stars increase the sky coverage, and (3) the FOV is wider. However, tip-tilt mirrors generally cannot correct rapid fluctuations in the wavefront, albeit provide medium or low bandwidth pointing in real time. These systems are limited to two Zernike modes (x and y-tilt), while a higher order system compensating many Zernike mode (see Appendix D) is required to remove high frequency errors.

4.4.1.2 Deformable Mirror

The residual phase errors are corrected by the use of a deformable mirror, a very thin flexible mirror, whose shape is changed by a force applied by actuators stacked behind it to create a conjugate surface. When some voltage, V_i, is applied to the ith actuator, the shape of deformable mirror is described by its influence function,[7] $r_i(x, y)$, multiplied by V_i. It resembles a bell-shaped (or Gaussian) function for deformable mirrors with continuous face-sheet (there is some cross-talk between the actuators, typically 15%). When all actuators are driven, the shape of the deformable mirror is equal to,

$$D(x, y) = \sum_i V_i D_i(x, y). \tag{4.99}$$

The primary parameters of deformable mirror based AO system are the number of actuators, the control bandwidth and the maximum actuator stroke. The response time, given by (4.48), is of the order of a few milliseconds. Although, it is relatively better at longer wavelengths, the 3-db bandwidth need not be that high; typically 100 Hz is enough for the visible wavelengths regime.

The wavefronts must be measured and corrected in a time less than the atmospheric coherence time, τ_0. The number of degrees of freedom needed for the deformable mirror, and hence the number of measurements which are required to be made in each interval, increases with the size of the telescope aperture. The number of actuators, N on the deformable mirror determines the number of degrees of freedom (wavefront inflections) the mirror can correct in an AO system, which is given by the following equation,

$$N = \left(\frac{D}{r_0}\right)^2 \propto \lambda^{-12/5}, \tag{4.100}$$

with D as the diameter of the telescope aperture; the required stroke is a few microns.

---

[7] Influence function of a deformable mirror is the characteristic shape corresponding to the mirror response to the action of one actuator (like a Greens function).

Several types of deformable mirrors based on the following concepts have been developed:

1. Segmented mirror: It consists of array of single tip-tilt elements that can be moved independent of each other and can be replaced easily. The motion of the individual mirrors is restricted to piston and tilt. But the disadvantages are the diffraction effects due to the discontinuity of such a mirror at the edges of the elements and the difficulty in achieving intersegment alignment. The gap between the segments may be the source of radiation in infrared wave band, which deteriorates the image quality. The actuators are normally push-pull type. The present generation piezoelectric actuators are no longer discrete, but ferroelectric wafers are bonded together and treated to isolate the different actuators.

2. Bimorph mirror: It consists of two piezoelectric wafers which are bonded together and oppositely polarized, parallel to their axes. A 50-mm bimorph deformable mirror and its measured influence function can be seen in Fig. 4.11.
   An array of electrodes is deposited between the two wafers. When a voltage is applied to an electrode, one layer expands in the $\mathbf{x}(= x, y)$ plane and contracts in the z direction, while the second element does the opposite. The changes in the $\mathbf{x}$ plane produce a local bending of the surface much like that of a bimetal strip in a thermostat; these deformable mirrors are called curvature mirrors. The local curvature, $\nabla^2\phi$, is proportional to the applied voltage, V;

$$\nabla^2\phi(\mathbf{x}) \propto \frac{-V(\mathbf{x})}{t^2}, \qquad (4.101)$$

with t as the total thickness of the bimorph mirror.

3. Membrane mirror: This mirror consists of a thin conductive and reflective membrane stretched over an array of electrostatic actuators. Applying voltages to these actuators, individual responses superimpose to form the necessary optical figure, can locally deflect electrically grained conductive membrane. Figure 4.12

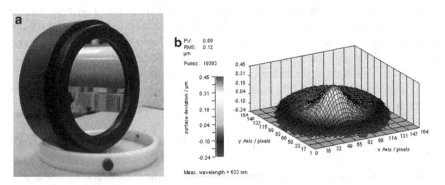

Fig. 4.11 (a) Deformable mirror of bimorph type and (b) its measured influence function of a 80 actuator mirror with keystone geometry; a single actuator is poked with a voltage of 150 V. The resulting peak has a height of 0.69 μm (Soltau 2009; courtesy: D. Soltau)

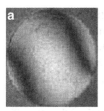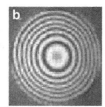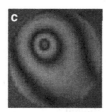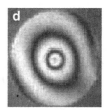

Fig. 4.12 Test patterns of micro-machined deformable mirror; (**a**) in absence of any voltage to its actuators depicting astigmatism (an off-axis point wavefront aberration, caused by the inclination of incident wavefronts relative to the optical surface) shape, (**b**) with equal voltage to all actuators showing spherical shape, (**c**) voltage is applied to one of the adjacent actuator to the central actuator displaying comatic shape, and (**d**) central actuator is applied voltage, depicting defocus shape Mohan et al. (2005). During testing and characterization of the mirror, voltages are applied to single and multiple actuators of the deformable mirror

depicts the test patterns of micro-machined deformable mirror at different conditions. The deflection of the membrane is given by,

$$\nabla^2 Z(\mathbf{x}) = \frac{-P(\mathbf{x})}{T_m(\mathbf{x})},$$ (4.102)

where $Z(\mathbf{x})$ is the deformation of the membrane, $P(\mathbf{x})$ the external pressure at position $\mathbf{x}$, and $T_m(\mathbf{x})$ the membrane tension; the effect of applying a local pressure is to change the local curvature of the membrane.

Wavefront correction can also be achieved using liquid crystal[8] deformable mirrors because they can be made into closely packed arrays of pixels, which may be controlled with low voltages.

4.4.1.3   Adaptive Secondary Mirror at MMT

Another way to correct the disturbance in real time is usage of adaptive secondary mirror that makes a compact system with high transmission (Bruns et al. 1997). Such a system may be suitable for infrared observation. The other notable advantages are: (1) enhanced photon throughput that measures the proportion of light which is transmitted through an optical set-up, (2) introduction of negligible extra IR emissivity, (3) causes no extra polarization, and (4) non-addition of reflective losses (Lee et al. 2000). Due to the interactuator spacing, the resonant frequency of such a mirror may be lower than the AO bandwidth.

---

[8] Liquid crystals are to a state of matter intermediate between solid and liquid and are classified in nematic and smectic crystals. The fundamental optical property of these crystals is their birefringence. They are suitable for high spatial resolution compensation of slowly evolving wavefronts such as instrument aberrations in the active optics systems.

An $f/15$ AO secondary with 336 actuators is installed on the 6.5 m (mono-lithic mirror) telescope of Multi Mirror Telescope (MMT) observatory, Mt. Hopkins, Arizona. It is a 642-mm diameter ultra-thin (2 mm thick) secondary mirror operating at 560 Hz. The actuators are basically like acoustical voice coils used in stereo systems. There is a 40-μm air space (with 16 μm stroke) between each actuator and a magnet that is glued to the back surface of the ultra-thin secondary mirror. The viscosity of the air is sufficient to damp out any unwanted secondary harmonics or other vibrations (Wehinger 2004).

The MMT adaptive secondary mirror system employs a Shack–Hartmann sensor (see Sect. 4.4.2.2), which adds two extra refractive surfaces to the wavefront sensor beam (Lloyd-Hart 2000). The corrected beam is relayed directly to the infrared science instrument, with a dichroic beam splitter passing light beyond 1 μm waveband and reflecting visible light back into the wavefront sensing and acquisition cameras. Owing to very low emissivity of the system, the design of this system is optimized for imaging and spectroscopic observations in the 3–5-μm band. The Large Binocular Telescope (LBT) secondaries having 672 actuators on each of the adaptive secondary mirror that has a diameter of 911 mm and a thickness of 1.6 mm with the operating gap of 100 μm, whose shape can be controlled by voice coil. The unit would observe the light from a target star and the wavefront from a close reference star a pyramid sensor (see Sect. 4.4.2.4). The signals from this sensor would permit to close a servo-loop with the LBT secondary mirror. Based on the 1,170 actuators a deformable secondary mirror for ESO-VLT UT-4 Telescope (see Sect. 7.3.6) is also being designed.

4.4.2 Wavefront Sensors

A wavefront sensor consists of front-end optics module and a processor module equipped with a data acquisition, storage, and sophisticated wavefront analysis programs. It should, in principle, be fast and linear over the full range of atmospheric distortions. It estimates the overall shape of the phase-front from a finite number of discrete measurements that are, in general, made at uniform spatial intervals. The phase of the wavefront does not interact with any measurable way, hence the wavefront is deduced from intensity measurements at one or more planes. The algorithms to unwrap the phase and to remove this ambiguity are also slow. The performance of a wavefront sensor depends on the effective measurement of the errors in the incoming wavefront. In what follows, a few wavefront sensors based on interferometric and geometrical optics concepts are enumerated in brief.

4.4.2.1 Lateral Shearing Interferometer

A lateral shearing interferometer can be used to measure wavefront phase through intensity measurements (Sandler et al. 1994) for AO applications. This can be

Fig. 4.13 Example of shear
in lateral shearing
interferometer

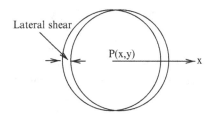

located at the pupil of the telescope, coinciding with the primary mirror. However, an interferometer of this kind has an intrinsic phase ambiguity of 2π, while phase distortions due to the atmosphere exceed typically 2π; therefore the wavefront sensors must be fast and linear over the full range of atmospheric distortions. A shearing device splits the incoming wavefront and shifts it laterally (see Fig. 4.13) and mixing it with itself, thereby, obtains the interference patterns which correspond to the wavefront tilt in the shear direction. These wavefronts are mutually displaced by a distance, s, called shear. They interfere in their overlap area. The interference fringes are a measurement of the phase difference over the shear distance. Placing the detector in a pupil image plane, the resultant intensity is given by,

$$I(\mathbf{r}) = \frac{1}{2} \left| e^{i\psi(\mathbf{r})} + e^{i\psi(\mathbf{r}+\mathbf{s})} \right|^2$$
$$= 1 + 1 \cos[\psi(\mathbf{r}) - \psi(\mathbf{r}+\mathbf{s})]$$
$$\approx 1 + 1 \cos\left[s \frac{\partial\psi(\mathbf{r})}{\partial s} \right], \tag{4.103}$$

where the phase in the cosine term is directly proportional to the slope of the wavefront at the measured position, which is expressed as,

$$\psi(\mathbf{r}) = \frac{2\pi}{\lambda}\delta(\mathbf{r}), \tag{4.104}$$

and $\delta(\mathbf{r})$ is the optical path difference (OPD) induced by the atmospheric turbulence.

If the shear distance is reduced and the wavefront deformation is small, the phase difference may be expanded by a Taylor series. For a shear along the x- direction, one may get

$$\psi(\mathbf{r}) - \psi(\mathbf{r}+\mathbf{s}) = s\frac{\partial\psi}{\partial x}(\mathbf{r}) + O(s^2), \tag{4.105}$$

where $O(s^2)$ denotes terms of order s^2 or higher of the expansion, which may be neglected in first approximation in signal analysis.

For compensating atmospheric turbulence the (4.105) needs a shear less than the size of the atmospheric coherence length, r_0. Further, the removal of 2π ambiguity from (4.104), a proper choice of s is required.

4.4.2.2 Shack–Hartmann Wavefront Sensor

Shack–Hartmann sensor measures the shape of the wavefront by a two dimensional lenslet array positioned in a plane conjugate to the telescope pupil. The incident wavefront is sampled into small sub-apertures by this array producing the Fraunhofer diffraction pattern (see Sect. 1.4.2), which corresponds to the wavefront sampling on to the detector plane. The sub-aperture size (typically of r_0 diameter) determines the spatial resolution of the sensor. A relay lens re-images these arrays of focal spots onto a high frame rate CCD camera (see Sect. 6.2). Shack–Hartmann wavefront sensor requires a reference plane wave generated from a reference source, in order to calibrate precisely the focus positions of the lenslet array. Due to aberrations, light rays are deviated from their ideal direction, producing image displacements (see Fig. 4.14). The positional offset of these images is used to derive mean gradient of the wavefront phase over each sub-aperture.

The centroid (center of gravity) displacement of each of these subimages provides an estimate of the average wavefront gradient over the subaperture. The position of the centroid of the focal spot is directly related to the phase mean slope over the sub-aperture. The centroid displacement of each of these subimages gives an estimate of the full vectorial wavefront tilt in the areas of the pupil covered by each lenslet. The centroid or the first order moments, C_x, C_y, of the image intensity with respect to x- and y-axes are given by,

$$C_x = \frac{\sum_{i,j} x_{i,j} I_{i,j}}{\sum_{i,j} I_{i,j}}, \qquad C_y = \frac{\sum_{i,j} y_{i,j} I_{i,j}}{\sum_{i,j} I_{i,j}}, \qquad (4.106)$$

in which $I_{i,j}$ are the image intensities on the detector pixels and $x_{i,j}$, $y_{i,j}$ the position coordinates of the positions of the CCD pixels, (i, j).

The sum is made over all the CCD pixels in the lenslet field and normalized over the signal. Because of the normalization by $\sum_{i,j} I_{i,j}$, the Shack–Hartmann sensor is

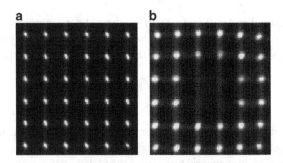

Fig. 4.14 Intensity distribution at the focal plane of a 6×6 lenslet array captured by an EMCCD camera: (**a**) for an ideal case at the laboratory and (**b**) an aberrated wavefront taken through a Cassegrain telescope (courtesy: V. Chinnappan)

insensitive to scintillation. Hence, a set of local slopes of the wavefront can be measured. Modal wavefront estimation using Zernike polynomials (see Appendix D) is one way in which the wavefront from the local slopes can be reconstructed. By integrating these measurements over the beam aperture, the wavefront or phase distribution of the beam can be determined. The intensity and phase information can be used in concert with information about other elements in the optical train to predict the beam size, shape, phase and other characteristics anywhere in the optical train. Moreover, it also provides the magnitude of various Zernike coefficients to quantify the different wavefront aberrations prevailing in the wavefront. The variance for the angle of arrival, $\alpha_x = C_x / fM$, in which M is the magnification between the lenslet plane and the telescope entrance plane, is given by,

$$\langle \sigma_x \rangle^2 = 0.17 \left(\frac{\lambda}{r_0} \right)^2 \left(\frac{D_{sa}}{r_0} \right)^{-1/3} \qquad \text{rad}^2, \qquad (4.107)$$

where D_{sa} is the diameter of the circular subaperture; (4.107) can be written for the y-direction as well.

However, the main drawback of the Shack–Hartmann sensor is a misalignment problem. Also, it suffers from low spatial frequency and the measurement noise arises from an uncertainty in the determination of centroid position of each spot and limited in terms of range and accuracy of measurement dictated by CCD performance and diffraction effects of the lenslet array. The wavefront estimation performance may also be limited by both the measurement technique and the noise propagation.

4.4.2.3 Curvature Wavefront Sensor

The curvature sensor, developed by Roddier (1988b), to make wavefront curvature measurements. It measures a signal proportional to the second derivative of the wavefront phase. The Lapacian of the wavefront, together with wavefront radial tilts at the aperture edge, are measured, providing data to reconstruct the wavefront by solving the Poisson equation with Neumann boundary conditions.[9] The intensity distribution in a defocused image of the reference star is measured (or the difference on either side of the focal plane as shown in Fig. 4.15), and this is proportional to $\nabla^2 \psi$. The displacement at the edge of the image provides the boundary condition needed to obtain ψ from $\nabla^2 \psi$. In geometrical approximation, the normalized intensity difference is related to the wavefront phase ψ in the pupil plane by,

$$
\begin{aligned}
C_n &= \frac{1_1(\mathbf{x}) - I_2(\mathbf{x})}{1_1(\mathbf{x}) + I_2(\mathbf{x})} \\
&= \frac{f(f-s)}{s} \left[\frac{\partial \psi(\rho, \theta)}{\partial \rho} \delta_c - \nabla^2 \psi(\rho, \theta) \right],
\end{aligned} \qquad (4.108)
$$

---

[9] Neumann boundary conditions specify the normal derivative of the function on a surface.

Fig. 4.15 Principle of a
curvature sensor

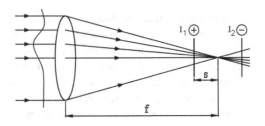

where ∇^2 denotes the 2-D Laplacian operator, (see 1.2), $I_1(\mathbf{x})$ and $I_2(\mathbf{x})$ the respective irradiance distribution of the light in the intra-focal (the one before the focus) and extra-focal images (the one after the focus), $\partial\psi/\partial\rho$ the first derivative of the wavefront, δ_c the linear impulse distribution around the pupil edge, f the focal length of a telescope, and s the distance between the focal point and the intra/extra-focal plane.

The computed sensor signals are multiplied by a control matrix to convert wavefront slopes to actuator control signals, the output of which are the increments to be applied to the control voltages on the deformable mirror. Subsequently, the Poisson equation is solved numerically and the first estimate of the aberrations is obtained by least squares fitting Zernike polynomials to the reconstructed wavefront. A conjugate shape is created using this data by controlling a deformable mirror, which typically compose of many actuators in a square or hexagonal array.

In curvature sensor, the choice of the distance s is very critical. Increasing this distance increases spatial resolution on the wavefront measurement albeit decreases sensitivity. On the contrary, a smaller distance yields a higher sensitivity to low order aberrations. The normalization of the difference in illumination makes the curvature sensor insensitive to scintillation. Such a sensor works well with incoherent white light, although there are articles which claim that curvature sensors work with the Sun as a target, no one has ever succeeded to do so (Berkefeld 2008). The main problem of a curvature sensor is the limited dynamic range in which it acts linear.

4.4.2.4 Pyramid Wavefront Sensor

Another wavefront sensor, called pyramid wavefront sensor, introduced by Ragazzoni (1996) for astronomical applications, consists of a four-faces optical glass pyramidal prism, which is placed with its vortex on the nominal focus of the optical system (see Fig. 4.16).

The four faces deflect the beam in four different directions, depending on which face of the prism gets the incoming ray. Using a relay lens located behind the pyramid, these four beams are then re-imaged onto a high resolution detector, obtaining four images of the telescope pupil. The intensity distributions in the $j(=1,2,3,4)$th pupil are represented by $I_j(\mathbf{x})$, in which $\mathbf{x} = x, y$ is the 2-D

Fig. 4.16 Pyramid wavefront sensor. A stands for light beam coming from telescope, B for Pyramid, and C for detector

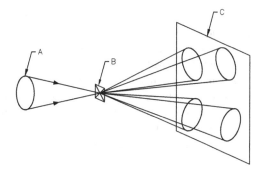

position vector. It is possible to retrieve the derivatives or slopes $\partial \psi(\mathbf{x})/\partial x$ and $\partial \psi(\mathbf{x})/\partial y$ of the wavefront along two orthogonal axes, i.e.,

$$\frac{\partial \psi(\mathbf{x})}{\partial x} = \delta \theta_x \frac{[I_1(\mathbf{x}) + I_4(\mathbf{x})] - [I_2(\mathbf{x}) + I_3(\mathbf{x})]}{\sum_{j=1}^{4} I_j(\mathbf{x})}, \qquad (4.109a)$$

$$\frac{\partial \psi(\mathbf{x})}{\partial y} = \delta \theta_y \frac{[I_1(\mathbf{x}) + I_2(\mathbf{x})] - [I_3(\mathbf{x}) + I_4(\mathbf{x})]}{\sum_{j=1}^{4} I_j(\mathbf{x})}, \qquad (4.109b)$$

where $\delta \theta_x$ and $\delta \theta_y$ are the modulation introduced in the x and y directions.

The pyramid sensor has been installed in the high order testbench at European Southern Observatory (ESO; Pinna et al. et al. 2008). One advantage of such a sensor is that of capability of changing the continuous gain and sampling, thus enabling a better match of the system performances with the actual conditions on the sky (Esposito and Riccardi 2001). However, the energy distribution of this sensor over the pupil is non-uniform. Most of the illumination is over the edges of the pupil leading to low S/N ratio. This adds complication in the closed loop AO system and procedure is time consuming.

4.4.3 Wavefront Reconstruction

The wavefront reconstructor receives the wavefront slope sensor output signals and calculate a phase estimate for each of the actuators of the deformable mirror. It measures any remaining deviations of the wavefront from ideal and sends the corresponding commands to the deformable mirror. The small imperfections of such a mirror like hysteresis or static aberrations are corrected automatically, together with atmospheric aberrations. The real-time computation of the wavefront error, as well as correction of wavefront distortion involves digital manipulation of wavefront sensor data in the wavefront sensor processor, the reconstructor and the low-pass filter, and converting to analog drive signals for the deformable-mirror actuators.

The functions that must be computed are: (1) sub-aperture gradients, (2) phases at the corners of each sub-aperture, (3) low-pass filter phases, and (4) to provide actuator offsets to compensate the fixed optical system errors and real-time actuator commands for wavefront corrections.

The measurements of wavefront sensor data may be represented by a vector, $\mathbf{S}$, whose length is twice the number of sub-apertures, n, for a Shack–Hartmann sensor because of measurement of slopes in two directions and is equal to n for curvature wavefront sensor. The unknowns (wavefront), a vector, ψ, specified as phase values on a grid, or more frequently, as Zernike coefficients[10] is given by,

$$\psi = \mathbf{B S}, \tag{4.110}$$

where $\mathbf{B}$ is the reconstruction or command matrix, $\mathbf{S}$ the error signal, and ψ the increment of commands which modifies slightly previous actuator state, known as closed-loop operation.

This kind of operation is generally being used in an adaptive optics system. The matrix equation between $\mathbf{S}$ and ψ can be read as,

$$\mathbf{S} = \mathbf{A}\psi, \tag{4.111}$$

in which $\mathbf{A}$, called the interaction matrix, is determined experimentally in an AO system.

The number of measurements is typically more than the unknowns, therefore the least-square solution is used, in which the wavefront phase, ψ, is estimated in order to minimize the error. The resulting reconstructor is recast as,

$$\mathbf{B} = \left(\mathbf{A}^t A\right)^{-1} \mathbf{A}^t, \tag{4.112}$$

with $\mathbf{A}^t$ as the transpose of $\mathbf{A}$.

4.4.4 Wavefront Controller

In an adaptive optics system, a wavefront control system is generally employed to correct beam aberrations. Such systems are broadly classified as open-loop systems and closed-loop systems. In the former case, the wavefront sensor measures the wave aberration before it is corrected by the deformable mirror, while in the case of the latter, the wavefront sensor measures the wavefront aberration after it is corrected by the deformable mirror. The adaptive optics system works in closed-loop systems, where parallel paths are employed whereby one channel controls lower order aberration modes, such as focus and tilt, while the other channel simultaneously

---

[10] Usually Karhunen Loeve modes that are a unique set of orthogonal functions having statistically independent weights, are used for wavefront reconstruction. These modes for the turbulence over a circular aperture can be expressed analytically in terms of Zernike polynomials (Noll 1976).

corrects the higher order wavefront errors with a deformable mirror. The functions of the controller may be described in two parts such as:

1. The wavefront is reconstructed from the input signals from wavefront sensor on the basis of the geometry of deformable mirror actuators. It estimates the control vector containing the control voltages, which are to be fed to the tip-tilt and deformable mirrors.
2. It involves the hardware using digital-to-analog converters, control voltages are generated and fed to the tip-tilt and deformable mirrors to enable the required corrections.

Although AO control systems often analyze and direct many parallel channels of information, most of the systems in reality are based on single channel linear processing algorithms. Advancement of digital computers permits the parallel operations to be performed at very high speed.

The error signal measured by the wavefront sensor as shown in Fig. 4.17 is given by,

$$e(t) = x(t) - y(t), \tag{4.113}$$

in which $x(t)$ is the input signal (e.g., a coefficient of some Zernike mode) and $y(t)$ the signal applied to the deformable mirror.

The error signal must be filtered before applying it to deformable mirror, or else the servo system would be unstable. In order to compute the reaction of the system in closed-loop, the control system in the frequency domain:

$$\widehat{y}(f) = \widehat{e}(f)\widehat{G}(f), \tag{4.114}$$

$$\widehat{e}(f) = \widehat{x}(f) - \widehat{y}(f) = \widehat{x}(f) - \widehat{e}(f)\widehat{G}(f), \tag{4.115}$$

where $\widehat{x}(f)$, $\widehat{y}(f)$, $\widehat{e}(f)$, are the Laplace transform[11] of the control system input, $x(t)$, output, $y(t)$, and the residual error, $e(t)$; in the frequency domain, $\widehat{G}(f)$ is known as open-loop transfer function of the control system.

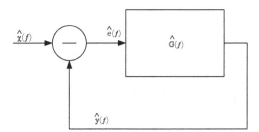

Fig. 4.17 Schematic diagram of the control system

---

[11] Laplace transform is an integral transform and is useful in solving linear ordinary differential equations. It simplifies the system analysis and is normally used because it maps linear differential equations to linear algebraic expression. Such a theorem has applications in mathematics, physics, optics, electrical and control engineering, and signal processing, etc.

Thus the transfer functions for the closed-loop error, χ_c, and for the closed-loop output, χ_o are deduced respectively as,

$$\chi_c = \frac{\widehat{e}(f)}{\widehat{x}(f)} = \frac{1}{1 + \widehat{G}(f)}, \tag{4.116a}$$

$$\chi_o = \frac{\widehat{y}(f)}{\widehat{x}(f)} = \frac{\widehat{G}(f)}{1 + \widehat{G}(f)}, \tag{4.116b}$$

By replacing $\widehat{G}(f)$ with g/f, in which $g(= 2\pi v_c)$ is the loop gain, v_c the 3-db closed-loop bandwidth of the control system, and $f = 2i\pi v$, the closed-loop error transfer function of the time frequency, v, one may write,

$$\widehat{E}(v) = \frac{iv}{v_c + iv}. \tag{4.117}$$

Thus the power spectrum of the residual error, $e(t)$ is derived as,

$$|\widehat{E}(v)|^2 = \frac{v^2}{v_c^2 + v^2}. \tag{4.118}$$

The response time to measure the wavefront signal by the wavefront sensor is represented by, $\widehat{G}(f)e^{-2i\pi\tau v}$, for a delay of time, τ. A certain time in computing the control signal is also required, since the response of the deformable mirror is not instantaneous due to its resonance and hysteresis. The transfer function, $\widehat{G}(f)$ accumulates additional phase delays with increasing frequency, hence the delay turns out to be larger than π. This implies that the servo system amplifies the errors. Such a system becomes unstable when the modulus of the closed loop transfer function exceeds 1.

4.4.5 Laser Guide Star

Implementation of adaptive optics system depends on the need for bright unresolved reference source, at reasonably close proximity in the sky, preferably within an iso-planatic patch for measuring the wavefront errors, as well as for mapping the phase on the entrance pupil. Thus, a notable drawback of AO system is that of non-availability of many such sources within the said patch to use for wavefront sensing. The problem becomes acute for observing obscuring source such as Young Stellar Objects (YSO) and dusty envelope stars where the visible source is often too faint.

A reference source may either be a component of the science target, e.g., the unresolved bright core of an active galaxy, or a non-variable distant star, referred to as natural guide star of known flux density, position, and polarization, with an angular size much less than the diffraction-limit of the telescope, or a fringe separation produced by an interferometer. The fluxes of such a star may change and should be monitored for absolute flux calibration. The natural guide star should be observed with the same instrumental set-up including detector integration time as for the target of interest. Many observatories are currently implementing the artificial laser guide star technique (Foy and Labeyrie 1985). The utility of the AO system is greatly enhanced by use of such a guide star source as reference. A laser would produce light from three reflections such as (1) resonance scattering by sodium (Na) in the Earth's mesosphere at an altitude of 90–105 km, (2) Rayleigh scattering from air molecules (oxygen and nitrogen) between 10 and 20 km altitude, and (3) Mie scattering from dust.

With a poor beam divergence quality laser, the telescope's primary mirror can be used as an element of the laser projection system, while with a diffraction-limited laser, projection system can be side-mounted and bore-sighted to the telescope (Tyson 2000). The beam is focused onto a narrow spot at high-altitude in the atmosphere in the direction of the scientific target. Light is scattered back to telescope from high altitude atmospheric turbulence, which is observed and used to estimate wavefront phase distortions. In principle, the laser guide star should descend through the same patch of turbulence as the target object.

The major advantages of a laser guide system are: (1) it can be put anywhere and (2) it is bright enough to measure the wavefront errors. Among the notable disadvantages are: (1) spread out of laser light by turbulence on the way up, (2) finite spot size ($0.5''$-$2''$), (3) increased measurement error of wavefront sensor, and (4) difficulties in developing such an artificial star with high power laser (laser beacon may be dangerous for aircraft pilots and satellites). There are other difficulties associated with laser guide stars too. Relatively, these stars are at a low height, due to which the path of light through the atmospheric turbulence from a laser guide star differs from that of a scientific object.

In order to produce backscatter light from Na atoms in the mesosphere, a laser is tuned to a specific atomic transition; the strongest laser beacon is from NaD_2 line at 589 nm. Higher altitude of sodium layer is closer to sampling the same atmospheric turbulence that a starlight from infinity comes through. Existence of layer in mesosphere containing alkali metals such as sodium (10^3–10^4 atoms cm^{-3}), potassium, calcium, permits one to use such a technique. The high altitude of the process makes it suitable for astronomical AO systems since it is closer to sampling the same atmospheric turbulence that a starlight from infinity comes through. However, the laser beacons from either of the Rayleigh scattering or of the sodium layer return to telescope are spherical wave, unlike the natural light where it is plane wave, hence some of the turbulence around the edges of the pupil is not sampled well.

Concerning the flux backscattered by a laser shot, Thompson and Gardner (1988) stressed the importance of investigating two basic problems: (1) the angular anisoplanatic effects and (2) the cone effect that arises due to the parallax between the remote astronomical source and artificial source. The problem due to the former occurs if an adaptive optics system uses natural guide stars to estimate the wavefront errors. The guide stars should be selected within the iso-planatic patch, θ_0, of the target. The mean square aniso-planaticity error, $\langle\sigma_{\theta_0}\rangle^2$, is given in (4.90).

However, the laser beacon, Rayleigh beacon in particular, suffers from the cone effect since it samples a cone of the atmosphere instead of a full cylinder of atmosphere since it is at finite altitude above the telescope, whereas the astronomical objects are at infinity. Although rays from the laser guide star and the astronomical source pass through the same area of the pupil, in reality they pass through a different patch of the atmosphere, and hence undergo different perturbations. A turbulent layer at altitude, h, is sampled differently by the laser and starlight. Due to this cone effect, the stellar wavefront may have a residual error although the laser beam is compensated by the AO system, i.e.,

$$\langle\sigma_c\rangle^2 = \left(\frac{D}{d_0}\right)^{5/3}, \tag{4.119}$$

with D as the diameter of the telescope aperture and d_0 the parameter characterizing the cone effect; the cone effect becomes severe with the increase of the telescope diameter.

4.4.6 Multi-conjugate Adaptive Optics

Due to severe iso-planatic patch limitations, a conventional AO system fails to correct the larger field-of-view (FOV), which can be achieved by employing Multi-Conjugate Adaptive Optics (MCAO) system (Beckers 1998). Such a system enhances the corrected FOV by reproducing the three dimensional (3-D) structure of the atmospheric phase aberrations in a series of altitude-conjugate deformable mirrors, generally conjugate to the most offending layers. Each deformable mirror is conjugated optically to a certain distance from the telescope. Several guide stars are used to reconstruct the 3-D atmospheric perturbations in the entire cylinder of air above the telescope mirror. This concept is known as turbulence tomography. Perturbations at various elevations with different shifts between the wavefront sensors; for Shack–Hartmann-sensing, all sensors are optically conjugated to the pupil (lenslet) and focus (camera).

Apart from correcting a larger FOV to image large objects, the MCAO is useful to obtain more uniform correction over the FOV, as well as to enhance the possibility of finding a natural guide star in a larger FOV. Increased FOV is due to overlap of fields toward multiple guide stars. It also takes care of the cone effect when using laser guide stars, which may increase the compensation performance on any large telescopes.

Figure 4.18 depicts a schematic of a MCAO system, in which, two wavefront sensors are looking at two off-axis laser guide stars. The information from these wavefront sensors is processed by a central processing unit (CPU), which feeds it to a reconstructor. To note, it is possible to measure the turbulence in three dimensions

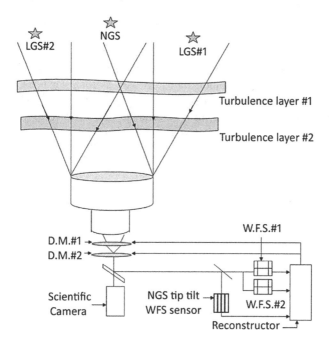

Fig. 4.18 A sketch of a multi-conjugate adaptive optics system

(3-D) not only with laser guide star but also with natural guide star or at solar telescopes at various elevation. However, this system enables near-uniform compensation for the atmospheric turbulence over considerably wider field-of-view.

However, its performances depends on the quality of the wavefront sensing of the individual layers. A multitude of methods have been proposed. Apart from solving the atmospheric tomography, a key issue, it is apparent that a diversity of sources, sensors, and correcting elements are required to tackle the problem. Atmospheric tomography is the measurement of the 3-D structure of turbulence in the atmosphere. The turbulence tomography, similar in its principle to medical tomography, is needed (1) to get the information for wavefront correction from distant natural guide stars, (2) to correct the cone effect when using laser guide stars, and (3) to drive several deformable mirrors.

This can be accomplished by measuring the integrated wavefront aberration from multiple probes, either stars or artificially generated laser guide stars. The integrated wavefront distortion along any line of sight contained by the beacons can also be calculated. A solution may be envisaged to mitigate the cone effect in employing a web of guide stars. Combining the wavefront sensing data from these stars, one can enable to reconstruct the 3-D structure of the atmosphere and eliminate the problem of aniso-planatism.

Ragazzoni et al. (2005) have developed the layer-oriented MCAO approach, in which the turbulence is sensed within a small volume where a few detectors can be placed in a variety of combinations. One may conjugate one wavefront sensor to each deformable mirror and optically co-add the light of many natural guide stars on the detector. The advantage of such an approach is that one may require as many wavefront sensors as deformable mirrors. MCAO is also useful for 8 m class telescopes, as has already been demonstrated, as well as for solar telescopes, where there is no guide star problem exists. However, the limitations are mainly related to the finite number of actuators in a deformable mirror, wavefront sensors, and guide stars.

Chapter 5
Diluted-aperture Stellar Interferometry

5.1 Methodology of Interferometry

Success in radio interferometry caught the attention of astronomers working at optical and infrared (IR) wavelengths, spanning the whole range from 0.35 to 20 μm. Thus, a Long Baseline Optical Interferometer (LBOI) came into existence offering unprecedented resolution. As stated earlier in Sect. 3.1.1, stellar interferometry can be achieved with a single telescope into which light from a distant object can pass through two (or more) apertures in a mask covering the telescope and then combine the two light beams to produce fringes. Another way to achieve stellar interferometry is to use two or more telescopes looking at the same star and reflecting light beams into a single receiver.

The standard technique in use is the Michelson (amplitude) stellar interferometry. This method provides better sensitivity in the optical band. The light from a source collected by several apertures that are separated by some distance may be combined coherently at the focus. These apertures can be either telescopes or Siderostats.[1]

A stellar interferometer measures the visibility (see Sect. 2.1.4) function at individual points in the u, v-plane (see 2.23), which decreases as the baseline, B increases. The rate of decrease depends on the angle subtended by the star. This function should, in principle, be normalized by dividing geometric means of the individual telescope responses such that $V = 1$ for a point source. The vector spacing between two of the apertures (the baseline vector), $\mathbf{B}$, with coordinates is given by,

$$\mathbf{B}_{ij} = \left(B_{x_0}, B_{y_0}\right) = (x_i - x_j, y_i - y_j). \tag{5.1}$$

---

[1] Siderostat is a single steerable flat mirror device that directs starlight in a fixed direction either directly through an aperture hole along an optical bench or first through a beam-compressor (reducing the beam diameter). It has limited sky coverage and has an image rotation, while in a coeleostat, a double mirror system, there is no image rotation. In this, the first mirror tracks the source and the beam is reflected to the second mirror kept at a fixed position. The rotational speed of both systems should be half of the Earth's speed to compensate the diurnal motion. However, both the systems are prone to have spurious polarization and therefore makes polarization measurement difficult due to the changing, non-normal reflection angle of the flat. Nevertheless, it offers a more stable structure for minimizing vibrations and pivot-point drifts for accurate astrometry.

S.K. Saha, *Aperture Synthesis*, Astronomy and Astrophysics Library,
DOI 10.1007/978-1-4419-5710-8_5, © Springer Science+Business Media, LLC 2011

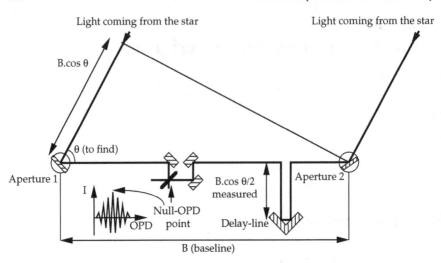

Fig. 5.1 Two-aperture optical interferometer

As evident from Fig. 5.1, the wavefront arriving from a celestial object at an angle θ from the baseline does not reach the apertures at the same time. The signal travels an extra path to reach one of the telescopes. There is a time difference, called delay or, equivalently, an optical pathlength difference (OPD) between the arms of the interferometer. The external optical delay, d, observed in a broad spectral range, is:

$$d = |\mathbf{B}| \times \cos\theta. \tag{5.2}$$

The light travels from each arm of the interferometer to the beam combiner passing through different paths in vacuum, as well as different dispersive pathlengths in air and glass. A wavefront, in general, arrives at one of the telescopes with an extra pathlength difference d_{ext}. In order to compensate the path difference, appropriate delay is introduced in the other arm (telescope) of the interferometer, providing an internal vacuum delay, d_{int}. When the delay-line is adjusted to provide an OPD of $\Delta d(= d_{ext} - d_{int}) = 0$, the peak intensity of the fringe pattern is centered in the envelope. If this path length difference is larger than the coherence length, l_c (see 2.1), the light from both apertures cannot interfere. Hence, it is essential to place both the telescopes at equal distance from the combining optics (see Fig. 5.1) so as to compensate the geometrical delay. However, it is a difficult task to maintain such an accuracy since both the coherence time, τ_c (see 2.2) and coherence length at optical wavelengths are small. For example, the coherence length in the case of the optical light at $\lambda = 0.5~\mu m$ with a bandwidth, $\Delta\lambda = 5~\text{Å}$ is 500 μm and for delay tracking, the resolution needed for 1% precision in amplitude measurements would be $l_c/16 = 30~\mu m$, while in the infrared band at 5 μm, it turns out to be 300 μm.

Amplitude interferometry relies on the visibility of interferometric fringes to provide measurement of the mutual degree of coherence between different telescopes.

As stated earlier in Sect. 2.1.2 that for a given baseline, the visibility, $\mathcal{V}$, is given by the contrast of the interference fringes. The general expression of the detected fringe signal is,

$$I_{int} = I_1(\mathbf{r}, t) + I_2(\mathbf{r}, t) + 2\mathcal{V}\sqrt{I_1(\mathbf{r}, t)I_2(\mathbf{r}, t)}\cos(2\pi\mathbf{B}\cdot\hat{\mathbf{s}}/\lambda + \psi), \qquad (5.3)$$

in which $I_1(\mathbf{r}, t)$ and $I_2(\mathbf{r}, t)$ are the intensities of the beams, ψ the phase difference between two signals given by (3.42), $\mathbf{B}$ the baseline vector, $\hat{\mathbf{s}}$ the unit vector indicating the direction of the source, and λ the interferometer operating wavelength.

It is required to obtain the visibility, $\mathcal{V}$, from $I_1(\mathbf{r}, t)$ and $I_2(\mathbf{r}, t)$ that can be measured either by tapping into each beam before combination to get simultaneous photometric measurements, or after fringe measurement by feeding the detectors with the light coming from only one telescope followed by the other. The delay can be determined from the position of the optical delay-line of the instrument set-up such that the central fringe of the interference pattern appears in a narrow observation window. The position, as well as $|\mathbf{B}|$, are measured by laser metrology. Hence, θ is deduced with a high precision. For ground-based interferometers, the baseline is fixed to the Earth and co-rotates with it. But in space the interferometer must reorient the baseline to measure both angular coordinates. For a space-borne interferometer, the issue is to find a reference for the angle measured. Usually, a grid of far objects like quasars are used as a reference frame. Light waves coming out of these two apertures are directed toward the beam combination table and superimposed at a common focal plane (see Fig. 5.1).

5.1.1 Resolving Power of an Interferometer

The angular resolution of an interferometer depends on the largest possible spacing between the individual telescopes. Such an instrument projects the fringes onto the source's intensity distribution and yields a measure of fringes between the corresponding apertures at the spatial frequency vector, $\mathbf{u}$ (see 2.23). The magnitude of the fringe amplitude is given by the structural content of the source at scales of the fringe spacing, while the phase of the fringe is given by the position of the fringe. If the size of the source is very small compared to fringe spacing, a high contrast in the fringes appear (see Fig. 5.2a), but as the size of source increases the contrast of fringes decreases.

To note, the modulus of the fringe visibility is estimated as the ratio of high frequency to low frequency energy in the average spectral density of the short-exposure. The distance between the fringes being directly proportional to the wavelength, one has to derotate the fringe phase in an interferometric experiment like in Mark III interferometer and Palomar Testbed Interferometer (PTI; Colavita et al. 1999) or to measure the group-delay using Real Time Active Fringe-Tracking (RAFT; Koechlin et al. 1996) system that is applied on Grand Interférométre à deux Télescope (GI2T; Labeyrie et al. 1986).

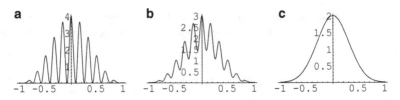

Fig. 5.2 Interference patterns from (**a**) a point source, (**b**) a partially resolved source, and (**c**) a uniform extended source

An extended source, whose centroid does not lie in the direction $\hat{s}$, produces fringes with lower visibility. In this case, the output of the interferometer is given by, $A \cos(2\pi \mathbf{B}_\lambda \cdot \hat{s} + \alpha')$, where A is the fringe amplitude, α' the fringe phase, and the quantity $2\pi \mathbf{B}_\lambda \cdot \hat{s}$ the fringe period. The period is a function of only the source-baseline geometry and independent of the source distribution. To note, fringe phase is proportional to the photocenter of the intensity distribution and here α' is because of the extended nature of the source; larger the ratio of the source size to the fringe separation, the smaller the observed amplitude as some parts of the source produce a positive output, while others produce a negative output. Thus, there would be some degree of cancellation in the total signal and the output signal is reduced compared to that obtained from a point source of the same total flux density having angular size less than fringe spacing. Figures (5.2b, c) depict the patterns of the fringes that are obtained from a partially resolved source and a uniform extended source, respectively. As seen from the Fig. 5.2c, it is difficult to obtain interference fringes from an extended source, since it has a large spectral bandwidth. As a result the coherence length is short.

Let the one-dimensional intensity distribution of an extended source be $I(x)$, where x is the displacement in angle from the field center, which means if $x = 0$, it corresponds to the angle θ to the normal with the field center. A signal from a point x has power $I(x)dx$. Since the fringes are spread over the source, the output of the interferometer is the sum of all the contributions,

$$I_{int}(\theta) = \int_{pupil} I(x) e^{i 2\pi \mathbf{B}_\lambda \sin(\theta + x)} dx. \qquad (5.4)$$

The integral in (5.4) extends over the whole source, which is a Fourier transform of source intensity, and the contribution from x is proportional to the intensity $I(x)$. The response to a point source in the direction of θ is given by the exponential term of (5.4). For small x, $\sin(\theta + x) = \sin \theta + x \cos \theta$, and on separating the term in $\sin \theta$, one gets,

$$I_{int}(\theta) = e^{i 2\pi \mathbf{B}_\lambda \sin \theta} \int_{pupil} I(x) e^{i 2\pi x \mathbf{B}_\lambda \cos \theta} dx, \qquad (5.5)$$

where $e^{i2\pi B_\lambda \sin\theta}$ represents the effects due to the path delay that is compensated in the correlation for white light fringes.

According to the van Cittert–Zernike theorem (see Sect. 2.2), the complex visibility, V of a source is related to the intensity distribution, through Fourier transform relationship, i.e.,

$$V(B_\lambda) = \int I_{int}(\theta)e^{i2\pi B_\lambda \cdot \hat{s}}d\hat{s},$$ (5.6)

By performing inverse Fourier transform of the source visibility function, the source morphology may be recovered,

$$I_{int}(\theta) = \int V(B_\lambda)e^{-i2\pi B_\lambda \cdot \hat{s}}d\hat{s},$$ (5.7)

In the Cartesian coordinate system, $\mathbf{x}(= x, y)$, are drawn on the the sky with origin at the field center, where x and y are the Eastward and Northern displacements respectively. In terms of celestial coordinates, $x = -\Delta\alpha \cos\delta$ and $y = \Delta\delta$, in which $\Delta\alpha$ and $\Delta\delta$ are the offsets in right ascension α and declination δ, respectively. Thus,

$$I(\mathbf{x}) = \int_{-\infty}^{\infty} \widehat{V}(\mathbf{u})e^{-i2\pi \mathbf{u} \cdot \mathbf{x}}d\mathbf{u}.$$ (5.8)

where $\mathbf{u}$ is the spatial frequency vector (see 1.61).

For small field-of-view (FOV), the intensity distribution can be derived from the measured visibility function by an inverse Fourier transformation. Since $I(\mathbf{x})$ is real, the complex visibility is Hermitian, $\widehat{V}(-\mathbf{u}) = \widehat{V}^*(\mathbf{u})$. The complex visibility measured by an interferometer can be obtained from the mean conjugate product or cross-correlation of the signals. While the two signals are identical in the case of a point source, they differ in phase for extended objects due to the extra propagation time to one of the telescopes.

The diffraction-limit of a telescope is dictated by an angle, θ_{tel} as given in (1.99). In the case of a two element interferometer, the light disturbance from a point source in the focal point is derived as,

$$U_{int}(\theta) = \int_{B/2-D/2}^{B/2+D/2} e^{i(2\pi x\theta/\lambda)}dx + \int_{-B/2-D/2}^{-B/2+D/2} e^{i(2\pi x\theta/\lambda)}dx$$

$$= 2D\cos(\pi\theta B_\lambda)\frac{\sin(\pi\theta D/\lambda)}{\pi\theta D/\lambda},$$ (5.9)

where D is the diameter of the aperture.

The intensity in the focal point is deduced as,

$$I_{int}(\theta) = 2I_{tel}(\theta)\left[1 + V\cos(2\pi\theta B_\lambda)\right],$$ (5.10)

in which I_{tel} is the envelope shape and V the visibility of the object.

The (5.10) is the product of the broad envelope of a single-telescope diffraction pattern and the interference term. This term depends on the separation between the telescopes. The angular resolution limit of the interferometer is described by the full width at half maximum (FWHM) of one of the narrow fringes as given by (2.30). The envelope of the fringe packet, N_{pac} has an angular width of $2\theta_{int}$ between first nulls. The number of fringes in a fringe packet is calculated as,

$$N_{pac} = 2.44\frac{B}{D}. \qquad (5.11)$$

If a star presents a circular disk of uniform brightness, the observed fringe visibility, $\mathcal{V}$, at a baseline, B, is given by (3.3). Similar to determining diffraction-limited resolution of a telescope, the Rayleigh criterion (see Sect. 1.5.3) can also be applied for defining the resolving power of a long baseline optical interferometer. An equal brightness binary star is resolved by such an interferometer provided the fringe contrast goes to zero at longest baseline. In order to measure the angular diameter of the source, the baseline between the two sub-apertures should be gradually increased until the fringes first vanish.

5.1.2 Astrometry

Astrometry is an important aspect to astronomy. It is useful to determine apparent dimensions and shapes of stars, their positions and time-variations, orbital motion of binary or multiple stars, and proper motion of star representing its apparent path on the sky; it helps in establishing cosmic distance scale as well. This technique is a complementary technique to the radial velocity method. It has a different detection bias, more suitable for tracking of near-Earth objects. The presence of the faint component is deduced by observing a wobble (oscillatory motion) in the proper motion of the bright component. Such a perturbation takes place due to gravitational influence from its unseen component (a star or a planet) on the primary star. Wobble is generally measured by looking at the variations of the radial velocity of a star. The combination of the motion around the center of mass between these two objects and the proper motion on the celestial sphere gives rise to a wobbling motion on the celestial sphere. It has also been employed to detect planets orbiting other stars, which are referred to as exoplanets, by measuring the displacement they cause in their parent star's apparent position on the sky, due to their mutual orbit around the center of mass of the system.

The primary objective of astrometry to derive their motion. They are analyzed in different ways such as kinematical approach describing motions and dynamical approach studying the forces governing them. Stellar kinematics deals with the relations or correlations between the components of stellar motions and some of their intrinsic properties like chemical compositions, age, spectral type etc. The dynamical approach illustrates the studies of the motion of planets and satellites in terms of various forces present in the Solar system and in galactic dynamics one tries to explain the structure of the Galaxy from the parameters of motions of stars (Kovalevsky 1995, 2004).

Various techniques, such as speckle interferometry, radio interferometry, *Hipparcos*, and LBOIs are employed to determine astrometric positions. *Hipparcos* (1997) uses the phase-shift measurement of the temporal evolution of the photometric level of two stars seen drifting through a grid. Very valuable astrometric results from space have already been obtained by the *Hipparcos* (Perryman 1998). The successor of *Hipparcos*, Gaia (Lindengren and Perryman 1996), may use the same technique with improvements, yielding more accurate results on a larger number of objects.

Astrometric interferometer estimates the relative angular position of celestial objects through measurements of the delay required to get stellar fringes on each source. These fringes must be observed at two or more baseline orientations to determine the two angular coordinates of an astronomical object. In a two-aperture interferometer, astrometry can be done with two modes of observations. They are (1) wide-angle mode and (2) narrow-angle mode. However, for a ground-based interferometer of this kind suffers from atmospheric turbulence that imposes severe limitations on the accuracy of astrometric measurement, wide-angle astrometry in particular.

Wide-angle astrometry determines accurate relative-positions of targets that are widely separated. This is employed at the Mark III (Shao et al. 1988a) and Navy prototype optical interferometer (NPOI; Armstrong et al. 1998). The accuracy that can be maintained is about 2 mas over large angles, but requires a very stable baseline laser metrology. The large angle difference between the reference and the studied object usually requires collector motions. However, it is difficult to have always a correct reference star within the field-of-view (FOV) for any studied object. In order to recover the two coordinates for each star from the observed delays, sufficient delay measurements are required. This would enable the baseline vectors of the interferometer to be derived along with the delay constants and the position of the objects (Hummel et al. 1998).

Quest for exoplanets (Wallace et al. 1998) is a challenging task for narrow-angle astrometry that requires a dual-feed system to observe two stars simultaneously, plus an extra delay line. Such a method can be applied for the measurement of parallaxes and more so for characterizing wobble in the position of a star, which is required for exoplanets search. In this technique, the two objects are in the FOV of the instrument, so that the collector motions are not required and the accuracy of the measurement can be improved. Moreover, the limitations due to atmospheric turbulence are far less severe for differential measurements over small field compared to wide-angle astrometry (see Sect. 9.3.2).

5.1.3 Nulling Interferometry

An interferometric technique, known as nulling interferometry, is gaining importance in optical astronomy. This technique was first proposed by Bracewell (1978) for applications in radio astronomy.[2] He opined that the light from the central star can be diminished by an interferometer, if the signals from some of the telescopes are delayed slightly. By precisely adjusting this delay, the central bright object is canceled out, allowing the faint, nearby faint object to stand out.

The purpose of such a technique is to enhance the ability to measure the relatively dim light of diffuse circumstellar material or faint companions such as planets near bright, nearby stars. Unlike coronagraphy (see Sect. 7.5.3) whose purpose is same as mentioned above, this technique cancels the image of the on-axis bright star due to a 180° phase delay on one of the interferometer arms before combination. The former method is possible only beyond several Airy radii from the on-axis stellar source and is useful for the nearest star, while the latter is expected to be effective within the core of a telescope's point spread function (PSF), so can be employed for stars at greater distances, where a larger sample is available (Serabyn 2000). When the two apertures are properly phased, the 180° phase shift introduced between the two beams produces a destructive interference on the optical axis. The separation between the two apertures can then be adjusted to place the first constructive fringe of the transmission pattern on a faint off-axis companion.

Nulling interferometry has been employed successfully in infrared (IR) wavebands in the range of wavelengths $\lambda \sim 8$–13 μm by Hinz et al. (1998). It is to be noted that the optical light from an exoplanet is hidden by the brightness of the star that it orbits. By nulling, such a planet may be detectable. The effect of nulling will suppress a bright star by a factor of $\sim 10^4$ or 10^5. The advantage of using IR over optical wavebands is that planets have their peak brightness in IR where the brightness of the star is reduced. This paves the way to detect a planet; the expected brightness ratio between the star and its planet in the IR region is $\sim 10^6$, while it is $\sim 10^9$ in the visible band. To note, the given contrast is for an Earth-like planet.

Figure 5.3 depicts the schematic of nulling interferometer at the MMT (Hinz et al. 1998). The diameter of each of the two primary mirrors is 1.83 m and the center to center separation of these mirrors is 5 m. The wavefronts from the two co-mounted telescopes, A and B, are translated via three mirrors M1, M2, M3 for superposition without relative rotation or tilt. The unmatched reflections from these telescopes are made to nearly normal incidence to minimize polarization differences. The folded beams are combined at the zinc selenide beam-splitter, Z_1. The secondary mirrors bring reflected collimated beam to a final focus via tertiary mirrors. The light beam from the telescope A is folded down at M_1 and is transmitted by the said beam-splitter Z_1 before reaching the focus. The light beam from the second telescope B is also

[2] van Albada (1956) did 'avant-la-lettre' nulling interferometry using photography at the Bosscha Observatory in Lembang near Bandung (Indonesia) in order to establish the orbits of Sirius B and other close binaries; a special technique was employed to eliminate the effects of contraction of the photographic emulsion. He used aperture diaphragms to obtain a diffraction pattern throwing most light from α CMa (Sirius A) in another direction than the location of Sirius B to make the latter visible. In addition, a five-wire grating was used on the aperture mask to obtain diffraction images of Sirius A of similar magnitude as Sirius B, for magnitude estimation and position calibration. van Albada established the required shape of the diaphragm analytically in a long Fourier derivation. Observations carried out during December 1955 to February 1958 by him are available in van Albada (1962, 1971).

Fig. 5.3 Schematic of the nulling interferometer that was used on the old Multiple-Mirror Telescope (MMT) in 1998 just before the MMT was converted from six 1.8 m mirrors to a single 6.5-m primary mirror (Hinz et al. 1998: courtesy: P. M. Hinz)

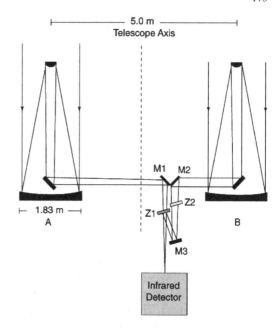

folded down at M_2 and passes through the second beam-splitter Z_2. It travels further and reflects off from another mirror M_3 before reflection from the beam-splitter Z_1.

A large space-based telescope may fail to detect the planet due to the photon noise. Since the contrast ratio of a star and its planet is very high, the signal from the latter gets obscured. The nulling interferometer yields a desired signal-to-noise (S/N) ratio at the output of the detector by canceling the light of the star arriving from the two apertures, while it enhances the image of the planet by constructive interference. Here, at the MMT experiment, the light collected from the two mirrors is combined in order to generate a deep destructive interference fringe at the stellar position, thus selectively nulling the star by many orders of magnitude relative to the surrounding off-axis environment, such as a planetary system. Basically, a 180° phase-shift is introduced in one wavefront segment, so that when it interferes with another segment of the same wavefront, a perfect cancellation is achieved. This phase-shift, $\delta\psi$, is a function of the baseline, but is independent of the aperture diameter. The wavefront from an oblique direction arrives at one of the primary mirrors earlier than at the other mirror, and therefore produces a path-length difference of $\mathbf{B}\sin\xi\cos(\eta - \theta)$, in which $\mathbf{B}$ is the baseline vector between the two primary mirrors, ξ, and η the angular distance with respect to the line of sight and azimuthal coordinates respectively, and θ the angle with which the interferometer is oriented. The relative phase of the light coming from the two telescopes is given by,

$$\delta\psi = \kappa\mathbf{B}\sin\xi\cos(\eta - \theta), \tag{5.12}$$

with $\kappa = 2\pi/\lambda$.

The corresponding disturbances of the light at the image plane is expressed as,

$$U_1 = \frac{1}{2}U_0\left(1 - e^{i\delta\psi}\right),\tag{5.13}$$

in which U_0 represents the peak of the interferometer's transmission, while U_1 provides the transmission as a function of phase shift. U_1 would thus be null at phase shift 0, and will then follow a square sinusoidal pattern for off-axis positions.

The resulting intensity is deduced as,

$$|U_1|^2 = |U_0|^2 \sin^2\left(\frac{\delta\psi}{2}\right)\tag{5.14}$$

$$= |U_0|^2 \sin^2\left[\frac{\pi B}{\lambda}\sin\xi\cos(\eta - \theta)\right].\tag{5.15}$$

For small values of ξ and the companion oriented in the direction of the baseline, this expression (5.18) can be simplified to obtain the square sinusoidal pattern. The central fringe of the interference pattern is dark, allowing the fringe pattern from a faint object to appear. The quality of a nulling is defined by the 'null depth', N_d, i.e., the ratio of the transmitted powers in the destructive and constructive states,

$$N_d = \frac{1 - \mathcal{V}\cos\varphi_e}{2} \approx \left(\frac{\pi\sigma_{\delta\psi}}{\lambda}\right)^2,\tag{5.16}$$

where φ_e is the phase error between the two recombined beams, $\mathcal{V}$ the fringe visibility modulus, and $\sigma_{\delta\psi}$ the standard deviation of the OPD between the two beams.

The null depth defined in (5.16) is valid on the optical axis and would capture the influence of imperfect cophasing of the beams. This is an instrumental effect and is generally referred to as "instrumental nulling ratio" or "null floor". The other instrumental effects contribute to the null floor are imperfect beam balance, polarization effects, etc. The term, null depth, is more often used in an astrophysical context, for the actual ratio of constructive to destructive output measured on a star. The null depth therefore also accounts for the finite angular size of the stellar photosphere (a contribution generally referred to as 'geometric stellar leakage').

To create the 180° phase-shift, a few important techniques may be employed:

1. using rooftop reflectors to achieve a reversal of sign of the electric vector and
2. introducing a precise thickness of glass whose index acts to retard all wavelengths by very nearly one-half wavelength.

The on-axis cancellation of stellar radiation demands a very symmetric and stable optical system. For a given polarization state, the intensities of the two beams, electric field rotation angles that can be achieved by adjusting the accompanying image rotation, and phase delays should be matched to $2\sqrt{N_d}$. Hinz et al. (1998) demonstrated the viability of nulling interferometry using two 1.8-m mirrors of the original six-mirror MMT. They detected the thermal image of the surrounding, circumstellar dust nebula around α Orionis. Figure 5.4 depicts constructive and destructive

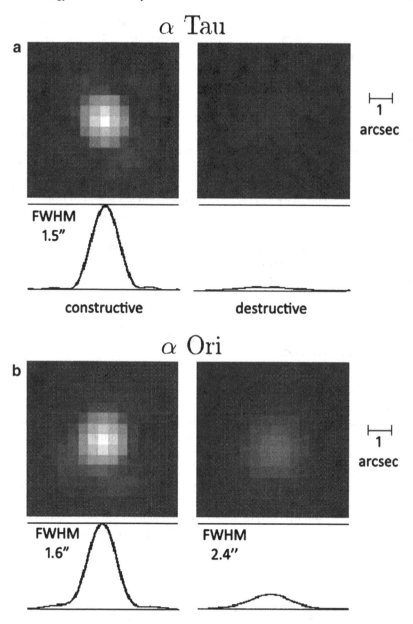

Fig. 5.4 Constructive and destructive interference fringes obtained by Nulling interferometer at the MMT (Hinz et al. 1998: courtesy: P. M. Hinz). The images of α Tauri and α Orionis were observed using two of the 1.8-m mirrors of the old MMT. α Tauri is a control image to create the PSF, while in the case of α Orionis, there is some suggestion of an extended envelope sourcing the red supergiant star

interferences of (a) α Tauri, which is unresolved at the said baseline (5 m) and (b) α Orionis. These single short-exposure images are obtained at 10.3 μm wavelength. Hinz et al. (2001) have estimated the disk sizes of a few Herbig Ae stars with the said interferometry and suggested that the extent of mid-IR emission around these stars may be much smaller than current models predict. They constrain $\sim$90% of the 10 μm flux to be within an approximately 20 AU diameter region.

Nulling interferometer could be employed at the upcoming large interferometers to observe faint structures close to non-obscured central sources. This technique can also be used in space to search for Earth-like exoplanets through their thermal emission and to determine the atmospheric signature of life with spectroscopic analysis (Angel et al. 1986; Hinz et al. 1998). The major difficulties may arise from the zodiacal light, which appears when the light from the Sun get scattered from the dust particles in the interplanetary space. The integrated thermal emission of this zodiacal dust is about 10^{-4} that of the sun's luminosity at a wavelength of 10 μm. This diffused radiation may enter a space-based interferometer and create a background noise. A similar situation may arise from exo-zodiacal dust grains congregating in and around the orbital plane of the targeted solar system, which may be imaged as a fairly uniform distribution across the instrument's FOV.

5.2   Baseline Geometry

Accuracy in measuring the position of a stellar source depends on the precision with which the parameters of the baseline of a two-aperture optical interferometer are known. The telescopes are located on the surface and track a source in the sky, which changes its position due to Earth's rotation inducing a variation of the length and orientation of the baselines; owing to wind, tide, precession, and nutation, the Earth rotates non-uniformly. The coverage of the spatial frequencies needs to be carried out by locating the apertures in order to take the advantage of the Earth's rotation. As Earth rotates, the baseline separation between the two telescopes change. When two telescopes are closer by half of a wavelength, the light waves from their output cancel each other, making the source disappear. Hence, it is useful to translate the delay into celestial coordinates system.

5.2.1   Celestial Coordinate System

The position of a source in the sky may be specified in various spherical coordinate systems in astronomy to determine the positions of the object on the celestial sphere. Such a sphere, an imaginary one with the observer at its center, represents the entire sky and rotates relative to the Earth's surface (Smart 1956). Celestial coordinate systems (summarized in Appendix E), namely, (1) equatorial coordinate system, (2) alt-azimuth coordinate system, (3) ecliptic coordinate system, and (4) galactic coordinate system are based on three common principles:

1. All celestial objects other than the Earth are imagined as being located on the inside surface of the celestial sphere.
2. Each coordinate axis, one of the fixed reference lines of a coordinate system, is a great circle on the celestial sphere. The great circle, perpendicular to the principal axis, is the fundamental great circle (FGC) along which one coordinate is measured. There are infinite number of secondary great circles corresponding to the y- axis on the plane surface, which are called vertical circles. These circles are perpendicular to the FGC, which meet at the poles of the principal axis. The principal axis is parallel locally to the direction of gravity. Extended upward, this axis intersects with the celestial sphere at a point, known as zenith (Z); the point downward is called nadir. These points are determined by the local plumb-line defining the direction of gravity at the place in question. The FGC at the center corresponding to the x-axis on the plane surface denotes the local horizon, which is formed by the intersection of the horizontal plane at a place on the Earth with the sphere. It is marked on the celestial sphere by a plane perpendicular to the zenith-nadir axis and tangent to the Earth at the point of the observer. There are four reference points on the astronomical horizon, North, East, South, and West. The two vertical circles passing through the North and South points together define the celestial meridian of the place, while those joining the East and the West define the prime vertical. The intersection of the celestial meridian and the celestial equator, Σ is down from the zenith by an angle equal to the latitude.
3. Coordinate measurements of an object to be located are made along two great circles, one a coordinate axis and the other perpendicular to it and passing through the object. Measurements are made either in degrees or in hours. Distances are measured on the celestial sphere in degrees along the arc of a great circle, which is drawn on the celestial sphere in a plane passing through the center of the sphere, dividing it into two parts. The coordinate system has a principal axis or a polar axis. Around this axis, the system rotates. The points of intersection of this axis and the celestial sphere are the poles of the system.

5.2.1.1   Equatorial Coordinate System

Equatorial coordinate system is commonly used astronomical coordinate system for indicating the positions of stars or other celestial objects on the celestial sphere. Every object in the sky has a position given by two celestial coordinates such as right ascension (RA) and declination (δ). The principal axis of this system coincides with Earth's rotation axis. Its poles, known as celestial poles are the points where the extended Earth's axis intersects the celestial sphere. The North and South celestial poles are determined by projecting the rotation axis of the Earth to intersect this sphere. The North celestial pole is directly above the Earth's North pole, while the South celestial pole is directly below its South pole. The great circle on the celestial sphere halfway between the celestial poles is called the celestial equator. It can be thought of as the Earth's equator projected onto the celestial sphere; it divides the celestial sphere into northern and southern skies. An important reference point on the celestial equator is the vernal equinox,[3] the point the Sun crosses the celestial equator in March.

In order to designate the star's position, the observer considers an imaginary great circle on the celestial sphere passing through the North celestial pole and the star in question. This is referred to as star's hour circle, analogous to a meridian of longitude on Earth, which is used in deriving its right ascension (RA) and declination (δ). By definition, the vernal equinox and is located at RA $= 0^h$ and $\delta = 0°$. The noted spherical coordinates are defined as:

1. Right ascension (RA): It is the angle between the vernal equinox and the point at which the hour circle intersects the celestial equator. RA is parametrized by α and is measured eastward from the 'First point of Aries' in units of time (hours, minutes, and time) rather than in the more familiar degrees of arc (1 h of right ascension is equivalent to 15° degrees of apparent sky rotation).

2. Declination (dec): It corresponds to latitude projected on the sky, which is the angle between the celestial equator and the object along the star's hour circle. The declination is measured in degrees, minutes, seconds North ($+\delta$) or South ($-\delta$) of the celestial equator. Thus the declination of the North celestial pole is $+90°$, and that of South pole is $-90°$. The declination is obtained by measuring the meridian zenith distance of the star. The declination of a star is given by $\delta = \Phi \pm z$, in which Φ is the latitude of the place (observatory) in question, z the zenith distance, depending on whether the star is North or South of the zenith.

3. Hour angle (HA): It is the arc of celestial equator between the object's hour circle and the meridian. HA is the angular distance measured westward along the object being located. Denoted by H, the hour angle is calculated by subtracting the right ascension from the local sidereal time. It is expressed in terms of time and is used to measure astronomical time.

4. Sidereal time (ST): It is the time as measured by the apparent motion of the stars across the sky. The hour angle of the First point of Aries is referred to as

---

[3] The celestial equator and the great circle through ecliptic intersect in two points. These points are referred to as equinoxes. One of them is called the vernal equinox and considered as a reference point in the sky. This point, also called the 'First point of Aries', symbolized by ♈, the astrological symbol for the sign of Aries, marks the celestial meridian. The Sun moves through this point on or about 21 March. On autumnal equinox, the Sun resides at the 'First point of Libra'.

Because of the effects of the precession on the Earth, neither the North celestial pole nor the First point of Aries are fixed. The former describes a small circle of radius 23.4° around the pole of the ecliptic and the latter regresses westwards along the ecliptic (the Sun's apparent path on the celestial sphere) in a period of 25,800 years. At the time of Hipparchus, who derived the precession of the Earth, during the second century BC, the vernal equinox was in the eastern boundary of the constellation Aries, from which ♈ takes its name. Over the centuries, it has moved right across the Aries and at present it is near the western boundary of Pisces and a few degrees away from Aquarius. In spite of its present position, it retains the name 'First point of Aries'. Roughly 23,000 years from now, the Sun would complete its circuit of the zodiac, and the point ♈ may lie among stars from which it takes its name.

sidereal time. The sidereal day is the period during which the Earth completes one rotation on its axis, which is 3 min 56.6 s shorter than the solar day.

It is obvious that the RA and δ of a celestial object remain fixed as the sky rotates. This makes it possible to catalog astronomical objects by RA and dec. Because of the precession, the First point of Aries slides westwards along the ecliptic very slowly relative to the stars, so that RAs of all the stars are increasing at a rate of about 0.008 s per day. Thus over a number of years, a star's RA and δ change. The cataloged positions are not accurate and are up-dated periodically to correct the precision. To note, in the catalog all source positions are specified in standard epochs (B1950 or J2000.0), in which B stands for Besselian year and J the Julian year. Therefore, it is necessary to state the epoch to which RA and δ are referred. For much of the twentieth century, equatorial coordinates were referred by the epoch B1950.0, while at present, catalogs and atlases refer RA and δ to the epoch J2000.0.

Since RA and HA are both measured along the celestial equator but with opposite direction, one gets

$$H_\Upsilon - H = \alpha - \alpha_\Upsilon, \qquad (5.17)$$

in which H_Υ and α are the hour angle of the First point of Aries and RA of the star respectively.

With $\alpha_\Upsilon = 0$, (5.17) turns out to be,

$$H = T_s - \alpha. \qquad (5.18)$$

When $H = 0$, it provides $\alpha = T_s$, which means the RA of a star is the ST at the moment of its transit across the meridian. At the observatory, sidereal time is kept by a special sidereal clock which runs faster than the normal mean time clock by about 3 m 56.6 s per day. To note, the RA of the Σ point is equal to the observer's local sidereal time. The angular distance from such a point to a star's hour circle is its hour angle; it is equal to the star's RA minus the local sidereal time. Because the Σ point is always visible, the hour angle is used in locating a celestial object.

5.2.1.2   Altitude–azimuth (alt-az) Coordinate System

In the alt-azimuth coordinate system, the position of a body on the celestial sphere is described relative to an observer's celestial horizon and zenith. Coordinates in such a coordinate system are defined as:

1. Altitude: It is defined as the angular distance of the celestial object (for example, star) above or below the horizon measured along a vertical circle. In other words, it is the angle made from the position of such an object to the point nearest on the horizon. The altitudes of the zenith and the nadir are $+90°$ and $-90°$, respectively. To note, the altitude of any point on the horizon is $0°$.
2. Azimuth: It is the angle from the northernmost point of the horizon to the point on the horizon nearest the object. In general, azimuth is measured Eastwards

Fig. 5.5 Altitude–azimuth
local coordinate system

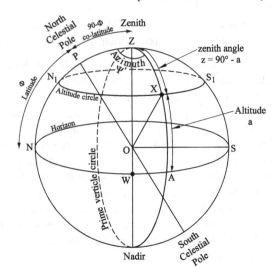

from due North. The azimuth of the North, East, South, and West points are 360°
or 0°, 90°, 180°, and 270° points, respectively. An object's altitude and azimuth
change as the earth rotates.

In an alt-az reference system (see Fig. 5.5), azimuth (Ψ') of the star at location X
is defined as the spherical angle P Z X or the great circle arc N A. The point on the
celestial sphere vertically overhead the observer (O) on the surface of the Earth is
the zenith (Z). The O Z axis is perpendicular to the plane of the horizon with positive
direction towards zenith. The direction in the local vertical half-plane defined by A
is given by the zenith angle $z(= 90° - a)$, where a is the altitude. Any two objects
with the same altitude lie on a small circle ($N_1 X S_1$) called the parallel of altitude.
The terrestrial reference system is defined by the direction of the pole, O Z and
the Greenwich meridian. The coordinates of the observation are the latitude, Φ and
longitude, L reckoned positively towards the East. However, this coordinate system
is not a general system. As Earth rotates, both the altitude and the azimuth of a
celestial object change continuously. Most of the interferometers that employ large
telescopes are using alt-az telescope designs.[4]

---

[4] Modern large telescopes are configured with the alt–azimuth mount, because of its compact
design. This system has two motions, altitude (up and down/vertical) and azimuth (side to
side/horizontal). The main axis is the vertical axis that is pointed straight at the zenith. The other
axis rotates at right angles to the vertical axis. Pointing in this axis is difficult as the transformation
for converting from the right ascension (RA), declination (δ) coordinate system to the alt–azimuth
coordinate system is more complex, and tracking requires moving both axis at the same time.
A computer is required to control and position the telescope.

5.2.1.3   Other Coordinate Systems

Two less frequently used systems of coordinates are elucidated in brief:

1. Ecliptic coordinate system: The principal reference circles for the ecliptic
   coordinate system is the ecliptic, i.e., the orbital plane of the Earth around
   the Sun. The revolution of the Earth around the Sun defines an orientation and
   the line per pendicular to the plane of the ecliptic through the center of the Earth,
   strikes the surface of the celestial sphere, defines the ecliptic poles. The North
   ecliptic pole lies in the constellation Draco, while the South ecliptic pole falls
   in Dorado constellation. The ecliptic latitude is defined as the angle between a
   position and the ecliptic, which takes values between −90 and +90°, while the
   ecliptic longitude begins from the vernal equinox and runs from 0 to 360° in
   the same eastward sense as RA. The obliquity of the Earth's equator against the
   ecliptic amounts to $23°26'21.448''$ (J2000.0); it changes very slowly with time
   due to gravitational perturbations of Earth's motion.
2. Galactic coordinate system: Established in the year 1958 by the International
   Astronomical Union (IAU), the galactic coordinate system is the key to understand the location of the objects within the Galaxy (the Milky Way). This system
   is useful for specifying an object's location relative to the Sun and the galactic
   core of the Galaxy. It is centered on the Sun and is aligned with the apparent
   center of the Galaxy. The principal axis is the galactic equator (the intersection
   of the plane of the Galaxy with the celestial sphere) and the reference points
   are the galactic North pole at RA $12^h51^m26.282^s$, δ $+27°07'42.01''$ (J2000.0)
   and the zero point on the galactic equator that has the equatorial coordinates,
   RA $17^h45^m37.224^s$ and δ -28°56'10.23'' (J2000.0), which lies in the direction
   of the center of the Galaxy (Reid and Brunthaler 2004). The inclination of the
   galactic equator to Earth's equator is 62.9°. The coordinates of a celestial object
   are its galactic longitude and galactic latitude. The former is measured with baseline the direction to the center of the Galaxy from the Sun in the galactic plane,
   while the latter is derived between the object and the galactic plane with origin
   at the Sun.

5.2.2   Coordinates for Stellar Interferometry

The principle behind the aperture-synthesis method may be considered as the operation of a conventional filled-aperture telescope composed of a large number of
telescopes connected pairwise with different spacings and orientations of the baselines. In such a technique, it is possible to synthesize a beam that would correspond
to a filled-aperture telescope having diameter equal to that of the array. Signals from
the different telescopes are transported simultaneously to a central point receiving
system and are cross-correlated to obtain the complex visibility pertaining to each
pair of telescopes. The positions of the telescopes are specified in a coordinate system such that their separation is the projected separation in a plane normal to the

phase center. In other words, in such a coordinate system the separation between the telescopes is as noticed by the observer sitting in the source reference frame. If the source brightness is specified on coordinates of right ascension (RA) and declination (dec), the inverse Fourier space has axes referred to as u, v, the respective East–West and North–South components of the effective baseline at the ground of an interferometer. In order to determine the fringe-source geometry the projected baseline, **b**, of the physical baseline, **B**, the direction of which is defined from the telescope 1 to the telescope 2 with coordinates, as viewed from a star, is required to quantified. It is perpendicular to the unit vector in the source direction, $\hat{s}$, and is given by,

$$\mathbf{b} = \hat{s} \times [\mathbf{B} \times \hat{s}] = \mathbf{B} - (\mathbf{B} \cdot \hat{s})\hat{s}. \tag{5.19}$$

This is equal to the difference between the physical baseline, **B**, and its component in the direction of the source. Here, $\mathbf{B} \times \hat{s}$ is the path delay and the projected baseline seen from the source has a magnitude $|\mathbf{B} \times \hat{s}|$, that is $|\mathbf{B}| \sin \theta = |\mathbf{b}|$. Such a projected spacing may be decomposed into a set of components u, v, w, in which w is in the source direction (center of the field-of-view), u in the E-W, and v in the N-S directions. Figure 5.6 depicts the right handed (u, v, w) coordinate system fixed on the surface of the Earth. The units of these components (u, v, w) are noted in wavelengths. The term, w, may be neglected for small FOV and for the East–West interferometers.

The effective length of a two-element interferometer baseline for observation in the direction of the source is $B \cos \theta$, in which θ is the angle between the stellar source and the baseline joining the telescopes. The degree of resolution is specified

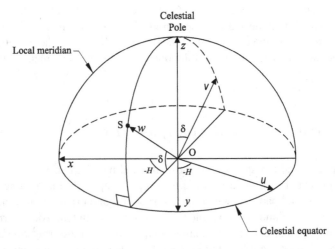

Fig. 5.6 Relation between the terrestrial coordinates (B_X, B_Y, B_Z) and the (u, v, w) coordinate system with the u, v-plane parallel to the tangent plane in the direction of the phase center on the celestial sphere and the w-axis in the direction of phase center

by the fringe separation, i.e., $B_\lambda \sin \theta$ in radians with $B_\lambda = B/\lambda$. The effective orientation of the baseline is determined by the position angle of the line normal to the fringes at the point of observation. The fringes cross the line between this point and the baseline pole (North) at right angles. In the local system of the coordinates, the components of the baseline vector, $\mathbf{B}$, are written as,

$$\begin{bmatrix} B_X \\ B_Y \\ B_Z \end{bmatrix} = \begin{bmatrix} B \cos \Psi \\ -B \sin \Psi \\ h \end{bmatrix} = \begin{bmatrix} B \cos \Delta_b \cos \Theta \\ B \cos \Delta_b \sin \Theta \\ B \sin \Delta_b \end{bmatrix} \tag{5.20}$$

in which Ψ is the azimuth of the baseline vector, $\mathbf{B}$, h the difference of height, Δ_b and Θ the declination and hour angle of the baseline respectively.

The components of the unit vector, $\hat{s}$ in the direction of the source given by its azimuth, Ψ' and zenith distance, z is,

$$\begin{bmatrix} s_X \\ s_Y \\ s_Z \end{bmatrix} = \begin{bmatrix} \sin z \cos \Psi' \\ -\sin z \sin \Psi' \\ \cos z \end{bmatrix}. \tag{5.21}$$

In the celestial direct local coordinates given by hour angle, H, and declination, δ, the components of the unit vector, $\hat{s}$ can be recast into,

$$\begin{bmatrix} s_X \\ s_Y \\ s_Z \end{bmatrix} = \begin{bmatrix} \cos \delta \cos H \\ -\cos \delta \sin H \\ \sin \delta \end{bmatrix} \tag{5.22}$$

The rotation matrix linking these two systems is,

$$\sin z \cos \Psi' = \sin \Phi \cos \delta \cos H - \cos \Phi \sin \delta, \tag{5.23a}$$
$$\sin z \sin \Psi' = \cos \delta \sin H, \tag{5.23b}$$
$$\cos z = \sin \Phi \sin \delta + \cos \delta \cos \Phi \cos H, \tag{5.23c}$$

where Φ is the latitude of the telescope location and H the hour angle related to the Greenwich sidereal time T_s.

Owing to the rotation of the Earth, the path difference to the telescope apertures changes with time, so that one moves the receiver or activate delay-lines. The difference of paths between the center of the telescope apertures, B_w, which is required to be compensated is given by,

$$\begin{aligned} \mathbf{B} \cdot \hat{s} &= B_X \cos \delta \cos H + B_Y \cos \delta \sin H + B_Z \sin \delta \\ &= B \left[\sin \delta \sin \Delta_b + \cos \delta \cos \Delta_b \cos(H - \Theta) \right]. \end{aligned} \tag{5.24}$$

It follows from the law of cosines that,

$$\cos\theta = \frac{\mathbf{B}\cdot\hat{\mathbf{s}}}{|\mathbf{B}|} = \sin\delta\sin\Delta_b + \cos\delta\cos\Delta_b\cos(H - \Theta)$$
$$= \cos(\delta - \Delta_b) - \cos\delta\cos\Delta_b[1 - \cos(H - \Theta)], \qquad (5.25)$$

and from the four-parts formula,

$$\cot\sigma = \tan\Delta_b\cos\delta\,\mathrm{cosec}(H - \Theta) - \sin\delta\cot(H - \Theta), \qquad (5.26)$$

with σ as the position angle of the telescope from the point of observation.

The diurnal motion of the Earth produces a continual change in the relative distance from the star to the two apertures, and hence in the phase relationship between the source at these apertures. The rate at which the fringes cross this source depends on its hour angle and declination, as well as the baseline parameters. This rate, called fringe frequency, v_f, varies as a function of position of the source and can be deduced as,

$$v_f = |\mathbf{B}|\cdot\frac{d\cos\theta}{dH} = -|\mathbf{B}|\cos\delta\cos\Delta_b + \sin(H - \Theta)$$
$$= \cos\delta\,[\mathbf{B}_Y\cos H - \mathbf{B}_X\sin H] \qquad Hz. \qquad (5.27)$$

The baseline vector, $\mathbf{B}$, is subject to small systematic changes arising from the pointing mechanism of the two telescopes. The derivative of (3.42) provides resulting change in phase, ψ,

$$d\psi = \frac{d}{dt}\mathbf{B}\cdot\hat{\mathbf{s}}. \qquad (5.28)$$

in which $\hat{\mathbf{s}}$ is the unit vector in the direction of the source.

The fringe phase, $\alpha'[= 2\pi B_\lambda\cos\theta(\tau)]$, varies as the Earth rotates with a rate,

$$\frac{d\alpha'}{dt} = -2\pi B_\lambda\cos\delta\cos\Delta_b\sin(H - \Theta)\frac{dH}{dt}, \qquad (5.29)$$

in which

$$\alpha' = 2\pi B_\lambda\{\sin\delta\sin\Delta_b + \cos\delta\cos\Delta_b\cos(H - \Theta)\}. \qquad (5.30)$$

The complex visibility may rotate at this rate in the complex plane and should be rotated backwards before averaging. This is referred to as fringe stopping. If the component telescopes have equatorial mounts,[5] the term, $d\mathbf{B}$, may be expressed in terms of coordinate system. Thus, one finds,

---

[5] The equatorial mount was widely used for astronomical telescopes, prior to the advent of the alt–azimuth mount; it is more common type mount for operating a moderate or a small telescope. In this, the vertical axis is aligned with the projection of the Earth's pole onto the sky. In the northern hemisphere, this point is marked by the pole star, Polaris. The other axis is the declination axis. In such a mount, tracking and pointing are greatly simplified. This system is suitable for astronomical observation for a long period of time, for astrophotographic observations in particular. As Earth

$$d\alpha' = d\,B_X \cos\delta \cos H + d\,B_Y \cos\delta \sin H + B_Z \sin\delta. \qquad (5.31)$$

Since the effect of $d\,B_Z$ is independent of hour angle, the variation of the path difference, B_w, with time is

$$\frac{d\,B_w}{dt} = \frac{d\,T_s}{dt}[-B\cos\Psi \sin\Phi \cos\delta \sin H + B\sin\Psi \cos\delta \cos H$$
$$-h\cos\Phi \cos\delta \sin H] = As. \qquad (5.32)$$

The dependent quantity, H, in the right hand side may be differentiated, where $d\,T_s/dt = dH/dt = s$ with $s = 15.041068$ arcsecs per second.

The phase α' is related to the instantaneous centroid offset of the brightness distribution from the axis of the interferometer beam. For a known position of a point source, the fringe phase is zero. On displacing the source by a fraction of the fringe and if the source moves in the interferometer plane, the fringe pattern would have the same period but the fringes may not occur at the expected time. The shift in the source position would give the corresponding fringe phase $\alpha' = 2\pi\Delta\theta/\Delta\psi$, in which $\Delta\theta$ is the fringe separation and $\Delta\psi$ the displacement of the source. As the direction of the source changes, the fringe amplitude oscillates (Burke and Graham-Smith 2002), and for a spacing of many wavelengths, if the source is close to transit, the variation is nearly sinusoidal. The resolution of the interferometer can be determined by the fringe angular spacing, i.e.,

$$\frac{d\alpha'}{d\theta} = \frac{d(2\pi B_\lambda \cos\theta)}{d\theta}$$
$$= -2\pi B_\lambda \sin\theta. \qquad (5.33)$$

The term $B\sin\theta$ in (5.33) is the component of baseline normal to the direction of the source. The separation of fringes is a phase change of $\Delta\psi = 2\pi$ radians, or an angular change of $\Delta\theta = 1/(B_\lambda \sin\theta)$. The interferometer can be arranged in either East–West or North–South orientation:

1. East–West orientation: A pair of telescopes operating along a horizontal East–West baseline provides the analog of a meridian circle. The baseline defines two diametrically opposite points on the celestial sphere. For such horizontal meridian, $\Psi = \pi/2$ and $H = 0$, therefore,

$$\frac{dx}{dt} = Bs\cos\delta \cos H. \qquad (5.34)$$

In this case, the variation of the path difference and consequently the movement of the fringes is maximum at the meridian.

---

rotates around its axis with a constant angular velocity, the stationary stars appear to move across the sky.

2. North–South orientation: In this case, the baseline is normal to the prime vertical plane. This may be suitable to observe the source near its transit over the prime vertical. The instrument becomes the equivalent to a prime vertical instrument. When a star crosses the great circle, the phase difference between the two apertures becomes zero. Such a crossing occurs if the source declination lies within the range, $0 < \delta < \Phi$. For a North–South interferometer, $\Psi = 0$ and $H = 0$, therefore,

$$\frac{dx}{dt} = -Bs \sin \Phi \cos \delta \sin H. \tag{5.35}$$

At the meridian in this case, the hour angle, $H = 0$, x and the fringes are stationary. This is the optimum configuration to analyze the fringes. With an introduction of a suitable delay-line, the source may be observed near meridian transit.

5.2.3 (u, v)-plane Tracks

The uv-coverage, also known as synthesized aperture, represents the spatial frequencies sampled by the array, in which for each baseline, the visibility of the fringe pattern provides the modulus of one component of the Fourier transformation. The shorter baselines, where the u, v-points are closer to the origin, provide the low resolution information about the source structure and are sensitive to the large scale structure of the source, while larger baselines give the high resolution information or sensitive to small scale structures in the source.

As the Earth rotates, the path of the projected baseline of a tracking interferometer describes an ellipse in a plane having Cartesian coordinates (u, v), which is given by (Fomalont and Wright 1974),

$$\frac{u^2}{a^2} + \frac{(v - v_0)^2}{b^2} = 1, \tag{5.36}$$

in which

$$a = \sqrt{B_X^2 + B_Y^2} = B \cos \Delta_b, \tag{5.37a}$$

$$b = a \sin \delta = B \cos \Delta_b \sin \delta, \tag{5.37b}$$

$$v_0 = B_Z \cos \delta = B \sin \Delta_b \cos \delta, \tag{5.37c}$$

and δ is the declination of the star.

Since Earth's rotation changes the orientation of the baseline, the measured Fourier component changes with time. It is worthwhile to mention that if a two-element interferometer is placed at the North pole, it describes a circle in the equatorial plane as the Earth spins. From a source at $\delta < 90°$, the track appears to be an ellipse. It is reiterated that in order to measure the angular details with finest distinguishability, the beam from separate mirrors is to be combined. The finer details of the image depend on the baseline of the telescope and on the range of baselines it can be adjusted to. These two conditions are together called as 'full coverage of the u, v-plane'. In an aperture-synthesis process, it involves a combination of physical movement of the telescopes in the array and the rotation of the Earth. Let the origin of the coordinate system be centered on one of the mirrors of the pair. For a tracking interferometer, the baseline vector, $\mathbf{B}$, in terms of astronomical coordinates can be written as,

$$
\begin{bmatrix} u \\ v \\ w \end{bmatrix} = \frac{1}{\lambda} \begin{bmatrix} \sin H & -\cos H & 0 \\ -\sin \delta \cos H & -\sin \delta \sin H & \cos \delta \\ \cos \delta \cos H & \cos \delta \sin H & \sin \delta \end{bmatrix} \begin{bmatrix} B_X \\ B_Y \\ B_Z \end{bmatrix}, \tag{5.38}
$$

where H is the hour angle of the star, and (B_X, B_Y, B_Z) the vector describing a baseline (single) in a coordinate system with $(0, 0, 0)$ at the Earth's center; the X and Y axes are in the plane of the equator towards the Greenwich meridian and the West, respectively, while B_Z is towards the North celestial pole.

With the rotation of the Earth, the hour angle of the source changes continuously, generating different sets of (u, v, w) coordinates for each telescope pair at each instant of time. The locus of projected telescope spacing components u and v (5.38) describes an ellipse with hour angle as the variable given by,

$$
u^2 + \left(\frac{v - (B_Z/\lambda) \cos \delta_0}{\sin \delta_0} \right)^2 = \frac{B_X^2 + B_Y^2}{\lambda^2}, \tag{5.39}
$$

where (H_0, δ_0) defines the direction of the phase center.

In the u, v-plane, this is an ellipse, referred to as u, v track with hour angle changing along the ellipse. The pattern generated by all the u, v-points sampled by the entire array of telescopes over the period of observation is known as the u, v-coverage and as found from the above transformation matrix, is different for different declination, δ. The celestial coordinates depend on the line of intersection of the ecliptic and equatorial planes. The u, v-coverage, in turn, depends on the position of the source in the celestial coordinate system. The reference line of such a coordinate system changes due to the precession of the Earth's rotation axis, hence the u, v-coverage turns out to be a function of the reference epoch for which the source position is specified. Each point in the u, v, w-plane measures a particular spatial frequency. This spatial frequency coverage differs from one epoch to another; therefore, it is advisable to process the source coordinates to the current coordinates prior to the observations. Processing of the visibility data for the purpose of mapping should be carried out with u, v, w evaluated for the epoch of observations.

The (u, v) coordinates correspond to a snapshot projection of the interferometer baseline on the plane of the wavefront of the incident stellar radiation from the source, in spatial frequency units (cycles/arcsec), thus,

$$
u = B_\lambda \sin \theta \sin \sigma = B_\lambda \cos \Delta_b \sin(H - \Theta), \tag{5.40a}
$$

$$v = B_\lambda \sin \theta \cos \sigma$$
$$= B_\lambda \left[\sin \Delta_b \cos \delta - \cos \Delta_b \sin \delta \cos(H - \Theta) \right]. \qquad (5.40b)$$

The law of sines provides $\sin \theta \sin \sigma = \cos \Delta_b \sin(H - \Theta)$; its multiplication with (5.24) gives,

$$\sin \theta \cos \sigma = \sin \Delta_b \cos \delta - \cos \Delta_b \sin \delta \cos(H - \Theta). \qquad (5.41)$$

The locus of these as hour angle, varies is an ellipse on the (u, v)-plane with center $(0, B \sin \Delta_b \cos \delta)$, semi-major axis $\cos \Delta_b$ and semi-minor axis, $\cos \Delta_b \sin \delta$.

The relation for the optical delay, τ_d, in terms of telescope and source parameters translates into;

$$\tau_d = -\frac{B}{c} \left[\delta \sin \Delta_b + \cos \delta \cos \Delta_b \cos(H - \Theta) \right]. \qquad (5.42)$$

In terms of (5.42), one may conclude that an interferometer operating at wavelength λ with projected baseline, $B \sin \theta$, measures the flux in the Fourier components of the source brightness at the spatial frequency, $B_\lambda \sin \theta$. It is pertinent to note that prior to the observations, the fringes should be located within a few millimeters zone around the dynamic position of the zero OPD. The optical path equalization is a must from a source to the interference location with an accuracy to a fraction of $c/\Delta v$ in which Δv is the bandwidth of the system, c the velocity of the light, and v the frequency of the beacon. Such an accuracy is obtained by means of a baseline cartography that is a set of triplets (B_X, B_Y, B_Z) corresponding to the sets of the positions of each telescope. At the Interféromètre à deux télescope (I2T; Labeyrie 1975), as well as at the Grand interféromètre à deux télescopes (GI2T; Labeyrie et al. 1986), the value B_X, B_Y, B_Z denote the location values along the rails, the horizontal, and vertical deviation with respect to the North–South direction respectively. The value of B_X is given to be 1 mm by employing a 'Geodimeter' before an observation. This is followed by the determinations of the values B_Y, B_Z, the accuracy of which should be less than 0.1 mm by interpolation through the pre-stored table. Besides, the apparent latitude of the baseline and declination of the source should be known to be 0.1 arcsecond. The right ascension and the setting of the computer clock should also be precise to the 0.1 s so that fringe drift is less than ~ 200 μm/hour for an average 20 m baseline.

5.3 Imaging Interferometry

Images of the source brightness distribution in the sky, using aperture-synthesis technique, can be obtained by Fourier synthesis of the observed complex visibility on different spacings. If the Fourier phases of the source distribution, which can be made possible to be recovered from the fringes that are detected from the sufficiently large number of baselines, are retrieved, the source image can be reconstructed. The expression for the image formation (1.74), can be recast into,

$$I(\mathbf{x}_1) = \int\!\!\!\int_{-\infty}^{\infty} K(\mathbf{x}_1 - \mathbf{x}_0')K^*(\mathbf{x}_1 - \mathbf{x}_0'')J_{12}(\mathbf{x}_0';\mathbf{x}_0'')d\mathbf{x}_0'd\mathbf{x}_0''. \qquad (5.43)$$

where $K(\mathbf{x}_1 - \mathbf{x}_0'')$ is the telescope transfer function (see 1.73), $\mathbf{x}_0'(= x_0', y_0')$, $\mathbf{x}_0''$ $(= x_0'', y_0'')$ are the position vectors of the fields, and $J_{12}(\mathbf{x}_0';\mathbf{x}_0'')$ the mutual intensities.

The (5.43) illustrates that in the general case, the system is linear with respect to the mutual intensities. This system has a 4-D transfer function and evaluation of this equation is generally complicated. The result depends on the respective width of the pupil function of the system, P, as well as the coherence area, A_c, of the incoming wave. For example, if $A_c \gg P$, it is in the coherent regime and if $A_c \ll P$, it is in the incoherent regime. According to the van Cittert–Zernike theorem, if the number of samples of the coherent function can be made large, the spatial frequency spectrum of the object can be reconstructed. The (2.70) may be modified as,

$$\mu_{ij} = \frac{\int\!\!\!\int_\sigma I(\boldsymbol{\alpha})e^{-i2\pi(\mathbf{u}_{ij}\cdot\boldsymbol{\alpha})}d\boldsymbol{\alpha}}{\int\!\!\!\int_\sigma I(\boldsymbol{\alpha})d\boldsymbol{\alpha}}$$

$$= \frac{\int\!\!\!\int I(\hat{\mathbf{s}})e^{-i\kappa\mathbf{B}_{ij}\cdot\hat{\mathbf{s}}}d\hat{\mathbf{s}}}{\int\!\!\!\int I(\hat{\mathbf{s}})d\hat{\mathbf{s}}}, \qquad (5.44)$$

where $I(\boldsymbol{\alpha})$ the intensity distribution of the observed source in angular coordinates, $\boldsymbol{\alpha}(= \alpha, \delta)$.

The degree of coherence between the parts of a telescope received at two separate points from a distant source is the Fourier transform of the brightness distribution of the source. It also implies that it may not be needed for all parts of a telescope mirror to be present at a time. Thus, it is possible to employ an array of small telescopes to synthesize a much larger telescope. The primary observable quantity in such an interferometer is μ_{ij}, called the visibility of the fringe pattern for the two sub-apertures situated in points p_i and p_j. Assuming that one looks at a point source, $S(\boldsymbol{\alpha}) = \delta(\boldsymbol{\alpha} - \boldsymbol{\alpha}_0)$, one has $\mu(\mathbf{u}) = 1$ for all spatial frequencies and the visibility of the fringe pattern is maximum for all baselines, $V(B) = 1$. The coherent addition of all the fringes provides a sharp concentration at the image position $\boldsymbol{\alpha}_0$. For larger pupil, i.e., $B > \sqrt{A_c}$, $|\mu(\mathbf{u})| \leq 1$, while in the limit, $B \gg \sqrt{A_c}$, the visibility of the fringes is null, which means the object is spatially resolved by the system.

5.3.1 Phase-closure Imaging

As stated earlier, the atmospheric turbulence perturbs the path of light rays randomly from the source to the different telescopes in an array, in spite of the complete correction of wavefront distortions across all the telescope apertures, the visibility phases may not contain any information about the astronomical object. Jennison (1958) devised the phase sensitive interferometer, which measures the amplitude and phase of the complex Fourier transforms of spatial brightness distribution of the source corresponding to the three interferometer baselines. This technique provides the phase information needed to reconstruct images and has been applied in the field of radio astronomy. Such a method is immune to the atmospherically induced random phase errors, as well as to the permanent phase errors introduced by the imaging systems.

Potential of the phase-closure technique lies in exploiting fully the resolution attainable with large optical telescope. With the introduction of the third sub-aperture, the Fizeau mask provides three pairs of sub-apertures and yields the appearance of three intersecting patterns of moving fringes. Baldwin et al. (1986) reported the measurements of the closure phases obtained at high light level with three hole aperture mask set in the pupil plane of the telescope and recorded the interference pattern of the object using CCD detectors.

The phase of the transform can be obtained from the phase of the fringe relative to a known datum. The observed phases, ψ_{ij}, on the different baselines contain the phases of the source Fourier components $\psi_{0,ij}$ and also the error terms, θ_j, θ_i, introduced by errors at the individual antennas and by the atmospheric variations at each antenna. The observed fringes are represented by the following equations,

$$\psi_{12} = \psi_{0,12} + \theta_1 - \theta_2, \tag{5.45a}$$

$$\psi_{23} = \psi_{0,23} + \theta_2 - \theta_3, \tag{5.45b}$$

$$\psi_{31} = \psi_{0,31} + \theta_3 - \theta_1, \tag{5.45c}$$

where the subscripts refer to the antennas at each end of a baseline, ψ_{ij} are the observed phases of the fringes produced by the antennas i and j on different baselines containing the phases of the source Fourier components, $\psi_{0,ij}$, and θ_i, θ_j, introduced by errors at the individual antenna, as well as by the atmospheric variations at each antenna and the subscripts refer to the antennas at each end of a particular baseline.

Figure 5.7 describes the concept of a three element interferometer. The sum of observed phases around a triangle of baselines, is the sum of phases of the source Fourier components and can be written as,

$$\beta_{123} = \psi_{12} + \psi_{23} + \psi_{31}. \tag{5.46}$$

This (5.46) displays the sum of the phases of the raw visibilities round a closed loop of interferometer baselines as the true source visibilities. The quantity, β_{123}, which

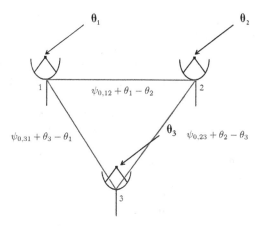

Fig. 5.7 Concept of closer phase measurements as applied to a three element interferometer

is free from all the systematic phase errors is called the 'closure phase'. It is free from systematic measurement errors irrespective of the separation and orientation of the interferometric baselines as well. The (5.46) is true for any interferometric system having more than three telescopes as well. Factorizing these into separate telescope and baseline dependent components, in which case the following expressions may apply,

$$\mathcal{V}_{12} = \left|\bar{\mathcal{V}}_{0,12}\right| e^{i\psi_{0,12}} G_1 e^{i\theta_1} G_2 e^{-i\theta_2} G_{12} e^{i\delta_{12}}, \tag{5.47}$$

with $\mathcal{V}_{12}$ as the measured complex visibility, $\left|\bar{\mathcal{V}}_{0,12}\right|$ and $\psi_{0,12}$ the true object visibility amplitude and phase, and G_1, G_2 the telescope related gain errors, θ_1, θ_2 the corresponding phase errors, and G_{12} and δ_{12} the baseline dependent gain and phase errors, respectively.

Thus, the observed visibility phases may be expressed as,

$$\psi_{12} = \psi_{0,12} + \theta_1 - \theta_2 + \delta_{12}. \tag{5.48}$$

If the errors are large and unknown, the measured fringe phases may not individually contain useful information about the source structure. In the case of small baseline dependent phase errors, the closure phase defined by (5.48), remains unaffected, regardless of the magnitude of the unknown telescope based errors, θ_1 and θ_2 and would preserve information about the structure of the source. Using the measured closure phases and amplitudes as observable quantities, the object phases are determined (mostly by least square techniques, viz., singular value decomposition, conjugate gradient method). From the estimated object phases and the calibrated amplitudes, the degraded image is reconstructed.

The amplitude function of the Fourier transform of brightness distribution across a source may be determined from the visibility of the fringes in the reception pattern of an interferometer. Similar to the closure phase, an analogus quantity, called closure amplitude, can also be defined for an interferometer comprising of at least four element arrays (Twiss et al. 1960, 1962). A four element interferometer forms six interferometric pairs to provide six visibility amplitudes, V_{12}, V_{23}, V_{34}, V_{41}, V_{31}, V_{24}. If a'_1 and a'_2 are the amplitude errors of the elements 1 and 2, respectively, the observed amplitude on baseline 12 is,

$$V_{12} = \bar{V}_{12}(1 + a'_1)(1 + a'_2), \tag{5.49}$$

where $\bar{V}_{12}$ is the true visibility amplitude for the baseline 12.

Similarly, one may write the visibility amplitudes for baselines 23, 34, and 41. The closure amplitude, A_{1234}, which should be error free, can be obtained by combining the observed visibility amplitude as,

$$A_{1234} = \frac{|V_{12}||V_{34}|}{|V_{23}||V_{41}|} = \frac{|\bar{V}_{12}||\bar{V}_{34}|}{|\bar{V}_{23}||\bar{V}_{41}|}. \tag{5.50}$$

5.3.2 Aperture-Synthesis Interferometry

Generally, three or more telescopes are required for aperture-synthesis technique. The direct measurements of the closure phase together with the measurements of visibility amplitude allow one to reconstruct an image of any object. The percent of phase information retained by the closure phases improves as the number of telescopes increases. To note, smooth reconstruction of the intensity distribution requires visibility measurements with many such baselines. For example, three-telescope interferometry can obtain about 33% of the phase information, while with 27 telescopes, the percentage of such information goes up to 92% (Downes 1988).

When the paths of the individual beams are matched to an accuracy of a sub-micrometer, the synthesis array acts as a single coherent telescope. The angular resolution of such an array is determined by the longest baseline. Aperture-synthesis method is based on,

- The measured visibility function, which is the spatial coherence function of the incoming wave field; it is modified by known characteristics of the interferometer elements, and
- The distribution of intensity of a source, which is computed over a limited FOV and with limited precision, provided spatial coherence function of the wave field is sufficiently well sampled.

Aperture-synthesis imaging technique with single telescope involves observing an object through a multi-aperture screen. Light can pass through a series of small holes (sub-apertures). A detector records the resulting interference patterns in a series of short-exposure. Each pair of sub-apertures introduces a set of overlapping two-holes interference fringe patterns at a unique spatial frequency in the image plane. Such patterns contain information about the structure of the object at the spatial frequencies from which an image of the same object can be reconstructed by measuring the visibility amplitudes and closure phases. According to the diffraction theory (Born and Wolf 1984), the image obtained in the focal plane of the system is the result of the summation of all such fringe patterns produced by all possible pairs of sub-apertures, i.e.,

$$I = \sum_{i,j} \langle A_i A_j^* \rangle. \tag{5.51}$$

The term, $A_i A_j^*$ in (5.51), is multiplied by $e^{i\psi}$, where ψ is the random instantaneous shift in the fringe pattern. Each sub-aperture is small enough for the field to be coherent over its extent. The aperture masks can be either placed in front of the primary or a secondary of a telescope or placed in a re-imaged its aperture plane. These masks are categorized either as non-redundant or partially redundant. The former category consists of arrays of sub-apertures in which no two pairs of holes have the same separation vector. This method produces images of high dynamic range (ability to detect a faint source near a bright one), but is restricted to bright objects. The other advantages are: (1) attainment of the maximum possible angular resolution by using the longest baselines and (2) built-in delay to observe objects at low declinations. However, the instantaneous coverage of spatial frequencies is sparse and most of the available light is discarded. The latter category is designed to provide a compromise between minimizing the redundancy of spacings and maximizing both the throughput and the range of spatial frequencies investigated (Haniff et al. 1989). The potential of this technique in the optical domain is demonstrated by the spectacular images produced with aperture-masking of a single telescope (Tuthill et al. 2001).

5.3.2.1 Aperture-Synthesis with Telescopes

In an interferometer with an array of telescopes, the light could be coherently combined. Each pair of telescopes yields a measure of amplitude of the spatial coherence function of the object. The visibility, $\widehat{\mathcal{V}}(\mathbf{u})$, can be measured simultaneously at various spatial frequencies, $\mathbf{u}$, corresponding to the different spacings between the pair of telescopes. Moreover for a given pair, as the Earth rotates, the projected spacing, as seen from the vantage point of the object being observed, changes continuously, measuring the visibility over a range of spatial frequencies in the u, v-plane. Therefore, a requisite number of measurements in the spatial frequency domain can be obtained in one single night. This is known as Earth-rotation aperture-synthesis method.

Table 5.1 Functional components of an optical imaging array

Components	Parameters to consider
Beam combiners	Number and nature of combiners
Beam transporter	Free or guided
Collectors	Size, number, array design
Delay compensator	Vacuum or air
Detector	Sensitivity, temporal and spectral resolution

The image of the object can be obtained by inverse Fourier transforming (5.6) using the visibilities obtained using all possibles pairs of baselines. Having the advantage of being cost effective, developing such an instrument avoids severe structural problems of developing very large optical telescope and allows large effective collecting area to be obtained with a minimum of engineering structure. The functional components of an optical imaging array is given in Table 5.1.

To reiterate, for n elemental aperture-synthesis, each pair of telescopes provides an ellipse on the u, v plane by tracking a source, although some part of it would be missed out due either to the source having set below the horizon or to hour angle limitations of the telescopes. The output of a synthesized array is a large collection of complex visibilities at points on the (u, v)-plane determined by the baseline ellipses. The quality of the image depends on the available u, v-plane data.

In such an arrangement, there are $n(n-1)/2$ independent baselines tracking in the u, v plane, with $n-1$ unknown phase errors. There are n amplitude errors related to the n telescope elements. Thus, data are sampled by a set of partial ellipses, with a data point taken at every integration point. However, the number of independent closure phases is dictated by $(n-1)(n-2)/2$, which is equivalent to holding one telescope fixed and forming all possible triangles with that telescope (Readhead et al. 1988); one obtains $n(n-3)/2$ independent closure amplitudes as well. They are free of all the telescope based errors. In such n interferometer elements, there are p closure relations of type given in (5.46), which can be represented by a matrix notation of kind,

$$\mathbf{H}_\psi \cdot \boldsymbol{\psi} = \boldsymbol{\psi}_c, \tag{5.52a}$$

$$\mathbf{H}_A \cdot \mathbf{A} = \mathbf{A}_c, \tag{5.52b}$$

in which $\mathbf{H}_\psi$ is the closure phase matrix of size $p \times q (p < q)$, $\mathbf{H}_A$ the closure amplitude matrix of size $r \times q (r < q)$; both matrices are composed of elements 0 or ± 1, $\boldsymbol{\psi}$ the column vector of q observed phases, $\boldsymbol{\psi}_c$ the column vector of p closure phases, and q the number of observed phases, $n(n-1)/2$.

The signal in the nth segment due to a source of emission is expressed by,

$$V_n = a_n cos(\omega t + \psi_n), \tag{5.53}$$

in which a_n is the amplitude of the signal and ψ_n the relative phase of the radiation.

If these signals are added together vectorily and time averaged, the intensity of the light I_n is derived as,

$$I_n \propto \frac{1}{2} \sum_{j=1}^{N} \sum_{k=1}^{N} a_j a_k cos(\psi_j - \psi_k)$$

$$= \frac{1}{2} \sum_{j=1}^{N} a_j^2 + \sum_{j=1}^{N-1} \sum_{k=j+1}^{N} a_j a_k \cos(\psi_j - \psi_k). \qquad (5.54)$$

where j and k represent the individual elements.

The first term on the right hand side of (5.54) is proportional to the sum of the power received from the individual elementary apertures, which does not contain high resolution information. The second term contains the cross product and the phase terms and represents the way in which the power changes with source direction. It is the cross terms that posses the response to high spatial frequencies, and hence determine the resolving power.

An image can be reconstructed from sequential measurements of all cross products using pairs of telescopes. Each term can equally be measured with two elementary areas in positions j and k. Following (3.42), the term $\psi_j - \psi_k$ is expressed by,

$$\psi_j - \psi_k = \frac{2\pi}{\lambda} \mathbf{B}_{j,k} \cdot \hat{s}, \qquad (5.55)$$

where $\mathbf{B}_{j,k}$ is the separation of the two elemental areas and $\hat{s}$, the unit vector defining the source.

5.3.2.2 Array Configuration

A single interferometer with a fixed baseline measures one Fourier component of sky brightness within the envelope pattern of the instrument. So by sampling the source's complex visibility function at particular intervals in the baseline one can obtain brightness distribution from the Fourier transform of the visibility function. A number of telescopes are usually placed in arrays in two ways such as:

1. Linear tracking arrays: In a two element array, separated by a distance, B, the net far field in the direction of θ is given by,

$$E(\theta) = E_1(\theta)e^{i\psi/2} + E_2(\theta)e^{-i\psi/2}, \qquad (5.56)$$

where $\psi = \kappa B \sin \theta + \delta$, $\kappa = 2\pi/\lambda$ is the wavenumber and δ is the intrinsic phase difference between the two sources, E_1 and E_2 the amplitudes of the electric field due to the two sources at the distant point under consideration; the phase center is taken halfway between two elements.

Fig. 5.8 Linear tracking
array with two elements

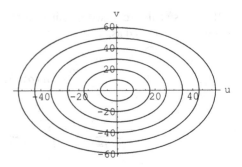

With the equal strength of both sources, (5.56) turns out to be,

$$E(\theta) = 2E_0 \cos(\psi/2). \tag{5.57}$$

The power pattern is obtained by squaring the field pattern. As stated earlier, the length and orientation of interferometer baseline changes due to the Earth's rotation. This causes u, v-plane to sample in an elliptical way during a 24-hr period. A series of ellipses are formed in u, v-plane when telescopes are placed in East–West direction. The ellipses (see Fig. 5.8) are centered at origin of (u, v). If the spacings between the telescopes increase uniformly, these ellipses are concentric with uniform increment in their axes.

The number of baselines that can be traced simultaneously with n telescopes are $n(n - 1)$. The main theme of array telescopes is to have maximum coverage of u, v-plane. But to have continuous Fourier components, all the intermediate spacings are required. If the source is constant over an extent of observations, it is not necessary that all the necessary baselines simultaneously exist. In the case of a uniform linear array of n elements of equal amplitude, the far field pattern is expressed as,

$$E(\theta) = E_0 \left[1 + e^{i\psi} + e^{i2\psi} + \cdots + e^{i(n-1)\psi} \right], \tag{5.58}$$

in which $\psi = \kappa B \sin \theta + \delta$, δ is the progressive phase difference between the sources.

The sum of this geometric series is,

$$E(\theta) = E_0 \frac{\sin(n\psi/2)}{\sin(\psi/2)} e^{i(n-1)\psi/2}. \tag{5.59}$$

If the array center is chosen as the phase reference point, the above result will not have the phase term of $(n - 1)\psi/2$. In order to obtain the total field pattern for non-isotropic but similar elements, E_0 is replaced by the element pattern, $E_i(\theta)$. The field pattern in (5.59) has a maximum value of nE_0, if $\psi = 0, 2\pi, 4\pi, \cdots$.

The interferometer spacings can be filled up by fixing a linear array of telescopes and having one or more movable telescopes (Booth 1985). In this case, one samples the interferometer output at several positions within the aperture size by displacing the telescopes to different stations along the baselines and collects data on two baselines simultaneously. This technique was applied in radio astronomy; the map of Cassiopeia A was made with three antennas at Cambridge (Rosenberg 1970). Let B^{EW} and B^{NS} be the orthogonal East–West and North–South components of the baseline vector at the ground of an interferometer located at the terrestrial latitude Φ, the (u, v) point sampled from a star of declination δ, when its hour angle is H, is given by,

$$u = B_\lambda^{EW} \cos H - B_\lambda^{NS} \sin \Phi \sin H \qquad (5.60a)$$

$$v = B_\lambda^{EW} \sin \delta \sin H + B_\lambda^{NS} (\sin \Phi \sin \delta \cos H + \cos \Phi \cos \delta), \quad (5.60b)$$

where Φ is the latitude, H the hour angle related to the Greenwich sidereal time, and δ the declination of the source.

2. Phased arrays and correlator arrays: Phased array is a combination of telescopes in which all the telescopes are connected to a single power combiner. The signals from an n-element phased array are combined by adding the voltage signals from the different telescopes after proper delay and phase compensation. This summed voltage is put through a detector, and the output is proportional to the power in the summed signal. For identical elements, this phased array provides a sensitivity, which is n times the sensitivity of a single element, for observing a point source. The beam of such an array is narrower than that of the individual elements. In certain cases, the voltage signal from each element of the array is put through a detector, which is followed by adding the powers from the elements to obtain the final output of the array. This corresponds to an incoherent addition of the signals from the array elements, while the first method gives a coherent addition. In the incoherent phased array operation, the beam of the resultant telescopes has the same pattern as that of a single element, since the phases of the voltage from individual elements are lost. This beam-width is wider than the coherent phased array telescope. The sensitivity to a point source is higher for the coherent array telescope as compared to the incoherent phased array telescope, by a factor of $\sqrt{n}$. However, such incoherent array method is useful for extended source. Another method, known as correlator array, in which the telescopes are connected pairwise in all possible combinations, is also employed.

Phased arrays of small telescopes can be used as single elements in correlator arrays. If voltages at all telescope outputs are $V_1, V_2 \cdots$, the output is proportional to the square of $V_1 + V_2 + \cdots + V_n$, in which n is the number of telescopes. In the case of n telescopes there are $n(n - 1)$ cross product terms which are of the form $V_i V_j$, where $i, j = 1, 2, \cdots, n; i \neq j$. If the electrical pathlength of the signal from each telescope to the detector is same, the signals are combined in phase, and the direction of the incoming radiation is given by,

$$\theta = \sin^{-1}\left[\frac{N}{l_\lambda}\right], \qquad (5.61)$$

where N is an integer and l_λ the spacing between the telescopes measured in wavelengths.

The outputs of the correlator array are cross products of voltages of two connected telescopes. These are equal to the cross product terms in a phased array (Mills and Little 1953). The loss of self product terms reduces the sensitivity of the correlator array to extended sources.

In a cross (+) shaped array aperture, the width of arms is finite but small compared with length of the arms. The outputs of two arms (which are usually in the East–West and North–South directions) go to a single cross correlating receiver, so that the spatial sensitivity is a square. A T-shaped array uses the East–West arm and half of North–South arm of a cross array. The equivalence between the spatial transfer function of cross and T shaped arrays is due to the fact that any pair of points in the aperture of a cross, for instance, one on East arm and one on North arm, there is a corresponding pair on the West and South arms for which the spacing vector is identical. Thus one of the four half length can be removed without effecting the (u, v) coverage. But the sensitivity would be less in a T array since the antennas in one half of one of the arms in a cross array is absent (Christiansen and Högbom 1985).

5.3.2.3 Beam Combination

It is to reiterate that the light from the same source may be combined either by refocusing the beams in a common focal plane yielding spatial fringes or by introducing a beam-splitter resulting in temporal fringes. However, the position of the reimaged telescope pupils dictate the type of interferometer (see Fig. 5.9):

1. Image-plane scheme: This approach involves adding the fields from each telescope in a focal plane, for example Fizeau interferometry (see Sect. 3.2.1). The advantage of this Fizeau mode is that the ratio aperture diameter/separation is constant from light collection to recombination in the image plane (homothetic pupil). In this case, the baseline length, B is equal to the combination baseline B_0

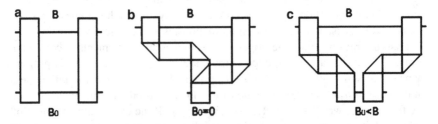

Fig. 5.9 Schematic diagrams of the optical set-up required for various types of beam combination. The exit pupil placement for a baseline B_0: (**a**) homothetic or image-plane (Fizeau interferometer; $B = B_0$), (**b**) co-axial or pupil-plane (Michelson interferometer; $B_0 = 0$), and (**c**) densified (Michelson stellar interferometer; $B_0 < B$), schemes; B and B_0 are the collection and combination baselines respectively. In the case of the pupil-plane configuration, each of the two complementary outputs from the beam-splitter is focused onto a single pixel detector and the intensity is measured as a function of a introduced OPD

(see Fig. 5.9a). The angular size of the image is governed by the diameters of the incoming beams. This sets the minimum size of the detector array that displays a superposed fringe pattern. On the other hand, the fringe period and the necessary pixel size depend on the physical separation of the incoming beams at the focusing optics. Image-plane scheme is generally used for multi-axial combiners, where the configuration of the telescopes, as seen from the target, is re-imaged to a smaller scale, maintaining orientation and relative separations, before the beams interfere. The fringe encoding relies upon using a non-redundant input pupil, so that every pair of beams corresponds to a identifiable vector separation and hence spatial fringe period on the detector. Taking a large number of images for many different configurations, one may superpose the Fourier transforms of these images and reconstruct the image.

2. Pupil-plane mode: Unlike Fizeau mode, in Michelson stellar interferometry, the ratio of aperture diameter/separation is not constant since the collimated beams have the same diameter from the output of the telescope to the recombination lens. In this case, the spatial modulation frequency in the focal-plane is independent of the distance between the collectors. The distance between pupils is equal to the baseline, B, at the collection mirrors and to a much smaller value just before the recombination lens (see Fig. 5.9b). The beams from an array of telescopes are simply overlapped or combined in the image plane without maintaining the input pupil configuration. The light coming from an off-axis direction has a different delay than the light coming on-axis, referred to as differential delay. At angles where the differential delay becomes higher than the coherence length, the fringes disappear and the high resolution information on the off-axis objects is lost. In this scheme, images of the apertures are given a virtual zero separation, re-imaged and combined by means of a beam-splitter or its equivalent fiber or integrated optics (see Sect. 6.1.3.4) components and the two resulting output beams are each focused on a single detector. In classical Michelson's experiment, a beam-splitter was used, by which the fields could be added at such a splitter and any optics thereafter would serve to deliver the resulting intensity to a suitable detector. By modulating the time delay of one beam with respect to the other, the interference can be modulated and detect interferometric fringes. Also, it is possible to disperse each of these beams to provide a one-dimensional array of spectro-interferometric measurements. The coherent addition is no longer spatially displayed, but is read point by point and the value depends on the OPD, $d(x) = c\tau$; the fringes are visualized by introducing an OPD between the beams as a function of time. The variation of the observed signal against $d(x)$ in the neighboring $d(x) = 0$ is comparable to an interference pattern obtained with Fourier spectrometer. The combined beam intensities is given by (Traub 1999),

$$I_{\text{int}} = 2I_{\text{tel}} \left[1 \pm \mathcal{V} \sin(2\pi \Delta d(t)/\lambda) \right], \qquad (5.62)$$

where $I_{\text{tel}} = \int I_{\text{tel}}(\theta) d\theta$ and $\mathcal{V}$ the visibility.

Here, a dephasing of $\pi/2$ at the beam-splitter occurs and leads to a sine modulation, providing an odd function. In multi-beam plane-plane combiners (Mozurkewich 1994), non-redundant modulation of the optical paths in each beam is used to give rise to a modulated intensity output that is separable into various temporal frequency components (Haniff 2007). A Michelson-type interferometer has the advantage of concentrating all the photons in one central peak in the fringe pattern. It works better with a few telescopes having longer baselines.

3. Pupil densification: This method of interferometric recombination was implemented by Michelson in his stellar interferometry with the focal plane image being crossed by fringes. In this case, the combination baseline, B_0 is less than the collection baseline, B (see Fig. 5.9c). In order to circumvent the 'golden rule' (Traub 1986) that an imaging interferometer must have an exit pupil identical to the entrance aperture, Labeyrie (1996) has imagined such a method for very large array. Conceptually, it differs from Fizeau interferometer, the S/N-ratio decreases with the number of telescopes, which limits the number of baselines that can be used in a system. Besides, when the size of the baseline is large compared to the single telescope diameter, a complex interference occurs and a dominant central peak, surrounded by many faint side-lobes filling the halo of diffracted light caused by the small sub-apertures, appears at the center of the image if the phasing is not disturbed by the atmosphere or by imperfections of the focusing optics. The relative amount of energy in the central peak becomes vanishingly small if the aperture becomes highly diluted. Densifying the exit pupil, for instance, distorting it to increase the relative size of the sub-pupils, in such a way that the pattern of sub-aperture centers is preserved, concentrates the halo and intensifies the image (Labeyrie 1996). In the recombination plane, the distance between two pupils corresponding to two telescopes is minimized to become about equal to their diameter. This principle can be a optimal solution for instruments with many telescopes (see Sect. 7.5.2) and short baselines.

5.3.2.4 Array Beams

The response of an aperture synthesis array can be calculated using the equation for the corresponding transfer function (3.50), since the analysis is identical to that of a single telescope. Of course, one needs a large number of telescopes in order to fill the (u, v) plane. Since the ability to make an image depends on the filled fraction of u, v plane, most arrays of telescopes are arranged in u, v-plane in an optimum manner; the more telescopes in an array, the better the observed image/map.

Since the visibility amplitude and phase measured by an interferometer is directly related to a single component of the Fourier transform of the source brightness distribution, the source distribution can be obtained if $\widehat{\mathcal{V}}(\mathbf{u})$ is measured over the entire (u, v)-plane. While reversing the baseline vector, the visibility is conjugated, i.e., $\widehat{\mathcal{V}}(\mathbf{u}) = \widehat{\mathcal{V}}^*(-\mathbf{u})$, and hence, no new information is obtained. For example, when the number of samples of the coherent function is large, the spatial frequency spectrum of the object can be reconstructed. When such measurements are carried out at several different separations of the apertures, they yield the Fourier transform of the distribution of intensity across the star, and hence the angular diameter of the star.

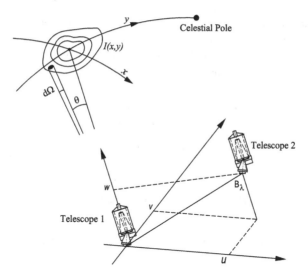

Fig. 5.10 Relation between baseline and image coordinates of a two element interferometer

It is reiterated that a tracking interferometer gives the information regarding amplitude and phase corresponding to the u, v, w value at a given instant of time. Significant variation of this value must not be allowed during the integration, or else results in loss of information. According to the van Cittert–Zernike theorem, this set of numbers provides one Fourier component of the source brightness distribution. Figure 5.10 shows the geometry for deriving the generalized 2-D Fourier transform relation between the visibility and the source brightness distribution in the u, v, w coordinate system. The correlator output is given by the Fourier transform of the sky brightness,

$$\widehat{\mathcal{V}}(u, v, w) = \int_{-\infty}^{\infty} \frac{I_{int}(x, y)}{\sqrt{1 - x^2 - y^2}}$$
$$\times e^{i2\pi\left[ux+vy+w\left(\sqrt{1-x^2-y^2}-1\right)\right]} dx\,dy. \tag{5.63}$$

Presume that in an array, where the aperture distribution is produced by many spatially distinct telescope apertures, all telescopes are precisely located in the (u, v)-plane, w is exactly zero, and (5.63) reduces to the 2-D Fourier transform relation between the source brightness distribution and the visibility. This holds good for a perfect East–West array. The integral in (5.63) is finite for a small portion of the sky. If the FOV is small, i.e., for small x and y, $\sqrt{1 - x^2 - y^2} - 1 \approx -(x^2 + y^2)/2$ and may be neglected. A simpler expression emerges by ignoring w,

$$\widehat{\mathcal{V}}(\mathbf{u}) = \int_{-\infty}^{\infty} I'_{int}(\mathbf{x}) e^{i2\pi \mathbf{u} \cdot \mathbf{x}} d\mathbf{x}, \tag{5.64}$$

where $I'_{int} = I_{int}/\sqrt{1 - x^2 - y^2}$ and $\mathbf{x} = x, y$.

It is pertinent to note that ignoring the w-term puts restrictions on the FOV that can be mapped without being effected by the phase error which is approximately equal to $\pi(x^2 + y^2)w$. This (5.64) shows the 2-D Fourier transform relation between the intensity and visibility, which may be inverted,

$$I'_{int}(\mathbf{x}) = \int_{-\infty}^{\infty} \widehat{\mathcal{V}}(\mathbf{u}) e^{-i2\pi \mathbf{u} \cdot \mathbf{x}} d\mathbf{u}, \tag{5.65}$$

Similarly, for an extended object, the observed visibility function, $\mathcal{V}$, may be expressed as,

$$\mathcal{V}(\mathbf{B}_\lambda) = \int_{-\infty}^{\infty} I_{int}(\theta) e^{i2\pi \mathbf{B}_\lambda \cdot (\hat{\mathbf{s}} + \alpha)} d\alpha, \tag{5.66}$$

with α is a small displacement from the field center, and hence the phase term in (5.66) may be expanded to first order accuracy,

$$\mathbf{B} \cdot (\hat{\mathbf{s}} + \alpha) \simeq \mathbf{B} \cdot \hat{\mathbf{s}} + \mathbf{B} \cdot \alpha$$
$$= \mathbf{B} \cdot \hat{\mathbf{s}} + \mathbf{b} \cdot \alpha + (\mathbf{B} \cdot \hat{\mathbf{s}}) \cdot (\hat{\mathbf{s}} \cdot \alpha). \tag{5.67}$$

The projected baseline $\mathbf{b}$ is composed into E-W and N-S components, u and v. This will lead the third term in (5.67) to be equal to zero, when $\hat{\mathbf{s}} \cdot \alpha = 0$, i.e., $\hat{\mathbf{s}}$ is perpendicular to α.

In an array, let a set of spacings be $\mathbf{B}_{j=1,2,\cdots,N}$ and the transfer function be $P_A(x)$, therefore the transfer function of the array is given by,

$$P(x) = \sum_{j=1}^{N} P_A(\mathbf{x} - \mathbf{B}_j), \tag{5.68}$$

and the intensity distribution of the array, $I(\alpha)$, is

$$I(\alpha) = A(\alpha) \sum_{j=1}^{N} e^{-i2\pi \mathbf{B}_j \cdot \alpha}, \tag{5.69}$$

where $A(\alpha)$ is the intensity distribution of the individual aperture.

By ignoring the intensity distribution of the individual elements, $A(\alpha)$, the synthesized beam of an array is recast in rectangular coordinates,

$$I(\mathbf{x}) = \sum_{j=1}^{N} e^{-i2\pi \mathbf{u}_j \cdot \mathbf{x}}. \tag{5.70}$$

This (5.70) is a set of visibility functions, $\widehat{\mathcal{V}}(\mathbf{u}_j)$, proving a resultant intensity distribution,

$$I(\mathbf{x}) = \sum_{j=1}^{N} \widehat{\mathcal{V}}(\mathbf{u}_j) e^{-i2\pi \mathbf{u}_j \cdot \mathbf{x}}. \tag{5.71}$$

The point spread function (PSF) or synthesized beam of the array is the intensity distribution obtained by observing a point source with unit visibility. The observed intensity distribution is the convolution of the real intensity distribution with the synthesized beam. Since there are finite number of telescopes in an aperture-synthesis array, the u, v-coverage is not continuous. The maximum baseline length, B_{max} provides the resolution λ/B_{max} and the gaps in u, v coverage give rise to sidelobes in the synthesized beam. At some angles, signals from parts of the aperture add up in phase to produce ripples in intensity distortion. The major causes of sidelobes are due to (1) diffraction at the edge of the aperture, (2) sparse aperture coverage in the (u, v) plane, and (3) discrete nature of the array. Errors in the synthesized pattern may also appear due to a phase offset, a phase variation, a baseline error or a gain variation, delay stepping, frequency dependence of the system, and time-constant effects. The beam of the synthesized telescope is also given by the Fourier transform of a set of unity values at each u, v-point, which is the same as the point source response.

The visibility functions are to be obtained by averaging the quasi-sinusoidal response of each interferometer pair in the array. It is necessary to average such function of all baselines at equal intervals. Data are required to be Fourier transformed in order to produce required aperture-synthesis map of source intensity distribution. Complex visibility function values are calculated for every point in the $u - v$ plane. Sky brightness distribution is obtained by taking the Fourier transform of the above results.

Chapter 6
Basic Tools and Technical Challenges

6.1 Requirements for the LBOI

Interferometric imaging requires detection of very faint signals and reproduction of interferometric visibilities to high precision; the time resolution of the detectors should reach 1 ms. Developing such an interferometer using Michelson technique is very challenging, which is linked with the advancement of required technology in the areas of opto-mechanics, opto-electronics, and computing. Apart from the design and construction of such an instrument, the physical stability, equalization of pathlength, fringe-tracking, vibration control, adaptive optics, dispersion, and calibration problems are necessary to look into. As stated earlier (see Sect. 5.1) the fringes can be obtained if the optical path difference (OPD) between the light from a stellar source through the two telescopes to the point of interference is smaller than the coherence length, $\leq l_c$. Over an extended field-of-view (FOV), this OPD needs to be continuously measured and corrected before the beam combination takes place. The contrast of the fringes tones down due to:

- non-zero OPD that may cause the interference fringes to tilt, which is due to a progressive phase variation of the fringes according to the wavelength; its effect can be revealed by a shift in the peak as compared to the central position;
- if the two interfering beams do not posses same intensity, which may be affected by a progressive defocalisation of the images due to the temperature variations, or a difference in transmission between the individual light paths; and
- if the polarization planes of the interfering field undergo an angular deviation during the diurnal movement of the source that is larger as it is observed further from the transit; interference occurs with a reduced efficiency, which in turn, attenuates apparent visibility.

This requires high optical quality and high-precision tracking instruments. Another difficulty arises from combining the beams in phase with each other after they have traversed exactly the same optical path from the source through each telescope down to the beam combination point. The situation is further complicated by the effects of atmospheric turbulence. As stated earlier (see Sect. 3.2.2) that one of the contributors to OPD is the geometric time delay, τ_g, associated with relative orientation of the source and the baseline B of the interferometer (see 3.40). An error in

S.K. Saha, *Aperture Synthesis*, Astronomy and Astrophysics Library,
DOI 10.1007/978-1-4419-5710-8_6, © Springer Science+Business Media, LLC 2011

delay of a few nanoseconds may cause the different parts of the observing band to partially interfere, leading to a drop in the amplitude; different frequencies in the band arrive with different phases. In order to determine τ_g, it is required to obtain time of the fringes accurately and also locate the telescopes. The pathlength from the astronomical source to the detector is determined by,

$$d_a(t) = \mathbf{B} \cdot \cos \delta \sin H(t), \tag{6.1}$$

in which δ and H are, respectively, the declination and the hour angle of the source.

The time delay, τ, between the waves arriving at x in the image plane is given by,

$$\tau = \tau_o + \frac{d(x)}{c}, \tag{6.2}$$

where $\tau_o = \tau_g - \tau_i$, τ_g is the geometrical delay due to the Earth's rotation, τ_i the compensating instrumental delay, and $d(x)$ the OPD depending on the imaging set-up.

If the delay τ_o is kept within the coherence time, τ_c (see 2.2), the interference term may appear as an intensity cosine-modulation within the composite image, that is, the superposition of two individual images on the final image plane (on the detector), along the spatial coordinate and its observation requires a two dimensional detector. The period of modulation is fixed by the separation of the images of each aperture. Failing to achieve the zero path difference, the amplitude of the cosine falls gradually to zero as τ_o increases and the modulation is not observable. Another requirement for developing an interferometer is to have a very complex real-time control system. The necessary hardware and software are to be developed. Several interferometers use a distributed control system and a mixture of real-time and non-real-time operating systems.

6.1.1 Delay-line

Delay-line system enables to maintain the same length of the light path from a star, independent of the telescope through which it has been caught. The delay-line mechanism is one of the few methods by which light beams from various telescopes of an optical interferometer can be combined in a proper manner to produce interferometric fringes at the focal point. The differences in the length of the light path from a star occur owing to (1) the static geometric path length difference between the telescopes in a certain configuration, (2) the diurnal motion of the source during observation due to Earth's rotation, and (3) the atmospheric disturbances and/or mechanical vibrations along the optical path length.

In order is to correct the path variation, a carriage carrying cat's eye retro-reflecting optics is translated; cat's eye reflectors allow to keep beams aligned automatically. This mechanism, called delay-line, switches in and out various path-lengths to compensate the geometric delay so that the light that left the star at the same time arrives simultaneously at the beam combining optics to produce interference fringes (Shao et al. 1988a). The position of the carriage is monitored using a laser metrology hardware to provide a precise delay to a fraction of the wavelength of observation divided by the fractional bandwidth used. The optical elements are, in general, vibrationally decoupled from the moving carriage and the motion of the entire assembly is controlled by a combining effort of a coarse motor stage and a precisely commanded electromagnetic actuators.

Beam combination is being carried out by means of delay-lines in many LBOIs. Although delay-line technology is a difficult process to implement, it is the most advanced and suitable technique since it permits by and large unlimited excursions in optical path within the range defined by the apparent motion of the source from East to West. Such a technology is also used for a phase interferometry observations. Figure 6.1 depicts a schematic diagram of a cat's eye based delay-line.

At the Mark III interferometer (Shao et al. 1988a), the retroreflection is produced by focusing the incoming beacon to a point coinciding with a small flat mirror attached to a piezo-electric stack, which reflects and re-collimates (see Fig. 6.1). The beams from the two arms are combined with a beam splitter, which is split again in two beams, one directed towards the optical correlator, the other towards the star tracker. The siderostats keep the two interfering wavefronts parallel to within a small fraction of an arcsecond. This was realized by a star tracker. In order to provide smooth tracking, a computer determines the corrections to the position of the siderostats and controls them by steps of 0.02 arcsecs. A receiving system that was employed to analyze the interference pattern, could be used to compute the corrections to the position of the mirror. This mirror can be driven on precision rails by a motor drive. Of course, the difficulty arises from avoiding various aberrations and vignetting, particularly when light is fed through long and narrow pipes. In this set-up, the laser interferometers measure the position of the delay-line carts.

Several present day interferometers, viz., US Navy Prototype Optical Interferometer (NPOI; Armstrong et al. 1998), Infrared Optical Telescope Array (IOTA;

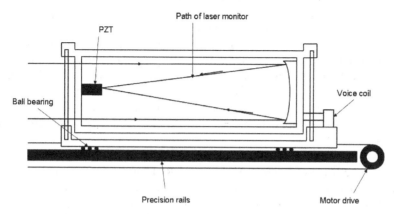

Fig. 6.1 Schematic of re-combination optics in the Mark III interferometer interferometer

Carleton et al. 1994), at Mt. Hopkins, Arizona, USA use the vacuum delay-lines. Since the geometric delay occurs in vacuum, its correction should be implemented using vacuum delay-lines. Delay-lines of IOTA, have a 28-m long delay-line which is moved each time a new object is observed. It does not move during fringe acquisition. IOTA has a second 'short' delay-line of about 2 m, which tracks the sidereal motion during fringe acquisition. Both the delay-lines use a dihedral (two plane mirrors at 90°) mounted on a carriage. For the long delay-line, the carriage is moved by a pulley-and-cable system powered by a stepper motor, while for the short delay-line a linear motor system allows precise motion of the carriage (10 nm steps). For both delay-lines, measurement of the carriage position carried out is by a laser metrology system using the Doppler–Fizeau effect of a laser beam that is sent to the carriage and bounced back.

Delay-line system at the Very Large Telescope Interferometer (VLTI), Chile and Keck Interferometer, Mauna Kea, Hawaii, consists of a retro-reflector mounted on a moving cart; the cat's eye used in this set-up is a Ritchey–Chretien type. The moving cart enables the cat's eye to travel along a 60-m long rail track with an accuracy of better than 10 nm, thereby providing corrections for the OPD of up to 120 m. This delay-line is based on linear motor technology, combined with high accuracy piezo-electric control elements. A fringe sensing unit provides a signal to the delay-line system via a fast link to the delay-line local control unit. An optical data-link to the cat's eye on the carriage ensures the transfer of data to the Piezo controller.

6.1.2 Spatial Filtering

Spatial filter is an optical device that uses the principles of Fourier optics to alter the structure of a beam of coherent light or other electromagnetic radiation. Often a light beam passes through a system, dust in the air or on optical components may disrupt the beam and create scattered light. The latter can leave unwanted ring patterns in the beam profile. A spatial filter is designed to remove unwanted multiple-order energy peaks and pass the central maximum of the diffraction pattern. This may be performed either by a micron-sized aperture or by a fiber optics system. The first application of spatial filtering was to the microscope, in Abbe's theory of image formation. It is also used to clean up the output of lasers, removing aberrations in the beam due to imperfect, dirty, or damaged optics, or due to variations in the laser gain medium itself. Spatial filtering in a Fourier transform spectrometer modifies the output by adjusting the amplitude and phase of the measured longitudinal correlation function.

In an interferometer, this approach helps to spatially filter the light being delivered to the beam-combiners (Shaklan and Roddier 1988), thus smoothing the turbulence-induced corrugated wavefronts and one finds that the contribution variations are much reduced. The incoming wavefronts propagate through a spatial filter that selects a single geometric mode of the beams. With an extended source, spatial

Fig. 6.2 Principle of wavefront smoothing by spatial filtering with a pin-hole (*top panel*) and with a single mode fiber (*bottom panel*)

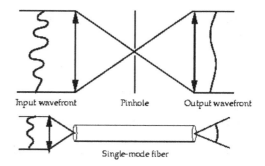

filtering is required to measure a proper contrast. The advantage of such a technique for interferometry is a reduction of the uncertainty on the measured visibility. However, the drawbacks of spatial filtering are a loss of optical coupling efficiency and larger photometric variations due to the turbulence. Figure 6.2 depicts the principle of wavefront smoothing by spatial filtering with a pin-hole and a single-mode fiber. Both pin-holes and single-mode optical fiber can act as suitable spatial filters though many interferometric implementations use fiber components. However, it is wise to use spatial filter, if the size of a telescope exceeds a few times than Fried's parameter, r_0.

Fibers are used for telecommunication because of their high efficiency in carrying light over long distances ($\sim$100 km). A spool of fibers that can be stretched or relaxed to increase or decrease the optical pathlength transmit light beam from one place to other. They are widely used to connect telescopes to spectroscopes, which enables the latter to be kept away from the former in a temperature-controlled room. This reduces the problems of flexure that occur within telescope-mounted spectroscopes as gravitational loads change with the telescope orientations.

An optical fiber eliminates the need for mirrors or lenses, and alignment required for these elements. It exploits total internal reflection by having an inner region of low refractive index and a cladding of higher index; light is confined by repeated reflection. Most of the research performed in recent years for this purpose is devoted to the development of efficient fibers and waveguides for the infrared.

Fiber-optics posses a refractive index structure that guides the beam. Such cables usually consist of a thin (10–500 μm) filament of glass encased in a cladding of another glass with a lower refracting index. One or more of these strands make up the cable as a whole. Light entering the core is transmitted by multiple internal reflections off the interface between the two glasses, provided that its angle of incidence exceeds the critical angle,

$$\theta_c = \sin^{-1} \sqrt{\mu_{co}^2 - \mu_{cl}^2}, \tag{6.3}$$

in which μ_{co} and μ_{cl} are, respectively, the refractive indices of the core and cladding, for total internal reflection.

These cables are, in general, characterized by their numerical apertures, $NA = \sin \theta_c$. The minimum focal ratio F_{min} that can be transmitted by the core is dictated by $\sqrt{1 - NA}/2NA$. The noted advantages of fiber can be envisaged in the form of

- having high throughput; 100 m silica fiber has a throughput of $\sim 99\%$ at $\lambda = 1.6\mu m$,
- offering flexibility since the degrees of freedom are located at its entrance and output,
- selecting the plane-wave part of a wavefront,
- splitting a guided wave into any desired intensity ratio, and
- combining two guided waves interferometrically.

The disadvantage of transmission by optical fibers is the signal attenuation due to propagation and the need for repeaters to reformat and amplify the optical signals after long distances. The other notable problems of beam transportation by fibers are:

- chromatic dispersion due to mismatch of the different fibers when light is made to pass through them,
- injection losses or function losses,
- mechanical and thermal sensitivity, and
- birefringence[1] of the material introducing elliptic polarization at the output, when linearly-polarized light is injected. This causes a loss of the measured visibility.

However, certain instruments like FLUOR (Fiber-Linked Unit for Optical Recombination; Coudé du Foresto and Ridgway 1992) or VINCI (VLT INterferometer Commissioning Instrument; Kervella et al. 2000) uses very limited length of fibers, so they are not affected by chromatic dispersion or birefringences. For sufficiently small core sizes, a few times the wavelength in question, the field inside the fiber has one single mode. Such a fiber converts phase errors across the telescope pupil into amplitude fluctuations in the fiber. Light is fed into two such fibers. Fiber coupler acts as beam combiner for coaxial beam combination. They are made by polishing and merging, at a given point and on a short length, the claddings of two fibers such that the distance between the cores is a few microns (for fibers that are single-mode in K band). Because of the common cladding, the fibers exchange energy from the electric fields that are carried.

Following coupling, the beam may be partially split in order to monitor the amount of photometric output, as well as can be interfered with beam from another fiber using a coupler, the fiber equivalent of a beam-splitter. The limitations associated with single-mode fibers arise from low coupling due to (1) poor coupling efficiencies, (2) high dispersion,[2] and (3) low polarization stability. Imperfections in

---

[1] Birefringence is the property by which certain materials have two different refractive indices for two orthogonal polarization components.

[2] Fiber-optic communication suffers mostly from dispersion causing a marked decrease in transmitted power. It occurs when the light traveling down a fiber optic cable, spreads out and becomes

the walls of the fibers and internal stresses cause the degradation of the focal ratio. Such a degradation may lead to loss of light if the incident angle exceeds the critical angle; it affects long focal ratio light beam the most. This gives rise to the mismatch between the cable output and the focal ratio of the instrument being used.

6.1.3 Beam Recombination in Reality

Role of the beam recombining system, also called 'fringe detector', is important in the measurement of central fringe intensity, a quantity that is used in amplitude interferometry, and its phase or displacement. The incident light collected by the telescopes must be directed to a central laboratory for beam combination. Two of the three beam combination processes that have been discussed in Sect. 5.3.2.3 are, in general, implemented at LBOIs. These are (1) image-plane scheme, where the focal plane image being crossed by fringes and (2) pupil-plane beam mode, which involves superposing afocal beams[3] from each telescope at a half-silvered plate. The VLTI has exploited both these schemes. However, for phased combination either in the image-plane or in the pupil-plane, the individual incoming beams from the arms of an interferometer should have identical pupil orientations, image orientations, and polarization characteristics (Traub 1999).

Decrease in fringe visibility is related to off-axis observation. The visibility loss is a function of the an-isoplanatic OPD, which has two effects on the observation of the faint object such as (1) reduction of the limiting magnitude on the faint object and (2) reduction of the accuracy with which the visibility of the object is measured. However, the visibility loss can be calibrated on reference stars (see Sect. 6.1.9). Another important point is to be noted that in the case of image-field scheme, the speckle patterns are dependent on the atmospheric coherence length, r_0; therefore, the visibility of the fringes is lowered randomly, and the quality of the estimate is generally poor.

6.1.3.1 Recombination Scheme at the Plateau de Calern

With an optical interferometer, it is difficult to observe objects much fainter than 5th magnitude, although 17th magnitude has been reached by speckle interferometry on a single moderate telescope (Foy et al. 1985). This limit is strongly seeing related;

---

longer in wavelength and eventually dissipates. Dispersion can be categorized into three main types such as (1) material or chromatic dispersion, (2) wave guide dispersion, and (3) modal dispersion. Two other major mechanisms of attenuation in optical fibers are absorption and scattering. The former occurs when the light beam is partially absorbed by lingering materials, namely water and metal ions, within the core of the fiber as well as in the cladding, while the latter occurs when atoms or other particles within the fiber spread the light.

[3] If an image is transferred as a beam of parallel light rays, the beam is termed an afocal beam.

at Mt Wilson, one could reach $m_v = 7.4$ (Mourard 2009). However, a better result can be envisaged if such an interferometer tracks fringes constantly to maintain coherence between the apertures.

Labeyrie (1975) utilized concept of merging speckles to obtain fringes of a star at the focal plane of his first prototype two aperture interferometer. The same principle has been implemented at the Grand Interféromètre à deux Télescope (GI2T; Labeyrie et al. 1986) as well. It is to reiterate that speckles are the results of the interference of the wavefronts diffracted by the pupils, thus one point of the image plane is illuminated by the whole (interferometric) pupil but the intensity receives from the various parts of the (interferometric) pupils could change depending on the seeing conditions. This creates a speckle figures crossed by fringes that can be witnessed by looking with an eyepiece at the direct focus of the GI2T. In this instrument, the beam-recombining optical devices are kept on a cart in the central laboratory, which move parallel to the baseline. The cart's displacements are controlled by computer and correct the optical path length drift induced by the diurnal rotation of the tracked source.

Large scale phase variations also limit fringe measurements to a confined coherence volume, encompassing spatial, temporal, and chromatic phase changes. The fringes are randomly phase-shifted across neighboring speckles, and therefore cannot be coherently combined (Bosc 1988). The average delay is derived from the integrated power spectra of individual fringed speckles, and the speckles must be processed individually in each frame. Figure 6.3 depicts the concept of acquiring fringes of a star using GI2T and Fig. 6.4 shows its optical table. The beam

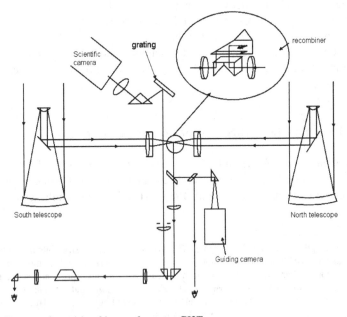

Fig. 6.3 Concept of acquiring fringes of a star at GI2T

Fig. 6.4 First optical table of GI2T (courtesy: D. Mourard)

recombining optical devices that were employed at both I2T and GI2T at the initial stage in a central laboratory are enumerated below:

- Collimator: The star beam forms a fixed collimating beam from each telescope to the central beam combiner, which provides a constant image size irrespective of the position of the telescopes on the tracks. The advantage of such an arrangement of transporting the beam is high throughput and low wavelength dependency. Of course, the main drawback is that of having thermal sensitiveness and mechanical disturbances. Aligning the beams from both telescopes require many degrees of freedom.
- Recombining element: The beams from the telescopes are recombined in an image plane after reconfiguring the pupils.
- Dispersion compensator: In the case of the interferometers, while transporting the beam through air from the telescope to the beam combining facility, significant differential chromatic dispersion takes place. Hence, a dispersion compensator in the form of wedges of glass are inserted into the beam for compensating the chromatic phase effect induced by the un-equal air travel that varies according to the limb of the interferometer. Of course, another alternative technique is to transport the beam through vacuum.
- Removable optical system: It enables the observer to control the interval between the pupils and hence the inter-fringe value, which is governed by the interval between the pupils images re-formed on the beam recombining optical device.
- Slit: On the image plane, the field is required to be diaphragmed by means of a slit, which corresponds to the diameter of Airy pattern and the length is equal to two such diameters. Such a slit allows the image of the star to pass on, where the beam undergoes anamorphosis, as well as dispersion. However, the fringe visibility gets reduced if the spectrograph slit is made larger than mono-speckle. The

advantage of a narrow-slit is that several adjacent slits can be used in the form of an image-slicer similar to those serving in conventional large spectrograph. The surrounding star field is re-imaged on to a photon counting camera for guiding, which allows the control of all the motions of each telescope. This pilot controlled tracking is essential to have accuracy of the order of 0.5 arcsecs, which is required for the superposition of the images.

- Anamorphosing: Sampling of the fringes is magnified in the vertical and horizontal directions by using two crossed cylindrical lenses. Owing to anamorphosis, the resulting focal length at the camera image plane would vary accordingly if it is considered as perpendicular to the fringe or parallel to the direction of the dispersion. This enables the inter-fringe being adapted to the resolution of the camera while fitting the spectral band within the field.

- Slicing: The image may be sliced in several strips (being oriented perpendicular to the fringes), whose slit-width matches the speckle size. GI2T had used an image slicer, comprising of a series of 10 wedges of different angles cemented on a field lens, for the fringe-tracker; approximately 12 dispersed speckles fit into the field of the detector. The fringes are contained in dispersed image slices and appear across speckles, each of which has a limited spatial extent and lifetime. The slicer was kept at the entrance of the spectrograph. However, the drawback of slicing the image is that it destroys part of the low resolution information, for example, for a binary star having separation more than the slide-width, the corresponding speckles for both companions may fall on adjacent slices, and hence the object cannot be resolved.

- Dispersion grating: Dispersion reduces the spectral interval for a given point in the image without loss of the total amount of collected light. Increase in coherence length induces a narrowing of the bandpass, and hence a decrease in the available photons. In order to get rid of this situation, Michelson and Pease (1921) devised this technique, known as 'dispersed fringe observation', for his stellar interferometer (see Sect. 3.1.2). The spectral analysis may be achieved either by dispersing with a dispersive component or by using OPD modulation (double Fourier transform mode) in coaxial mode. In the case of the former, the fringe light is focused on the spectrograph slit with a cylindrical optics to concentrate the flux along the slit. A dispersion grating can be employed to observe fringes simultaneously in several adjoining spectral channels. It enables to record the fringes with longer integration time, and selects different spectral channels for differential visibility measurements.

In the case of observing in dispersed light, bi-dimensional (2-D) image should be converted to one-dimensional (1-D) pattern. A 1-D distribution is composed of putting all the strip along the same line. The final strip is dispersed like in the mono-speckle case. The sampling of the fringes needs $\simeq 6$ pixels per speckle. However, in the case of non-zero OPD, the position of a fringe varies according to the wavelength yielding tilted fringes, and therefore, requires to be analyzed in 2-D instead of one. With multi-r_0 apertures, the recombined image has dispersed fringe pattern within speckles as shown in Fig. 6.5.

Fig. 6.5 A dispersed image
of a fringe pattern across a
star taken with GI2T
(courtesy: D. Mourard)

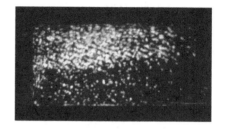

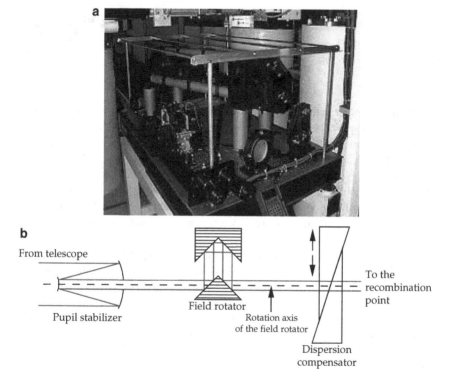

Fig. 6.6 (**a**) General view of the GI2T/REGAIN visible spectrograph, installed at the visible coherent focus (courtesy: D. Mourard) and (**b**) optical processing of a beam from one telescope of GI2T by the REGAIN recombiner (Saha and Morel 2000)

Subsequent improvement on the beam recombining optics, telescope guiding, and data collection permitted accurate computation of the fringe visibility in digitized images by a Fourier transformation. In the new recombiner (see Fig. 6.6a), called REcombineur pour GrAnd INterféromètre (REGAIN; Mourard et al. 2000, 2001), the 76-mm Coudé beams coming from the telescopes meet a pupil stabilizer, a field rotator, a wedge prism, and the beam combiner. These beams are first compressed to 5 mm in order to stabilize the pupil image in a fixed plane. Then, field rotators consisting of four plane mirrors are used for each beam to

Fig. 6.7 GI2T/REGAIN
dispersed fringes of R Cas
recorded in K-band (courtesy:
D. Mourard)

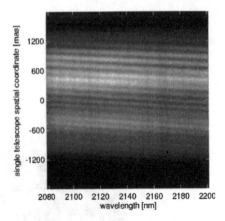

compensate the polarization difference affecting the visibility measured. The different chromatic dispersion between the two beams (due to operation at atmospheric pressure) is compensated by using two prisms for each beam which can slide on their hypotenuse, forming therefore a plate with adjustable thickness. This thickness is modified every 4 min, following the variation of the altitude of the observed object. Figure 6.6b depicts the process performed by an arm of the REGAIN table prior to recombination. REGAIN uses a delay-line named LAROCA (Ligne A Retard de l'Observatoire de la Côte-d'Azur) featuring a cat's eye reflector with a variable curvature mirror.

GI2T/REGAIN is a spectrometer working either in dispersed fringes mode or in Courtès mode. In dispersed fringes, the spectral range is 480–750 nm and the spectral resolution can reach $R = 30,000$. Separated recombination of two orthogonal polarizations may be possible. The Courtès mode consists in forming images at different wavelengths of speckles with fringes given by the recombination. The spectrometer of REGAIN in Courtès mode may provide 16 images at the same time. Figure 6.7 displays the GI2T/REGAIN dispersed fringes of R Cas recorded in near-IR K-band.

6.1.3.2 Phase-referencing Scheme

Two different approaches such as closure-phase methods (Jennison 1958) and phase-referencing scheme that was implemented at Mark III interferometer are employed to deal with the problem of atmospheric, as well as instrumental phase corruption. Generally, three or more telescopes are required for the former method, while in the case of the latter, two can suffice if the interferometer has the ability to utilize a nearby bright reference source of the target star, preferably within isoplanatic patch, to measure and correct for atmospheric time delays. The brighter star is also called phase-reference. It is reiterated that if these stars are far away, the profiles of the atmospheric turbulence along the individual line of sight to each star are not correlated to each other. In principle, the separation between these two stars should be within 1°, in order to keep the S/N-ratio losses smaller than 20%. A large phase variation during an interferometric observation reduces the fringe S/N-ratio, i.e.,

$$\mathcal{V}_{av}^2 = \mathcal{V}_0 e^{-\langle\sigma_D\rangle^2}, \qquad (6.4)$$

where $\mathcal{V}_{av}$ and $\mathcal{V}_0$ are the time-averaged and original visibility amplitudes and $\langle\sigma_D\rangle^2$ the phase variation over the detector integration time.

Phase-referencing can be used primarily for astrometric applications, e.g., to search for the reflex motion of stars orbited by planets, and to determine the positional offset of a circumstellar envelope from the central star. This method has been implemented in radio interferometry to permit long coherent integrations on targets, via a fast switching technique. Another utility of this method is that reference phase can be used to extend the effective atmospheric coherence time, allowing longer coherent integrations on the target source.

A different approach may be pursued by the implementation of dual-field interferometry and phase-referencing to cope with the sensitivity limiting effects of the turbulent atmosphere. The dual-field module helps reaching these limits on fainter objects, where observation of fringes from two close stars are obtained simultaneously; one of these stars should be bright enough to measure fringe and preset the delay-lines precisely. This fringe stabilization that enables longer coherent integration times and higher S/N-ratio on the second star is referred to as fringe-tracking based on phase-referencing. Palomar testbed interferometry (PTI) employed such a dual-star module for phase referencing and narrow-angle astrometry. This has been implemented at VLTI. The phase-referenced technique with a fringe-tracking channel was also used at Mark III interferometer to recover the phase information (Quirrenbach et al. 1996).

Another method is to observe a target at multiple wavelength for differential visibility measurements (continuum and spectral line) and use data from one part of the spectrum such as the continuum emission to calibrate another part (Mourard et al. 1989). This method, known as spectro-interferometry, has the ability to compare the size scales of one spectral region with another. In this set up, the different spectral channels are employed; the two spectral windows should be within the allowed spectral bandwidth. The continuum channel is supposed to originate from the unresolved region of a star, e.g., photosphere and the spectral line centered on a part of the spectrum created in an extended region, such as circumstellar medium.

6.1.3.3 Fiber-linked Recombination

An interesting innovation was introduced to interferometry in the late 1980s by employing optical fibers. Shaklan and Roddier (1987) demonstrated in the applicability of single-mode fibers, based on which the second recombiner named FLUOR was developed (Coudé du Foresto and Ridgway (1992). The implementation of such a beam combiner was a real breakthrough; however, fiber combiners are very sensitive

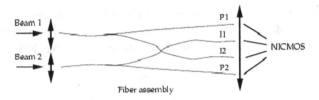

Fig. 6.8 Schematic of the FLUOR recombiner. P1 and P2 are the photometric output fibers. I1 and I2 are the interferometric output fibers; temporal fringe pattern can be measured in I1 and I2. These outputs are imaged by a lens on a NICMOS infrared array detector

to external environmental conditions like temperature and mechanical stresses. In other part of physics they are used as sensors for monitoring fluctuations of these observables. FLUOR recombination unit was used for diluted aperture interfero-meter operation at the McMath solar tower of the Kitt-Peak National Observatory (Arizona), with a 5-m baseline in 1991 by these authors. This unit used at the IOTA (Coudé du Foresto et al. 2001).

FLUOR consists of single-mode fiber optics interfering beams (see Fig. 6.8). These fibers have been designed to propagate infrared light at Near-IR (K band, 2.2 μm in TEM mode. Therefore, only plane-waves perpendicular to the axis of the fiber may propagate over long distance. The FLUOR bench used a 180-μm stroke piezo-electric transducer (PZT) for scanning the fringes. The detector used is the NICMOS3 of IOTA. Four pixels are read (two for the interferometric fiber outputs, two for the photometric fiber outputs). The significance of the FLUOR at IOTA is that visibilities can be calibrated with sub-1% precision, which is important for many astrophysical applications like stellar atmosphere studies, Cepheid pulsation measurements for distance determinations, and detection of angular anisotropies arising in disks around young stars.

FLUOR uses spatial filtering and photometric monitoring to allow calibration of fringe visibilities. The beams including atmospheric turbulence effects are char-acterized by two parameters such as the amplitude and the phase of the outgoing electric field. Such a system is implemented at the with CHARA (Center for High Angular Resolution Astronomy) interferometer, Mt. Wilson, USA. The fiber based beam combining technique for multi-telescope interferometers may face some dif-ficulties since light has to be split several times, combined together many times.

6.1.3.4 Integrated Optics

Integrated Optics (IO) is an optical technology, photonics analog to integrated chips in micro-electronics, which allows to reproduce optical circuits on a planar sub-strate. Such hardware potentially allows large tables of bulk optics to be replaced by miniature devices (Haguenauer et al. 2000) and provides easy access to spatial filtering and photometric calibration. IO system is proved to be viable alternative (Malbet et al. 1999) to the bulk optics for a multi-telescope aperture synthesis.

IOs are compact, provide stability, low sensitivity to external constraints like temperature, pressure or mechanical stresses, no optical alignment except for coupling, simplicity, and an intrinsic polarization control. Unlike fiber optics, which introduces decisive inputs with a reduction of the number of degrees of freedom of instrumental arrangement and the modal filtering, planar optics introduces additional arguments. However, its application is restricted to the combiner and is not suitable for beam transportation and large optical path modulation.

IO represents a higher level of integration for optical functionality, the principle of which is that light beam propagates inside a planar substrate. The optical communication employing single mode fiber optics for long distance connection impose periodic signal amplification. In a planar waveguide, the wave propagation of a collimated incident beam is guaranteed by the three step-index infinite planar layers. A high index layer is sandwitched between two low-index layers providing the range of acceptable incident angle; the core layer thickness ranges between $\lambda/2$ and 10λ depending on index difference.

IO on planar substrate is based on glass ion exchange: Na^+ ions from a glass substrate are exchanged with ions (Ag^+, K^+, or Ti^+) of molten salts. The local modification of the glass chemical composition increases the refractive index at the surface of the glass. A three-layer structure (air/ions/glass) is created and the light is confined. The implementation of the optical circuit is created by photo-masking techniques to ensure the horizontal confinement of light. The additional process of embedding the guide is carried by applying an electric field to force the ions to migrate inside the structure (Kern et al. 2005). The waveguide core is the ion-exchange area and cladding the glass substrate and air. Another method consists of etching layers of silicon of various indices. These layers may be either phosphorus doped silica or silicon nitride.

Light may be observed at the structure output if total internal reflection occurs between two successive reflected wavefronts. The main part of the carried energy lies in the waveguide core, but evanescent field propagates in lateral layers and contribute to the mode propagation. A guiding structure with a given thickness and layers refractive index is characterized by a cut-off wavelength, λ_c, separating the single mode propagation, ($\lambda > \lambda_c$), in which the primary mode propagates. A single mode waveguide propagates in the direction parallel to the waveguide.

Integrated optics are used in telecommunication applications with multiple optical functions like power dividing, coupling, multiplexing, etc. Such an approach is well suited for interferometrically combined numerous beams to achieve aperture synthesis imaging, as well as for space-borne interferometers where stability and a minimum of optical alignments are expected. In IO based combiners, many fibers may be matched to a small planar substrate with miniature waveguides etched in place to split or combine the light. Several beam splitting or combinations can be fitted into a small area of a few square centimeters. IO has several limitations, such as (1) low transmission, (2) limited wavelength coverage, (3) birefringes, and (4) dispersion. Results with first generation devices was found to be promising (LeBouquin et al. 2006). The integrated beam combiners on chips (Kern et al. 2008) may lead to new direction for combining beams in aperture synthesis interferometry.

6.1.3.5   Beam Combination at VLTI

The Very Large Telescope Interferometer (VLTI), situated at Cerro Paranal, Chile, has several beam combining systems. Their salient features are enumerated below:

1. VINCI (VLT INterferometer Commissioning Instrument; Kervella et al. 2000): It is a single-mode optical fiber recombiner operating at 2.2 μm like FLUOR (Coudé du Foresto et al. 1998) and is intended to be used for debugging the upstream sub-systems of VLTI. The VINCI instrument coherently combines the light from two telescopes in the IR K-band, where the fringe pattern is obtained by temporal modulation of the optical path difference. The light is injected into the fibers, followed by combining in a single-mode coupler. First fringes were observed with this instrument by Glindemann et al. (2001). VINCI is no longer in use.

2. IONIC: An integrated optics near-IR combiner was employed to permit operation in the H-band (IONIC; LeBouquin et al. 2004). Figure 6.9a depicts the IONIC-3 system, which is capable of measuring closure phase, while Fig. 6.9b shows the

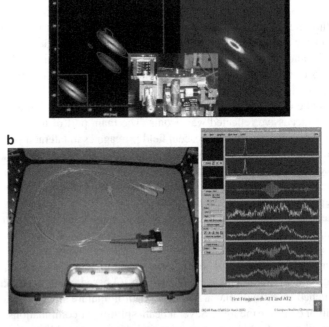

Fig. 6.9 (a) The IONIC-3 beam combiner in front of the first aperture-synthesis images from three element IOTA interferometer, depicting λ Virgo based on 1.65 μm observations (Monnier et al. 2004); the binary components are resolved from each other but are unresolved individually and (b) the integrated optics beam combiner (IONIC2T-K) at the VLTI (*left*) using two auxiliary telescopes (LeBouquin et al. 2006). Courtesy: F. Malbet

integrated optics beam combiner (IONIC2T-K) at the VLTI using two auxiliary telescopes (AT). Light from each telescope is focused into a single-mode fiber, and three fibers are aligned using a silicon V-groove array mated to the planar waveguides on the IO device (Berger et al. 2003). The optical circuit acts to split the light from each telescope before recombining each telescope pair at three IO couplers. This pair-wise combination scheme leads to six interferometric channels (two for each baselines). The optical path difference (OPD) between the two light beams is modulated to sweep through the fringes, which appears as temporal modulation. A fringe packet centroiding algorithm is applied during observations for removing the instrumental OPD drift. The present system, suitable for IR K-band, is connected at the input by single-mode fibers that maintain polarization (LeBouquin et al. 2006). The notable results from this equipment was the measurement of (1) diameters of very low mass stars, (2) oblateness of the fast rotating star, and (3) calibration of the brightness-color relation of Cepheids.

3. MIDI (MID-Infrared Interferometric Instrument): This instrument is based on the pupil-plane concept (see Sect. 5.3.2.3), built around a Mach-Zehnder/Michelson type two-beam optical recombiner, which operates in the mid-infrared N-band (8–13 μm). It uses as light-collectors either two VLT unit-telescopes, or two VLT auxiliary-telescopes, and the whole VLTI infrastructure, viz., delay-lines, M16, switch-yard, beam-compressors, and fringe tracker. MIDI combines two beams to provide visibility moduli in the u, v plane and features spectroscopic optics to provide visibilities at different wavelengths within the N-band. The beams are dispersed after combination, allowing to measure visibility, $\mathcal{V}(\lambda)$ at two possible resolution, either R = 30 (prism) or R = 230 (grism; Leinert et al. 2003). The delay-lines are equipped with variable curvature mirrors, that relay the pupil image correcting for the distance of the delay-line carriage. The detector image of MIDI may show up to four channels that are spectrally dispersed in the horizontal direction: two interferometric channels containing fringes at the center of the detector, and two optional photometric beams, which sample the photometric level (at a given wavelength) of each beam coming from a telescope.

4. AMBER (Astronomical Multiple BEam Recombiner; Petrov et al. 2007): This is an near-IR instrument based on image-plane mode (see Sect. 5.3.2.3), which performs recombination of three light beams from three of the telescope apertures (either UTs or ATs as shown in Fig. 6.10a) in order to enable imaging through closure-phase techniques. Since the fringe encoding relies on using non-redundant input pupil, the AMBER instrument is arranged with 1-D configuration of beams allowing to produce spectrally dispersed fringes (Robbe-Dubois et al. 2007) with mas angular scales operating in the range between 1.3 μm and 2.3 μm (in J-, H-, and K-bands). The beams are dispersed before combination by focusing the parallel beams onto the detector plane. The interferograms are spectrally dispersed with three possible spectral resolutions ($R = 35$, $R = 1,500$, $R = 12,000$). With the lowest one, one may simultaneously observe the J-, H- and K-bands; observations are being made in K-band with the rest. AMBER can measure visibilities and a closure phase in a few hundred different spectral channels.

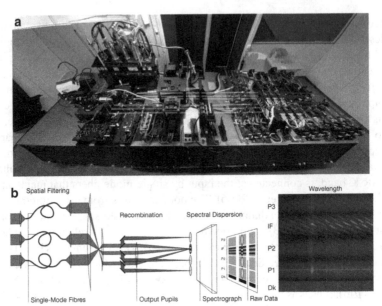

Fig. 6.10 (**a**) The VLTI instrument AMBER; the incoming beams have been superimposed on the photograph. (**b**) The basic concept of AMBER. To note, the *vertical brighter line* indicating the Brγ emission line. In this figure, the detector image contains a view, rotated by 90 degrees, of the AMBER real time display showing three telescope fringes, in medium resolution between 2,090 nm and 2,200 nm, on the bright Be star α Arae. The three superimposed fringe patterns form a Moiré figure because, in that particular case, the three OPDs were substantially different from zero during the recording (Malbet et al. 2007b). Courtesy: F. Malbet

AMBER features a set of single-mode optical fibers; each beam is spatially filtered by such a fiber (Petrov et al. 2003). After each fiber, the beams are collimated so that the spacing between the output pupils is non redundant. The multi-axial recombination consists of common optics that merges the three output beams in a common Airy-disk containing Young's fringes. With a cylindrical optics anamorphosor, this fringed Airy disk is fed into the input slit of a spectrograph (see Fig. 6.10b). In the focal plane of this spectrograph, each column (in the figure, but in reality each line) of the detector contains a monochromatic image of the slit with three photometric (P1, P2, P3) zones and one interferogram (IF). On the detector plane of AMBER, the light is dispersed along the vertical axis.

5. PRIMA (Phase-Referenced Imaging and Microarcsecond Astrometry; Quirrenbach et al. 1998): The primary objective of this instrument is the detection and characterization of exoplanets including their precise mass, orbital inclination and a low resolution spectrum. It is a recombiner dedicated to narrowangle astrometry in band H or K down to the atmospheric limit of 10 μas. PRIMA is composed of four major sub-systems: star separators, differential

delay-lines,[4] laser metrology (to monitor the PRIMA instrumental optical path errors), and Fringe Sensor Units (FSU; Cassaing et al. 2000).

PRIMA picks up two stars in the Coudé and feeds it into the delay-lines, which produces the fringes in the VLTI laboratory. Each FSU recombines light from both telescopes for the two sources separately to form interferometric fringes. This unit uses spatial phase modulation in bulk optics to retrieve real-time estimates of fringe phase after spatial filtering; a R = 20 spectrometer across the K-band makes the possibility of retrieval of the group delay signal. Integrated recently at the VLTI, this unit yields phase and group delay measurements at sampling rates up to 2 kHz, which are used to drive the fringe-tracking control loop (Sahlmann et al. 2008). PRIMA is a dual feed system adding a faint object imaging and an astrometry mode to the VLTI, whose objective is to enable simultaneous interferometric observations of two sources on the sky – each with a maximum size of 2 arcsec having separation by up to 1 arcmin. The brighter (reference) source, in the field-of-view of the telescope, can be used to control the fringe position, permitting the interferometer to look at the fainter scientific object. When the fringes of this star are stabilized, the fringes of the fainter science object can be studied. Recently, starlight from two ATs was fed into the PRIMA system, and interference fringes were detected on PRIMA's Fringe Sensor Unit (van Belle 2008).

In addition, the second generation instruments are also being developed in order to enlarge the scientific possibilities of VLTI. These are:

- MATISSE (multi-aperture mid-infrared spectroscopic experiment; Lopez et al. 2006): It is a mid-IR spectroscopic interferometer combining the beams of up to four UTs or ATs of the VLTI. MATISSE is going to be the successor to MIDI and plans to measure closure phase relations, which will offer a capability for image reconstruction. This instrument would provide imaging capability in three spectral bands of the mid-infrared domain like L, M, and N-bands. The possible principle of this instrument is a multi-axial combination concept like REGAIN on GI2T or AMBER on VLTI. In spectroscopic mode, it might be a dispersed fringes instrument operating in the image plane.

- GRAVITY (general relativity analysis via VLT interferometry; Eisenhauer et al. 2008): This is an interferometric imager with 10 μas astrometry and interferometric imaging with the VLTI. It will have the ability perform extremely accurate astrometric measurements on very faint sources; it may enable to measure interaction of celestial sources through gravity.

- VSI (VLTI Spectro-Imager; Malbet et al. 2008): This instrument will provide spectrally resolved near-IR images of astronomical sources at angular resolutions down to 1.1 mas and spectral resolutions up to $R = 12,000$. With 6 telescopes, it will measure simultaneously 15 baselines and 10 closure phases.

---

[4] The function of differential delay-lines is to compensate for slight differences in path lengths due to the separate positions on the sky of the two sources. The difference of the white light fringe position of these stars can be expressed as a differential OPD that must be adjusted with the differential delay-lines with a high precision of 5 nm.

6.1.4 Phase and Group Delay Tracking

A method, called group-delay tracking, was used by Michelson and Pease (1921)
to discern fringes (by eye) with the stellar interferometer (see Sect. 3.1.2) employing a direct view prism. Labeyrie (1975) used an identical approach while
acquiring fringes using I2T. This practice continued until 1984 at I2T and until 1995 at GI2T. Group-delay tracking, based on the integration of the moduli
of all the computed Fourier transforms (FT; Lawson 1994), has been employed
with the interferometers like Sydney University Stellar Interferometer (SUSI),
and Cambridge Optical Aperture-Synthesis Telescope (COAST). A combination
of phase and group delay estimation provides precise delay measurements, which
is useful for astrometric observations. The phase estimation can be used to provide better resolution, while the group delay estimation is employed so that the
phase is unwrapped with reference to the peak of the coherence envelope (Lawson
1999).

Let a number of harmonic waves with different frequency be superimposed to
form a composite disturbance, the field amplitude varies in the form of beats (see
Fig. 6.11). Considering two harmonic waves of equal amplitude (see Sect. 1.2),

$$U(\mathbf{r}, t) = A(\mathbf{r})e^{-i\omega_1 t} + A(\mathbf{r})e^{-i\omega_2 t}, \tag{6.5}$$

in which $\mathbf{r} = x, y, z$ is the position vector, (2.12) for the self-coherence function,
$\Gamma(\mathbf{r}, \tau)$, is written as,

$$\Gamma(\mathbf{r}, \tau) = \lim_{T_m \to \infty} \frac{1}{T_m} \int_{-T_m/2}^{T_m/2} U^*(\mathbf{r}, t)U(\mathbf{r}, t + \tau)dt. \tag{6.6}$$

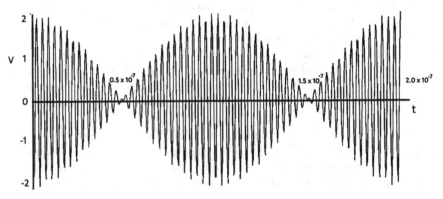

Fig. 6.11 1-D plot of the cosine factor of the field amplitude in time domain; the high frequency
carrier wave is 300 MHz, while the low frequency is 10 MHz

Thus, in this case,

$$
\begin{aligned}
\Gamma(\mathbf{r}, \tau) &= \lim_{T_m \to \infty} \frac{1}{T_m} \int_{-T_m/2}^{T_m/2} \left[A^*(\mathbf{r}) e^{i\omega_1 t} + A^*(\mathbf{r}) e^{i\omega_2 t} \right] \\
&\quad \times \left[A(\mathbf{r}) e^{-i\omega_1(t+\tau)} + A(\mathbf{r}) e^{-i\omega_2(t+\tau)} \right] dt \\
&= \lim_{T_m \to \infty} \frac{|A(\mathbf{r})|^2}{T_m} \int_{-T_m/2}^{T_m/2} \left[e^{-i\omega_1 \tau} + e^{-i\omega_2 \tau} \right] \\
&\quad + \left[e^{-i\omega_1 \tau} e^{-i(\omega_1-\omega_2)t} + e^{-i\omega_2 \tau} e^{-i(\omega_2-\omega_1)t} \right] dt.
\end{aligned}
\tag{6.7}
$$

Since the contributions from the third and fourth terms of the right hand side of (6.7) are null, the self-coherence function can be recast into,

$$
\Gamma(\mathbf{r}, \tau) = |A(\mathbf{r})|^2 \left[e^{-i\omega_1 \tau} + e^{-i\omega_2 \tau} \right].
\tag{6.8}
$$

With the normalized coherence function (see 2.16), and $\Gamma(\mathbf{r}, 0) = 2|A(\mathbf{r})|^2$, one finds

$$
\gamma(\mathbf{r}, \tau) = \frac{1}{2} \left[e^{-i\omega_1 \tau} + e^{-i\omega_2 \tau} \right].
\tag{6.9}
$$

The visibility function, $\mathcal{V}$ is considered to be equal to $|\gamma(\mathbf{r}, t)|$; hence, one derives it in the form of,

$$
\begin{aligned}
\mathcal{V} &= \frac{1}{2} \left| e^{-i\omega_1 \tau} + e^{-i\omega_2 \tau} \right| \\
&= \frac{1}{2} \left[4 \cos^2 \frac{\omega_1 - \omega_2}{2} \tau \right]^{1/2} \\
&= \left| \cos \left(\frac{\omega_1 - \omega_2}{2} \tau \right) \right|.
\end{aligned}
\tag{6.10}
$$

This (6.10) implies that the visibility function takes a periodic dependence on the time delay, τ. By extending this to a sum of many harmonic waves of different frequency, ω_m, (6.5) is,

$$
U(\mathbf{r}, t) = \sum_{m=1}^{N} A_m(\mathbf{r}) e^{-i\omega_m t}.
\tag{6.11}
$$

In the limit of closely spaced harmonic waves, it appears to be similar to (1.12b) and the self-coherence function is given in (2.12).

The phase velocity of a wave is the rate at which the phase of the wave propagates in space. Each small peak in the carrier travels with the phase velocity, v_p, i.e.,

$$
v_p = \frac{(\partial \psi / \partial t)_x}{(\partial \psi / \partial x)_t} \cdot
\tag{6.12}
$$

Since the wavefront is a plane normal to the wave vector, κ, $d\mathbf{r}/dt$ is directed along κ, the phase velocity, v_p, of the planar wavefronts is derived as,

$$v_p = \left| \frac{d\mathbf{r}}{dt} \right| = \frac{\bar{\omega}}{\bar{\kappa}} \cong \frac{\omega}{\kappa}. \tag{6.13}$$

Thus, one finds that the phase velocity $v_p \cong \omega/\kappa$ is an approximation in the case for neighboring frequencies and wavelengths in a continuum. A waveguide, based on its geometry, imposes a cutoff frequency, ω_c on any propagating electromagnetic wave and cannot sustain waves of any lower frequency. The dominant wave pattern of a propagating wave with a frequency of ω will have a wave number κ given by, $\kappa = \sqrt{\omega^2 - \omega_c^2}/c$, in which c is the speed of light. In this case, the phase velocity is,

$$v_p = \frac{c}{\sqrt{1 - \left(\frac{\omega_c}{\omega}\right)^2}}. \tag{6.14}$$

The propagation of information (or energy) in a wave occurs as a change in the wave. Modulation of the frequency (and/or amplitude) of a wave is required in order to convey information. Each amplitude wave contains a group of internal waves, the rate at which the modulation envelope advances is known as the group velocity. Assume that the differences between the frequencies and propagation constants are small; thus,

$$v_g = \frac{\omega_g}{\kappa_g} \cong \frac{d\omega}{d\kappa}. \tag{6.15}$$

The group velocity in a waveguide with cutoff frequency, ω_c is,

$$v_g = c\sqrt{1 - \left(\frac{\omega_c}{\omega}\right)^2}. \tag{6.16}$$

The relation between group and phase velocities is derived as,

$$
\begin{aligned}
v_g = \frac{d\omega}{d\kappa} &= \frac{d}{d\kappa}\left(dv_p\right) \\
&= v_p + k\left(\frac{dv_p}{d\kappa}\right) = v_p - \lambda\left(\frac{dv_p}{d\lambda}\right),
\end{aligned} \tag{6.17}
$$

where the spectroscopic wavenumber $k = 1/\lambda$.

In an idealized dispersion-free medium, in which v_p is independent of λ, $dv_p/d\kappa = 0$, hence (6.17) turns out to be $v_g = v_p$. This is the case of light propagating in vacuum, where $v_g = v_p = c$. While in the case of light waves through dielectric media, the dependence of the index of refraction, $n(\omega)$ on the wave frequency causes dispersion. Therefore, the phase velocity is given by,

$$v_p = \frac{c}{n(\omega)} = \frac{\omega}{\kappa}, \qquad \text{then,}$$

$$\frac{dv_p}{d\kappa} = \frac{d}{d\kappa}\left(\frac{c}{n(\omega)}\right) = \frac{-c}{n^2(\omega)}\left(\frac{dn(\omega)}{d\kappa}\right). \tag{6.18}$$

Invoking (6.17), one obtains,

$$v_g = v_p\left[1 - \frac{\kappa}{n(\omega)}\left(\frac{dn(\omega)}{d\kappa}\right)\right]$$

$$= v_p\left[1 + \frac{\lambda}{n(\omega)}\left(\frac{dn}{d\lambda}\right)\right], \tag{6.19}$$

where $\kappa = 2\pi/\lambda$ and $d\kappa/d\lambda = -(2\pi/\lambda^2)$. If $dn(\omega)/d\kappa = 0$, the group velocity turns out to be equal to the phase velocity, i.e., $v_g = v_p$.

It is to reiterate that light travels to the beam combiner from each arm of an interferometer through different dispersive elements. The OPD between the wavefronts may be expressed in terms of the indices of refraction of the different media, n_j, and the path lengths in each arm of the interferometer, x_{1j} and x_{2j},

$$x(k) = \sum_{k=0}^{K}(x_{2j} - x_{1j})n_j(k). \tag{6.20}$$

Putting $x_{2j} - x_{1j} = x_j$, the phase of the fringes may be written as,

$$2\pi x(k) = 2\pi k\left[x_0 + \sum_{k=1}^{K} x_j n_j(k)\right], \tag{6.21}$$

with x_0 stands for a vacuum delay and K is for dispersive media.

The group-delay, k_0, is proportional to the rate of change of phase as a function of wavenumber, evaluated at the center of the band,

$$k_0 = \frac{d}{dk}kx(k) = x_0 + \sum_{k=1}^{K}\frac{d}{dk}kx_j n_j(k). \tag{6.22}$$

The group delay can be measured if the combined beams from an interferometer are dispersed in a spectrometer. The group-delay tracking yields a peak whose position is proportional to the OPD.

6.1.5 Coherence Envelope

Substituting $x = s_2 - s_1$, in which $s_{j=1,2}$ are the two optical path lengths, (2.24) can be recast into,

$$I(v, x) = I_s\left[1 + |\gamma(\mathbf{r}_1, \mathbf{r}_2, \tau)|\cos(2\pi vx - \varphi(\mathbf{r}_1, \mathbf{r}_2, \tau)\right] + I_b, \tag{6.23}$$

with $I_s = 2\sqrt{I_1(\mathbf{r}, t) + I_2(\mathbf{r}, t)}$ and $I_b = I_1(\mathbf{r}, t) + I_2(\mathbf{r}, t) - I_s$.

Using a finite bandwidth, the recorded intensity is expressed as,

$$I(\bar{v}, x) = \int_{-\infty}^{\infty} \widehat{\Psi}(v - \bar{v}) I(v, x) dv, \tag{6.24}$$

where $\bar{v}$ is the mean frequency of the beam, $\widehat{\Psi}(v - \bar{v})$ the filter function that includes both the shape of the bandpass and the frequency response of the detector.

By setting the relation, $v' = v - \bar{v}$, and performing the integration in (6.24) with respect to v', (6.23) is written as,

$$I(\bar{v}, x) = I_s \left\{ 1 + |\gamma| \left[\cos(2\pi \bar{v} x - \varphi) \int_{-\infty}^{\infty} \widehat{\Psi}(v') \cos(2\pi v' x) dv' \right. \right.$$
$$\left. \left. - \sin(2\pi \bar{v} x - \varphi) \int_{-\infty}^{\infty} \widehat{\Psi}(v') \sin(2\pi v' x dv') \right] \right\}. \tag{6.25}$$

It is to be noted that the background irradiance, I_b is ignored here. Defining $\Psi(x) = |\Psi(x)| e^{i\varphi_\Psi}$ as the Fourier transform of $\widehat{\Psi}(v)$, it is found,

$$|\Psi(x)| \cos \varphi_\Psi = \int_{-\infty}^{\infty} \widehat{\Psi}(v) \cos(2\pi v x) dv \tag{6.26a}$$

$$|\Psi(x)| \sin \varphi_\Psi = \int_{-\infty}^{\infty} \widehat{\Psi}(v) \sin(2\pi v x) dv. \tag{6.26b}$$

Thus, (6.24) is recast into,

$$I(\bar{v}, x) = I_s \left[1 + |\gamma_x| \cos(2\pi \bar{v} x - \varphi + \varphi_\Psi) \right], \tag{6.27}$$

in which the apparent visibility $|\gamma_x| = |\gamma||\Psi(x)|$ is the product of the true visibility and the modulus of the Fourier transform of the filter function.

The term $\Psi(x)$ describes the coherence envelope. When $\widehat{\Psi}(v)$ is symmetric, $\Psi(x)$ is real valued, $\varphi_\Psi = 0$. The true visibility can be observed at zero delay, where the envelope is at its peak. With a rectangular bandpass, the coherence envelope resembles a sinc function (see 1.66),

$$\Psi(x) = |\Delta v| \frac{\sin \pi x \Delta v}{\pi x \Delta v}, \tag{6.28}$$

and

$$\widehat{\Psi}(v) = \begin{cases} 0 & |v| > \Delta/2 \\ 1 & |v| < \Delta/2. \end{cases} \tag{6.29}$$

The sinc function is characterized by the location of its first zero crossing, in which the distance x is equal to $1/\Delta v$. This distance is thought to be the coherence length of the radiation of the source under observation.

6.1.6 Fringe Acquisition and Tracking

It is reiterated that an interferometer pair is able to detect fringes as long as the cross correlation performed on the light beam arriving at the two telescopes is done within the coherence length, l_c (see 2.1). Fringes are searched by adjusting the delay-line position; however, the error on the OPD must be smaller than l_c. In an interferometer, this correction is carried out for the pointing zenith direction of the telescope, but for any other angle the OPD becomes different. In the case of an interferometer having two moderate (or large) apertures, the mean value of the delay between the two sampled wavefronts is measured. The position of the delay-line (or any other delaying device in the optical path) must be adjusted in order to keep the fringes within the 'observation window'. This real-time control is called fringe-tracking, which is essential to increase the efficiency and sensitivity of the LBOIs. Such a method predicts the position and motion of the center of a fringe packet. Like adaptive optics systems, such a system reduces the blur in interferometric observations. It allows longer acquisition times as well.

Time dependent behaviors affect the position of the central white-light fringe; sidereal motion, instrument flexures, and fine telescope pointing decay on smaller scales. Atmospheric turbulence also affects ground-based observation inducing 'piston' at the interferometer baseline scale. Piston is the concept of zero-order atmospheric turbulence, equivalent to an optical path offset of the whole pupil. Being constant over the aperture, the piston does not affect single telescope imaging. In interferometry, one talks about the difference of piston between the pupils, or 'piston error', which changes the effective OPD and biases a phase (in astrometric) measurement. The magnitude and velocity of the differential piston variation is roughly correlated with the magnitude of seeing and the atmospheric coherence time. The fringe position is continuously varied by the differential piston, hence the integration time needs to be shorter than the timescale of the piston variation to avoid a blurring of the fringe pattern. The variation in piston phase across an aperture has been derived (Fried 1965; Noll 1976). If the spatial variations in phase are due to Kolmogorov turbulence, the phase structure function is given by (4.41). If the wavefront tilt is not compensated, the mean squared variations after piston phase is removed, measured over a diameter, D is shown in (4.43). In the case of GI2T, its star tracker corrects tilt up to 0.5 Hz.

Interferometer signal should be analyzed faster than the atmospheric coherence time, τ_c, to prevent visibility losses due to fringe blurring. With a perfectly corrected Airy disk, the sensitivity is limited by the fringe motion, the exposure time may be reduced to about 10 ms. The remaining fringe motion must be removed by fringe-tracker. The fringe-tracker ensures the fringe stability with an appropriate delay-line controlled with a proper sampling of the OPD fluctuations at a frequency compatible with the considered time scale. It is important to note that fringe stabilization is a must to enhance the sensitivity. This fringe-tracker analyzes the fringe position and actively control a small delay-line to compensate the atmospheric delay. It avoids visibility losses due to fringe blurring as well.

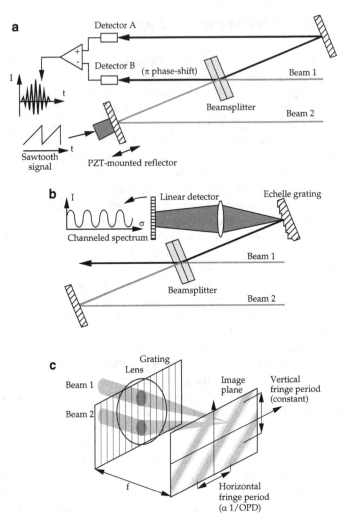

Fig. 6.12 Three possible set-up for beam recombination and fringe acquisition: (**a**) white-light fringes, (**b**) channeled spectrum, and (**c**) dispersed fringes (Saha and Morel 2000)

There exist three possible set-ups for fringe acquisition for visible spectrum (see Fig. 6.12):

1. White fringes: The OPD is temporally modulated in this category by a sawtooth signal, using a fast- and short-travel delaying device (usually a reflector mounted on PZT). The intensity of the recombined beams, therefore describes over time a fringe pattern that is recorded by single-pixel detectors. Natural OPD drift due to the Earth-rotation may also be used for acquiring fringes, as it was done by the Synthèse d'Ouverture en Infra Rouge avec DEux TElescopes (SOIRDÉTÉ) interferometer (Rabbia et al. 1990) at Observatoire de Calern, France. At Mark III

interferometer, each phase estimate was made as an average between a measurement made on the up-stroke and a measurement made on down-stroke (Lawson 1999). Several cycles of up- and down-strokes were averaged subject to the condition of atmospheric turbulence.

2. Channeled spectrum: This method consists of imaging the dispersed recombined beam on a linear detector such as CCD or photon-counting camera. The spatial information contained in the fringes is generally modified through the process of detection. Hence, the response in the direction of dispersion turns out to be a convolution of the optical point spread function of the spectrometer and the detector. The power spectrum of this response determines the sensitivity to fringes as a function of path-difference. If the resolution is limited by sampling, the coherence length is determined by the bandwidth per pixel. If the light is dispersed in a spectrometer in order to separate the bandwidth $\Delta \nu$ into N smaller bands of width $\Delta \nu / N$. The coherence length for fringe detection changes from $1/\Delta \nu$ to $N/\Delta \nu$. Such a length turns out to be N times larger.

3. Dispersed fringes: In this technique, beams are dispersed prior to the recombination. Unlike the two previous techniques, recombination is not done by overlapping the beams, but by focusing them with a common lens, like in Michelson stellar interferometer. The detector used in this case is a bi-dimensional photon counting camera. In the absence of infrared (IR) photon-counting camera, it is essential to use as little pixel as possible in order to reduce global readout noise. Hence, the 'white' fringes set-up is preferably used for IR observations.

Techniques, to compensate the OPD drift between the two beams of a standard optical interferometer, are characterized by their ability to monitor the OPD drifts and re-positioning the delay-lines so as to reduce the delay error to a small fraction of the coherence length, l_c, or to monitor the OPD fluctuations and correct the OPD to a small fraction of a wavelength. These modes are, respectively, referred to as coherencing and co-phasing (Beckers 1988). The former lets phase fluctuations blur the fringes at long exposures, while the latter keeps stabilized fringes visible at any exposure time. The fast compensation of the OPD variations due to the differential piston mode of the turbulence is, therefore, done in order to freeze the fringes.

6.1.6.1 Coherencing

In coherencing, it is required to disperse the fringes and map them onto a 2-D space of wavenumber against fringe phase, and examine the 2-D spatial frequency of the resulting fringes (Basden and Buscher 2005). The value of this spatial frequency measures the center of the coherence envelope, what is referred to as group delay tracking, by a scan (OPD modulation) and applies this measurement to a delay-line in order to compensate the OPD; the resolution in delay should be equal to the coherence length of the total bandpass, which is of the order of 12 μm for the near-IR K-band. The coherencing with non-white-light fringes is compared with the active optics. In white light it was carried out, as for IOTA, by scanning the OPD while acquiring signals to find the null-OPD point in the fringe pattern (Morel et al.

2000). This coherencing yields the OPD correction to apply to the delay-line, at a few Hz servo-loop rate. With a channeled spectrum, one may rewrite (6.27) as,

$$f_n(x) = I_n \left[1 + |\gamma_n| \cos(2\pi \bar{\nu}_n x - \varphi_n)\right], \tag{6.30}$$

in which n is an index number that counts the pixels across the spectrometer, I_n the intensity of the spectrum, $|\gamma_n|$ the fringe visibility amplitude, and φ_n the fringe phase.

Here, φ_n is set to zero. The (6.30) reveals that the OPD is proportional to the fringe frequency in the channeled spectrum. If p fringes are counted between wavelengths, λ_{min} and λ_{max}, then, one gets,

$$p = \left(\frac{x}{\lambda_{min}} - \frac{x}{\lambda_{max}}\right), \quad \text{or} \tag{6.31}$$

$$x = \frac{p}{\Delta \nu}, \tag{6.32}$$

with $\Delta \nu = 1/\lambda_{min} - 1/\lambda_{max}$.

The optical difference may be derived by estimating the frequency or number of fringes in the channeled spectrum. Such a number determines the distance from a zero group delay, but not the offset if it is positive or negative (Lawson 1999). The minimum requirement to remove the ambiguity in sign is to have two measurements of channeled fringes at different delays.

Another method, viz., the Real Time Active Fringe-Tracking (RAFT) system that is based on photon counting detector and real time image processing unit has been applied to dispersed fringes on GI2T using 2-D FT (Koechlin et al. 1996). The advantage of RAFT over group-delay tracking is the knowledge of the sign of the OPD to measure and the possibility to be used with apertures larger than Fried's parameter, r_0, where overlapping wavefronts would blur the fringes. However, dispersed fringes with large apertures require a complex optical system to rearrange the speckles in the image plane before dispersion and recombination (Bosc 1988). Both group-delay tracking and RAFT allow a slow servo-loop period (up to a few seconds) by multiplying the coherence length by the number of spectral channels used. To note, their common use of the Fourier transform make them optimal in the sense that they yield the same OPD than a maximum likelihood estimator.

6.1.6.2 Co-phasing

Co-phasing, which may be compared with adaptive optics, is a more demanding technique and is performed with white-light fringes, using the synchronous detection method. To note, a dual-feed system can be used for phase-referenced interferometry, by which one measures the atmospheric delay on the brighter star, and uses this to track fringes on the fainter star. This permits extended coherent integration of fringes on the fainter star. This method allows for direct Fourier inversion of the visibility data provided if an unresolved off-axis phase reference is utilized

(Haniff 2007). It is designed to track quickly the OPD over a wavelength range and hence requires faster response of several orders of magnitude. Using trigonometric identity, one may rewrite the fringe (6.23) as,

$$I(v) = \frac{1}{\tau} [N + X \cos(2\pi v x) + Y \sin(2\pi v x)], \qquad (6.33)$$

with

$$N = \tau I_s, \quad X = \tau I_s |\gamma(\mathbf{r}_1, \mathbf{r}_2, \tau)| \cos(\Delta\varphi), \quad Y = \tau I_s |\gamma(\mathbf{r}_1, \mathbf{r}_2, \tau)| \sin(\Delta\varphi), \qquad (6.34)$$

in which $\Delta\varphi = \varphi(\mathbf{r}_1, \mathbf{r}_2, \tau) - \varphi_0$, τ is the measuring period and the background intensity I_b is assumed to be zero.

Shao and Staeline (1977) implemented a system where two beams are combined pairwise, while a mirror is stepped at quarter-wavelength intervals. The broadband white-light fringe is detected at one of the beam-splitter outputs. The signal acquired from the detector is then processed in order to yield the phase-shift and the visibility, which can be done by integrating signal over four $\lambda/4$ bins, namely, A, B, C, and D, i.e.,

$$A = \frac{(N + X)}{4}, \quad B = \frac{(N + Y)}{4}, \quad C = \frac{(N - X)}{4}, \quad D = \frac{(N - Y)}{4}. \quad (6.35)$$

Thus the phase-shift and visibility modulus are then given by,

$$\Delta\varphi = \arctan\left(\frac{B - D}{A - C}\right) \qquad (6.36)$$

$$\mathcal{V} = \frac{\pi \sqrt{(A - C)^2 + (B - D)^2}}{\sqrt{2}(A + B + C + D)}. \qquad (6.37)$$

In fringe-tracking system, it is possible to obtain high spectral resolution data, since a broad band white-light fringe can be used for tracking the fringe, while the remaining output channels may be dispersed. However, its sensitivity limit is low compared to the fringe envelope scanning interferometer. At NPOI interferometer, a compound method, based on the synchronous detection applied to signals from several spectral channels, has been used (Benson et al. 1998). Fringe-tracking is usually done from data acquired for scientific purpose (i.e., visibility extraction), in order not to 'share' the photons between two instruments. Hence, the fringe S/N ratio is optimal. Such a ratio determines the precision with which the visibility can be measured as a function of $n_p \mathcal{V}^2$, in which n_p is the number of photons detected per sub-aperture and $\mathcal{V}$ the fringe visibility. The fringe S/N ratio is given by the expression (Tango and Twiss 1980),

$$S/N = n_p \Delta t \bar{\mathcal{V}}^2 \left[\frac{(T/\Delta t)}{2(1 + 2n_p \Delta t \bar{\mathcal{V}}^2)} \right]^{1/2}, \qquad (6.38)$$

where, the bar represents time average.

To note, the S/N ratio at low count rates depends on both the overall integration time T and the basic sampling period, Δt. In order to avoid phase smearing, it is desirable to make the sampling period as short as possible, although it has the effect of lowering the S/N ratio. The dependence on $n_p \mathcal{V}^2$ implies that interferometry becomes increasingly difficult for faint sources, particularly for those with complex structures. The $n_p \mathcal{V}^2$ limit can be addressed in various ways, viz., (1) using larger sub-apertures, (2) slicing of the image at the entrance of the spectrograph, (3) bootstrapping, and (4) tracking fringes on a point source to increase integration time on the target.

Fringe-tracking methods may be enhanced by the introduction of *a priori* information, in order to allow observations at fainter visibility, $\mathcal{V}$ or fainter magnitudes. By filtering data with a function computed to reduce the photon noise, one may improve white light cophasing.

6.1.6.3 Bootstrapping

It is possible to stabilize the fringes using bootstrapping methods with the interferometers. These methods can be categorized into two systems:

1. Wavelength bootstrapping: It requires two recombiners, one for visibility measurement, the other for fringe-tracking. For example, at long baselines, when the expected fringe visibility is too low for tracking, it is possible to use a longer wavelength where the fringe contrast, for a white object, is higher. Meanwhile, fringes for computing the visibility are acquired at shorter wavelength. The main advantage of wavelength bootstrapping is that the fringe-tracking data may be used to stabilize the fringe. Hence, it is possible to overcome the atmospherically induced limit on the integration time, which in turn, allows one to push the data from the photon-starved regime into the photon-rich regime and can shorten the observation time considerably (Mozurkewich 1999). The limitation of such a technique is that since the fringe-tracking is performed at the longest wavelength, it should be carried out with high precision and speed to stabilize fringes at the shortest wavelength. Moreover, the atmosphere introduces a relative piston between the apertures that are co-located but of different diameters. Therefore, in the absence of an adaptive optics system, one must limit the size of the apertures to that which is permitted at the shortest wavelength. The limitations of the aperture size as well as the integration time limit the sensitivity. In addition, uncertainty of accurate value of the refractive index of air prevents to predict the fringe position from a measurement at a different wavelength (Mozurkewich 1999). Thus, it may be difficult to apply wavelength bootstrapping over more than a factor of two or three in wavelength. This method was applied at Mark III interferometer (Quirrenbach et al. 1996).

2. Baseline bootstrapping: It estimates the phase on the longest baselines with high precision to stabilize the fringe. The integration time may also be extended beyond the atmospheric limit. Such a method is performed by tracking fringes over a connected series of short baselines to allow low visibility fringes to be measured on the longest baseline. Figure 6.13 depicts a four-element, linear, redundant

Fig. 6.13 Principle of
baseline bootstrapping;
apertures are represented by
circles (Saha and Morel 2000)

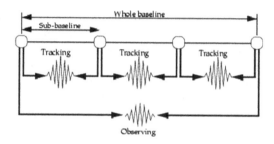

array. In this method the photons are shared and it works well because of
the closure-phase relationships. Here, the baseline is divided into sub-baselines
by adding apertures along the baseline. Baselines connecting the apertures are
identical, hence the number of Fourier phase is reduced to three from six. Fringe-
tracking is performed on each sub-baseline simultaneously, where the visibility
is higher than with the entire baseline. Hence, fringes are tracked on the whole
baseline as well. However, if N short baselines are strung together to phase a
long baseline, the phase noise on all the short baselines contributes to the uncertainty in the long baseline phase. The phase noise on the latter may be $\sqrt{N}$
times the phase on the former. This gives rise to the lower system visibility on
the longer baselines (Mozurkewich 1999).

NPOI employs such a system that enables one to reach spatial frequencies beyond
the first visibility null (Pauls et al. 1998). A related idea has been developed for the
Keck interferometer, as well as for the VLTI.

6.1.7 Effect of Polarization

In a two element long baseline interferometer, the signals are affected by instrumental polarization. The interference laws of Fresnel–Arago state that (1) two waves
linearly polarized in the same plane can interfere and (2) two waves linearly polarized with perpendicular polarizations cannot interfere. In the case of the latter, the
situation remains the same even if they are derived from perpendicular components
of unpolarized light and subsequently brought into the same plane, but interfere
when they are derived from the same linearly polarized wave and subsequently
brought into the same plane (Collett 1993). Since the two waves are propagating
in the z direction and linearly polarized with the x-component of the electric field
vector. The optics of one arm of the interferometer can introduce, (1) polarization
containing s and p linearly polarized components, and left and right circularly polarized components, which is large at optical wavelengths (about 1–10%), while it
is less at the near-IR wavelengths, (2) a phase-shift effect between reference polarization directions, and (3) a rotation of the reference system of the polarization
directions (Rousselet-Perraut et al. 1996). To note, polarization mismatches between
the arms of a feed system reduce the observed scalar visibilities by up to a few tenths
of a percent (Elias II 2001). Two components of one element, i.e., E_{x1}, E_{y1}, of such

an interferometer are correlated with the components of other element, E_{x2}, E_{y2}, to give four complex correlations or the visibility coefficients corresponding to the four elements of the brightness matrix (see Sect. 1.3). On simplification, one may obtain the visibilities which are related to the Stokes parameters, I, Q, U, and V, implicitly through a Fourier transform relationship. The observables, which are related to Stokes parameters are (Elias II 2004), (1) the complex normalized Stokes visibilities, $\mathcal{V}_I$, $\mathcal{V}_Q$, $\mathcal{V}_U$, and $\mathcal{V}_V$, and (2) normalized Stokes parameters, Q_{12}/I_{12}, U_{12}/I_{12}, and V_{12}/I_{12}. Both sets approach the normalized uncorrelated Stokes parameters as the baseline length approaches zero. In the case of an array system, if all the four correlations as described in Sect. 1.3 are measured by all the interferometers, one can obtain a spatial Fourier coverage which is the same for all the polarization components, and therefore, the same synthesized beam for all four polarization images may be obtained.

It is imperative to consider the state of polarization of the light as, according to the law of Fresnel–Arago, orthogonal states of polarization do not interfere. Optical designs of interferometers change the state of polarization differently in different arms. The geometry of the relay optics should not corrupt the relative polarization of the two beams coming from these arms. Instrumental differential relations and phase shifts between the polarization directions may affect the visibility of fringes, and therefore, instrumental errors which require calibration.

In general, the Coudé trains are kept symmetrical in order to prevent differential polarization relations and phase shifts, but leads to a large number of mirrors which gives low throughput; the differential optical coatings may provide residual polarization effects. Each reflection from a mirror surface introduces a phase shift between polarization states. If the corresponding mirrors at each reflection are of same type, both beams that come from two arms of an interferometer would experience the same phase-shifts; the respective s and p components combine independently in the focal-plane and produce identical fringe packets. If the sequence of reflections is different, the visibility $\mathcal{V}_{pol}$, of an interferogram is,

$$\mathcal{V}_{pol} = \left| cos\frac{\psi_{sp}}{2} \right|, \tag{6.39}$$

where ψ_{sp} is the $s - p$ shift between the two beams.

The extent of optics may be reduced to a minimum and the resulting polarization effects should be calibrated and corrected inside the interferometer. The loss of coherence due to misalignment of optical train, aging of coatings, and accumulation of dust can also be analyzed (Elias II 2001). To note, in spite of utmost care in designing the optical path to have the most symmetric path for each beam, in reality the incident angles may not be the same and the mirrors do not have the same coatings. Another strategy for reducing polarization errors is to separate the two polarization states in the beam-combining area. In the SUSI instrument, one polarization is employed for tip-tilt correction, while the other is used for visibility measurement. One may also use one polarization state for fringe-tracking and other for fringe measurement, although there is a possibility of a small phase difference between s and p

polarization states, even in the case of symmetric paths (ten Brummelaar 1999); the center of the fringe envelopes may not be in the same location.

Differential rotations may be compensated by rotator devices, while differential phase-shifts can be corrected by Babinet compensators. At GI2T, the REGAIN uses the field rotators consisting of four plane mirrors for each beam to compensate the polarization difference (Rousselet-Perraut et al. 1996). In the COAST, starlight passing through the central siderostat undergoes an additional two reflections so that its s and p polarizations experience the same reflections as light from the other siderostats (Baldwin et al. 1998).

Fibers have a natural birefringence and a solution for compensating such a birefringence consists in winding the fibers into one or two loops (Lefèvre 1980). The supplementary birefringence introduced by this system depends on the radius of the loops; it cancels the effects of the natural birefringence. With this system, the polarization plane can be rotated by twisting the fiber (by displacing the loops around the main fiber axis). Another solution to minimize birefringence effects consists in using a Babinet compensator (a birefringent quartz crystal consisting of two thin prisms cemented together to form a thin parallel plate) at the input of each fiber.

Since LBOIs employ mirrored feed systems, they are prone to have instrumental polarization. It is probable to integrate polarizing optics along with the rest in the new generation beam combiners, for instance, CHARA/VEGA (Visible spEctroGraph and polArimeter) instrument (Mourard et al. 2009). In addition, it is also necessary to remove interstellar polarization caused by the forward scattering off interstellar dust grains from the data. These grains are oblong (or needle) shaped, which are aligned perpendicular to the galactic magnetic field. Therefore, the induced interstellar polarization is parallel to the galactic magnetic field (Serkowski et al. 1975).

6.1.8 Dispersion Effect

Owing to the atmospheric refraction the real optical path deviates from the theoretical position of the source. This results in an offset at a given wavelength. This effect is chromatic as is the optical index of the atmosphere and the refracted beacon undergoes a dispersion along the vertical direction inducing a spread in the zero-OPD position against the wavelength, which in turn, reduces the apparent visibility. The decreasing factor is evaluated from the wavelength dependence $dn/d\lambda$ of the refractive index at the mean wavelength, $\bar{\lambda}$, as well as from the baseline component, B_e projected on the direction of dispersion. In the case of the North–South baseline, the attenuation factor is dictated by, $\sin x/x$, where

$$x = \pi B_e \tan z \frac{dn}{d\lambda} \frac{\Delta\lambda}{\bar{\lambda}}, \qquad (6.40)$$

in which z is the zenith distance and $\Delta\lambda$ the bandwidth.

Differential dispersion may arise from the unequal path lengths within the instrument through vacuum, air and glass, which gives rise to differential dispersion, and thus a reduction in visibility. A compensating system for correcting atmospherically induced dispersion is essential at the recombiner. The bias introduced by differential chromatic dispersion between the two beams coming out of the interferometer due to the optical components are necessary to be compensated or calibrated. In the REGAIN, such compensation is taken care of by using two prisms which can slide on their hypotenuse, forming therefore, a plate with adjustable thickness. This thickness is modified every 4 min, following the variation of the altitude of the observed object (Rousselet-Perraut et al. 1996). At NPOI, dispersion compensation of the internal optical path observations of the white light source, which are effected by delay fluctuations due to air paths between the siderostats and the vacuum feed system can be corrected to 0.3 μm (Hutter 1999).

In the fiber-linked recombiner, the dispersion of a fiber optics coupler made by two fibers of 1 and 2 is expressed by the phase curvature:

$$\frac{d^2\psi}{dk^2}, \tag{6.41}$$

where ψ is the phase of the spectrum of the interferogram.

It was demonstrated by Coudé du Foresto et al. (1995) that the phase curvature can be given by:

$$\frac{d^2\psi}{dk^2} = -2\pi c \lambda^2 (D_2 \Delta L + L_1 \Delta D), \tag{6.42}$$

where D_2 is the dispersion coefficient of fiber 2, $\Delta L = L_2 - L_1$ the difference of length between the two fibers, and $\Delta D = D_2 - D_1$ the difference of dispersion.

New fiber designs, such as reduced-slope fibers, have low dispersion over a very wide range because the slope of dispersion is less than 0.05 ps/nm²-km, which is significantly lower than standard fibers (Hecht 2000). The dispersion coefficient depends on the refractive indices of both the core and the cladding of the fiber. One problem of fiber dispersion is the 'flattening' of the interferogram, reducing the fringe contrast. To minimize the dispersion, the length of each fiber must be calculated from the dispersion coefficients of each fiber.

6.1.9 Calibration

Long baseline interferometric measurements need precise calibration with reference stars. This requires an assessment of the phase errors associated with each sub-aperture in the array. A general calibration requirements for an interferometer are: (1) calibration of the visibility, (2) flux calibration to wipe out the thermal background from the signal, (3) internal instrument calibrations such as characterization of the optics and of the detector, and (4) checking the instrument, as well as monitoring of important fluctuating parameters.

In order to optimize a synthesized array, it is essential to make initial calibrations apart from delay corrections (see Sect. 6.1.3). The telescope tracking center must follow the same intended position for all telescopes; errors in tracking may reduce sensitivity and distort the image of the extended objects. The telescope pointing error is the difference between the actual pointing position and the desired one. It has a complicated directional dependence mainly due to (1) misalignment of telescope rotation axes, (2) differential heating of the structure, and (3) wind loading.

A perfect synthesis array preserves a linear relationship between the measured visibilities, $\mathcal{V}'(\mathbf{u})$ and the actual visibilities, $\mathcal{V}(\mathbf{u})$, which, in general, is expressed as,

$$\mathcal{V}_{ij}(t) = G_{ij}(t)\bar{\mathcal{V}}_{ij}(t) + \epsilon_{ij}(t), \tag{6.43}$$

where $\mathcal{V}_{ij}(t)$ are the visibilities measured between telescopes i and j, $\bar{\mathcal{V}}_{ij}(t)$ the true visibilities for the baseline, ij, $G_{ij}(t)[= g_i(t)g_j^*(t)]$ the gain factor representing amplitude and phase errors introduced by the interferometer itself and during the propagation of the signal through space and atmosphere, $g_i(t)$ and $g_j(t)$ the complex gain factors for the individual telescope i and j respectively, and $\epsilon_{ij}(t)$ the additive noise.

The calibration equation (6.43) for such a source turns out to be,

$$\mathcal{V}_{ij}(t) = g_i(t)g_j^*(t)S + \epsilon_{ij}(t), \tag{6.44}$$

with

$$g_i(t)g_j^*(t) = \bar{g}_i(t)\bar{g}_j(t)e^{i[\psi_i(t)-\psi_j(t)]}, \tag{6.45}$$

$\bar{g}_i(t)$ as the multiplicative gain of the ith telescope, $\psi_i(t)$ the phase of the ith telescope, and S the strength of the unresolved point source.

Since the expected amplitude of such a source is constant, and the expected phase is zero, any differences in the $\bar{g}_i$ and any non-zero phase differences $\psi_i - \psi_j$ should be due to the telescopes themselves. One may solve for complex gain $g_i(t)g_j^*(t)$ on $n(n-1)/2$ baselines for n-telescopes. Hence, for large n one may use a least squares algorithm in order to solve for the telescope-based complex gains g_i, in which, $i = 1, 2, \ldots n$.

The measured visibility, $\mathcal{V}$, differs from the true visibility, $\bar{\mathcal{V}}$, owing to the several sources of perturbation. The measurements of $\mathcal{V}$ are biased by a random factor depending on the seeing quality. The imperfect seeing provides low value of visibility. The other sources of perturbation such as mechanical vibrations, the polarization effects in the instrument, instrumental flaws leading to optical aberrations and non-balanced flux between the two beams modify the measured visibility modulus as well. The surface irregularities of the reflecting mirror distort a plane wavefront with phase errors. Wavefront errors causes fluctuations of the beam overlap or of the flux that is injected in the single-mode fibers. Strehl's ratio, $\mathcal{S}_r$, concept (see Sect. 1.5.2) is applicable to the fringe modulation in an interferogram. Such a ratio for fringes,

$$\mathcal{S}_r = \frac{\mathcal{V}_{max}}{\bar{\mathcal{V}}_{max}}, \tag{6.46}$$

can be employed to estimate the combined effect of different sources of visibility from the star and the instrument.

The complex gains are usually derived by means of observation of a calibration source at the phase center before and after the target source, and the change in interferometer orientation due to the diurnal rotation of the Earth permit one routinely to obtain synthetic images of great quality. For high quality measurements, it is therefore important to obtain the impulse response in the effective modulation transfer function (MTF) of the interferometer. This can be achieved in principle by calibrating each measure on an object by measuring visibility on a calibrator (unresolved star or a star of known angular diameter), V_0, for which its true visibility, $\bar{V}_0$ is known *a priori*. This allows one to obtain the interferometric transfer function, $T_{int} = V_0/\bar{V}_0$. Hence, the estimated visibility of target of interest is $\bar{V}' = V/T_{int}$.

In principle, a calibration source is typically chosen to be unresolved, whose true visibility, $\bar{V}_0 = 1$, but its flux may be too low to measure V_0 precisely. In such a situation, the point source model for a very short baseline can be used. However, it is realistic to use as model a Uniform Disk (UD) function having as parameter the angular diameter, θ_{UD} of the calibrator:

$$I(r) = \begin{cases} I_0 & \text{if } r < \theta_{UD}/2, \\ 0 & \text{otherwise.} \end{cases} \tag{6.47}$$

The visibility model for an unresolved source can be obtained by using (3.3). in general, the diameter, θ_{UD}, is derived from the photometric observation of the calibrator. However, the problem arises from the uncertainty on the effective temperature (see Sect. 9.1.4) that can be determined from a direct measurement of θ_{UD} in some cases (Perrin et al. 1998). Another problem comes from the limb-darkening (see 8.1.1) of the stars observed. The radial intensity profile of a star may be given (Hestroffer 1997) by,

$$I(r) = I(0) \left(1 - \frac{r^2}{R^2}\right)^{\alpha/2}, \tag{6.48}$$

where R is the radius of the star and α the limb-darkening factor depending on the stellar atmosphere.

An LBOI experiences visibility reductions due to instrumental and environmental limitations as well. It is required to differentiate astrophysical and instrumental effects on the data. To reproduce the instrumental conditions, the calibrator must roughly be as bright as the object to calibrate; it should have ideally a spectral type and a magnitude similar to the target object. Observations of both these sources should be interleaved for recording the fringes back and forth a few times during the observing run. Hence, one can interpolate the transfer function for each object-observation period. Also, the calibrator should, in principle, be in the neighborhood in both time and space, preferably within $1°$ of the science target, to minimize motions of telescopes and to get identical turbulence condition (Shao et al. 1988b).

The visibility loss experienced by fringes observed in the target source and calibrator must be the same. The fringes must be measured at the same fixed position on coherence envelope. Such an envelope may move during an observation due to the errors in the astrometric model and random path variations ($\sim$10 μm/m RMS of baseline) introduced by the atmospheric turbulence. Indeed, observation of this kind calls for some form of phase- or group-delay tracking.

Another effect may arise from diffraction, which is unavoidable since the long paths are required in an interferometer. It is required to introduce differential paths in order to compensate for the external path implemented by the projected baseline, and hence yields differential diffraction in the reduced visibilities. To note, the magnitude limit of a source within an isoplanatic patch restricts the observations. However, a few objects are themselves providing the reference, for example, at some wavelengths an object may get resolved, while for the remaining part of spectrum, it may act as point source. In this technique, known as spectro-interferometry (see Sect. 6.1.3.2; Mourard et al. 1989), the various effects undergone by the visibility are expected to be same in and out of the emission line.

In general, a photometric correction is applied on the recorded interferometric signal taking into account the variations of the flux for each incoming beam. The variation of the telescope fluxes are monitored together with the interferometric signal. The principle of photometric calibration associated to the modal filtering, which permits improvement of the fringe visibility estimation, has been applied with accuracy down to 0.3% with the FLUOR instrument (Coudé du Foresto et al. 1995).

An *a priori* baseline vector, **B**, is used to fit the interferometer response to a computed fringe pattern. In order to determine the baseline precisely, the process of calibration is employed. Point sources of accurately known positions are observed over a wide range of hour angle and declination, and the systematic change of the visibility phase response with hour angle and declination is used to derive a more accurate value of the baseline. Determination of the ratio of the flux density of the calibrators to the correlated amplitude of the response is also a necessity.

6.1.10   Role of Adaptive Optics Systems

Interferometry with mirrors larger than a meter or so would be less than fully effective if the images from the individual mirrors are less sharp than the diffraction limit. Enough photons are required to phase the telescopes into a coherent aperture, therefore, to enhance the instrumental visibility. Without any active wavefront correction on the individual telescopes and photon-counting detectors, the limiting magnitude for such an interferometric technique is similar to that of bispectral imaging (see Sect. 8.2.2) at single telescopes.

The coherence radius of the perturbed wavefront due to the influence of the atmospheric turbulence exhibits a $\lambda^{6/5}$ dependency. Image distortion varies across the sky, and the angle over which such distortions are correlated is the isoplanatic angle, θ_0 (see Sect. 4.3.1.1). There are a few more aspects of the atmosphere, viz., (1) existence of atmospheric tilts that may persist over a long period of time and

(2) outer scale of turbulence. Atmospheric tilts provide unequal paths through the atmosphere to the telescopes of an interferometer, thus changing the apparent position of a star. The root-mean-square variation between the stellar positions during observations on successive days may be attributed to such a tilt. Of course, pressure gradients in the atmosphere parallel to the Earth's surface, as well as the gradients of humidity may provide modest tilts in the optical path length through the atmosphere. If two wavefronts of width, D, are tilted by an angle, α, the interference pattern gets smeared and visibility reduced. Following (3.3), the visibility factor from a 2-D circular aperture of diameter, D can be expressed as,

$$\mathcal{V}_{tilt} = \frac{2J_1(\pi D\alpha/\lambda)}{\pi D\alpha/\lambda}. \tag{6.49}$$

To note, an interferometer works well if the wavefronts from the individual telescopes are coherent. The maximum useful aperture area is proportional to $\lambda^{12/5}$. These telescopes should, in principle, not be larger than a few times the atmospheric coherence length, r_0. Corrections for atmospheric turbulence may be avoided if the diameter of the telescopes is much smaller than r_0 (see 4.58), in which case the S/N ratio is low. Moreover, the measurement of the contrast of the fringe pattern should be made in a time τ_c or else the pattern would become blurred. In order to make the coherence length longer than the atmospherically induced optical path fluctuations, a narrow optical bandwidth defined by interference filter is used. These constraints restrict LBOIs to bright targets.

In order to improve the sensitivity of an interferometer, each telescope will have to be reasonably large. In that case, an adaptive optics system (see Sect. 4.3) is required to be fitted at each telescope for improving the resolution, and thereby allows better interferometric imaging with telescope array. For a small telescope, correction of the tip-tilt components of the perturbations may be adequate. If the value of D/r_0 exceeds three or more, higher order adaptive optics systems are essential. For example, an aperture of 2-m class produces more than 100 speckles in the image and the fringe pattern within each speckle is randomly phased. The main errors can be corrected by incorporating such a system or by an appropriate optical design adjusting the entrance pupil diameter to the local value of r_0 for the considered wavelength; the remaining phase errors on the incoming wavefront can be removed using a spatial or modal filtering. However, the limiting magnitude of reference source required for adaptive optics correction sets an upper limit on the limiting magnitude for interferometric arrays.

By employing a low order adaptive optics system in the form of tip-tilt correction, which allows the aperture size to be increased to a few r_0 without reducing the contrast of the interference fringes, astronomers are able to target the faint sources. Fast tip-tilt guiding system tracks fast jitter of the stellar image. It corrects the first order term of the wavefront perturbations, aligning the wavefronts to permit for stable beam combination. Most of the existing or planned separate element interferometers have some adaptive optics corrections (often only the image position or tip-tilt component). For example, GI2T and IOTA are using tip-tilt control system. This correction is sensed in the visible, using CCDs, and fringe detection is done in the Near-IR. IOTA tip-tilt correction system uses a 32×32 pixel CCD for each beam. The maximum rate is about 200 Hz. A computer reads each frame and computes the centroid. The value of the centroid position is sent to a piezo mirror placed downstream the secondary mirror of each telescope. The Keck interferometer and the VLTI employ large telescopes fitted with adaptive optics systems.

6.2 Limitations and Constraints

The major limitations of optical interferometry are atmospheric turbulence and limited response time of interferometer. The turbulence, both external turbulence as well as internal turbulence within the interferometer, can made pathlengths to fluctuate rapidly over timescales of the order of milliseconds. The fluctuating amount of integrated atmospheric pathlengths above each aperture introduces random delays in the light arrival time resulting in phase shifts in the measured fringes. This makes the source appear and disappear, making measurement more difficult. The other factor is the limited response time of the interferometer. Due to the small wavelengths of the signals the interferometer should be fast enough to respond to them. Stringent mechanical tolerance and limited fluxes from the stellar sources also add to the misery to implement such an interferometer. The optical surface errors, the aberrations in the design, manufacture and alignment of optical trains in the system can reduce its performance greatly as well; a root-mean-square surface error of η_{sur} reduces the efficiency of the mirror by a factor of $\cos^2(4\pi\eta_{sur}/\lambda)$. The random variations in MTF of atmosphere and instrument deteriorates the visibility of the fringes.

Operating a LBOI, for instance, finding fringes, measuring their visibility, interpreting the result, is a long and difficult process. It requires accurate alignments, high stability, full control of any effect decreasing visibility. The limitations come from the parameters, viz., (1) precise determination of visibility, (2) sensitivity in measuring weak sources, (3) accurate measurement of fringe phases, and (4) availability of range of baselines. High speed photo-detectors with high-level storage and processing capabilities and frequency-stabilized lasers to measure continually changing delay-line lengths are also required.

Light detection in LBOI is a difficult task since the light from the source must be gathered within a few milliseconds, or the source will move from the field of view. The interference pattern should also be detected in a few milliseconds to avoid smearing due to turbulence. Such an interferometry requires a high quality sensor that transforms the energy distribution into a flow of interpretable information. Most of the available detectors are based on the principle of 'photo-electric effect' (Einstein 1905b). Among these detectors the (1) CP40 (Blazit 1986), (2) Precision Analog Photon Address (PAPA; Papaliolios et al. 1985), (3) Resistive anode position sensing detector system, (4) Low light level CCD (L3CCD; Jerram et al. 2001) and (5) Near Infrared Camera and Multi-Object Spectrometer (NICMOS) are in use.

Although many advances in photo-detection have been made (Morel and Saha 2005 and references therein), some important aspects are required to be considered. These are (1) the spectral bandwidth, (2) the quantum efficiency[5] (QE; Janesick 2001), (3) the number of pixels that can be put into an array, (4) the detector noise that includes dark current, read-out and amplifier noise; photo-conductors have more noise by a factor of $\sqrt{2}$, (5) the time lag due to the read-out of the detector, and (6) the array size and the spatial resolution. Together, they may be gauged by the astronomical capability metric (Bahcall et al. 1991), which is proportional to the number of pixels multiplied by the square of the sensitivity per pixel. Such a metric indicates the number of positions on the sky that is measured in a given time to a given detection limit.

6.2.1 Instrumental Constraints

For precise measurement of phase, an important aspect for astrometry, or mapping of regions that are asymmetric with respect to inversion, the pathlengths to the individual telescopes in an interferometer should be maintained to an accuracy of better than a wavelength of the beam. A correct determination of the baseline vector is necessary to be established, following which, one knows, to observe a given object, how to set the position of the optical delay-line to get fringes within, usually, a few hundred micron (μm) interval around the expected null optical path difference (OPD) point.

In order to avoid various aberrations and vignetting, optics must be adjusted, which may be difficult to avoid when light is fed through long and narrow pipes. It is reiterated that fringes are searched by adjusting the delay-line position. However, once they are located, mechanical constraints on the instrument, errors on the pointing model, thermal drifts, various vibrations and atmospheric turbulence make the null-OPD point changing.

Optical surface figure aberrations reduce the quality of the optical imaging system. If the combining wavefronts, each having a root-mean-square perturbation of σ with respect to a perfect wavefront, and if such perturbations are randomly distributed across the wavefront, and uncorrelated between the two wavefronts, Strehl's ratio, S_r, as well as fringe visibility may degrade. The (1.93) for Strehl's ratio (see Sect. 1.5.2), can be recast for N mirror surfaces with root-mean-square of $\langle \sigma_0 \rangle$,

$$S_r \simeq e^{-\kappa^2 N \langle \sigma_0^2 \rangle}, \tag{6.50}$$

with $\kappa = 2\pi/\lambda$ and $\langle \sigma \rangle = \sqrt{N} \langle \sigma_0 \rangle$.

---

[5] Quantum efficiency (QE) is defined as the ratio of the electrons generated and collected to the number of incident photons and it determines the sensitivity of the detector. Generally, it is wavelength dependent.

The cumulative effect of N optical surfaces may strongly affect the performance of an interferometer. It is also necessary to apply the corrective factors in order to get rid of certain other constraints that come from field rotation, asymmetrical flux effect from apertures, differential atmospheric dispersion, detection effects (arising from insufficient sampling, i.e., number of pixel per inter-fringe), and non-zero OPD. Since the light beam coming out of the telescope undergoes several reflections on its way to the optical correlator, the optical set-up should be devised when pointing at zenith in order to preserve identical orientation of the fields. But the fields are rotated relative to other when pointing off-zenith in a fashion that projections of the fields contribute to the interference. This results in decreasing the measured visibility. Proper optical components or design may be used in order to compensate such an effect (Shao and Colavita 1987). Alternately, a corrective factor, $1/\cos 2\theta'$, in which θ' is the rotation angle undergone by each field vector such that $\sin\theta' = \sin H_\star \cos\delta_\star / \sin\theta$, where $H_\star$ denotes hour angle of the astronomical source, θ the angle between the baseline direction and the source direction, and $\delta_\star$ the declination of the source, should be applied on the measured visibility.

Due to the photometric asymmetry between the two arms of the interferometer, attenuating factor appears, which affects the visibility. Hence, the fluxes are required to sample for each beam beside the recording of interference fringes so as to restore the corrected visibility. Measuring individual fluxes during the recording fringed image enables to correct partial residual effects of the atmospheric turbulence. The corrected visibility is expressed as,

$$V_c = V\frac{1+\alpha}{2\sqrt{\alpha}}, \tag{6.51}$$

where α is the flux ratio (small/large).

Such a correction is applied to the mean value of the fluxes which may be affected by a slow de-focusing due to temperature variations, or a difference in transmission between one way or the other, etc. The other notable problems arise from aberrated imaging the apertures, poor metrology, and instability of path lengths:

1. Aberrated imaging of apertures at the optical correlator induces the axial separation of pupils owing to asymmetry in length of the arms, which causes a mismatch of beams geometry inducing a spurious modulation of the fringes leading to decrease the visibility. Another effect, appearing at large baseline, comes from the de-magnification factor from input to output pupils, the effect of which is a reduction of the effective FOV; this distorts the fringe patterns and falsifies the visibility.

2. Metrology is governed by the coherence length, l_c; the mechanical constraints on the instrument, errors on the pointing model, thermal drifts, various vibrations and atmospheric turbulence make the null-OPD point changing. In such a situation, the visibility is affected by factor, $\sin(\pi d'/l_c)/(\pi d'l_c)$, in which d' is the optical path difference, which results from incoherent addition of several fringe patterns over the spectral bandwidth. A poor metrology increases the uncertainty about the location of the fringes.

3. Instability of the pathlengths that arises from vibrations associated to moving components such as telescopes, delay-lines, horizontal propagation through air, affects the measurements. In addition, the effect of microseismicity should also be taken into account for the large baselines.

6.2.2   Field-of-view

The field-of-view (FOV) is the angular extent of the object that is imaged by the system. It is determined by magnification and the focal lengths of the objective and eyepiece lenses; the more magnification the less FOV. In an angular field it may be expressed in degrees. In an interferometer, when the mirrors' images are completely parallel, the interference fringes appear circular. If the mirrors are slightly inclined about a vertical axis, vertical fringes are formed across FOV. These fringes can be formed in white light if the path difference in part of the FOV is made zero, albeit a few fringes appear.

The interferometric FOV is the product of the spatial and spectral resolutions, which is limited by the spectral resolution, $\lambda/\Delta\lambda$. The limited FOV is the main constraint for an aperture synthesis imaging. It introduces correlations in the complex visibility in the u, v-plane. A low spectral resolution also produces an effect called 'bandwidth smearing' (Thomson 1994). Short coherency envelope for wide bandwidth observations give rise to a limitation of the FOV. This limits the FOV, i.e.,

$$FOV \sim \frac{\lambda^2}{\mathbf{B}\Delta\lambda}, \tag{6.52}$$

where $\mathbf{B}$ is the baseline vector.

This (6.52) reveals that the FOV can be baseline dependent. However, this may be avoided by employing a spectrometer in order to limit the bandwidth of individual observing channel. The primary beam of an individual telescope used in an interferometer may limit the FOV. The limit is dictated by λ/D, in which D is the diameter of the largest telescope. The flux outside the diffraction-limited beam is rejected for many beam combiner. In order to map the entire primary beam, the bandwidth smearing should be small for any imaging interferometer, the requirement of which is determined by the longest baseline divided by the diameter of an individual telescope.

The interferometer observes a finite bandwidth, but the external geometrical OPD is compensated for the central wavelength. The averaging of the visibility over the bandwidth produces a radial blurring of the image. In order to enlarge the FOV, it is necessary to increase the spectral resolution, but a high spectral resolution may be incompatible with the observation of faint objects. Hence, it is desirable to keep a moderate spectral resolution. A wide FOV is required for observing extended or multiple objects. It is useful for the fringe acquisition and tracking on a nearby unresolved object as well. Beckers (1986) opined that the FOV should at least be equal to the isoplanatic patch for ground-based arrays.

Depending on the type of interferometer, there are specific methods to deal with the FOV problem. Two notable methods are generally employed to observe a large FOV:

1. Homothetic mapping: A Fizeau-type instrument is intrinsically a homothetic mapper, where the beam combination scheme has a natural wide FOV. A Michelson-type telescope array can be used as a homothetic system as well provided the images are recorded in the focal plane and if the exit pupil after the telescopes is an exact demagnified replica of the input pupil as seen by the incoming wavefront. Let M be the angular magnification of the telescope, hence the effective baseline at the exit pupil must be $B_0 = B/M$, in which B and B_0 are the collection and combination baselines respectively. Pupil rotation is required to be precisely controlled to maintain the orientation of the exit pupils. An instrument based on homothetic system is being developed in the Delft Testbed Interferometer (DTI; van Brug et al. 2003). However, such a system requires a highly complex positioning system and demands very accurate calibration of the optical parameters of the instrument (Montilla et al. 2005).

2. Wide-field mosaic imaging: This technique, analogus to the mosaicing method that scan the object in steps of one Airy disk, employed in radio astronomy (Cornwell 1994), is adapted to a Michelson pupil-plane beam combiner with detection in the image plane. In order to image a region of the sky in more detail, multiple overlapping scans are recorded, the individual images are put together to form the large image. A delay-line is used to scan the optical path length through the sky and an array detector records simultaneously the temporal fringe patterns from many adjacent telescope fields. The data needs to be collectively deconvolved to reconstruct the image. With a large format detector, as well as with high spatial frequency sampling of the image plane, this method may be used to multiply the field size considerably, reaching a field of several arcmin. This method demands long observation time since wide FOV is not acquired in a single shot.

To note, the OPD should not vary more than $\lambda/10$ over the FOV in order to have the white light fringe on the individual stars. The accuracy requirements for the pupil reimaging should also be dynamic due to the Earth rotation, as well as for the scale factors.

6.2.3 Sensitivity

Sensitivity of an optical interferometer is essentially determined by three factors such as (1) atmospheric turbulence restricting the coherent aperture size and coherent integration time, (2) optical transmission, and (3) detector/background noise. The relative pathlengths for the signals from the two telescopes should necessarily be tracked precisely. In addition to the precise baseline vectors, the accurate determination of stellar position depends on the signal-to-noise (S/N) ratio obtainable in measurement of the interference pattern. An adequate sensitivity is necessary to determine the phase of interference during such timescales for signals from a narrow solid angle in an astronomical FOV. If S/N ratio approaches unity or less for these times, no fringe measurement can be made. The other notable major problems consist of:

- the λ-dependency of diffraction and interference phenomena, which demands sub-micron quality and stability of the optical instrumental design,
- low optical throughput due to the large number of reflections between the telescopes and the detector,
- the lack of detector linearly sensitive to the electromagnetic field in the optical domain, which forbids post-detection amplification and imposes the requirement to encode information prior to detection, and
- noise that is generated at each stage of the process (Mariotti 1988).

Sensitivity of an interferometer may go down further with the diffractive losses during transmission of light. Of course, coatings of these mirrors, which deteriorate over time may add to this misery. New generation interferometers may have to look for a simplified beam trains with a few reflections; the use of integrated optics may provide some solution in this respect.

The fundamental noise for an ideal amplitude interferometer arises due to thermal radiation incident on the sensor. For an ideal sensor, this produces a noise power fluctuation of,

$$N_d = h\nu \left[\frac{2\Delta\nu}{t} \frac{1-\epsilon}{e^{h\nu/k_B T} - 1} \right]^{1/2}, \tag{6.53}$$

with h as Planck's constant, $h\nu$ the quantum energy, k_B the Boltzmann's constant, T the temperature of optics and atmosphere through which the signal is received (in K), $\Delta\nu$ the bandwidth, t the post-detection averaging time (s), and ϵ the fractional transmission of light reaching the telescope.

The (6.53) is valid for a photo-diode. To note, noise is due to fluctuations in the number of quanta in the received light. For a two-aperture interferometer, the signal-to-noise (S/N) ratio is limited by the statistical fluctuations in visibility squared $\mathcal{V}^2$ (see 6.38), in which $\mathcal{V}$ is the fringe visibility.

It is improbable to tilt the baseline of an interferometer in the vertical plane, and the transit of the source is employed to produce the interference fringe pattern on the detecting system. For a baseline oriented in the East–West direction, it produces a fully fringed pattern, but as the baseline is rotated in azimuth, the period of the fringes increases. On the other hand, for a North–South baseline, the fringe frequency remains zero at transit. The source may occupy any position on the lobes of the fringe pattern, however it is not possible to get a direct reading of the visibility at any point along the baseline. A continuously variable length of baseline can be arranged in order to produce fringes. The source distribution in the direction of North–South can be computed from the variation of fringes with the aperture spacing.

6.2.4 Bandwidth Limitations

In reality, the LBOI works for finite bandwidth where the pass-band defines the effective source spectrum. Indeed large bandwidths are required to get high sensitivity on continuum sources. However, a significant limitation comes from this bandwidth, the effect of which is to produce a frequency dispersion. The optical path delay is different for the different frequencies within the band, which can lead to signals being out of phase. The time over which the radiation is considered, the actual time coherency is derived from the Fourier transform of the pass-band used. In order to observe continuum sources over a finite bandwidth, $\Delta \nu$, one considers the effect of this on the interferometer response, since the sensitivity of the system increases with $\Delta \nu$. As stated earlier, the path difference between the two rays arriving at two apertures of an interferometer is, $\tau_g = B \sin \theta / c$. Replacing $\mathbf{B} = \lambda \mathbf{u}$, $\mathbf{u}$ the spatial frequency vector, and $c / \lambda = \nu$, the delay can be transformed into $\tau_g = \mathbf{u} \sin \theta / \nu$. Hence, the frequency dispersion $\Delta \nu / \bar{\nu}$ gives rise to a corresponding equivalent dispersion in the baseline of $\Delta \mathbf{u} / \mathbf{u}$. The resulting effect upon the aperture beam is to multiply with the fringe pattern by the Fourier transform of the bandpass function. For an interferometer with a bandwidth, $\Delta \nu$, the output, averaged over the frequency band is given by,

$$I = \int_{\nu_1}^{\nu_2} U_1 U_2 \cos \frac{2\pi \nu l_c}{c} d\nu, \qquad (6.54)$$

in which l_c is the coherence length.

. The tolerance for matching pathlengths in an interferometer depends on such spectral bandwidth. For an extremely narrow bandwidth, like a laser, interference takes place even if pathlengths are not matched, while for a wide bandwidth, the number of fringes in an interferometer equals to the inverse of the fractional bandwidth, i.e., $N_{fr} \sim \lambda / \Delta \lambda$. Hence, a stringent requirement in the case of wide bandwidth observation, the pathlengths should be matched within a wavelength. It is quite natural that the received quantity of light should be increased in order to enhance the signal-to-noise ratio, therefore, increase $\Delta \lambda$. This is apparent that with increasing baseline the fringe pattern is limited by the transform of the frequency response. If the bandpass of width extends over $\Delta \nu$, in which $\Delta \nu > 1 / \tau$, the change of pathlength across the frequency band is sufficient to cause a loss of correlation. To note, the delay response for a rectangular bandwidth function, $\Delta \nu$, is proportional to, $\sin \pi \Delta \nu \tau_c / \pi \Delta \nu \tau_c$.

The compensating delay should be adjusted to an accuracy of about 0.1% of the reciprocal bandwidth to avoid loss of no more than 2% of the signal amplitude. Figure 6.14 depicts the response of an interferometer to a narrow band source. A narrow band is useful in permitting a relatively easy variable relative delay between beacons from the two telescopes in order to get the white light fringe. Conversions to low frequencies provides in easing a delay-time precision required to determine the phase precisely. It is essential to maintain the delay, τ, close to zero or

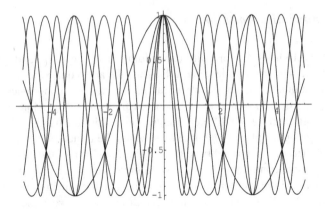

Fig. 6.14 Response of an interferometer to a narrow band source

else different frequency components would cancel. The amount of delay should be changed to compensate for the variation of the geometrical delay as the source is tracked.

6.2.5 Limitations due to Atmospheric Turbulence

Decrease in visibility is due to the incoherent addition of randomly shifted independent fringe patterns. The same effect may arise from mechanical instabilities that modify the instrumental OPD. It is necessary to maintain the instrumental stability so as to make the maximum expected OPD standard deviation a fraction of the typical one induced by the turbulence. According to Roddier (1981), the standard deviation of turbulence induced OPD, is given by,

$$\sigma_z = 0.417\lambda \left(\frac{L}{r_0}\right)^{5/6},\tag{6.55}$$

in which L is the smallest of either the baseline or the outer-scale of turbulence, L_0, according to the expected saturation of the phase structure function.

Another constraint, the spectral coherence, is the maximum usable bandwidth, $\Delta\lambda$ is proportional to $\lambda^2 \sigma_z^{-1}$, i.e.,

$$\Delta\lambda \propto \left(\frac{r_0}{L}\right)^{5/6},\tag{6.56}$$

at which coherence is 50%.

The signal-to-noise (S/N) ratio is proportional to the number of photons received per integration (exposure) time, Δt at low light level. For a given bandwidth, $\Delta\lambda$ and a given exposure time, it is proportional to the square of the atmospheric coherence length, i.e., r_0^2. The visibility of fringes decreases if they are allowed to move during the integration time, Δt; in general, the exposure time, Δt should be less than $0.36(r_0/\Delta v)$, in which Δv is the wind velocity dispersion with altitude (typically 5 m/s). Measurements of MTF in speckle work, suggest that fringes should be obtained in a time as short as atmospheric fluctuations ($\sim$0.01 s).

Incomplete superposition of images randomly decreases the coherent energy contribution in the image energy spectrum. In the pupil plane technique air tilt fringes appear, which reduces the efficiency of the flux modulation. With the increase of telescope size, the importance of image motions reduces. Performing image tracking on a large and separate bandwidth may enhance the sensitivity. In the image-field scheme, in each area the effect of non-zero OPD and of bad overlapping occur at random through the entire focal path whole angular extension is about $\sim \lambda/r_0$. Therefore, 2-D analysis is required for the fringe pattern.

6.2.6 Atmospheric Phase Errors

The instantaneous phases of fringes from the reference source act as a probe of the atmospheric conditions. This can be used to correct the corrupted phases on the target. The delay measured with an interferometer can be related to the interferometer baseline vector, $\mathbf{B}$, and the unit vector, $\hat{\mathbf{s}}$ to the star in the absence of the atmosphere. The basic astrometric equation is given by,

$$d = \mathbf{B} \cdot \hat{\mathbf{s}} + C, \tag{6.57}$$

where the delay, d, is the position of the central white light fringe and C the delay offset, which depends on the details of the interferometer optical path and the zero point of the laser metrology system.

For interferometer in the visible range, compensation of the delay variations due to the atmospheric turbulence can be obtained by measuring interferences at two different wavelengths, spaced widely enough so that the dispersion in the atmospheric refractive index can be used for correcting atmospheric seeing and refraction. Colavita et al. (1987) had applied two-color method in the visible region to astrometric data from the Mark III interferometer. In this, one observes one star at two wavelengths and uses narrow-angle astrometry to measure the shift in photocenter as a function of wavelength. It exploits the dispersion of the atmosphere in order to estimate simultaneously the instantaneous stellar position as well as the instantaneous atmospheric error. The two color estimate of a delay is written as,

$$d = d_1 - D^a(d_2 - d_1), \tag{6.58}$$

where d_1 and d_2 are the respective delays measured at two wavelengths λ_1 and λ_2 and D^a the dry air dispersion, which can be related to,

$$D^a = \frac{n(\lambda_1) - 1}{n(\lambda_2) - n(\lambda_1)} \tag{6.59}$$

with $n(\lambda)$ is the refractive index of the air.

The change of path in the atmosphere above an array of element is equivalent to replacing a length of the vacuum path with air. The observed fringe phase, ψ, is the error of the fringe servo and the actual fringe position is given by,

$$d(t) = l(t) + \kappa^{-1}\psi(t), \tag{6.60}$$

where l is the position of the laser-monitored delay-line and $\kappa = 2\pi/\lambda$ the wavenumber of the interfering light.

The two-color phase difference $\psi'(t)$ is

$$\psi'(t) = \psi_2 - \frac{\lambda_1}{\lambda_2}\psi_1, \tag{6.61}$$

in which ψ_2, ψ_1 are the fringe phase in the blue and red channels, respectively.

Owing to the correlations between ψ_1 and ψ_2, the coherence time of ψ' is much larger than that of either ψ_1, or ψ_2. In the case of observing a star at zenith, such phase difference ψ' turns out to be,

$$\psi' = \frac{1}{D^a}\frac{\lambda_1}{\lambda_2}\psi_1. \tag{6.62}$$

The atmospheric coherence time, τ_{01}, of one color phase, ψ_1, is proportional to $\lambda_1^{6/5}$, hence the coherence time, τ_0' for the two color phase difference may be written as,

$$\tau_0' = \left[\frac{D^a\lambda_2}{\lambda_1}\right]^{6/5}\tau_{01}. \tag{6.63}$$

This technique has been applied to narrow-angle astronomy in which fringe phase information is used to determine precise relative positions of nearby stars (Shao and Colavita 1992).

Chapter 7
Discrete-Element Interferometers

7.1 Direct-Detection Interferometers

A number of Long Baseline Optical Interferometers (LBOI), providing submilli-arcsec resolution, are in place on Earth, while space-based telescope arrays are being designed (Saha 2002; Monnier 2003; Labeyrie et al. 2006). Most of these instruments combine the features of the Michelson design and the radio interferometers. To note, interferometers with two telescopes provides the angular size and shape of the source, but cannot reconstruct the angular distribution across an asymmetrical source without ambiguity. For a binary component, such an instrument produces double images with 180° ambiguities. This limitation is possible to circumvent in systems with a larger number of telescopes, where sampling across all possible pairs may permit the reconstruction of high resolution stellar images.

While at an interesting stage of development, currently of more limited imaging capabilities, LBOIs have become routine instruments for scientific investigations, although some of them, namely (1) Interféromètre à deux Télescope (I2T; Labeyrie 1975), (2) Mark III interferometer (Shao et al. 1988a), (3) SOIRDÉTÉ, an IR interferometer (Assus et al. 1979; Gay and Mekarnia 1988), and (4) Infrared Optical Telescope Array (IOTA; Carleton et al. 1994) are no longer in operation. With the instruments as powerful as the current generation of working or planned LBOIs, the element of serendipity may bring many surprises to astronomy. In what follows, the salient features of the LBOIs are enumerated.

7.1.1 Interféromètre à deux Télescope

Labeyrie (1975) had built a direct-detection interferometer with two independent telescopes, called Interféromètre à deux Télescope (I2T), in optical wavelengths using a 12-m baseline at Nice Observatory in 1974, which combines directly the electric fields before photon detection. He had successfully obtained interferometric fringes of α Lyrae (Vega) in the visible band from this interferometer. Following the success of this experiment, it was moved to a relatively better site, at the Centre

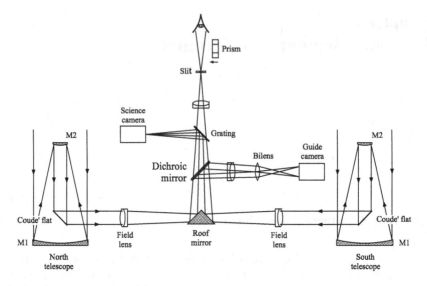

Fig. 7.1 Schematic drawing of I2T

d'Etudes et de Recherches Géodinamiques et Astronomiques (C.E.R.G.A.), in the Observatoire du Plateau de Calern, presently known as Observatoire de la Cote d'Azur, France (Labeyrie 1975).

The I2T consists of a pair of 26 cm telescopes on altitude–altitude (alt-alt) mounts[1] having a long coudé[2] focus (see Fig. 7.1). These telescopes track the same source (star) and send the collected light to the central laboratory where the star images are superposed at the focii in order to produce Young's fringes similar to those observed by Michelson and Pease (1921). Both the telescopes are kept on tracks so that inter-telescope spacing can be varied between 5 m and 67 m. This allows to obtain a variety of different resolving powers, providing flexibility for measurements of stellar diameter and observations of binary stars. When the telescopes are movable, the interferometer can cover many different baselines with different geometrical configurations. Each telescope has Cassegrain afocal system providing an angular magnification of 23 and the coudé beam is obtained by means of a single reflection on a flat tertiary mirror.[3] Due to such a magnification, air turbulence in the horizontal part of the beam has little effect on the image quality.

---

[1] An altitude–altitude (alt-alt) mount is sort of half-way between the equatorial mount and the alt-az mount. The design incorporates an image rotator into the main axes of the telescope, eliminating the need for an independently rotating coudé platform. The beam can be directed to an instrument with three reflections.

[2] A conventional coudé telescope has a very long focus that is useful mainly for high resolution spectroscopy.

[3] Usage of an additional flat mirror, placed on the altitude axis, in a conventional Cassegrain system, called Nasmyth telescope, enables to keep the focal point instruments at a fixed position.

The telescopes are positioned suitably along the North–South track, which allows a coarse balancing of optical paths at the time of stellar transit. The finer compensation is done by translating the optical table along the said direction in order to maintain constant zero optical path difference (OPD) within the coherence length between the two beams (see Sect. 6.1.3). Of late, a delay-line that is movable is inserted in one of the arms of the I2T, by means of a cat's eye system; the other arm was equipped with a fixed delay-line (Robbe et al. 1997).

The performance of this instrument was limited to a visual magnitude of $m_v = 4$, a limitation essentially due to the sensitivity of the photon counting camera system used for fringe monitoring. This instrument was used for measurements in the visible (Blazit et al. 1977) and in the near-IR (2.2 μm) waveband (di Benedetto and Conti 1983). However, the observations with this interferometer were limited to the stars brighter than fourth magnitude.

7.1.2 Grand Interféromètre à deux Télescope (GI2T)

A novel interferometer, known as Grand Interféromètre à deux Télescope (GI2T), with two 'boule' telescopes that run on North–South tracks with variable baseline of 12–65 m, was developed soon after the success of I2T at the same site with larger telescopes. This instrument is notable for its use of 1.52 m telescopes having specially developed spherical drives with altitude–altitude (alt-alt) coudé train, broad multi-channel spectral coverage, and beam combination in the image plane (Labeyrie et al. 1986). It uses a spectrograph as fringe detector to achieve simultaneously high spatial and spectral resolution. This interferometer has been found to be suitable to meet the requirements in terms of (1) vibrational stability; it has a good vibration absorption capacity, (2) mobility, and (3) the coudé train. Apart from the low cost, because of little mechanical structure, the other noted advantages of this type of telescope are (1) choice of rotation axis; the tracking mode can be computer controlled in order to change alt-az, alt-alt, equatorial etc., according to the type of experiment, (2) low thermal sensitivity owing to good insulator, and (3) low flexure. Figure 7.2 shows the GI2T at Plateau de Calern, France.

The domeless boule telescope is composed of a reinforced concrete spherical structure having 3.5 m diameter, a steel rotating support, and an electronic control unit. Each telescope has three mirrors directing the horizontal afocal coudé beam to the recombiner optics. The primary mirror sits on astatic cell at the bottom of the sphere. The secondary mirror has a mechanical focusing system, while the tertiary mirror is in a fixed position in order to obtain a focal point inside the sphere; the tertiary is motorized to have a fixed beam towards the laboratory. The sphere can accommodate auxiliary instruments like photometer, spectroscopes, etc., as well.

The driving system of each sphere consists of a pair of rings that are movable with respect to a third ring; an encoder measures their motion. Each ring is motorized by 3 actuators acting in 3 orthogonal directions within 2 different tangential planes. The two rings alternately carry the sphere, which in turn, produce continuous motion.

Fig. 7.2 Grand interféromètre à deux télescope (GI2T) at Plateau de Calern, France (courtesy: A. Labeyrie)

The clutching of the rings is performed by computer-controlled electro-valves and hydraulic pistons, which carry each of the 60-mm balls that raise the rings by 2 mm to carry the sphere. However, the pointing speed of the telescopes is slow (3°/mm), which is a major limitation for rapid change of targets.

The original driving system of the GI2T was computer driven hydraulic pistons, which was producing vibrations preventing to obtain fringes. This system was subsequently replaced with small DC-motors of 25 watts. Each telescope was equipped with 2 sets of 3 such motors and 3 linear encoders. The fringe movement was found to be by and large limited to the atmospheric path fluctuations. The resolution obtained here was of the order of 1 μm – equivalent to $0.1''$ on the sky. Further, improvement on the development of software specifically adapted to the driving system, made the interferometer workable. However, the main drawback comes from the slow pointing of the telescopes; 4 or 5 stars can be tracked during the night.

Like I2T, here too, the finer compensation is carried out by translating the optical table as discussed in Sect. 6.1.3.1. The configuration of the boule telescopes, the baseline, the choice of a Michelson pupils reconfiguration, and the main science drivers have defined the necessary functions of the interferometric system such as, (1) beam combiner and pupils remapping, (2) pupil stabilization, (3) image stabilization, (4) optical path difference (OPD) compensation, and (5) atmospheric dispersion compensation. The fringes are recorded at short-exposure using a photon counting detector (Labeyrie et al. 1986). The path-difference is derived from the slope of the dispersed fringes. However, the measures obtained were based on visual estimation of fringe contrast resulting in values that were strongly affected by the atmospheric turbulence. High error bars prevented full exploitation of the astrophysical capacity of these interferometers. The GI2T was used to observe the Be stars, Luminous Blue Variables (LBV), spectroscopic and eclipsing binaries, wavelength dependent objects, diameters of bright stars, and circumstellar envelopes. However, the scientific programs are restricted by the low limiting visible magnitude down to 5 (seeing and visibility dependent), which is mainly related to the number of photons per coherence cell of the atmosphere per coherence time. The initial visual estimate of visibilities was then replaced by a software version (Mourard et al. 1994). Adaptive optics in each sub-aperture can in principle extend the magnitude limit. The GI2T has now been stopped for a year (Labeyrie 2009).

7.1.3 Mark III Interferometer

Another interferometer built with smaller apertures for astrometry purposes, the Mark I, was able to measure and track atmospheric turbulence in real time (Shao and Staelin 1977; Shao and Staelin 1980). Further improvement had been witnessed in the form of developing Mark III interferometer at Mt. Wilson, USA, which was the first fully automated interferometer. Measurements of precise stellar positions and motions of the stars are the major programmes carried out with this instrument. Though the basic programme was wide-angle astrometry (Shao et al. 1990), this set-up has also been used to derive the fundamental stellar parameters, like the orbits for spectroscopic, as well as eclipsing binaries, structure of circumstellar shells etc. Fairly large samples of cooler stars, mostly G, K, and M-type giants, have also been observed in the visible with this interferometer with formal errors of a few percents.

Mark III interferometer was conceived to allow observations in amplitude, as well as in phase mode. It was preceded by two prototype instruments, and was operational from 1986 to 1993. Two well-separated siderostat mirrors, that divert the light from the distant object in a single direction, track the target star as the Earth rotates and direct its light into a series of mirrors. Mark III set up had four siderostats that can be used in pairs forming four different interferometers; the maximum size of the aperture was 5 cm. Since the central detection and fringe tracking system is the same for all, only one could be used at a time. The baseline of this interferometer could be configured to lengths ranging from 3.0 m to 31.5 m. The fringes of the objects are obtained with automated operation.

Unlike the interferometers like I2T and GI2T, this interferometer uses (1) an optical delay-line to control the internal optical path lengths (see Fig. 7.3) and (2) a laser metrology system to monitor the position of the siderostat mirrors. The 25-mm optical flat (siderostat mirror), is attached to a polished hemisphere. Light from the active siderostats is directed toward the central building by fixed mirrors. The two mirrors feeding light into the delay-lines are mounted on piezo-electric actuators and are part of the angle-tracking servo loop. A set of four laser interferometers measure the position of the sphere relative to a set of four corner cubes embedded in a reference plate.

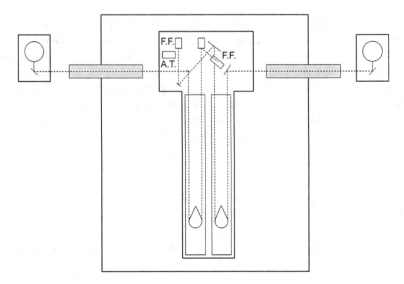

Fig. 7.3 Schematic layout of the delay-line arrangement at Mark III interferometer

Light beacons from the two arms of this interferometer travel in vacuum tubes
to avoid atmospheric turbulence between the siderostats and the central building.
Tilt-correction mirrors are used to adjust for disturbances caused by the turbulence.
They are directed towards the two delay-lines. The delay-lines set-up was kept in
a vacuum chamber. The delay-line has a peak to peak range of 20 m with a small
signal bandwidth. The phase of the signal was computed in real time. This phase
information permits one to determine the path difference accurately and is used for
astrometry. It is also used to control the position of the mirrors of the first delay-line
in order to realize the most accurate phase tracking. The delay was modulated with
a 500-Hz triangle wave of amplitude 800 nm. If the difference of the path length between the two interferometer arms is within the coherence length, the intensity at the
beam combiner output would vary sinusoidally with time. Four wavelength channels were made available by dividing the light in each of the two outputs. A broad
band channel centered around $\lambda = 700$ nm was used for fringe-tracking.

For the astrometric observations, the measurement of the central fringe is performed in two colors to correct for the error caused by atmospheric turbulence,
while for stellar diameter programmes, it is done through narrow band filters.

7.1.4 Sydney University Stellar Interferometer

After the successful venture of Narrabri intensity interferometer (see Sect. 3.3.3),
the Sydney University had built a 12.4-m prototype stellar interferometer and determined the angular diameter of α CMa (Davis and Tango 1986). Based on this

success Davis et al. (1999a) built the present Sydney University Stellar Interferometer (SUSI), which has a very long baseline ranging from 5 m to 640 m (North–
South) that are achieved with an array of 11 input stations equipped with a siderostat
and relay optics, located to give a minimal baseline redundancy. The intermediate
baseline forms a geometric progression increasing in steps of ~40%. SUSI employs
small apertures (14 cm; single r_0), adaptive optics for wavefront tilt correction, and
rapid signal sampling to minimize the effects of atmospheric turbulence.

Starlight is steered by two siderostats of 20 cm diameter using an Alt-Azimuth
mount placed upon large concrete piers, via relay mirrors into the evacuated pipe
system that carries the light to the central laboratory (Davis et al. 1992). It also
uses the dynamic path length compensation system to equalize the optical paths
to the point where the two beams of starlight are combined. At the central laboratory, the light enters a beam reducing telescope followed by the atmospheric
refraction corrector systems consisting of pair of counter-rotating Risley prisms.
It proceeds towards either the optical path length compensator (OPLC) or is di-
verted towards the acquisition camera. On leaving OPLC, the beams from the two
arms of this interferometer is switched to one of the optical tables (blue or red) for
recombination.

Beams from the two arms of the interferometer are directed from opposite sides
towards a beam-splitter, where the beams are combined. Two beams emerge out of
beams-splitter, and each is a superposition of the two input beams. The emerging
beams have irradiances (Davis et al. 1999b),

$$I_1(t) = I_0[1 + \mathcal{V}\cos\psi(t)], \tag{7.1a}$$
$$I_2(t) = I_0[1 - \mathcal{V}\cos\psi(t)], \tag{7.1b}$$

where $\mathcal{V}$ is a seeing affected fringe visibility, ψ the fringe phase that is the sum of
the complex degree of coherence and phase fluctuations introduced by pathlength
variations in the atmosphere; the difference of sign between these sub-equations is
a consequence of the π phase shift of one of the reflected beams at the beamsplitter.

At visible wavelengths the phase ψ is corrupted. With sufficient integration time,
the average of $[I_1(t) - I_2(t)]^2$, can be taken, the quantity $\cos^2\psi(t)$ turns out to
be $1/2$. Davis et al. (1999b) measured the mean correlation and fringe visibility by
means of,

$$\bar{\mathcal{C}} = \bar{\mathcal{V}^2} = 2\frac{\overline{[I_1(t) - I_2(t)]^2}}{\overline{[I_1(t) + I_2(t)]^2}}, \tag{7.2}$$

in which the bars represent time averages.

Two beam-combiners such as (1) blue beam-combination system having spectral
range between 430 and 530 nm and (2) red beam-combination system having spectral range between 530 and 950 nm are used for the scientific program. The former
is used to observe early-type stars, early-type binaries, while the latter is employed
to observe late-type stars, binaries, and Cepheids. SUSI is being upgraded for remote operation with a new beam-combination system (PAVO) that has several (10)
parallel spectral channels and spatial filtering.

This instrument has been employed (1) to measure the stellar diameters with high precision, effective temperatures, and luminosities, (2) to interferometric-spectroscopic studies of binary systems; it has determined the masses of β CMa stars, (3) to derive the angular diameter variations and, in combination with spectroscopic radial velocities, and (4) to determine the distances and mean radii of Cepheids. In addition, it has been employed for measuring the outer scale of turbulence, as well as for interferometric-asteroseismological studies that enable to determine the mass of a single star; interferometric spectropolarimetry was also carried out.

7.2 Spatial Interferometry in the Infrared (IR) Region

Operating an interferometer in IR band is easier rather than in visible wavelength, although it was limited by the suitable detectors at the initial stage. Recent advances in the development of IR sensitive detectors having the ability to detect the small number of photons collected by telescope in the allowed integration times made success to built a few two or multi-element long baseline IR interferometers. In what follows, the development and the prospects of the spatial interferometry in the infrared region are enumerated.

7.2.1 Heterodyne Detection

Like radio interferometry (see Sect. 3.2), heterodyne technique, which helps to detect radiation by non-linear mixing with radiation of a reference frequency has been employed in the infrared interferometry. However, paths of the signal from the stellar source to such a detection should be maintained to an accuracy $(\Delta\psi/\kappa)$, in which $\Delta\psi$ is the desired limit in phase uncertainty, and κ the wave number. This technique is sensitive for a very narrow bandwidth and following detection, the resulting signals are in the radio range, which may be amplified and used flexibly in various ways without loss of the S/N ratio (Townes 1984). A variable delay-line may be useful so that an interference between the two signals is obtained directly. This would enable to control the delay-line to a small fraction of a wavelength for determining the phase precisely.

Heterodyne signals from two separate telescopes are combined to provide an interference. A coherent local oscillator (LO) is employed in order to mix the energy received by the telescope in a single diffraction mode. Unlike homodyne technique in which the LO has the same frequency as the input, in a heterodyne system, the LO is frequency shifted. In the case of the IR region, heterodyne method selects the components of the radiation that are in phase with the wavefront of the laser local oscillator, thereby tries to prevent this difficulty. In direct interferometry, a similar result may be achieved by employing spatial filter with a glass fiber in order to obtain one coherent core, of course, with some loss of signal. It is to reiterate that in the optical interferometry, the dispersion in the refractive index is useful for possible correction to atmospheric seeing and refraction, which has been performed simultaneously at two wavelengths in order to correct some of the apparent variation in stellar position at a single wavelength (see Sect. 6.2.6), while in the IR region, such a mechanism may not be feasible to provide a correction.

Heterodyne technique in IR region uses a CO_2 laser as a local oscillator that mixes the energy received by the telescope. This method detects one polarization (that of the local oscillator) and has the following advantages in the case of beam recombination. These are (1) a larger coherence length and (2) a simplification of the transport of the signal from the collector to the recombiner (coaxial cables instead of mirrors).

Let $U_s(t)$ and $U_l(t)$ be respectively, the signal coming from a star and of an artificial source (laser) which are expressed as,

$$U_s(t) = a_{s0}e^{-i[\omega_s t - \psi]}, \tag{7.3}$$
$$U_l(t) = a_{l0}e^{-i\omega_l t}. \tag{7.4}$$

The laser is the phase reference. A detector like a photo-diode, illuminated by the sources (star + laser) yields an electrical signal corresponding to the light intensity:

$$
\begin{aligned}
I(t) &= |U_s(t) + U_l(t)|^2 \\
&= \left(a_{s0}e^{-i[\omega_s t - \psi]} + a_{l0}e^{-i\omega_l t}\right)\left(a_{s0}e^{i[\omega_s t + \psi]} + a_{l0}e^{-i\omega_l t}\right) \\
&= a_{s0}^2 + a_{l0}^2 + 2a_{l0}a_{s0}\cos[(\omega_l - \omega_s)t + \psi].
\end{aligned}
\tag{7.5}
$$

The optical path difference (OPD) is discretely adjusted by means of a set of commutable cable lengths. If ω_l and ω_s are close, the frequency of I is low enough to fit in the bandwidth of the detector and its electronics (a few GHz) and I carries the phase information from the radiation of the star. By correlating (multiplying) the signals I_1 and I_2 yielded by two apertures with heterodyne systems, one can extract a visibility term. However, the lasers must have the same phase for the two apertures.

As discussed in Sect. 6.2.3, noise is due only to fluctuations in the number of quanta in the radiation received in an amplitude interferometer. These fluctuations are also present in heterodyne detection, but are much smaller at IR frequencies than the uncertainty principle noise, and therefore are ignored from (6.53). The fundamental noise power for a heterodyne detector is equivalent to an average of single quantum per second per bandwidth in the polarization as the local oscillator. For an ideal sensor like a photo-diode, the noise power fluctuation (Townes 1999) is,

$$N_h = h\nu\sqrt{\frac{2\Delta\nu}{t}}, \tag{7.6}$$

in which t is the post-detection averaging time (seconds), $\Delta \nu$ the bandwidth (single sideband, Hz^{-1}), and $h\nu$ the quantum energy.

The noise described in (7.6) is inescapable result of quantum mechanics and the uncertainty principle. In IR interferometry, the noise is due to heat radiation impinging on the detector from the sources associated with Earth, atmosphere and the instruments. There is a substantial amount of noise due to quantization of the radiation field and the uncertainty principle in heterodyne detection. This sets the limit $\Delta\psi\Delta n \geq 1/2$, in which $\Delta n = \sqrt{n}$ and $\Delta\psi \geq 1/(2\sqrt{n})$, to the uncertainty of numbers of quanta, n and of phase, ψ (Townes 1984). In the case of linear detection that permits phase determination of a wave within limits, $\Delta\psi < \infty$, hence there is an uncertainty in the measurement of intensity. For a heterodyne system, uncertainties in $\Delta\psi$ are of a magnitude making this noise correspond to a single photon per mode of the radiation field Hz^{-1}. The noise power in a diffraction-limited telescope is $h\nu\sqrt{2\Delta\nu}$, in which $\Delta\nu = \Delta n/t$ is the bandwidth of the receiver and t the time of observations. For a heterodyne detector employing ideal detectors with 100% quantum efficiency, the S/N ratio for the detection of a point source with a heterodyne technique, as well as for the direct detection are given (Townes 1999) respectively by,

$$(S/N)_h = \frac{P(\nu)\sqrt{2\Delta\nu t}}{h\nu}, \qquad \text{and} \tag{7.7}$$

$$(S/N)_d = \frac{P(\nu)}{h\nu}\left[\frac{2\Delta\nu t\left(e^{h\nu/k_B T} - 1\right)}{1 - \epsilon}\right]^{1/2}, \tag{7.8}$$

where k_B is the Boltzmann's constant, ϵ the fractional loss, T the temperature, and $P(\nu)$ the power in each polarization per unit bandwidth of source.

If a direct detection has narrow bandwidths like heterodyne detection, it has advantage by the square root of $(e^{h\nu/k_B T} - 1)/(1 - \epsilon)$. Under ideal condition, if the heterodyne signals from two separate telescopes are combined, the S/N-ratio for the combined signal turns out to be the square of (7.7). For a single telescope, the S/N-ratio increases as the square root of t, while for an interference fringe, it increases as t. Since the phase of the fringe is subjected to arbitrary fluctuations due to the imperfect atmosphere, the S/N-ratio is poorer than that given by the square of the said equation. The S/N-ratio for the measurement of fringe power in a heterodyne interferometer is,

$$(S/N)_{fr} = \frac{\mathcal{V}^2 P^2(\nu)\Delta\nu}{(h\nu)^2}\sqrt{\tau_c\, t}, \tag{7.9}$$

in which $\mathcal{V}$ is the visibility of the source power, t the observing time period, and τ_c the length of time atmospheric fluctuations do not change the fringe phase by more than a radian.

As in radio astronomy, the determination of amplitude and phase of interferometric fringe patterns may provide complete mapping of distributions of infrared intensity. The high resolution of the order of $\sim 0.002''$ is adequate to map convection cells on the surface of the larger stars (Townes 1984).

7.2.2 Plateau de Calern IR Interferometer

The IR interferometer, called as Synthèse d'Ouverture en Infra Rouge avec DEux TElescopes (SOIRDÉTÉ) at Observatoire de Calern, France, was designed for multiwavelengths observations within given spectral ranges located in K, L, M, and N bands, respectively, centered at λ: 2.2 μm, 3.5 μm, 5 μm, and 10 μm. It comprised of a pair of 1-m telescopes with a 15-m East–West horizontal baseline (Gay and Mekarnia 1988). The effective baseline as seen by the object varied with the hour angle from 3.5 m up to 15 m. The corresponding interval of spatial frequencies, B/λ, in which B is the baseline, thus covered is about, in cycles per arcsecs (7.9, 33) at K-band and at N-band. In this project, the telescopes were on equatorial mounts with afocal coudé beam of diameter 25 mm (Rabbia et al. 1990).

Beams were received in the central laboratory on a double cat's eye delay-line on a step by step movable carriage. Figure 7.4 depicts the schematic of the set up for the mixing of incoming waves from the telescopes. Unlike the conventional devices, where the motion of the carriage should, in principle, be continuous, in order to follow the zero OPD or white-light fringe as the source's position varies, this mode of observation relied on recording an Earth-scanned complete interferogram, therefore carriage moved step by step, rolling on rails and under computer-control coupled to a metrological laser. The accuracy of this metrology was about 0.1 μm. For a given position of the carriage, one of the beams was given an excess pathlength, d_i' and the other beam had an astronomical excess pathlength, $d_a(t)'$. If these pathlengths satisfied the condition, $d_i = d_a(t)'$, the transit of the white light fringe occurs; transit of fringe takes less than 100 ms.

An active optical set up performed both the real time wavefront tilt correction and the control of the position of aperture's image. The tilt corrections were made

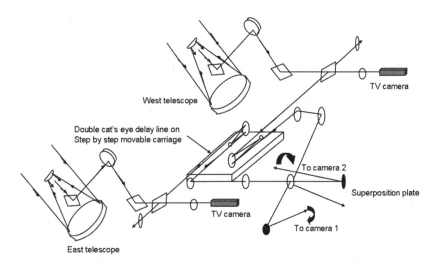

Fig. 7.4 Set-up for combing two incoming waves from the telescopes at SOIRDÉTÉ

by means of a two-axis tilting mirror onto which is imaged the aperture. The second feature prevents a reduction of contrast originating in Fresnel diffraction and occurring when images of pupils were separated along the virtual common propagation axis. The natural OPD drift due to the Earth-rotation was used for acquiring fringes (Rabbia et al. 1990). Two similar detectors were employed in order to increase the S/N ratio and to eliminate atmospheric fluctuations.

7.2.3 Infrared Spatial Interferometer

Heterodyne interferometry is also employed on the mid-IR Spatial Interferometer (ISI), Mt. Wilson, California, USA (Townes et al. 1998), which is well suited to study of circumstellar material around bright evolved stars. The major advantage of ISI's narrow heterodyne detection bandwidth is that one may tune the detection wavelength to be in or out of known spectral lines if required. This permits to determine the diameter of the target star; the distribution of hot water vapor in the circumstellar envelope can also be measured. This instrument is suited for spectral line research as well.

ISI features three Pfund-type movable telescopes. Each telescope comprises a stationary 1.65-m parabolic mirror and a 2 m siderostat on a steerable altitude-azimuth mount equipped with an automated guiding and tip-tilt control system at 2 μm (Lipman et al. 1998). The latter is used to redirect light from a star onto the former mirror that focuses the starlight through a hole in the flat mirror, onto an optical table mounted behind it. These telescopes are compact and kept on the semi-trailer that allows to move them to different discrete baselines. The maximum baseline between the two apertures can be made upto 56 m, which enables to resolve the diameters ($\phi \sim 0.02''$) of several stars in the mid-IR range. Such a range is best suited for measuring stellar diameters, since it is less affected by limb-darkening phenomenon (see Sect. 8.1.1.2).

Each telescope possesses a stable CO_2 laser, which acts as local oscillator. Starlight collected by the telescope comes into a Schwarzschild optical system (Hale et al. 2000). This optical system consists of a pair of spherical front-surface mirrors that provide a less divergent beam to optical components and allow matching of the incident stellar wavefront with that of the local oscillator. Emerging beam from Schwarzschild system converges to a focus, then split by a beam-splitter that reflects the mid-IR radiations and transmits the near-IR wavebands (see Fig. 7.5). The near-IR bands are sent to the guide camera, while the mid-IR radiation is sent to another beam splitter, where it is mixed with laser oscillator, transforming the signal to microwave frequencies, followed by pathlength matching and fringe detection in a correlator. An infrared sensitive photo-diode (liquid nitrogen cooled HgCdTe) signal detector is employed to record the intensity. On ISI, two CO_2 lasers are used, the phase of one being controlled by the other. Here, the interferometer noise is dominated by shot-noise of the laser and thermal background is negligible for setting the sensitivity limit (Hale et al. 2000).

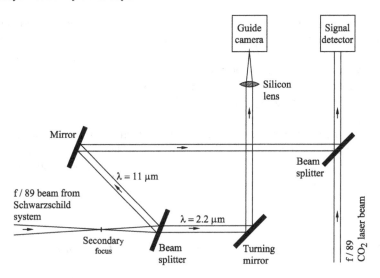

Fig. 7.5 Schematic of the ISI instrument

ISI utilizes Earth's rotation and periodic discrete changes of the baseline to obtain a wide range of effective baselines and map the visibility functions of the stellar objects. The visibility is measured by integrating power spectrum of the recorded fringe signal, followed by normalizing it with an IR power measurement obtained by position switching and chopping. The ratio of the two is proportional to the square of the visibility defined in (2.26). ISI has been outfitted with a filter-bank system to allow visibility measurements on spectral lines between 9 and 12 μm with spectral resolution of $\lambda/\Delta\lambda \sim 10^5$ (Monnier et al. 2000). The recent addendum to the ISI is the third telescope (Hale et al. 2004), which provides the advantage of observing in a closure phase (see Sect. 5.3.1) fashion. In this, a master laser is used, which splits the beam in three ways and sent to each telescope via the periscopes; the telescope lasers can lock on to the master laser.

7.3 Arrays with Multiple Telescopes

An array with multiple telescopes is a large and complex instrument, comprising of many active and passive systems. As the number of telescopes increases the complexity of the mirror system also increases as it is necessary to compensate for the numerous beams which contribute to the interference pattern. Interferometer of heterogeneous nature, such as Very Large Telescope Interferometer (VLTI), consists of one or more large telescopes and a few small ones, face problems that arise from the recombination of two telescopes, one large and one small, since the signal-to-noise (S/N) ratio would be the one given by the small telescopes.

For all imaging arrays, the combination of the light collected by the telescopes in the array needs to be performed, which can be done in a number of ways. The beam combiner ensures visibility and phase coding of the interference pattern. The image reconstruction implies precise control between the interferometer arms. Variations of the OPD may lead to loss of phase relationship, and thus reduce the capability of image reconstruction, although techniques such as speckle interferometry remain usable with non-phased sub-apertures. The opto-mechanical stability of the instrument affects the fringe complex visibility as well. In this respect, Integrated Optics (IO) may offer a better solution for beam combiner for large interferometric arrays (Kern et al. 2005). Moreover, for more than three aperture interferometry, a fringe-tracker is essential for each baseline.

7.3.1 Cambridge Optical Aperture Synthesis Telescope

As stated earlier in Sect. 5.3.1, combining more than two telescopes helps in measuring closure phase of any object. Baldwin et al. (1998) measured closure phase using arrays with several beams in visible band at the Cambridge Optical Aperture-Synthesis Telescope (COAST). The maximum baseline of this instrument is up to 100 m to provide images with a resolution down to 1 mas. It uses multi-stage four-way combiner. Light from the telescopes is combined co-axially (see Fig. 5.9b for co-axial method), where a beam splitter is used to combine two beams pairwise, of which one or both beams was already the combination of two others. The combined beam is focused onto one or more detectors for intensity readout.

Four telescopes, each comprising a 50-cm siderostat mirror that feed a fixed horizontal 40 cm Cassegrain telescope (f/5.5) with a magnification of 16 times, are arranged in a Y-layout with one telescope on each arm, movable to a number of fixed stations and one telescope at the center of the Y. Light from each siderostat passes through pipes containing air at ambient pressure into the beam combining laboratory. The laboratory accommodates the optical path compensation delay-lines, the beam combining optics, the detectors, the fringe acquisition systems etc. (Baldwin et al. 1994). The laboratory is a tunnel (32 m length × 6 m width × 2.4 m height), covered sufficiently with Earth having thick insulating end walls that provides a stable thermal environment internally for the path compensator delay-lines, combining optics and detectors.

Each of the four telescope beams is provided with a variable delay-line (maximum delay is 37 m); the equalization of the path delays for each of them is carried out by a movable trolley, carrying a roof mirror running on a rail track. The four beams emerging from the path compensator are each split at a dichroic; the longer wavelength ($\lambda > 650$ nm) of the visible band passes into the beam combining optics and the shorter ones are used for acquisition and auto-guiding. A cooled CCD detector system is used for both acquisition and guiding. The 4-way beam combiner accepts four input beams, one from each telescope provides four output beams, each having equal contributions from the light from the four telescopes. Each output beam passes through an iris diaphragm and is focused by a long focus lens on to a fiber-fed single-element avalanche photo-diode detector for fringe detection (Baldwin et al. 1994). A similar beam combiner has also been added for the near IR capability.

7.3.2 Infrared Optical Telescope Array (IOTA)

The IR-Optical Telescope Array (IOTA) at Mt. Hopkins, Arizona, USA, consisted of three 0.45-m collector assembly movable among 17 stations along two orthogonal linear arms. IOTA synthesized a total aperture size of 15×35 m (Traub et al. 2003). Each assembly comprises a siderostat feeding a stationary afocal Cassegrain telescope, which produces a 10× reduced parallel beam, and a piezo-driven active mirror that corrects for tip-tilt motion introduced by atmospheric turbulence. Identical reflections were applied to the beams from the collectors to the recombiner in order to avoid any loss of fringe contrast due to different polarization states between the two beams at the recombination point.

Following compression, each beam was directed vertically downward by the active mirror. The tip-tilt corrected afocal star light beams (4.5 cm diameter) from the telescopes are brought to the beam combining table through an evacuated envelope. Optical differences were compensated by the fixed and variable delays and the beams were recombined onto a beam splitter, producing two complementary interference signals (Carleton et al. 1994). These were focused onto a pair of InSb photovoltaic detectors through a K-band filter. Fast auto-guiding system was used to correct the atmospheric wavefront tilt errors and a precise optical delay-line was employed to compensate the effect of Earth's rotation. Two active delay-lines for three telescopes were provided. A scanning piezo mirror was used to modulate the optical path difference between the two telescopes. A SITe 512×512 CCD with a quantum efficiency of 90%, was used as sensor for IOTA's visible light detector. A supplementary optical path delaying system (consisting of a long-travel delay-line fixed while observing and a short-travel delay-line tracking the star), a control system for the long delay-lines had also been implemented (Traub et al. 2000). Figure 7.6 depicts the overview of IOTA interferometer.

There were three combination tables at the IOTA, and all of them implement pupil-plane beam combination. In two cases, at visible and near-IR wavelengths, they use classical beam-splitter (see Fig. 6.12a). The two recombined beams are focused on two pixels of a 256×256 HgCdTe NICMOS3 IR camera. The third table housed the monomode fiber-fed FLUOR experiment, which is suitable for stellar atmosphere studies. Later, an integrated optics combiner, IONIC-3 (Berger et al. 2003) was employed to measure precise visibilities and closure phases.

Fig. 7.6 Overview of IOTA interferometer

7.3.3 Navy Prototype Optical Interferometer

The astrometric array of Navy Prototype Optical Interferometer (NPOI), a joint project of the US Naval Research laboratory and the US Naval Observatory in co-operation with the Lowell Observatory, is designed to overcome the effects of the atmosphere and instrumental instabilities, which are required to achieve wide-angle astrometry. Located on Anderson Mesa, Arizona, USA, this interferometer is capable of measuring positions with precision comparable to that of *Hipparcos* (1997). The anticipated wide-angle astrometric precision of the NPOI is about $\sim$2 mas (Armstrong et al. 1998). NPOI plans to measure the positions of some radio stars that would help in matching radio sources with their optical counterparts.

This interferometer includes sub-arrays for imaging and astrometry and is developed at Y-shaped (Very Large Array-like) baseline configuration. The light beams are passed through vacuum pipes to the central laboratory. A pair-wise recombination technique is employed at NPOI, where a beam combiner uses a different detector for each baseline. For astrometric mode, 4 fixed siderostats (0.4 m diameter) are used with the baselines extendable from 19 m to 38 m (Armstrong et al. 1998). The shared backend covers 450–850 nm in 32 channels. The other notable features are the delay system, active group-delay fringe-tracking, etc. The use of vacuum delay-lines renders the interferometer insensitive to plane parallel atmospheric refraction and permits simultaneous fringe-tracking in spectral channels over the entire optical bandpass.

The astrometric sub-array has a laser metrology system to measure the motions of the siderostats with respect to one another and to the Earth-fixed reference system (bedrock). This metrology system, consisting of 56 laser interferometers is configured in three subsystems, which allow continuous measurement of the time evolution of the baselines relative to the bedrock. Five such interferometers measure the position of a 'cat's-eye' retroreflector near the intersection of the rotation axis of each siderostat with respect to the adjacent reference plate.

NPOI is well suited for imaging stars, but not for the faint sources. For imaging mode, six transportable siderostat flats (0.12 m diameter) are used with the base-lines from 2 m to 432 m; the best angular resolution of the imaging subarray is 0.2 mas. The position of three such flats are kept with equal space for each arm of the Y. Coherence of imaging configuration is maintained by phase bootstrap-ping. Observations in visible spectrum with 3-elements have been carried out using avalanche photo-diode as detector (Hummel et al. 1998). The dynamic range in the best of the NPOI images exceeds 100:1 (Armstrong et al. 1998).

7.3.4 Palomar Test-bed Interferometer

Palomar Test-bed Interferometer (PTI), is an infrared phase-tracking interfero-meter in operation situated at Palomar Observatory, California. It was developed by the Jet Propulsion Laboratory and California Institute of Technology for NASA as a test-bench for the Keck interferometer. Figure 7.7 depicts the overview of Palomer Test-bed Interferometer. The main thrust of this interferometer is to develop techniques and methodologies for doing narrow-angle astrometry for the purpose of detecting exoplanets (Wallace et al. 1998).

Three 40-cm siderostats coupled to beam compressors (reducing the beam dia-meter) can be used pairwise to provide baselines up to 110 m (Colavita et al. 1999). This interferometer tracks the white light fringes using an array detector at 2.2 μm (K band) and active delay-lines with a range of ±38 m. Among others, the notable feature of this interferometer is that of implementation of a dual-star astrometric

Fig. 7.7 Overview of Palomer test-bed interferometer (courtesy: Peter Lawson)

ability where light from two stars are selected and observed simultaneously using different delay-lines, allowing to achieve both relative astrometry and phase referencing. An end-to-end heterodyne laser metrology system is used to measure the optical path length of the starlight (Wallace et al. 1998).

7.3.5 Keck Interferometer

The Keck Interferometer, a joint programme by Jet Propulsion Laboratory (JPL) and California Association for Research in Astronomy (CARA), built at the Mauna Kea observatory, combines the light from the twin 10-m apertures main telescopes with a fixed baseline of 85 m. These large telescopes in the world are being used for optical and near infra-red astronomy. The primary mirror is composed of 36 hexagonal mirror segments, each 1.8 m across mounted together in a honeycomb pattern, which work in concert as a single piece of reflective glass. This instrument has a spatial resolution of 5 mas at 2.2 μm, and 24 mas at 10 μm. The system transports the light to a laboratory located between the main telescopes, where a beam combiner and infrared camera combine and process the light. In order to equalize the path length between the arms of this interferometer, the delay-lines consisting of movable mirrors, are used to enable interference.

It was planned to have a few outrigger telescopes at the Keck interferometer, but abandoned because of building permit. Lack of such telescopes makes this instrument unsuitable for interferometric imaging. However, this interferometer can be used in nulling mode (Colavita et al. 2008) to remove the effects of bright stars and study the much dimmer surrounding areas (see Sect. 5.1.3). Potential on interferometric nulling was demonstrated when a recurrent nova, RS Oph was observed. It combines phased pupils provided by adaptive optics for the main telescopes (up to V = 9, 39 mas FWHM, Strehl's ratio = 30%) and fast tip/tilt correction on the outriggers. Beam recombination will be carried out by 5 two-way combiners at 1.5–2.4 μm for fringe-tracking, astrometry, and imaging. The astrometric accuracy is expected to be of the order of 20–30 μas/$\sqrt{\text{hour}}$.

Effort is on to develop a new instrument, called ASTRA (astrometric and phase-referenced astronomy) to broaden the astrophysical applications (Pott et al. 2008). This instrument is based on the dual-field technology, which will facilitate to observe two objects simultaneously, and measure the distance between them with an accuracy better than 100 μas. The astrometric functionality will be an additional tool for the research as diverse as exo-planetary kinematics, binary astrometry, and the galactic center.

7.3.6 Very Large Telescope Interferometer (VLTI)

An array of large aperture telescopes, known as Very Large Telescope Interferometer (VLTI), was built to perform interferometric measurements (Derie et al. 2000).

Located at a height of 2,635 m in Cerro Paranal, Chile, it was built by the European Southern Observatory (ESO). VLTI consists of a system of four 8.2 m fixed separate telescopes (unit telescopes; UT); the Antu telescope, the Kueyen telescope, the Melipal telescope, and the Yepun telescope). Each of them is a Ritchey–Chrétien type and can operate in Cassegrain, Nasmyth, and coudé focus. These telescopes are accompanied by four mobile auxiliary telescopes (AT) of 1.8 m diameter. The star separators have also been installed on two VLTI Auxiliary Telescopes. The star separator is an opto-mechanical system, designed to separate the light of two distinct stars, for the unit telescopes (UT) and the auxiliary telescopes (AT) located at the coudé foci feeding two arbitrary objects from the coudé field-of-view into the delay-lines of the VLTI. Such a system may compensate for field rotation and stabilize the beam tip-tilt and adjust the lateral alignment of the pupil. Figure 7.8 displays the recent photograph of VLTI system. The complete array is rotated by Earth rotation. This array allows access to baselines between 47 m and 130 m with the UTs, and between 8 m and 202 m with the ATs. The best resolution can be obtained for the large telescopes is 1.5 mas and for the auxiliary telescopes is 1 mas.

In VLTI, the unit telescopes are mainly used for independent astronomical observations. Its first fringes were obtained in 2001 (Glindemann et al. 2003). The auxiliary telescopes are mounted on tracks, which can be moved over an array of 30 stations and placed at precisely defined parking observing positions on the observatory platform. From these positions, their light beams are fed into the same common focal point via a complex system of reflecting mirrors mounted in an underground system of tunnels, where the beams from two or more chosen telescopes are brought together and combined coherently. The unit telescopes, as well as the AT stations are connected by a network of underground light ducts. coudé beams from these apertures are sent through delay-lines operating in rooms at atmospheric pressure but at a thoroughly controlled temperature in order to avoid turbulence.

Fig. 7.8 VLTI system at Paranal (ESO PR 51c/06; courtesy: F. Malbet); four auxiliary telescopes are also visible

The optical path differences are continually adjusted to correct for both long-range effects due to sidereal motion and fast, short-range variations due to differential atmospheric piston. Light from the telescopes is sent to the underground interferometric facility comprising the delay-lines tunnel and the VLTI coherent laboratory. Beams reach an optical switch-yard to be directed to one of the four recombiners (see Sect. 6.1.3.5). VLTI is equipped with a multi-application curvature adaptive optics (MACAO; Arsenault et al. 2002) system, which is located at the coudé focus of the UTs and makes use of a 60-element curvature wavefront sensor and a 60 actuator deformable mirror. Although the correction of turbulence does not provide a large gain on 8-m apertures at the wavelengths of MIDI beam combiner under good seeing conditions, in practice MACAO is needed for the operation of the fringe-tracker, FINITO (fringe-tracking Instrument of Nice and Torino; Mario et al. 2004), an optical 'instrument' operating in H-band.

VLTI has a good u, v-coverage with the power of its large collecting area. It has an angular resolution of 0.002 arcseconds at a wavelength of 2 μm. Using the big telescopes, the faintest object the VLTI can observe is $m_v = 7$ in the near-IR for broadband observations. Because of the large number of mirrors in VLTI, a significant fraction of light is lost before reaching the detector. This interferometric technique is efficient only to objects that are small enough so that all their light is concentrated at almost a single point, which helps to observe stars in the solar neighborhood and bright active galactic nuclei.

7.3.7 Center for High Angular Resolution Astronomy Array

The Center for High Angular Resolution Astronomy (CHARA) array is an interferometric facility operated by the Georgia State University and is situated at Mt. Wilson, California, USA (ten Brummelaar et al. 2005). Equipped with six telescopes of 1 m in diameter, this interferometer is arranged on a Y-shaped configuration with 15 baselines ranging from 31 to 331 m. It operates at optical (470–800 nm, 0.15 mas limiting resolution) and IR (2.0–2.5 μm, 0.6 mas limiting resolution) wavebands (McAlister et al. 1998). This instrument is very well laid out for binary star astrometry, observations of stars with well-determined spectroscopic elements, and determination of the metal abundance. Although it has large baselines, relatively large apertures for sensitivity, it lacks good (u, v)-coverage since the number of telescopes versus resolution is not very large.

Light from the individual telescopes, each of them is an afocal beam reducer with a 12.5-cm output beam, is sent through vacuum transport tubes to the centrally located beam synthesis facility, a L-shaped long building (McAlister et al. 1998) that houses the Optical Path Length Equalizer (OPLE) and the beam combination laboratory. This instrument employs Earth rotation aperture-synthesis (see Sect. 5.3.2.1). Since the starlight beams from the telescopes are directly combined, the delay-line in this interferometer require a series of adjustable mirrors to bounce the beams back-and-forth in order to delay them in time (see Fig. 7.9).

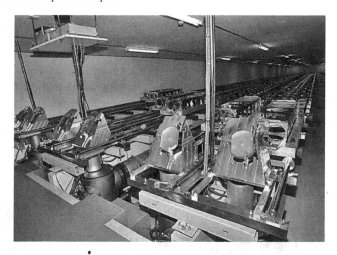

Fig. 7.9 The Optical PathLength Equalizers (OPLE) are placed on rails 50 m long, giving up to 100 m of delay controlled to an RMS of 10 nm (courtesy: S. Golden)

An image plane combiner, the Michigan Infra-Red Combiner (MIRC), has been developed to light from four telescopes of CHARA array simultaneously (Monnier et al. 2006a). These beams are filtered by single-mode fibers, which are rearranged into a 1-dimensional non-redundant pattern and brought to a focus. These overlapping beams create six interference fringes, each with a unique spatial frequency. The pattern is focused by a cylindrical lens into a line of fringes, which are subsequently dispersed by a spectrograph with spectral resolution $\Delta\lambda \sim 0.035 \ \mu$m. MIRC can measure six visibilities, four closure phases, and four triple amplitudes simultaneously over eight spectral channels spanning the astronomical H-band.

A newly developed combiner, called VEGA (Visible spEctroGraph and polArimeter), operating at visible wavelengths ($0.45 - 0.85 \ \mu$m) in a spectrally dispersed fringe mode, is used to record three-telescope fringes on two different triplets of baselines ranging from 34 to 211 m (Mourard et al. 2009). This allows it to combine high spectral resolution (up to 30,000) and high angular resolution (better than 1 mas with the longest baselines). VEGA spectrograph, similar to a classical spectrograph with a collimator, a slit, a grating, and a camera for reimaging the spectrum on the detector, is fed by four CHARA parallel beams vertically aligned so that fringes appear as horizontal lines at the coherent focus. Equipped with two photon counting detectors, it allows simultaneous recording of data, in medium spectral resolution, of the spectral regions around Hα and Hβ. The width of the slit is defined in a fashion so that almost two speckles are selected in the horizontal direction, while it covers a field of 4 arcsecs on the sky in the vertical direction.

VEGA recorded its first three-telescope fringes on two different triplets of baselines S1S2W2 (corresponding baselines 34 m 177 m 211 m) and S2W2W1 (107 m 177 m 251 m). The three pupils are combined in a redundant configuration and in the dispersed fringe mode. Two fringe patterns present the same interfringe (the high frequency peaks corresponding to the fringes are at the same vertical position in the spectral density) but are tracked and locked at different OPD so as to be horizontally shifted in the spectral density. Once the two fringe-trackers are on, the third fringe pattern (corresponding to the longest pupil distance), appears in the spectral density as a fringe peak with highest frequency. Visibilities and differential phases[4] can be simultaneously recorded on three baselines of various length and orientation and closure phase can be determined. VEGA may also measure dispersed visibilities for each polarization state.

Figure 7.10a presents the spectral density recorded on HD 3,360, whose visual magnitude is 3.7, around Hα line with a spectral resolution of 6,000. The

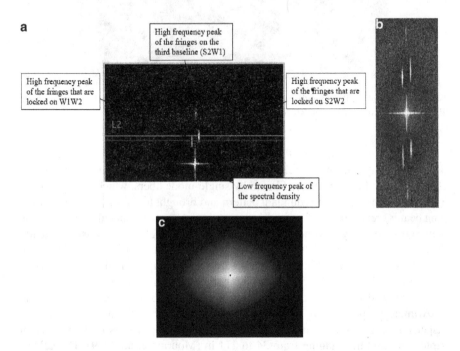

Fig. 7.10 (**a**) Real-time fringe-tracking image of HD 3,360 taken with CHARA/VEGA instrument on 10 October, 2008 at 03:00 UT, (**b**) integration of 3,000 individual spectral densities (spectral band of 20 nm around 610 nm), and (**c**) autocorrelation of 3,000 short exposures images. Courtesy: D. Mourard

---

[4] The cross-spectrum method can be used to measure differential interferometric quantities between two different spectral channels (Berio et al. 1999a, b; Tatulli et al. 2007). The averaged cross-spectrum between two bands of the interferogram is given by,

$$W_{12}(\mathbf{f}) = \langle \widehat{I}_{\lambda_1}(\mathbf{f})\widehat{I}^*_{\lambda_1}(\mathbf{f})\rangle,$$

in which $\widehat{I}_{\lambda_j}$ is the Fourier transform of the interferogram in the spectral band centered at λ_j.

The product of the complex visibility of each spectral band, $\mathcal{V}_{\text{diff}}$ (complex differential visibility),

$$\mathcal{V}_{\text{diff}} = \mathcal{V}_{\lambda_1}\mathcal{V}_{\lambda_2}e^{i(\psi_{\lambda_1}-\psi_{\lambda_2})},$$

where $\mathcal{V}_{\lambda_j}$ and ϕ_{λ_j} are the absolute visibility and the phase of the fringes in the spectral band centered at λ_j respectively, and the argument of $\mathcal{V}_{\text{diff}}$ is the differential phase $\Delta\psi_{12}$.

two other Fig. 7.10b, 7.10c, respectively, correspond to off-line processing of the data: Integration of 3,000 short-exposure such data with the configuration S2W2W1, corresponding to the Fig. 7.10a and sum of the autocorrelation of 3000 short-exposure images of the same having configuration S2W2W1. The S2W2 and W2W1 fringes have the same fringe spacing but their inclination being defined by the optical path difference (OPD), they create the honeycomb pattern in the center of the image. It is the first instrument, which has recorded interferometric data in the visible wavelength on Young Stellar Objects.

7.4 Interferometers Under Development

Success of obtaining interferometric fringes from the starlight by the interferometers using large telescopes made an enormous impact on developing future large optical arrays. Several such instruments will be in operation soon.

7.4.1 Large Binocular Telescopes

In an interferometer like VLTI, the apertures are some distant apart on separate mountings, therefore, they are unable to cover full u, v-plane; these interferometers can cover only a part of the sky. This limitation is because the telescopes baseline is not exactly perpendicular to the viewing object. This configuration restricts the field-of-view (FOV) as the baseline length decreases depending on the orientation of telescopes.

Although the interferometers using two apertures are limited to measurements of visibility amplitude, a different kind of interferometer called Large Binocular Telescope (LBT), currently being commissioned on Mt. Graham (elevation 3,200 m), Arizona, USA, can provide information in the u, v-plane, which can be continuously combined or co-added. From this the complex visibility can be recovered. LBT consists of two 8.4 m primary mirrors, each with f/14.2 (Hill 2000), mounted in a single mounting, spanning 22.8 m baseline. The equivalent circular aperture of LBT is 11.7 m. The primary mirrors are co-mounted on a fully steerable alt-az mounting, where variable delay-lines for the path equalization are not needed. The effective field is limited only by the atmosphere (2 arcminutes to 2.2 μm), while other interferometers are restricted to few arcseconds. LBT has a diverse range of characteristics such as sensitivity, spatial resolution, and wide field; it has a continuous FOV of 1 arcmin operating at 2.2 μm. Such an interferometer can be used in almost all fields of astronomical research.

The telescopes can be used separately or, by sending the light to a single camera between the telescopes. The configuration allows essentially complete sampling of all spatial frequencies in the image within a variable baseline of 0–22 m using interferometric imaging between the two pupils. This arrangement provides

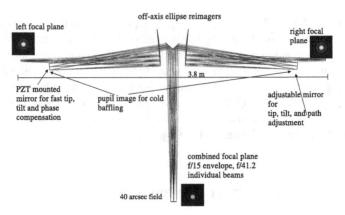

Fig. 7.11 Schematic of LBT beam-combining optics (courtesy: P. Hinz)

unique capabilities for high resolution imaging of faint objects. In the near-IR, LBT would exceed the HST, at its optimum wavelength by a factor of three. At such wavelengths, a field of view of one arcminute or more is expected with unprecedented spatial resolution of order of 8–9 mas at $\lambda \sim 1~\mu$m.

The binocular structure allows storing of secondary mirrors on-board the telescope and so one can interchange them in 10–15 min. This configuration helps in changing observing conditions. The beam-combining optics is shown in Fig. 7.11. When combined with adaptive optics system, LBT interferometric mode offers high signal-to-noise imaging on even the faintest objects, over a relatively wide field. The adaptive optics system may employ two 91 cm diameter 1.7-mm thick f/15 secondary mirrors with 672 actuators on each mirror operating at $\sim$900 or 1000 Hz (Wehinger 2004). Wavefront sensors at the f/15 bent Gregorian focii[5] would provide inputs for correction of the Gregorian adaptive secondaries. The quality of image is optimized due to the large field view of the LBT.

7.4.2 *Mitaka Optical and Infrared Array*

The Mitaka optical and InfraRed Array (MIRA) project (a collaboration between the University of Tokyo and the National Astronomical Observatory of Japan) consists of several interferometers built one-by-one, each instrument being an upgrade of the previous one. The first of the series was MIRA-I (Machida et al. 1998). It had

---

[5] Gregorian telescope, conceived by J. Gregory in 1663, is a reflecting telescope having a paraboloidal primary mirror with a hole in the center and a concave ellipsoidal secondary mirror placed beyond the focus of the former (prime focus). The latter forms the final focus by re-focusing diverging light cone coming from the primary. The telescope produces an image but with a small field of view, whose quality is determined by the particulars of optical configuration of the two mirrors.

25-cm siderostats and a 4-m baseline. The fringe detector was designed for 800 nm wavelength. Its successor, MIRA-I.2 (Sato et al. 1998) has the same baseline and slightly larger siderostats (30 cm). The main objectives of such an instrument is that of determining the diameters, as well as of obtaining accurate orbital parameters of binary systems.

MIRA 1.2 is a Michelson interferometer, and has two siderostats placed separately with a baseline length of 30 m, forming two pupils of the interferometer. It features the equipment encountered on many operating interferometers: beam compressors (yielding 30 mm beams), delay-line consisting of a micro-stepping motor cart and a cat's eye cart which operates in vacuum, tip-tilt correction system and laser metrology. After being reflected from the siderostats, the light beams from the observed star are transferred, guided through 50-m vacuum pipes, to the delay-line unit. This unit is housed in a vacuum chamber, which can be moved smoothly on a rail. The optical path lengths of the starlight coming via two arms of the interferometer can be adjusted accurately. The first stellar fringes with α Lyrae was obtained on a 30-m baseline employing MIRA 1,2 instrument (Naoko 2003), although MIRA-I and MIRA-I.2 instruments are specially designed for practicing interferometry and testing devices.

7.4.3 Magdalena Ridge Observatory Interferometer

The Magdalena Ridge Observatory Interferometer (MROI) is a high-sensitive imaging array being developed at a 10,500 ft altitude in the Magdalena mountains of Socorro county, central New Mexico (Buscher et al. 2006). This array is composed of ten 1.4-m diameter telescopes having focal ratio of roughly $f/2.5$. These telescopes are designed in an altitude-altitude configuration with an active secondary and driven tertiary. Each telescope contains the optical tip-tilt and tracking systems. The interferometer is arranged in a 'Y' configuration, along a baseline distance of up to 400 m operating at wavelengths between 0.6 and 2.4 μm. The telescopes are movable between sets of discrete foundations. The longest baseline may provide an angular resolution of 0.6 milli-arcsecond at an wavelength of 1 μm.

Light collected by the telescopes is being transported through evacuated beam relay pipes to the beam combining facility building. The paths traveled by the light from various telescopes are equalized by the vacuum delay-lines before they are interfered on optical table in the beam combining area. The other necessary components on this table are: (1) atmospheric dispersion compensators, (2) a beam reducer, (3) beam turning mirrors, and (4) pick-off dichroics to send light to various beam combiners and then to science and fringe-tracking cameras. There are fifteen reflections from telescope primary to detectors.

7.5 Interferometry with Large Arrays

The next generation imaging interferometers with at least 15 or more elements should have the snapshot capability in an instantaneous mode. Beams from separated telescopes of such an interferometer must be recombined in the focal point as in the case of a Fizeau interferometer that is optically equivalent to single large telescope masked with a multi-aperture screen so as to reproduce exactly the ensemble of collecting telescopes (Traub 1986).

7.5.1 Optical Very Large Array (OVLA)

The Optical Very Large Array (OVLA) concept, consisting of 27 mobile 1.5-m telescopes, was proposed by Labeyrie a couple of decades ago. It was conceived to be an array along an ellipse, which is the intersection between the ground plane and a virtual giant parabola pointing towards the star. As the star follows its diurnal path in the sky, the virtual parabola has to track it, so the telescopes are moving continuously on the ground to follow the ellipse distortion. Thus, no delay-line is required.

Each telescope structure was planned to be 2.8-m diameter fiberglass sphere, also serving as a dome (see Fig. 7.12), which may be oriented in any direction with three motors featuring 'barrel-caster' cabestans, mounted on their shafts. The sphere rests on these three barrel-casters: when one of them is steering the sphere, no friction occurs from the other two. A 1.5- m prototype telescope has been constructed at the Observatoire de Haute-Provence (OHP; Dejonghe et al. 1998). The active thin primary mirror, which enable to get high-quality wavefronts for interferometric

Fig. 7.12 Schematic view of the 1.5-m prototype telescope built at Observatoire de Haute Provence (OHP; courtesy: O. Lardière)

purpose, is supported by 32 actuators. A secondary mirror makes the beam afocal and compressed. A third steerable flat mirror sends this beam out through a slit located on the sphere to the central station where all other beams coming from other telescopes are combined into a single high resolution image. A motorized shutter can partially close this slit when a barrel-caster and the slit are in coincidence. It has been projected to mount the sphere and its motorization on a six-leg robot (hexapode) able to move on the ground while fringes are acquired (in order to compensate the OPD between beams). OVLA has also been considered for different possible aperture diameters including 12–25 m (Labeyrie 1998). A new telescope structure has been imagined for this class of very large collectors: the 'cage telescope', in which the sphere is replaced with an icosahedral truss steerable by a different mechanical system.

A Moon-based version of OVLA, called LOVLI (Lunar Optical Very Large Interferometer) was also proposed to the space agencies, but using small mirrors instead of telescopes. It was planned to build a fragmented giant telescope. Each mirror is shaped as if it were a segment of the paraboloid mirror of the synthesized giant telescope. The array is arranged on the ground to form an ellipse. Images are then directly obtained at a recombination station located at a focus of the ellipse. The synthetic aperture of the telescope array arranged in this configuration reproduces in a down-scaled version at the recombination optics, thus providing a Fizeau interferometer configuration. However, controlling the particular shape of each mirror must be variable, which requires a few actuators.

7.5.2 Hypertelescope Imaging

In recent years, using long baseline interferometers images have been reconstructed from Fourier components recorded with multiple exposures, obtained with different baseline settings. By applying the concept of hypertelescope imaging (Labeyrie 1996) with many diluted apertures, one may get direct images in a more efficient and sensitive way. Recently, Labeyrie et al. (2006) showed that there is a large gain in imaging performance, including sensitivity, if a given collecting area is divided in many small apertures, used in a hypertelescope mode, rather than a few larger ones used with aperture synthesis. Such a mode may use either image plane or pupil plane interferometry, as well as work through optical fibers.

As described earlier in Sect. 5.3.2.3, in a Fizeau mode with many apertures, the energy is spread over the sidelobes of the interference pattern limiting the sensitivity of the instrument. Most energy goes in a broad diffractive halo rather than in a narrow interference peak, which precludes obtaining usable snapshot images with kilometric or megametric arrays in space. However, several stars in the field would provide superposed interference patterns, identical but laterally shifted and adding their energy, analogus to the conventional image formation in monolithic telescopes (see Sect. 4.1.6.1). Although the deconvolution technique can be applied to reconstruct an image of the source, this becomes difficult with the complicated

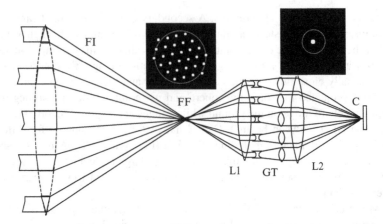

Fig. 7.13 Principle of the hypertelescope (courtesy: A. Labeyrie). The segmented lens at *left* provides a Fizeau image in its focal plane. The focal image FF provided by a Fizeau interferometer FI is re-imaged by lenses L1 and L2 on camera C, through an array of miniature and inverted Galilean telescopes GT. They densify the exit beam, thus shrinking the diffractive envelope (*dotted circle*) of the focal pattern with respect to the interference peak. For an off-axis star, they also attenuate the local tilt of the flat wavefront transmitted from L1, while preserving the global tilt. The wavefront from an off-axis star thus acquires stair steps while becoming densified at L2. The interference peak is displaced more than the envelope, but remains within it if the step is below one wavelength

spread function of a highly diluted Fizeau interferometer. The pupil (exit) densification (see Sect. 5.3.2.3) shrinks the halo and intensifies the direct image formed by the peaks (Labeyrie 1996), thus providing the answer to this problem.

Figure 7.13 depicts the concept of the hypertelescope, which is a giant diluted telescope at the scale of 100 m to kilometers, and possibly 100,000 km in space, made of many mirrors, located far apart with respect to their size. It is capable of providing direct images of faint objects at the focal plane with more sensitivity than the aperture synthesis procedure with its visibility measurement, phase calibration, and Fourier synthesis etc. By combining amplitude signals from all apertures, it thus avoids its inefficient combination of fringe intensities recorded in succession with different baselines. Such a telescope may provide an image with full luminosity in a narrow field of $\approx \lambda/B_s$, where B_s is the distance between the sub-apertures of the array. This favors using more apertures, at given array size, and these are then smaller for a given total collecting area. On a point source, such a system has a broad halo of interference features, formed as the coherent sum of the broad diffraction peaks contributed by each sub-aperture. With a perfect phasing a narrow central peak appears, which can be intensified by shrinking the halo with an array of small Galilean telescopes, operating backwards. This also shrinks the field of view since the wavefront in the exit pupil becomes stair-shaped if the star moves off-axis. This is verified with a miniature 100-aperture hypertelescope (Labeyrie 2009). Pedretti et al. (2000) have derived the integrated intensities of the central peaks of the images on the star Capella (α Aurigae) that are obtained by taking two separate exposures of 100 s in the Fizeau and densified-pupil mode of the hypertelescope. The comparison of these values showed an intensity gain of 24 ± 3 times of the densified with respect to the Fizeau configuration.

There are various practical ways of building hypertelescopes, with different optical architectures. For ground-based arrays, an elliptical track is one way of compensating the Earth's rotation, but with delay-lines, a periodic or non-redundant dilute aperture can also be built at the scale of 10 kilometers. Alternately, a spherical crater site can be equipped with many fixed mirrors, in which case no delay lines are needed if a focal corrector is inserted near the suspended camera for correcting the inherent spherical aberration. In space, it would be a hypertelescope version with non-redundant or periodic hexagonal paving. The periodic version has a Fizeau focus followed by a small pupil densifier, providing a fully densified exit pupil which is of interest for feeding a coronagraph (Boccaletti et al. 2000), and it can be globally pointed to avoid the use of delay lines, or the restriction of an elliptical, or else the use of a spherical locus with a focal corrector.

It can be shown that the maximal number of point sources to be imaged by a non-redundant multi-aperture interferometer is, per sub-aperture lobe, equal to the square of its number of apertures. The number increases beyond this limit if the aperture pattern is modified during a long exposure, using Earth rotation for example (Labeyrie 1996). One difficulty is cophasing all the beams, since $27 (= 3^3)$ telescopes are expected, the cophasing of the whole array may be done hierarchically by cophasing triplets of beams (yielding a honeycomb pattern in the image plane), then triplets of triplets, etc. Piston errors are measurable from the triplet images. The limitation of the cophasing procedure by the photon noise would not be very important. According to numerical simulations, the expected limiting magnitude of a hypertelescope imaging technique is found to be 8.3 m_v if 10-cm apertures are used and 20 m_v for 10-m apertures (Pedretti and Labeyrie 1999). The limit is expected to increase with the Carlina array (Labeyrie 1999b), a 100-element hypertelescope with a diameter of 200 m.

Another way of analyzing the piston errors (Labeyrie 1999a) is an extension of the classical dispersed-fringes used since Michelson: a set of monochromatic images recorded with a spectro-imager is organized as an x, y, λ data cube, and its 3-dimensional FT is calculated to extract piston errors in pairs or triplets of apertures. Obtaining snapshot images with milli-arcsecond angular resolution, and possibly micro- or even nano-arcsecond with 100 or 100,000 km arrays in space, may provide rich information on stellar photospheres and stellar oscillation modes expected on various types of stars. The proposed large hypertelescope arrays, equipped with a coronagraph and adaptive optics system, offer an access to get information on the faint and complex sources such as, circumstellar matter in the form of disks, jets, coronae, exoplanets, galaxies, etc., with a high angular resolution showing much finer detail carrying more information than multi-telescope LBOIs.

7.5.3 Carlina Array

Carlina[6] array is an alternate type of design, shaped like Arecibo radio telescope dish,[7] but in diluted form, which can accommodate numerous small apertures (Labeyrie 2005). The individual elements are supposed to be positioned in a spherical arrangement, exploiting a natural depression, (see Fig. 7.14) with an emphasis on interferometric imaging. To note, the spherical geometry reduces the amount of pathlength compensation required in re-pointing the interferometer array.

Carlina can have a broad primary field exploited with several focal gondolas, which can be panoramic when implemented in space, but requires correcting optics near the primary focal plane to compensate the spherical aberration. The main advantage of such a system is that it does not require delay-lines, and is therefore cost effective when utilizing many sub-apertures. The focal point optics contains a Mertz corrector as depicted in Fig. 7.14 and a pupil densifier. As the diluted primary mirror is spherical, the aberrated image is corrected by such a corrector. A CCD

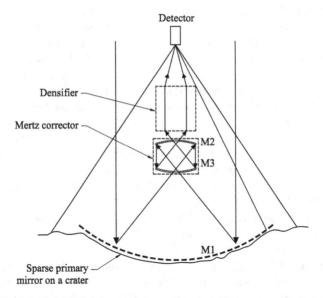

Fig. 7.14 Principle of the Carlina hypertelescope. Many small mirrors are distributed on a spherical surface. A camera and focal corrector of the spherical aberration are suspended at the focal sphere

---

[6] Carlina is name of a large ground-hugging thistle flower.

[7] The Arecibo radio telescope is located in a natural depression that is 14 km south-southwest of Puerto Rico. It is the largest single-dish telescope of 305 m diameter whose collecting area is about 73,000 m^2. This spherical reflector dish is made out of almost 39,000 perforated aluminum panels, each measuring 1 m by 2 m, supported by a mesh of steel cables. A platform is suspended 150 m above the reflector which houses the receiving equipment and can be positioned anywhere over the dish with the use of the drive system. The rotation of the Earth causes it to scan the sky once each day. The band of sky that can be observed is 39° in diameter.

camera is placed at the densified focus. With a simple metrology system, using a white laser source at the curvature center of the diluted primary mirror, its elements can be co-spherized with sub-micron accuracy. In addition, an adaptive optics system may be used within the densifier for correcting the atmospheric phasing errors. However, building such a giant system requires a crater-shaped site or, for observing near transit time, a deep glacial valley oriented East–West. A delicate suspension of the focal optics for tracking the star image along the focal surface, is also needed, as achieved at Arecibo. Le Coroller et al. (2004) had tested at the Observatoire de Haute Provence a proto-type system, using a tethered balloon to carry the focal camera 35 m above the mirrors. They verified the feasibility of tracking the diurnal motion of a star image equatorially by pulling a cable with a computer-driven winch. They have obtained interference fringes with a small baseline. The proto-type hypertelescope may operate initially without any adaptive optics system, and would observe by speckle interferometry, using either a redundant or non-redundant many-element aperture (Labeyrie 2009). On moderately complex sources, up to rather faint stellar magnitudes, image processing algorithms such as the bispectrum technique (see Sect. 8.2.2) are proving usable.

Figure 7.15 shows the point spread function (PSF) of a randomly distributed 500 apertures in a disk. The distributions are such that the aperture dimension was very

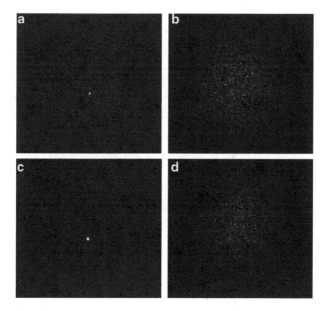

Fig. 7.15 (a) Densified image of an object located on-axis obtained with a simulated 200 random diluted apertures in disk having random phases in the range between 0 and π and (b) the speckle pattern obtained with the said arrangement having random phases in the range of $0-2\pi$, (c) and (d) represent, respectively, the same as (a) and (b) but with 500 apertures, which one may notice that the PSF contains more energy in the central peak. The ratio of disk diameter to densified pupil diameter D/d is 100:10

small compared to the distribution dimension, so essentially each individual aperture was considered as a delta function. An array of diluted apertures can be thought of as an array of delta functions, convolved with a disk, mathematically the aperture function term of such an array is given by,

$$\sum_{j=0}^{N} \delta(x - x_j, y - y_j),$$

in which n is the number of apertures and x_j, y_j provides the position for the jth diluted aperture.

If all diluted apertures are considered to be co-phased, the mod. squared of direct Fourier transform of this function provides the PSF of the distribution. The Fourier transform of aperture distribution,

$$\mathcal{F}\left[\sum_{j=0}^{N} \delta(x - x_j, y - y_j)\right] = \sum_{j=0}^{N} e^{-i2\pi(ux_j + vy_j)}, \tag{7.10}$$

thus, the PSF of such a distribution essentially becomes,

$$PSF = \left|\sum_{j=0}^{N} e^{-i2\pi(ux_j + vy_j)}\right|^2. \tag{7.11}$$

In the case of the random phase, for example, if the individual apertures are not co-phased, each of these apertures can have their own random phases. So, the aperture function term for the array consisting of delta functions modifies to,

$$\sum_{j=0}^{N} \delta(x - x_j, y - y_j) e^{i\psi_j},$$

where $e^{i\psi_j}$ is the random phase term associated with the jth diluted aperture.

Hence, the going back to the previous analysis one finds the Fourier transform,

$$\mathcal{F}\left[\sum_{j=0}^{N} \delta(x - x_j, y - y_j) e^{i\psi_j}\right] = \sum_{j=0}^{N} e^{i2\pi(ux_j + vy_j)} e^{i\psi_j}, \tag{7.12}$$

and the PSF is,

$$PSF = \left|\sum_{j=0}^{N} e^{i2\pi(ux_j + vy_j)} e^{i\psi_j}\right|^2. \tag{7.13}$$

To note, this analysis holds good for an on-axis star. In the simulation, two cases are considered:

1. ψ is a random variable with domain between 0 and π, in which the PSF showed mild speckle structures, albeit most of the energy of the PSF lie in a central peak (see Fig. 7.15a, c) and
2. ψ is between 0 and 2π, where enhanced speckle structures are envisaged (see Fig. 7.15b, d).

Here, the pupil densification is treated as convolving a large aperture to the initial diluted aperture distribution. Mathematically, the exit pupil after pupil densification can be represented by,

$$\left[\sum_{j=0}^{N} \delta(x - x_j, y - y_j) \right] \otimes circ\left(\frac{x}{R_0}, \frac{y}{R_0} \right),$$

with R_0 as the radius of densified aperture in exit pupil.

The Fourier transform of this term may be written as,

$$\left[\sum_{j=0}^{N} e^{i2\pi(ux_j + vy_j)} \right] \cdot \mathcal{F}\left[circ\left(\frac{x}{R_0}, \frac{y}{R_0} \right) \right],$$

where $\mathcal{F}[circ(x/R_0, y/R_0)]$, is the diffraction function (see Sect. 1.4.3) from the R_0 aperture.

On multiplying the PSF from diluted apertures with the diffraction function, one obtains the PSF for the densified case. For the random phase case, one multiplies this term with the random phase term.

7.5.4 High Resolution Coronagraphy

A coronagraph is an on-line telescopic attachment that blocks the central Airy peak and a few rings in the diffraction pattern of the brighter source, while allowing light from surrounding sources to pass through relatively undisturbed. It also greatly attenuates the outer diffraction rings. Such a mechanism, invented by Bernard Lyot (Lyot 1939), can record the emission component of the corona. It enables astronomers to study comets, sub-stellar companions (brown stars), and debris disks around stars as well.

The principle of a coronagraph is depicted in Fig. 7.16. It has an occulting disk in the primary image plane of the telescope, which absorbs most of the light from the center of the field of view. It also has a mask, the Lyot stop, in the following pupil plane. By using masks with a different effective shape, the diffraction pattern can be controlled so that the starlight is much dimmer closer to the center in some areas, and brighter in others. The primary objective lens (first lens), specially fabricated to

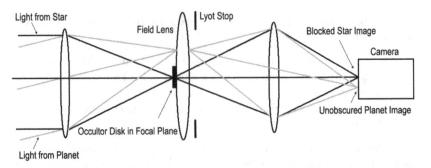

Fig. 7.16 Schematic of a Lyot coronagraph. The objective lens (first lens) makes a real image of the object on to a blocking disk that creates artificial eclipse. The field lens (second lens), located just after such a disk makes a real image of the objective lens onto a Lyot stop (after Labeyrie et al. 2006)

minimize scattered light, makes a real image of object on to this disk. The field lens, located after such a disk, makes a real image of the objective lens onto a matched re-imaged pupil plane, called Lyot stop. This stop has an aperture slightly undersized than the geometric pupil to account for aberrations and diffraction effects caused by other stops and baffles in optical systems. It blocks out the remaining diffraction rings of light from the central star and relays the hidden image by a third lens onto final observing plane. The star's Airy pattern is attenuated, while image of an off-axis object (e.g., planet) is little affected. While using the occulting stop, the size of the pupil should be chosen carefully in order to find the best trade-off between the throughput and image suppression.

However, the limitations come from the light diffracted by the telescope, instrument optics, and atmospheric turbulence. These effects substantially degrade the ability of a coronagraph to remove the starlight. To note, the wavefront aberrations throughout the optical train produce a speckle halo noise in the image plane preventing direct detection of the planet if these speckles (see Sect. 4.2) are bright. Speckles may be canceled by employing an adaptive optics system to measure and compensate wavefront aberrations (Bordé and Traub 2007). In a high dynamic range coronagraph regime, a problem related to the non-common path error, that is the difference between wavefront sensor (see Sect. 4.4.2) optics and the main path for the star and the planet images, should in principle be considered in the focal plane wavefront sensor (Guyon 2006). Speckle nulling featuring focal plane speckle measurement and iterative wavefront control may also solve the non-common path problem (Balasubramanian et al. 2006). Nishikawa et al. (2008) proposed a wavefront correction with an unbalanced nulling interferometer for an imaging coronagraph, with which they showed that in the image plane, the central star's peak intensity and the noise level of its speckled halo are reduced.

Direct imaging of planetary and other stellar companions may provide information on the planet's atmospheric composition. However, the contrast between planets and their host stars make direct imaging method technically challenging, particularly in the visible light; a Jupiter-like object is a billion time fainter in this waveband than their host stars. Several coronagraph designs using advanced focal plane masks have been reported (Kuchner and Traub 2002; Riaud et al. 2003; Mawet and Riaud 2005).

Coronagraphic method is important in the K-band where thermal radiation from the parts off the edge of the primary mirror would increase the background very significantly. However, an adaptive optics system in conjunction with a corona-graph is required in order to effectively image exoplanets and debris disks (Serabyn et al. 2006). Nakajima (1994) estimated that imaging with coronagraphy equipped with an adaptive optics system in a 6.5 m telescope could detect Jupiter-size exo-planets at separation $\sim 1.5''$ with signal-to-noise (S/N) ratio of 3 in 10^4 s. According to Rouan et al. (2000) a detection at a contrast of more than 8 mag difference between a star and a planet may also be feasible. However, the required dynamic range of 10^9–10^{10} for optical detection of Earth-like planets by a coronagraph can be achieved with a high quality wavefront of $\lambda/10,000$ root-mean-square (RMS) and an intensity uniformity of $1/1000$ RMS (Kuchner and Traub 2002; Lowman et al. 2004).

With improved technology, the interferometric arrays of large telescopes fit-ted with high level adaptive optics system that applies dark speckle[8] coronagraph (Boccaletti et al. 1998) may provide snap-shot images at their recombined focus using the concept of densified-pupil imaging (Pedretti and Labeyrie 1999), and yield improved images and spectra of objects. One of the key areas where the new technology would make significant contributions is the astrometric detection and characterization of exoplanets. A space-borne large hypertelescope interferometer may detect an image of an Earth-like planet (see Fig. 7.17).

High resolution stellar coronagraphy is a powerful tool in (1) detecting low mass companions, e.g., both white and brown dwarfs, dust shells around AGB and post-AGB stars, (2) observing nebulosities leading to the formation of a planetary system, ejected envelops, accretion disk, (3) understanding of structure (torus, disk, jets, star forming regions), and dynamical process in the environment of AGNs and QSOs, and (4) direct imaging of exoplanets (Aime and Vakili 2006). By employing such a coronagraphy equipped with image-motion compensation system for the stabiliza-tion of the telescope field, Golimowski et al. (1992) reported that it may be feasible to detect circumstellar objects 2 magnitudes fainter than those detectable with a con-ventional coronagraph. They have achieved the maximum attainable resolution gain factor of 2.2 by stopping the telescope aperture to $D \equiv 4r_0$.

---

[8] Dark speckle method uses the randomly moving dark zones between speckles, known as dark speckles; highly destructive interferences may occur occasionally depicting near black spots in the speckle pattern (Labeyrie 1995). Such a method exploits the light cancellation effect in ran-dom field, whose aim is to detect faint objects around a star when the difference of magnitude is significant. If a dark speckle is at the location of the companion in the image, the companion emits enough light to reveal itself. Dark speckle method, which features the combination of both speckle interferometry and adaptive optics system, may improve the possibility of detecting faint companions of stars.

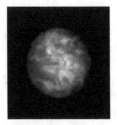

Fig. 7.17 Simulated image of an Earth-like Planet at the focus of a 100 km hypertelescope (courtesy: A. Labeyrie). An image of the Earth picture was convolved with the spread function from 150 point apertures, arrayed in three circles; it was multiplied by the hypertelescope envelope, i.e. the spread function of the sub-pupils as well, and then pushed somewhat the contrast to attenuate the diffractive halo

7.6 Space-borne Interferometry

Although impressive advances have been made in Michelson interferometry, baselines much longer than a few hundred meters as well as the ones with large apertures, encounter serious issues in atmospheric turbulence. In order to avoid such problem, it is prudent to deploy the interferometers into space. The other noted advantage is that observation can be made at any wavelength ranging from the ultraviolet (Lα; 1,200 Å) to the near-IR, and for longer duration. However, the difficulty comes from developing a technology featuring high precision positioning, as well as toughness required for space operation. A new generation of ultra-lightweight active mirrors may resolve the problems of size and weight. Several space-based small interferometers, capable of astrometric detection of planets, are being developed.

7.6.1 Space Interferometry Mission

Space Interferometry Mission (SIM) lite astrometric observatory is a space interferometer being developed by the NASA. The design of this mission consists of one free-flyer having a 6-m baseline science (Michelson) interferometer with 50-cm apertures operating at visible spectrum (0.4–0.9 μm). In order to get an accurate knowledge of the baseline vector, **B**, for wide-angle astrometry without collector motions, it would feature two auxiliary interferometers with 7.2 m baselines, aimed at reference stars (grid-locking). The sensitivity for astrometry is $m_v = 20$ after a 4-h integration and a few tens of observations per star to get the orbit. SIM lite would operate in an Earth-trailing heliocentric orbit, which may drift away from the Earth at the rate of 0.1 AU per year until it reaches a distance of 95 million kilometer.

The expected angular accuracy is 1 μas in narrow-angle mode (with a 1° field of view), collecting the new high-precision astrometry results (Unwin et al. 2008), including the possibility of Jovian planet detection around stars up to 1 kpc distant and terrestrial planet detection around nearby stars by detecting the astrometric wobble relative to reference stars (Catanzarite et al. 2006) and 4 μas in wide-angle mode. These two modes may enable to (1) collect data to help pinpoint stellar mass for specific types of stars, (2) determine the matter makeup of the Galaxy and the local cluster, (3) assist in determining the spatial distribution of dark matter in the Universe, (4) decisively define the cosmic distance-scale, and (5) undertake a comprehensive search for planets down to the mass of Earth (Unwin 2005 and references therein); it may detect the presence of 1 Earth mass planets at 1 AU around about 65 GV stars. It is expected to be launched soon.

7.6.2 Terrestrial Planet Finder

The Terrestrial Planet Finder (TPF) is a concept for a formation-flying interferometer, currently persued by NASA. It would take the form of two separate and complementary observatories, viz., (1) a coronagraph operating at visible wavelengths and (2) a large-baseline interferometer operating in the infrared. TPF interferometric mode would rely on a simultaneous and coordinated use of five separate spacecraft flying in formation, i.e., four aligned 3.5-m free-flying telescopes each on separate spacecraft and a central spacecraft that houses the beam-combining apparatus. The baseline from the two most separated telescopes can span 40 m or more. The collectors of TPF will rotate around the recombiner for planet detection.

TPF is designed to detect and characterize the mid-IR (6–20 μm) spectra of the atmospheres of Earth-like exoplanets (Lawson et al. 2008). Its main scientific goal is the detection and study of exoplanets by means of nulling interferometry and study other aspects such as (1) their formation and (2) development in disks of dust and gas around newly forming stars. Evidence of biological activity on planets around nearby stars would also be studied. TPF should be able to reveal individual star-formation regions in distant galaxies and in the star-forming regions of galaxies as well. In addition, it would explore the physical processes of icy cores of comets, the winds of dying stars, as well as the cores of distant luminous galaxies.

The nulling interferometer with, in the case of the interferometer concept, a rejection ratio (inverse of null depth) of at least $\geq 10^5$ would be able to suppress starlight below the background level of other sources of photon in the 7–12 μm range over a waveband of $\Delta\lambda/\lambda \geq 25\%$. To note, the Sun–Earth flux ratio is $\sim 10^7$ at 10 μm. For planet imaging, telescopes will move along parallel straight lines. Another important instrument is a coronagraphic spectrograph, which would be used to simultaneously characterize all planets, gas and dust within the coronagraphic field. Heap et al. (2006) envisaged a spectrograph with two optical channels, i.e., (1) blue spectral region (0.4–0.8 μm) and (2) red region (0.7–1.0 μm). Each channel has individual integral field unit, a set of dispersers (prisms and grating), and a photon-counting detector system optimized for the red or blue passband.

7.6.3 Darwin Mission

Darwin mission, a space mission (Penny et al. 1998; Fridlund et al. 2004) patronized by the European Space Agency (ESA), is a flotilla of four free-flying spacecraft, each containing 3-m light collectors. These collectors plan to redirect light to the central hub spacecraft, which would contain beam combiner, spectrographs, and cameras for the interferometer array. These collectors would operate together, with the separation of the satellites controlled to a few centimeters, and active optical path compensation systems, maintaining the path difference between the beams of the various telescopes to an accuracy of about 20 nm. Figure 7.18 depicts the conceptual design of Darwin mission.

The aim of Darwin is the discovery and characterization of terrestrial planet systems orbiting nearby stars (closer than 15 pc) by nulling (see Sect. 5.1.3) the light of the star with a rejection ratio of at least 105 (Ollivier et al. 2001). This involves the detection of photons from the planet and not from the star as it is done with Doppler–Fizeau effect detection or wobble detection. Basically, Darwin should overcome two major difficulties for achieving Earth-like planet detection. The first one concerns stellar light quenching. Interferometric nulling techniques may address this issue. Severe requirements about the optical quality of the nulling device might involve spatial filtering, by pinhole or single-mode fibers, to smooth the beam wavefronts. The second difficulty is the expected presence of exo-zodiacal light (infrared emission from the dust surrounding the observed star). The direct interferometric imaging combining the light of the telescopes in Fizeau mode (D'Arcio et al. 2003) is also planned.

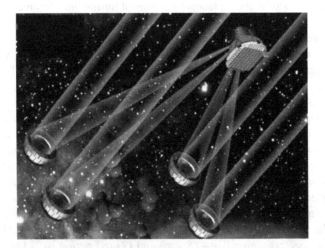

Fig. 7.18 Artist view of the configuration of the Darwin interferometer. Four mirrors collect the stellar and planetary lights, defining the bases of the interferometer, and sending them to the re-combining laboratory. The instrument stops the stellar light and analyzes spectroscopically the planetary light, in order to determine the planetary atmospheric composition, and search for possible biosignatures. The interferometer bases are adjusted to the star–planet separation by navigating the free flying spacecrafts (courtesy: A. Legér)

Another important aim of this mission is to detect rocky planets similar to the Earth and perform spectroscopic analysis of their atmospheric composition, for example water, ozone, and carbon-dioxide in the mid-IR spectrum (between 6 and 20 μm), where the most advantageous contrast ratio between star and planet occurs. It may also feasible to perform the spectroscopy of telluric exoplanets to obtain information on the composition of their atmosphere. One of the goals is the search for biosignatures (Leger 2009); the presence of oxygen, O_2, or ozone, O_3, reveals a photosynthetic activity on the planet. The name of the mission is for stressing the exobiology aspect of the mission.

7.6.4 Long-term Perspective

Space-borne interferometry projects for years spanning from 2020 to 2050 already exist. However, such projects are at the preliminary stage. These missions are capable to achieve in astronomical observations by increasing the achievable angular resolutions of 0.1 mas or more, over a wide range of wavelengths from X-ray to sub-millimeter. Hence, spectral imaging from such interferometers would be a major breakthrough in understanding the Universe. For the post-TPF era, NASA has imagined an enhanced version featuring several very large telescopes and a $R \geq 1,000$ spectrometer, which would be able to detect on an exoplanet lines of gases directly produced by biochemical activity. The next step would be an array of 25 telescopes, 40 m diameter each, that would yield 25×25 pixel images of an Earth-like planet at 10 pc, revealing its geography and eventually oceans or chlorophyll zones.

Space-based telescope clusters flying as phase interferometers have been proposed to attain baselines up to kilometers such as with Stellar imager (SI; Klein et al. 2007; Carpenter et al. 2009) and Luciola (Labeyrie et al. 2009). SI is a ultraviolet-optical interferometer designed to enable 0.1 mas spectral imaging of stellar surfaces and of the Universe and asteroseismic imaging of stellar interiors (Carpenter et al. 2009). Its primary goal is to understand the formation of planetary systems, of the habitability and climatology of distant planets, and of dynamos and stellar magnetic activity. However, the technological challenges, such as (1) precision formation flying of $\sim$30 spacecraft, (2) wavefront sensing and control and closed-loop optical path control of many-element sparse arrays, and (3) positioning mirror surfaces to 5 nm, need to be looked into. The maximum baseline of this interferometer ranges between 100 and 1,000 m, having mirrors of 1–2 m.

A comparable project, known as 'Exo-Earth Imager' has been proposed by Labeyrie (1999a). It consists of a hypertelescope employing 150 mirrors of 3 m each, forming an interferometer with a 150 km maximum baseline. Such an instrument would give a 40×40 pixel image of an Earth-like planet at 3 pc (see Fig. 7.17), providing the same information as described previously. Recently, Labeyrie et al. (2009) had proposed the Luciola project, a large (1 km) interferometer capable of imaging stellar surface details even in the ultra-violet, employing many small apertures for obtaining direct snapshot images with a high information content.

A diluted collector mirror, deployed in space as a flotilla of small mirrors, focuses a sky image, involving many large free-flying spacecraft, utilizing a densified pupil beam combiner. The optical path differences between the sub-apertures would be compensated by a Mertz corrector (see Fig. 7.14). It is as an observatory designed for different forms of observing capabilities from exoplanets to stellar physics and deep-field cosmology, with its spectro-imaging and coronagraphic attachments. It also covers a broader spectral range, from the deep ultra-violet to the mid infra-red.

Such an interferometer in space is expected to reach the limiting magnitude $m_v = 32$ with the 63 m^2 collecting area offered by 1,000 mirrors of 25 cm. Even smaller mirrors, as small as 3 cm, for example, would further increase the science efficiency at given collecting area. This may be feasible in the form of a laser-driven hypertelescope flotilla (LDHF), consisting of a highly diluted giant mirror equipped with a small pupil densifier light concentrator, for which laboratory testing is initiated (Labeyrie 2009). A laser driven hypertelescope for obtaining efficient high-resolution imaging in space can be developed by the usage of small laser-trapped mirrors[9] with nanoparticles trapped in standing waves of laser light (laser

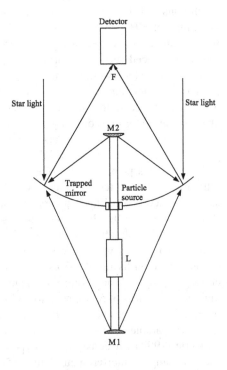

Fig. 7.19 Concept of space telescope having a gaseous or nanoparticle mirror. Two diverging beams of polychromatic light, emitted by a laser L through small mirrors M1 and M2, produce interference pattern having standing waves where nanoparticles can be trapped by radiation pressure, thereby generating a Bragg hologram or a parabolic mirror. A focused image of the stars is formed in the focal plane F

[9] A gaseous mirror (see Fig. 7.19) consists of atoms, molecules or larger particles that are trapped by the standing waves of polychromatic or tunable laser light (Labeyrie 1979). When such a laser strikes two deflectors kept on its either side, a pair of diverging beams propagate in opposite directions. Their interference pattern has standing waves where particles can be trapped by radiation pressure, thereby generating a Bragg hologram or a paraboloid, so as to focus starlight towards its

trapped mirror) and of the hypertelescope in the form of a large diluted mirror having thousands of small (few centimeters) mirrors driven by a common laser. They can be trapped by radiation pressure, axially in polychromatic standing waves and transversely with the 'laser tweezer' effect.

The future projects, such as (1) Black hole imager (BHI; Gendreau et al. 2004) to enable X-ray imaging of black hole event horizons, (2) TPF-I (Lawson et al. 2007) in optical/IR, and (3) Submillimeter Probe of the Evolution of Cosmic Structure (SPECS)/ SPace InfraRed Interferometric Telescope (SPIRIT; Leisawitz 2004) in far-IR/sub-mm, would provide similar capabilities. The technology developments needed for these missions are challenging.

7.7 Reviving Intensity Interferometry

Intensity interferometry in visible waveband, introduced by Hanbury Brown and Twiss (1956a, b), may be useful to derive the diameter and limb-darkening coefficients of stars, albeit unable to produce aperture synthesis images since the visibility phase information is not preserved. The phase behavior may be recovered with multi-element intensity interferometry. Such a method was suggested by Gamo (1963), which permits model-independent reconstruction of the object image allowing simultaneous measurement of the degree of coherence between any two-telescope pair of an array. Although it was not used for astronomical purpose, but was applied to spectral analysis, i.e., using time domain information. It is reiterated that the frequency power spectrum and the self-coherence function of an incident beam form a Fourier transform pair (2.39). The self-coherence function of a beam having an asymmetric spectrum profile is, therefore, a complex quantity, and the determination of its phase is required for the analysis of the asymmetric profile. The triple correlation between the photo-current fluctuations, Δi, receiving light at three detectors at times, t, $t + \tau$, and $t + \tau - \delta\tau$ is,

$$
\begin{aligned}
\Gamma(\mathbf{r}_1, \mathbf{r}_2, \mathbf{r}_3, \tau) &= \langle \Delta i(\mathbf{r}_1, t) \Delta i(\mathbf{r}_2, t + \tau) \Delta i(\mathbf{r}_3, t + \tau - \delta\tau) \rangle \\
&= K |\Gamma(\tau - \delta\tau)| |\Gamma(\tau)| |\Gamma(\delta\tau)| \cos[Q(\tau)\delta\tau],
\end{aligned}
\tag{7.14}
$$

focal plane. The pair of laser wavefronts produces a series of parabolic nodal surfaces. Through radiation pressure forces, dielectric particles are attracted toward bright fringes, and metallic particles towards dark fringes. The particles may be pumped to the central fringe by repeatedly shrinking the trap. Shifting the laser wavelength blueward shrinks the trap symmetrically on each side of the central fringe. They are subsequently cooled below $1°$ K by the laser action. This results a reflective surface in the shape of a mirror of almost arbitrary size. In order to generate a giant diluted mirror capable of focusing star light, many small semi-mirrors may be trapped on a single standing wave, concave in shape and generated by a single tunable laser. A word of caution, the trapped mirror must be protected from sunlight. The noted advantages of such a diluted mirror are: (1) potential for very large aperture mirrors with very low mass, (2) deployment without large moving parts, potential to actively alter the mirror shape, and flexibility to change mirror coatings in orbit, and (3) potential for fabricating naturally co-phased arrays of any shape. A laboratory experiment with a laser-trapped small mirror is in progress (Labeyrie et al. 2009).

where

$$\pm Q(\tau)\delta\tau = \psi(\tau) - \psi(\tau - \delta\tau) - \psi(\delta\tau), \tag{7.15}$$

and K is a constant dependent on the photo-electric conversion efficiency the area of photosensitive surface, $|\Gamma(t)|$ and $\psi(t)$ the magnitude and phase of the self-coherence function of the incident beam respectively.

The phases can be determined by using the same procedure described in Sect. 5.3.1 for non-redundant array. Sato et al. (1978) have proved experimentally that an intensity interferometer that correlates the intensities from three independent detectors can reconstruct the phase of the asymmetrical object. Fontana (1983) generalized intensity interferometry to N-detectors correlating all N currents to form a single output. The multiple correlation function, according to him,

$$C_N(\tau_1, \tau_2, \cdots, \tau_N) = \int_{-\infty}^{\infty} \langle I_1(t\tau_1) \cdots I_N(t\tau_N)\rangle dt. \tag{7.16}$$

where $\tau_{j=1,2,\cdots}$ are the electrical delays added to each beam.

The advantage of a multi-detector intensity interferometer is its independent ability to measure the distance to each source by searching for the maximum signal. Sensitivity improvements of an intensity interferometer can result from observing the same star with an array of telescopes, providing measurements over several baselines simultaneously. Recent technological developments of detectors and electronics have increased the band pass of detectors, and consequently, the potential sensitivity of intensity interferometry. Also, various gamma ray detection observatories, based on collection of Cerenkov radiation produced in the atmosphere, are potentially making available large collecting area mirrors.

Cosmic rays and gamma rays reaching the Earth's atmosphere trigger extensive particle showers, which produce Cerenkov light. Some high energy charged particles, through an optically transparent medium, travel greater than the speed of light in that medium.[10] These particles emit faint flashes of blue light, known as Cerenkov radiation. This phenomenon is named after Pavel A. Cerenkov, who discovered it in 1934. Cerenkov radiation can be detected with large telescopes equipped with fast photo-detectors and electronics. Gamma ray astronomy, particularly after the detection of Cygnus X-3, is one of the fast growing fields of research at current time. The recent advances in the field of ground-based gamma ray astronomy have led to the discovery of more than 60 sources of very high energy (VHE) gamma ray emitters (Aharonian et al. 2005; LeBohec 2009). This success is due to the sensitivity achieved with imaging Atmospheric Cerenkov Telescope (ACT) (Weekes 2003) used in stereoscopic arrays. The emission mechanisms and the regions of VHE gamma ray radiation in these sources are yet to be explored.

In the light of these new developments during the past decades, a renewed enthusiasm for the revival of intensity interferometry by which to achieve

---

[10] The cosmic rays are not traveling faster than the speed of light, but travel faster than the phase velocity of light in the medium.

diffraction-limited imaging over a square-kilometer synthetic aperture is envisaged (LeBohec and Holder 2006; Dravins and LeBohec 2008; LeBohec et al. 2008). Upcoming ACT projects would consist of up to 100 telescopes, each with $\sim$100 m^2 of light gathering area, and distributed over $\sim$1 km^2. These facilities can offer thousands of baselines from 50 m to more than 1 km and an unprecedented (u, v) plane coverage. A telescope array with baselines up to 1 km would provide an angular resolution better than sub-milliarcsecond in the visible band.

Using a precise electronic timing of arriving photons within intense light fluxes, combined with digital signal storage, may enable digital intensity interferometry with high degrees of freedom. These telescopes can be kept fixed and the electronic delays can enable the tracking of stars across the sky. Ofir and Ribak (2006a, b) examined a linear array of many detectors with a uniform spacing in order to combine signals off-line from each component telescope. This relaxes the need for spacecraft orientation and orbital stability for the space-based such instrument (Klein et al. 2007). If a reference star within the field of view is available the intensity interferometry may be employed for astrometry (Hyland 2005). LeBohec and Holder (2006) opined that existing ACT arrays can be employed as the receivers of intensity interferometers to study stars of magnitude as large as five at wavelengths shorter than 400 nm with baselines close to $\sim$200 m that provides an access to angles as small as $\sim$0.1 mas.

Narrabri interferometer (see Sect. 3.3.3.3) telescopes were employed to detect atmospheric Cerenkov light from extensive air showers by Grindlay et al. (1975a, b). They were looking for sources of very high energy ($E > 300$ GeV) gamma rays. Various gamma ray detection observatories, such as (1) Collaboration of Australia and Nippon for a Gamma Ray Observatory in the Outback (CANGAROO; Kubo et al. 2004), (2) High Energy Stereoscopic System (HESS; Bernlöhr et al. 2003; Cornils et al. 2003), (3) Major Atmospheric Gamma Imaging Cerenkov (MAGIC; Lorenz 2004), and (4) Very Energetic Radiation Imaging Telescope Array System (VERITAS; Holder et al. 2006), are potentially making available large collecting area mirrors having diameters ranging from 10- to 17 m. A recent addition is that of the High Altitude GAmma Ray (HAGAR) telescope array, which has seven telescopes, each with seven mirrors of total area 4.4 sq m. Installed at the Indian astronomical observatory (IAO) Hanle, India, at an altitude of 4300 m, it offers an advantage in lowering the threshold of energy of gamma ray observation. The total light gathering area of such a system is 31 sq m, which are deployed on the periphery of a circle of radius 50 m with one telescope at the center. Each of the seven mirrors in each telescope would be looked at by a UV sensitive photo-multiplier tube.

Arrays such as, CANGAROO, HESS, MAGIC, VERITAS, extend over 200 m, which are comparable to the Narrabri interferometer from the point of resolving power. Barring MAGIC, these arrays consist minimum of four telescopes, allowing measurements along up to six baselines simultaneously. This corresponds to a sensitivity gain of $\sim$2.5, which is a gain of one magnitude. Ofir and Ribak (2006c) opined that with a many telescope array, by using higher order correlation, it might be possible to obtain more than double the S/N ratio, and therefore improve sensitivity

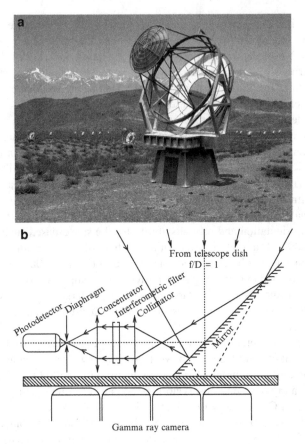

Fig. 7.20 (a) Artist's concept of the AGIS array operating in the range 40 GeV–100 TeV consisting of about 50 imaging ACTs (courtesy: J. Buckley and D. Braun) and (b) schematic of a possible implementation of stellar intensity interferometer on a Cerenkov telescope (LeBohec and Holder 2006; courtesy: S. LeBohec)

by almost one magnitude. Figure 7.20a shows a conceptual design of the Advanced Gamma-ray Imaging System (AGIS) array consisting of about fifty 12 m ACTs. The project is aimed at providing an order of magnitude improvement in sensitivity to gamma-rays from 50 GeV to 20 TeV by increasing the effective area (with a square kilometer array), improving the gamma-ray angular resolution (by a factor of 3) and increasing the angular field of view (from 3 to 8° diameter). The two-mirror demagnifying optical design would provide a wide field of view, isochronous optics, and a much better optical PSF than traditional Davies–Cotton optical design (Davies and Cotton 1957), providing better angular resolution and faster timing important for intensity interferometry (Buckley 2009). To note, Davies–Cotton Configuration is an arrangement of several mirrors of spherical form to produce what is effectively a single large mirror. This particular configuration has good off-axis performance to give an undistorted image; VERITAS array used Davies–Cotton optical design.

LeBohec and Holder (2006) considered the feasibility of implementing intensity interferometry on imaging ACT arrays, which requires the installation of a narrow band filter that is displayed in Fig. 7.20b. The light from the mirrors is reflected side way toward a collimator and analyzer before being focused on the photo-detectors. Such a system is planned to be mounted on the shutter of the atmospheric Cerenkov camera so it does not block the view for gamma ray observations. To note, the sensitivity of such an interferometer is proportional to the square root of the signal bandwidth. According to these authors, interferometer of VERITAS class, with $|\gamma(\mathbf{r}_1, \mathbf{r}_2, 0)|^2 = 0.5$, 100^2 m light collecting area, stars of visual magnitude 5 could be measured in 4 h; the parameters for the photo-detectors and electronics are assumed to be respectively $\Delta f = 100$ MHz and $\eta_d = 25\%$, in which η_d is the quantum efficiency. Signals from this channel may be transferred to a central location via an analog optical fiber. After digitizing at high rate, the signals can be duplicated and aligned in arrival time for each pair in the array before being multiplied and integrated over time.

An important aspect is to be noted that since the time of the Narrabri interferometer, many advances in photo-detection (Morel and Saha 2005; Saha 2007) have been made which could benefit the implementation of a modern intensity interferometer. Until recently, large area photo-diodes, required for practical use in intensity interferometry did not offer competitive electronic noise and bandwidth when compared to photomultiplier tubes, which provide an electronic bandwidth of over 1 GHz. Photo-multipliers offer a high bandwidth and a high gain, but suffer from a relatively low quantum efficiency (QE). The quantum efficiency of such detectors is typically less than 30%. Since the sensitivity of an intensity interferometer is proportional to the quantum efficiency, arrays of Geiger-APDs (see Sect. 6.2.5.3) on a single substrate can be used as detector, which offer detection efficiencies close to 60%. These devices are now becoming available with high signal bandwidth and gains from 10^5 to 10^6. Such photo-detectors are being considered for ACT high energy imaging cameras (Otte et al. 2007) and should be investigated for intensity interferometry as well.

However, unlike Narrabri interferometer, these arrays are at fixed locations and do not have a single pair of telescopes close together permitting measurement of the degree of coherence from an unresolved source. Signals from different telescopes should, in principle, be aligned in time for the correlation to be measured. Analog or digital programmable delays may be employed for this purpose. As in Michelson interferometers, because of the geometrical projection effect, the effective baseline between two telescopes changes during the observation of a star. This can be taken into account at analysis time. However, in ACT arrays, while there are no close pairs of telescopes with which to obtain almost zero baseline measurements, individual mirror within the telescope (for example, HAGAR) are fairly close and can be used for such measurements. Also it is possible to obtain a zero baseline measurement from single telescopes by splitting the beam of collimated and filtered light. The correlation of the fluctuations in the two beams from one telescope then provides a zero baseline measurement of the coherence (LeBohec et al. 2008).

With the advantages and limitations, the intensity interferometry appears to be a viable option to achieve diffraction-limited imaging by electronically combining multiple sub-apertures of extremely large telescopes, spanning over a square-kilometer synthetic aperture, for observations at short optical wavelengths in particular. If interferometry capability is not implemented directly in the high energy camera, all the necessary collimation and filtering optics with associated photo-detectors could be mounted on a plate to be installed in front of the high energy camera (Deil et al. 2008) for interferometric observations. The next generation of ACT arrays for very high energy gamma ray astronomy would offer an unprecedented (u,v) plane coverage.

The ACT intensity interferometry capable of working over a range of short visible bands, may contribute more to the measurements of the stellar diameter wavelength dependence. Predictions from three-dimensional dwarf star atmospheric models can be checked from limb-darkening curves observed at different wavelengths (Dravins and LeBohec 2008) by means of such an interferometry. The other possible stellar programmes that can be made are (LeBohec et al. 2008): (1) angular diameters and limb-darkening of single stars, (2) resolving Pre-Main-Sequence (PMS) stars, (3) determining atmospheric structures of super-giant stars including large-scale stellar convection, (4) calibration of Cepheid period-luminosity relation, (5) oblateness of fast rotating stars, (6) binary stars, double-lined spectroscopic binaries in particular, and (7) emission-line observations. Most of these programmes have been addressed by the existing LBOIs. In addition, Be stars phenomena can also studied in depth, although LBOIs such as I2T, GI2T, and NPOI were used to study the gas envelope of a few Be stars, γ Cas in particular (Mourard et al. 1989), in Hα emitting line with baselines of 40 m and less, using intensity interferometer with an Hβ emission-line filter may reveal finer details considerably smaller (Tycner et al. 2006).

Chapter 8
Image Recovery

8.1 Data Processing

The diffraction-limited phase retrieval of a degraded image is an important subject. Apart from the post-processing speckle interferometry, the adaptive optics systems may also require image-processing algorithms since the real-time corrected image is often partial. For a long baseline interferometry, the hypertelescope in particular, the image processing technique using bispectrum method (see Sect. 8.2.2) is needed. From the results of reconstructed image applying such a method for a (prototype) hypertelescope, where different aperture distributions and various objects were attempted, it appears that bispectrum algorithm is useful even before it becomes equipped with adaptive phasing. The natural aperture rotation, relative to celestial North, during hours of observing (possibly intermittent) may improve the reconstruction. Also, with a Carlina geometry, the pupil drift across a fixed and non-redundant array of mirrors tends to vary the aperture pattern, which improves the reconstructed image. Prior to using such algorithms, the basic operations to be performed are dead pixel removal, debiasing, flat fielding, sky or background emission subtraction, and suppression of correlated noise. The fringes are necessary to be examined and the visibility and contrast can be converted using Fourier transforms, so that the object under observation is mapped effectively. The visibility measured by such an instrument is characterized by the amplitude and phase of the fringe at different instants.

Sources producing fringes with an amplitude equal to the full received power of the source are known to have unit normalized visibility amplitude, and therefore are unresolved by the interferometer (i.e., point sources). In contrast, sources are said to be resolved if the fringes with amplitudes less than the received power have normalized visibility less than one (i.e., extended sources).

In radio astronomy, the Fourier transform of the visibilities obtained on different interferometer baselines in an array of antennas yields the first order map (or the dirty image)', $I'(\mathbf{x})$, which is the convolution of the true brightness distribution $I(\mathbf{x})$ with the synthesized array beam $S'(\mathbf{x})$, known as dirty beam. The latter is analogus to the point spread function (PSF) in optical astronomy,

S.K. Saha, *Aperture Synthesis*, Astronomy and Astrophysics Library,
DOI 10.1007/978-1-4419-5710-8_8, © Springer Science+Business Media, LLC 2011

$$I'(\mathbf{x}) = \mathcal{F}\left[\widehat{\mathcal{V}}(\mathbf{u})\widehat{S}'(\mathbf{u})\right]$$
$$= \mathcal{F}\left[\widehat{\mathcal{V}}(\mathbf{u})\right] \star \mathcal{F}\left[\widehat{S}'(\mathbf{u})\right]$$
$$= I(\mathbf{x}) \star S'(\mathbf{x}), \tag{8.1}$$

where the inverse Fourier transform of the observed visibility gives an image of an object; $\mathcal{F}$ represents the Fourier operator, the symbol, $\star$, stands for convolution parameter, and $\widehat{\mathcal{V}}(\mathbf{u})$ the voltage response function.

The optimal integration time required for measuring a visibility point is a trade-off between the number of photons to be collected and the Earth rotation shifting the sampled point in the (u, v) plane. In radio astronomy, the sampling function, $\widehat{S}'(\mathbf{u})$, is known precisely, and the dirty beam, $S'(\mathbf{x})$, is obtained from observed function $\widehat{S}'(\mathbf{u})$. Thus the image reconstruction in synthesis telescope reduces to deconvolution equation (8.1). However, in practice, this equation may be more complicated due to observational constraints, sky geometry, and computational limitations. Ideally, when the modulus and phase of a sufficient number of components in the u, v-plane are retrieved, the intensity distribution of the source can be derived by means of a deconvolution, obtaining interferometric images with high angular resolution.

8.1.1 Recovery of Visibility Functions

Theoretically, using the Fourier transform to extract information from the observed fringes would give an optimal estimate of the visibility modulus, as demonstrated by Walkup and Goodman (1973). A set of measured visibilities obtained (i.e., samples in the Fourier plane (u, v) corresponding to the image) allows partial reconstruction of image of the observed object. However, white-light fringes obtained from coherencing are corrupted by the intervening atmosphere which a times that modulates their frequency. Techniques used in radio-interferometry, like fitting a sine wave through the fringe data, are therefore, not suitable for the optical aperture synthesis imaging technique. Perrin (1997) has proposed a method to remove the piston from fringes. However, this method requires a high fringe S/N ratio and may only be applied when fringe S/N ratio is important.

Visibilities of several astrophysical systems such as (1) stars with uniform disk, (2) limb-darkened diameters, (3) binary stars, and (4) circumstellar envelops can be obtained. These systems have relatively simple visibility profiles, which may be sampled by a two-telescope interferometer. Such interferometers have limited possibilities for image reconstruction due to the absence of phase visibility recovering. Objects assumed with circular symmetry (standard stars) may be reconstructed with two-aperture interferometers. However, a problem comes from the limb-darkening of the stars observed (see Sects. 8.1.1.1 and 8.1.1.2 for details).

8.1.1.1 Visibility of a Star having Uniform Disk

For a circular star, its observed intensity decreases monotonically. A star has a finite diameter, hence each photon emitted from its surface is independent of rest of the photons. For a uniformly bright disk, one adds up the incoherent fringe patterns as well.

$$
\begin{aligned}
I_{UD}(\theta) &= \int_{disk} I_{int}(\theta - \theta_0) d\theta_0 \\
&= \int_{-\theta_{UD}/2}^{\theta_{UD}/2} 2I_{tel}(\theta - \theta_0)\left[1 + \cos(2\pi(\theta - \theta_x)B_\lambda)\right] d\theta_0 \\
&\simeq 2I_{tel}(\theta)\left[1 + \mathcal{V}_{UD}\cos(2\pi\theta B_\lambda)\right],
\end{aligned}
\tag{8.2}
$$

where $\theta = \lambda/D$ is the diffraction-limit of the individual telescope, $B_\lambda = B/\lambda$, λ the effective wavelength of the observed spectral pass band, D the diameter of the telescope, and $\mathcal{V}_{UD}$ the fringe visibility of a uniform disk in one dimension.

The model visibility amplitude $\mathcal{V}_{UD}$ for a uniform source of diameter, θ_{UD} (in radians), is given by,

$$
\mathcal{V}_{UD} = \pm\frac{1}{F_0} \int_{-\theta_{UD}/2}^{\theta_{UD}/2} B \cos[2\pi\theta B_\lambda] d\theta,
\tag{8.3}
$$

where F_0 and B are the respective total flux and the brightness distribution of the source.

The (8.3) reduces to,

$$
\mathcal{V}_{UD} = \frac{\sin[\pi B_\lambda \theta_{UD}]}{\pi B_\lambda \theta_{UD}}.
\tag{8.4}
$$

For objects with circular symmetry, the visibility function is expressed as,

$$
\mathcal{V}_{UD} = \left|\frac{2J_1[\pi B_\lambda \theta_{UD}]}{\pi B_\lambda \theta_{UD}}\right|,
\tag{8.5}
$$

where $J_1[\pi B_\lambda \theta_{UD}]$ is a Bessel function of the first kind.

Figure 8.1 depicts the visibility curves for a few uniformly circular disks. The visibility amplitude for a point source at a small angle to the line of sight is, in general, constant, but the angular displacement may provide a phase gradient. Prior to such fits, one may derive actual visibilities, which involves a calibration process using other stars.

8.1.1.2 Visibility of Limb-Darkened Diameter

Angular measurement that resolves the apparent angular diameter of a star only to the first zero of the visibility function is biased by limb-darkening; in a way it

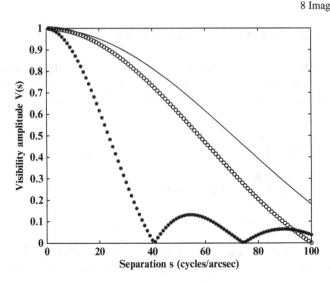

Fig. 8.1 Visibility curves for uniformly bright circular disks of various diameters such as (i) 30 mas (the lower profile with ∗ symbols, (ii) 12.2 mas (the middle curve with *o* symbols, and (iii) 10 mas (the *solid line* at the top)

depends on the wavelength (Mozurkewich et al. 2003). This diameter is known as apparent diameter, in contrast to the true apparent diameter, which is larger in the case of limb-darkening and smaller in the case of limb brightening. The true apparent angular diameter can be obtained by correcting limb-darkening by adequately resolving the center-limb brightness distribution. The limb-darkening occurs as a consequence of the following effects:

1. The emitted light passes through an absorbing atmosphere whose thickness depends on the radius because of increasing angle of incidence. The density of the star diminishes as the distance from the center increases. The photons arrive, on average, from optical depth, $\tau \sim 1$. Stars are spherical, thus the photons from the center of the disk include some from a hotter zone than those at the limb of the disk.
2. The temperature of the star decreases for an increasing distance from the center to of the star. Since surface brightness scales as T^4 for thermal radiation, the limb of stars appear fainter than the central regions of their disks.

To note, the temperature for the Sun does not uniformly drop as the radius increases, and for certain spectral lines, the optical depth is unity in a region of increasing temperature. In such a situation, one observes the phenomenon of limb brightening.

Observations of limb-darkening measurements may provide the atmospheric temperature structure. One requires to collect data in the vicinity of and beyond the first zero or minimum of the visibility function (see Fig. 8.2). The realistic description of true angular diameters needs to consider a radial profile of the source,

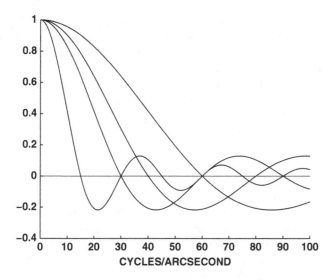

Fig. 8.2 Visibility curves for limb-darkening effect of a source, where the amplitude of the secondary maximum is also depicted. The abscissa represents the cycles per arcseconds and the ordinate the visibility functions

taking into consideration the center-to-limb variation of intensity. Such a profile of a star may be expressed as a polynomial function:

$$I(\mu) = I(0)\left[1 - \sum_{j=1}^{N} \alpha_j (1-\mu)^j\right], \tag{8.6}$$

where $I(\mu)$ is the disk brightness at angle $\mu (= \cos\theta)$ that is the cosine of the azimuth of the emissive point upon the stellar surface, θ the angle between the normal to the stellar surface and the line of sight to the observer, $I(0)$ the intensity of light in the center of the stellar disk ($q = 0$), and $\alpha_j [= \alpha_j(\lambda)]$ the limb-darkening coefficients for given band-pass, depending on the stellar atmosphere.

The coefficients α are provided by a least square fit to the limb-darkening, computed by the model atmospheres. A knowledge of the physics of the atmosphere of the interferometric observations in the second lobe of the visibility function is required, in order to know α. The visibility function of the limb-darkened source, $\mathcal{V}_{LD}$, is then taken by the Hankel transform (see Appendix B) of the selected profiles (Hanbury Brown et al. 1974b). The measurements, $\mathcal{V}_{LD}$ are, therefore, fitted by the new visibility function, leading to a sample of true angular sizes, ϕ_{LD} for each star. The visibility function in this case may be derived as,

$$\mathcal{V}_{LD} = \Gamma(n+1)\frac{|2J_n[\pi B_\lambda \theta_{LD}]|}{[\pi B_\lambda \theta_{LD}/2]^n}, \tag{8.7}$$

where $n = (\alpha_j + 2)/2$ and θ_{LD} is the limb-darkened diameter of the source.

Many interferometers cannot measure low visibilities existing at high angular frequency (i.e., when $\sqrt{u^2 + v^2}$ is large), beyond the first minimum of the visibility function. Reconstructions are, therefore, ambiguous and neither the diameter nor the limb-darkening factor may be accurately determined. Usually, the function for a uniform disk is fitted to the measured value of fringe visibility and a correction, obtained from model stellar atmospheres, is applied to obtain the true limb-darkened angular diameter (Tango and Booth 2000; Tango and Davis 2002).

8.1.1.3 Visibility of Binary Star

A binary system changes the laws of projection in accordance with the fact that the measured parameter, i.e., the Fourier transform of the brightness distribution across the source is a radically different function along the major and minor axes. These functions are approximately cosinusoidal and Gaussian, respectively. The former varies more rapidly with the spacing of the telescopes than the latter. As the baseline is rotated in azimuth, a gradual transition may occur between the two functions. In this case, the intensity at the focal point of a telescope or an interferometer is given by the superposition of appropriately shifted and scaled intensity patterns (Traub 1999).

For two unresolved sources, e.g., a binary system, the expression is formulated to,

$$\mathcal{V}_{bin} = \frac{\sqrt{1 + \beta^2 + 2\beta \cos[2\pi B_\lambda \rho_{1,2}]}}{(1 + \beta)}, \tag{8.8}$$

with β as the flux ratio of two binaries, B the projected baseline length, $\rho_{1,2}$ the separation between the two unresolved sources, and $\mathcal{V}_{1,2}(B)$ the visibilities arising from the sources 1 and 2, and

$$\mathcal{V}_{1,2} = \frac{2J_1[\pi B_\lambda \rho_{1,2}]}{\pi B_\lambda \rho_{1,2}}. \tag{8.9}$$

The visibility amplitude for a binary source or a double point source would depict two peaks. The minimum visibility amplitude is related to the component brightness ratio as the phase changes between them. The properties of this first minimum is used to determine the separation and brightness ratio for a binary stars. Figure 8.3 depicts plots of two unresolved components, in which the spacing and brightness ratio for these two sources are varied; the closer the stars, more the cycles occur in the visibility amplitude function. For equal brightness, the first minimum visibility amplitude comes down to zero. Such a minimum becomes shallower as the brightness ratio differs from unity.

It is then useful to use super-synthesis technique (see 5.71). After a large variation of hour angle H, several visibility moduli are, therefore, measured at different (u, v) points to determine the parameters, such as $\mathcal{V}_1$, $\mathcal{V}_2$ and ρ of the system by fitting the function.

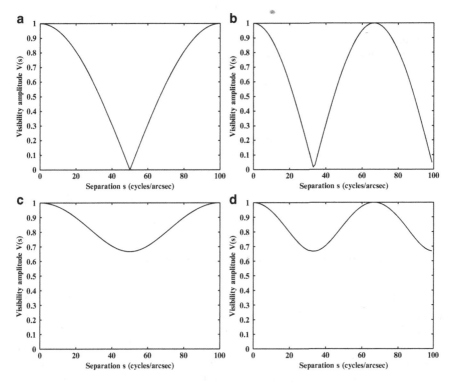

Fig. 8.3 Visibility curves of the binary systems. *Top*: two binaries have equal brightness with separation $\rho_{1,2} = 10$ mas (**a**) and $\rho_{1,2} = 15$ mas (**b**). *Bottom*: the brightness of one of the binaries is equal to 0.2 times the other one having separation $\rho_{1,2} = 10$ mas (**c**) and $\rho_{1,2} = 15$ mas (**d**). The position and depth of the minima change with separation and brightness ratio

8.1.1.4 Visibility of Circumstellar Shell

The spatial distribution of circumstellar matter surrounding objects which eject mass, particularly late type giants or supergiants are important to look at.

From the visibility function curve of the thin radial structures of the circumstellar shell, one may characterize the mass loss phenomena. Such a function may be computed by considering a coaxial uniform disk and a point source; therefore, the function is written as,

$$\mathcal{V}_{shell} = V_p + (1 - V_p) \left| \frac{2 J_1 [\pi B_\lambda \theta_{shell}]}{\pi B_\lambda \theta_{shell}} \right|, \tag{8.10}$$

where θ_{shell} is the diameter of the shell and V_p the ratio of power radiated by the star; the computer simulated visibility curve of a circumstellar shell surrounding a point source is depicted in Figure 8.4.

Fig. 8.4 Visibility curve
of the circumstellar shell
surrounding a point source.
The size of the disk is 30 mas
and $V_p = 0.5$

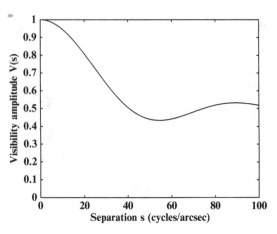

8.2 Reconstruction of Objects from Speckles

As stated earlier in Sect. 4.2.1 that from the spatial power spectrum function,
$|\widehat{O}(\mathbf{u})|^2$, in which $\widehat{O}(\mathbf{u})$ is the object transfer function and $\mathbf{u}$ the 2-D space vector,
one is able to derive the spatial auto-correlation. However, the autocorrelation technique falls short of providing true image reconstructions. The diffraction-limited
phase retrieval of a degraded image is a challenging job. Soon after publication
of the first results, astronomers tried to find algorithmic method to extract image.
Adaptive optics systems may also require image-processing algorithms since the
real-time corrected image is often partial. The following sub-sections outline a few
methods to obtain components of the object Fourier transform, as well as to reconstruct an image from these components.

8.2.1 Knox–Thomson Method

The Knox–Thomson (KT) method (Knox and Thomson 1974), which is a minor
modification of the autocorrelation technique, involves the centering of each specklegram with respect to its centroids and in finding the ensemble autocorrelation of
the Fourier transform of the instantaneous image intensity. Unlike the autocorrelation technique (see Sect. 4.2.1), KT technique defines the correlation of $I(\mathbf{x})$ by itself
multiplied by a complex exponential factor with a spatial frequency vector, $\Delta\mathbf{u}$,
which is larger than zero. The approximate phase-closure is achieved by two vectors
(see Fig. 8.4), $\mathbf{u}$ and $\mathbf{u} + \Delta\mathbf{u}$, assuming that the pupil phase is constant over $\Delta\mathbf{u}$. Let
the general second-order moment be the cross spectrum, $\langle \widehat{I}(\mathbf{u}_1)\widehat{I}^*(\mathbf{u}_2)\rangle$. It takes
significant values only if $|\mathbf{u}_1 - \mathbf{u}_2| < r_0/\lambda$; the typical value of $|\Delta\mathbf{u}|$ is ~0.2 -
0.5 r_0/λ. A 2-D irradiance distribution, $I(\mathbf{x})$ and its FT, $\widehat{I}(\mathbf{u})$, is defined by (4.71a).
In image space, the correlations of $I(\mathbf{x})$, is derived as,

$$I(\mathbf{x}_1, \mathbf{\Delta u}) = \int_{-\infty}^{+\infty} I^*(\mathbf{x})I(\mathbf{x} + \mathbf{x}_1)e^{i2\pi\mathbf{\Delta u \cdot x}}d\mathbf{x}, \qquad (8.11)$$

where $\mathbf{x}_1 = \mathbf{x}_{1x} + \mathbf{x}_{1y}$ are 2-D spatial co-ordinate vectors.

The KT correlation may be defined in Fourier space as products of, $\widehat{I}(\mathbf{u})$,

$$\widehat{I}(\mathbf{u}_1, \mathbf{\Delta u}) = \widehat{I}(\mathbf{u}_1)\widehat{I}^*(\mathbf{u}_1 + \mathbf{\Delta u}), \qquad (8.12a)$$
$$= \widehat{O}(\mathbf{u}_1)\widehat{O}^*(\mathbf{u}_1 + \mathbf{\Delta u})\widehat{S}(\mathbf{u}_1)\widehat{S}^*(\mathbf{u}_1 + \mathbf{\Delta u}), \qquad (8.12b)$$

where $\mathbf{u}_1 = \mathbf{u}_{1x} + \mathbf{u}_{1y}$, and $\mathbf{\Delta u} = \mathbf{\Delta u}_x + \mathbf{\Delta u}_y$ are 2-D spatial frequency vectors; the term, $\mathbf{\Delta u}$ is a small, constant offset spatial frequency.

In Knox–Thomson (KT) method, the transform $\widehat{I}(\mathbf{u})$ interferes with itself by a small shift vector $\mathbf{\Delta u}$) (see Fig. 8.5). It consists of evaluating the three sub-planes in Fourier space corresponding to $\mathbf{\Delta u} = \mathbf{\Delta u}_x$; $\mathbf{\Delta u} = \mathbf{\Delta u}_y$; $\mathbf{\Delta u} = \mathbf{\Delta u}_y + \mathbf{\Delta u}_x$. If digitized images are used, $\mathbf{\Delta u}$ normally corresponds to the fundamental sampling vector interval. A number of sub-planes is used by taking different values of $\mathbf{\Delta u}$. The argument of (8.12) provides the phase-difference between the two spatial frequencies separated by $\mathbf{\Delta u}$ and is expressed as,

$$arg|\widehat{I}^{KT}(\mathbf{u}_1, \mathbf{\Delta u})| = \psi(\mathbf{u}_1) - \psi(\mathbf{u}_1 + \mathbf{\Delta u}). \qquad (8.13)$$

Therefore, (8.12b) translates into,

$$\widehat{I}(\mathbf{u}_1, \mathbf{\Delta u}) = \left|\widehat{O}(\mathbf{u}_1)\right|\left|\widehat{O}(\mathbf{u}_1 + \mathbf{\Delta u})\right|\left|\widehat{S}(\mathbf{u}_1)\right|\left|\widehat{S}(\mathbf{u}_1 + \mathbf{\Delta u})\right|$$
$$\times e^{i[\theta_O^{KT}(\mathbf{u}_1,\mathbf{\Delta u})+\theta_S^{KT}(\mathbf{u}_1,\mathbf{\Delta u})]}. \qquad (8.14)$$

The difference in phase between points in the object phase-spectrum is encoded in the term of (8.14),

$$e^{i\theta_O^{KT}(\mathbf{u}_1,\mathbf{\Delta u})} = e^{i[\psi_O(\mathbf{u}_1)-\psi_O(\mathbf{u}_1+\mathbf{\Delta u})]}. \qquad (8.15)$$

a

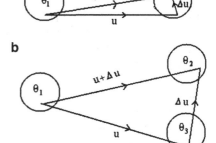

b

Fig. 8.5 Diagrammatic representation of pupil sub-apertures of diameter r_0; (a) approximate phase-closure is achieved in Knox–Thompson method, and (b) complete phase-closure is achieved in Triple correlation method

In a single image realization, the object phase-difference is corrupted by the random phase-differences due to the atmosphere-telescope OTF,

$$e^{i\theta_S^{KT}(\mathbf{u}_1,\Delta\mathbf{u})} = e^{i[\psi_S(\mathbf{u}_1)-\psi_S(\mathbf{u}_1+\Delta\mathbf{u})]}. \tag{8.16}$$

If (8.14) is averaged over a large number of frames, the feature $(\Delta\psi_S) = 0$. The term $|\widehat{O}(\mathbf{u}_1 + \Delta\mathbf{u})|$ is approximated to $|\widehat{O}(\mathbf{u}_1)|$, if $\Delta\mathbf{u}$ is small, $|\widehat{O}(\mathbf{u}_1 + \Delta\mathbf{u})| \approx |\widehat{O}(\mathbf{u}_1)|$, etc., and so,

$$\langle\widehat{I}(\mathbf{u}_1,\Delta\mathbf{u})\rangle = |\widehat{O}(\mathbf{u}_1)|\,|\widehat{O}(\mathbf{u}_1 + \Delta\mathbf{u})|e^{i[\theta_O^{KT}(\mathbf{u}_1,\Delta\mathbf{u})]}$$
$$\times\,\langle\widehat{S}(\mathbf{u}_1)\widehat{S}^*(\mathbf{u}_1 + \Delta\mathbf{u})\rangle, \tag{8.17}$$

from which, together with (4.75), the object phase-spectrum, $\theta_O^{KT}(\mathbf{u}_1,\Delta\mathbf{u})$, can be determined. The scheme for phase reconstruction consists of setting one point equal to unit amplitude and zero phase. By carrying out this process for two orthogonal values of $\Delta\mathbf{u}$, for instance $(\Delta u, 0)$ and $(0, \Delta v)$, the values of ψ_O may be derived at points on grid $(u, v) = (m\Delta u, n\Delta v)$ by summing the $\Delta\psi_S$ out from a reference point (say, $\mathbf{u} = 0$). This can be carried out by various routes in (u, v) plane, and the results can be averaged out in order to improve the accuracy. Once the phases are determined, one gets the complete function, $\widehat{O}(\mathbf{u}_1)$, from which the true image $O(\mathbf{x})$ is reconstructed.

8.2.2 Triple Correlation Technique

The Triple Correlation (TC) method is a generalization of closure phase technique (see Sect. 5.3.1) where the number of closure phases is small compared to those available from bispectrum (Lohmann et al. 1983). The latter is the Fourier transform of the triple correlation. In the second order moment or in the energy spectrum, phase of the object's Fourier transform is lost, but in the third order moment or in the bispectrum, it is preserved. The bispectrum planes, obtained as an intermediate step, contain phase information encoded by correlating $I(\mathbf{x})$ with itself multiplied by complex exponentials with higher spatial frequencies.

The TC method is insensitive to (1) the atmospherically induced random phase errors, (2) the random motion of the image centroid, and (3) the permanent phase errors introduced by telescope aberrations; any linear phase term in the object phase cancels out as well. The images are not required to be shifted to common centroid prior to computing the bispectrum. The other advantages are: (1) it provides information about the object phases with better S/N ratio from a limited number of frames and (2) it serves as the means of image recovery with diluted coherent arrays (Reinheimer and Weigelt 1987). The disadvantage of this technique is that it demands severe constraints on the computing facilities with 2-D data since the calculations are four-dimensional (4-D). It requires extensive evaluation-time and data storage requirements, if the correlations are performed by using digitized images on a computer.

Unlike in shift-and-add method,[1] where a Dirac impulse at the center of gravity of each speckle is put to estimate the same, a triple correlation is obtained by multiplying a shifted object, $I(\mathbf{x} + \mathbf{x}_1)$, with the original object, $I(\mathbf{x})$, followed by cross-correlating the result with the original one. For example, in the case of a close binary star, the shift is equal to the angular separation between the stars, masking one of the two components of each double speckle. The calculation of the ensemble average TC is given by,

$$I(\mathbf{x}_1, \mathbf{x}_2) = \left\langle \int_{-\infty}^{+\infty} I(\mathbf{x})I(\mathbf{x} + \mathbf{x}_1)I(\mathbf{x} + \mathbf{x}_2)d\mathbf{x} \right\rangle, \tag{8.18}$$

where $\mathbf{x}_j = \mathbf{x}_{jx} + \mathbf{x}_{jy}$ are 2-D spatial co-ordinate vectors.

The ensemble average of bispectrum is given by,

$$\widehat{I}(\mathbf{u}_1, \mathbf{u}_2) = \langle \widehat{I}(\mathbf{u}_1)\widehat{I}^*(\mathbf{u}_1 + \mathbf{u}_2)\widehat{I}(\mathbf{u}_2) \rangle, \tag{8.19a}$$
$$= \widehat{O}(\mathbf{u}_1)\widehat{O}^*(\mathbf{u}_1 + \mathbf{u}_2)\widehat{O}(\mathbf{u}_2)\langle \widehat{S}(\mathbf{u}_1)\widehat{S}^*(\mathbf{u}_1 + \mathbf{u}_2)\widehat{S}(\mathbf{u}_2) \rangle, \tag{8.19b}$$

with

$$\mathbf{u}_j = \mathbf{u}_{jx} + \mathbf{u}_{jy}, \tag{8.20a}$$

$$\widehat{I}(\mathbf{u}_j) = \int_{-\infty}^{\infty} I(\mathbf{x})e^{-i2\pi\mathbf{u}_j \cdot \mathbf{x}}d\mathbf{x}, \tag{8.20b}$$

$$\widehat{I}^*(\mathbf{u}_1 + \mathbf{u}_2) = \int_{-\infty}^{\infty} I(\mathbf{x})e^{i2\pi(\mathbf{u}_1 + \mathbf{u}_2) \cdot \mathbf{x}}d\mathbf{x}. \tag{8.20c}$$

The object bispectrum is given by,

$$\widehat{I}_O(\mathbf{u}_1, \mathbf{u}_2) = \widehat{O}(\mathbf{u}_1)\widehat{O}^*(\mathbf{u}_1 + \mathbf{u}_2)\widehat{O}(\mathbf{u}_2)$$
$$= \frac{\langle \widehat{I}(\mathbf{u}_1)\widehat{I}^*(\mathbf{u}_1 + \mathbf{u}_2)\widehat{I}(\mathbf{u}_2) \rangle}{\langle \widehat{S}(\mathbf{u}_1)\widehat{S}^*(\mathbf{u}_1 + \mathbf{u}_2)\widehat{S}(\mathbf{u}_2) \rangle}. \tag{8.21}$$

The modulus $|\widehat{O}(\mathbf{u})|$ and phase $\psi(\mathbf{u})$ of the object FT $\widehat{O}(\mathbf{u})$ can be evaluated from the object bispectrum $\widehat{I}_O(\mathbf{u}_1, \mathbf{u}_2)$. The argument of (8.19) gives the phase-difference and is expressed as,

$$arg|\widehat{I}^{TC}(\mathbf{u}_1, \mathbf{u}_2)| = \psi(\mathbf{u}_1) + \psi(\mathbf{u}_2) - \psi(\mathbf{u}_1 + \mathbf{u}_2). \tag{8.22}$$

---

[1] Shift-and-add method (Lynds et al. 1976; Worden et al.. 1976) involves calculation of the differential shifts of the short-exposure images. The image data frame is shifted so that the pixel with maximum S/N ratio in each frame can be co-added linearly at the same location in the resulting accumulated image. This provides an image with higher resolution (higher signal-to-noise at high spatial frequencies) than a long-exposure image.

The object phase-spectrum is encoded in the term $e^{i\theta_O^{TC}(\mathbf{u}_1,\mathbf{u}_2)}$. It is corrupted in a single realization by the random phase-differences due to the atmosphere-telescope OTF,

$$e^{i\theta_S^{TC}(\mathbf{u}_1,\mathbf{u}_2)} = e^{i[\psi_S(\mathbf{u}_1)-\psi_S(\mathbf{u}_1+\mathbf{u}_2)+\psi_S(\mathbf{u}_2)]}. \tag{8.23}$$

The S/N ratio of the phase recovery contains S/N ratio for bispectrum, as well as a factor representing improvement due to redundancy of phase-information stored in the bispectrum. If sufficient number of specklegrams are averaged, one can overcome this shortcoming. Let $\theta_O^{TC}(\mathbf{u}_1,\mathbf{u}_2)$ be the phase of the object bispectrum; then,

$$\widehat{O}(\mathbf{u}) = |\widehat{O}(\mathbf{u})|e^{i\psi(\mathbf{u})}, \tag{8.24}$$

$$\widehat{I}_O(\mathbf{u}_1,\mathbf{u}_2) = |\widehat{I}_O(\mathbf{u}_1,\mathbf{u}_2)|e^{i\theta_O^{TC}(\mathbf{u}_1,\mathbf{u}_2)}. \tag{8.25}$$

Equations (8.24) and (8.25) may be inserted into (8.21), yielding the relations,

$$\widehat{I}_O(\mathbf{u}_1,\mathbf{u}_2) = |\widehat{O}(\mathbf{u}_1)||\widehat{O}(\mathbf{u}_2)||\widehat{O}(\mathbf{u}_1+\mathbf{u}_2)|$$
$$\times e^{i[\psi_O(\mathbf{u}_1)-\psi_O(\mathbf{u}_1+\mathbf{u}_2)+\psi_O(\mathbf{u}_2)]} \rightarrow, \tag{8.26}$$

$$\theta_O^{TC}(\mathbf{u}_1,\mathbf{u}_2) = \psi_O(\mathbf{u}_1) - \psi_O(\mathbf{u}_1+\mathbf{u}_2) + \psi_O(\mathbf{u}_2). \tag{8.27}$$

Equation (8.27) is a recursive one for evaluating the phase of the object FT at coordinate $\mathbf{u} = \mathbf{u}_1 + \mathbf{u}_2$.

The reconstruction of the object phase-spectrum from the phase of the bispectrum is recursive in nature. The object phase-spectrum at $(\mathbf{u}_1 + \mathbf{u}_2)$ can be written as,

$$\psi_O(\mathbf{u}_1+\mathbf{u}_2) = \psi_O(\mathbf{u}_1) + \psi_O(\mathbf{u}_2) - \theta_O^{TC}(\mathbf{u}_1,\mathbf{u}_2). \tag{8.28}$$

If the object spectrum at $\mathbf{u}_1$ and $\mathbf{u}_2$ is known, the object phase-spectrum at $(\mathbf{u}_1 + \mathbf{u}_2)$ can be computed. The bispectrum phases are mod 2π, therefore, the recursive reconstruction in (8.19b) may lead to π phase mismatches between the computed phase-spectrum values along different paths to the same point in frequency space. However, according to Northcott et al. (1988), phases from different paths to the same point cannot be averaged to reduce noise under this condition. A variation of the nature of computing argument of the term, $e^{i\psi_O(\mathbf{u}_1+\mathbf{u}_2)}$, is needed to obtain the object phase-spectrum and (8.28) translates into,

$$e^{i\psi_O(\mathbf{u}_1+\mathbf{u}_2)} = e^{i[\psi_O(\mathbf{u}_1)+\psi_O(\mathbf{u}_2)-\theta_O^{TC}(\mathbf{u}_1,\mathbf{u}_2)]}. \tag{8.29}$$

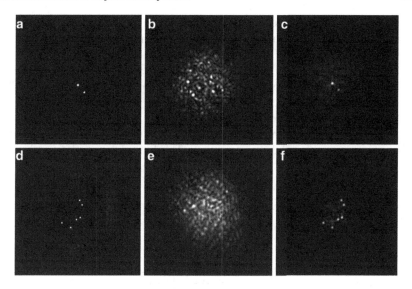

Fig. 8.6 Bispectrum recovery from simulated speckle images of a binary system and a cluster of six objects of a hypertelescope (see Sects. 7.5.2–7.5.3); 40 frames of speckle simulations were taken for analysis: (**a**) binary system (brightness ratio 3 : 1, (**b**) a convolved speckle frame of the binary, (**c**) reconstructed image through speckle masking, (**d**) cluster of six objects (equal brightness), (**e**) a convolved speckle frame of the cluster, and (**f**) reconstructed cluster image. The simulation used 200 random apertures inside a disk, which was further densified with densified aperture diameter: disk diameter ratio is taken as 1 : 5

The values obtained using the unit amplitude phasor recursive re-constructor are insensitive to the π phase ambiguities. Saha et al. (1999b) have developed a code based on this re-constructor. Application of this method is described in detail in a recent book (Saha 2007). Figure 8.6 demonstrates the reconstructed images of a few simulated objects.

8.2.3 Blind Iterative Deconvolution (BID) Technique

Most deconvolution techniques, in which 'a priori information' plays an essential role, can be simplified to the minimization /maximization of a criterion by using a constrained iterative numerical method (Gerchberg and Saxton 1972). Essentially, the Blind Iterative Deconvolution (BID) technique consists of using very limited information about the image, like positivity and image size, to iteratively arrive at a deconvolved image of the object, starting from a blind guess of either the object or both the object and convolving function. The iterative loop bounces back and forth and is repeated enforcing image-domain and Fourier-domain constraints until two images are found that produce the input image when convolved together (Ayers and Dainty 1988). The image-domain constraints of non-negativity is generally used in iterative algorithms associated with optical processing to find effective supports of the object and or PSF from a specklegram. The implementation of the algorithm of BID runs as follows.

Let the degraded image, $I(\mathbf{x})$, be used as the operand. An initial estimate of the PSF, $S(\mathbf{x})$, has to be provided. The image is deconvolved from the guess PSF by Wiener filtering (see Sect. 4.2.1), which is an operation of multiplying a suitable Wiener filter (constructed from $\widehat{S}(\mathbf{u})$, of the PSF) with $\widehat{I}(\mathbf{u})$. The technique of Wiener filtering damps the high frequencies and minimizes the mean square error between each estimate and the true spectrum. This filtered deconvolution takes the form,

$$\widehat{O}(\mathbf{u}) = \widehat{I}(\mathbf{u}) \frac{\widehat{O}_f(\mathbf{u})}{\widehat{S}(\mathbf{u})}. \tag{8.30}$$

The Wiener filter, $\widehat{O}_f(\mathbf{u})$, is derived as,

$$\widehat{O}_f(\mathbf{u}) = \frac{\widehat{S}(\mathbf{u})\widehat{S}^*(\mathbf{u})}{|\widehat{S}(\mathbf{u})|^2 + |\widehat{N}(\mathbf{u})|^2}. \tag{8.31}$$

The noise term, $\widehat{N}(\mathbf{u})$ can be replaced with a constant estimated as the root-mean-square fluctuation of the high frequency region in the spectrum where the object power is negligible. The Wiener filtering spectrum, $\widehat{O}(\mathbf{u})$, takes the form:

$$\widehat{O}(\mathbf{u}) = \widehat{I}(\mathbf{u}) \frac{\widehat{S}^*(\mathbf{u})}{\widehat{S}(\mathbf{u})\widehat{S}^*(\mathbf{u}) + \widehat{N}(\mathbf{u})\widehat{N}^*(\mathbf{u})}. \tag{8.32}$$

The result, $\widehat{O}(\mathbf{u})$, is transformed back to image space, the negatives in the image and the positives outside a prescribed domain (called object support) are set to zero. The average of negative intensities within the support are subtracted from all pixels. The process is repeated until the negative intensities decrease below the noise. A new estimate of the PSF is next obtained by Wiener filtering $I(\mathbf{x})$, with a filter constructed from the constrained object, $O(\mathbf{x})$; this completes one iteration. This entire process is repeated until the derived values of $O(\mathbf{x})$ and $S(\mathbf{x})$ converge to sensible solutions.

The comparative analysis of the recovery reveals that both the morphology and the relative intensities are present in the retrieved diffraction-limited image and PSF. This scheme of BID has the chief problem of convergence. It is indeed an art to decide when to stop the iterations. The results are also vulnerable to the choice of various parameters like the support radius, the level of high frequency suppression during the Wiener filtering, etc. The availability of prior knowledge on the object through autocorrelation of the degraded image is very useful for specifying the object support radius (Saha and Venkatakrishnan 1997; Saha 1999). In spite of this care taken in the choice of the support radius, the point spread function (PSF) for each star contains residual signatures of the binary sources. Although suggestions have been made for improving the convergence (Jefferies and Christou 1993), however, these improved algorithms require more than a single speckle frame.

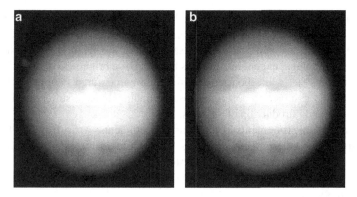

Fig. 8.7 (**a**) A specklegram of the Jupiter obtained on 24th. July 1994, through the green filter centered at 5,500 Å, with FWHM of 300 Å during its collision of comet, Shoemaker–Levy 9 (SL 9), at the Nasmyth focus of the 1.2 m telescope of Japal–Rangapur Observatory, Hyderabad (Saha and Venkatakrishnan 1997), and (**b**) its reconstructed image with BID

A major advantage of the BID is its ability to retrieve the diffraction-limited image of an object from a single specklegram without the reference star data. Often, it may not be possible to gather a sufficient number of images within the time interval over which the statistics of the atmospheric turbulence remains stationary. However, such an algorithm requires high S/N ratios in the data. This technique is indeed an ideal one to process the degraded images of extended objects. Figure 8.7 displays the impact of the collision of Jupiter with the fragments of the comet Shoemaker–Levy 9 (SL 9; 1993e[2]).

8.3 Aperture Synthesis Mapping

Aperture synthesis technique is the most powerful indirect imaging technique, which has been employed for many years in radio interferometry for obtaining high resolution images. Instead of measuring the image directly, this technique measures its Fourier components which form the required image after Fourier inversion. However, this image degrades mainly due to two reasons:

1. Inadequate sampling of the Fourier plane (u, v-plane): In order to obtain an image of decent quality and reliability, the coverage of the u, v-plane has to fulfill the

---

[2] During the period of July 16–22, 1994, the fragments of the comet Shoemaker–Levy 9 after breaking up under the influence of Jupiter's tidal forces, collided with Jupiter with spectacular results. The fragments closest to Jupiter fell with a greater acceleration, due to the greater gravitational force. Observations of the crash phenomena have been carried out extensively. Speckle interferometry was also employed to get the finer details. Saha and Venkatakrishnan (1997) have identified the complex spots due to impacts by the fragments using the Blind Iterative Deconvolution (BID) technique.

Nyquist sampling theorem, requiring a high number of baselines; all the Fourier coefficients up to infinite spatial frequency are required to be sampled. However, owing to finite size and non-uniform sampling of the synthesized aperture, the synthesized image, in general, deviates from the true brightness distribution. The reason being that while Fourier inverting, the unmeasured Fourier coefficients are naturally assumed to be zero. Thus, the obtained image suffers from loss of resolution and introduction of artifacts.
2. Errors in the measurements of the Fourier coefficients: Apart from the fact that the aperture synthesis technique does not sample all the Fourier coefficients, the measurements are affected by systematic and random errors. In the presence of these errors a simple deconvolution does not provide a good image reconstruction.

Reconstruction of complex images involves the knowledge of complex visibilities. Process after data acquisition consists of phase calibration and visibility phase reconstruction from closure-phase terms (see Sect. 5.3.1) by techniques similar to bispectrum processing. From complex visibilities acquired from a phased interferometric array, it is possible to reconstruct the image by actually interpolating the function in the (u, v) plane. This has been done in radio-interferometry for a few decades. Let the output of a synthesis array be a set of visibility functions $\mathcal{V}(u_j, v_j)$ that are obtained by averaging the quasi-sinusoidal response of each interferometer pair and hence, the resultant brightness distribution be given by the equation:

$$I'(x, y) = \sum_{j=1}^{N} \mathcal{V}(u_j, v_j) w_j e^{[-2\pi i (u_j x + v_j y)]}, \tag{8.33}$$

where $I'(x, y)$ is the dirty image which is the resultant of the true image $I(x, y)$ convolved by the response to an unresolved (point) source, $S'(x, y)$, and w_j the weight associated with $j^{th} N$.

The map may show sufficient details when the (u, v) coverage is good, but it is not the best representation of the sky and contains many artifacts notably the positive and negative side lobes around bright peaks. If the sampling of the (u, v) plane is irregular and uneven, the dirty beam will have large sidelobes and confuse and obscure the structure of interest in the image.

Image restoration is a kind of deconvolution problem, which can be carried out in the spatial domain based on (4.70) or in spatial frequency domain (4.72) or in both domains. A variety of non-linear image retrieval methods have been introduced to restore the unmeasured Fourier components in order to produce a physical map since the direct inversion method is generally not feasible. These methods work on the constraint that the image must be everywhere positive or zero and they make use of a priori knowledge about the extent of the source and statistics of the measurement processes. In reality, a kind of model fitting along with a prior knowledge about the image such as positivity, makes the deconvolution possible. The most commonly used approaches to deconvolve dirty image fall into two groups: (1) CLEAN algorithm that was introduced by Högbom (1974), and was later mathematically modified by Schwarz (1978) and (2) Maximum Entropy Method (MEM), which was introduced by Ables (1974); its subsequent development brought it to a satisfactory level (Wernecke and D'Addario 1976; Gull and Daniell 1978; Cornwell and Evans 1985). To some extent, CLEAN and MEM are complimentary in their application. It is found that CLEAN generally performs well on small compact sources, while the MEM does better on extended sources.

8.3.1 CLEAN

The CLEAN algorithm is the most routinely used, particularly in radio astronomy, because it is both computationally efficient and intuitively easy to understand. CLEAN deconvolves the images that consist of a few point like sources and its success lies in the 'empty' nature of the field-of-view, and therefore, is not suited for extended sources. This procedure is a non-linear technique that applies iterative beam removing method. Mathematically, it is a least-square data fitting technique and operates in the map plane and which uses the known shape of the dirty beam to distinguish between real structure and sidelobe disturbances of the dirty map.

CLEAN subtracts iteratively, from the image given by the inverse Fourier Transform of the visibilities measured on the object, the dirty map, a fraction of the image given by the array from an unresolved source, the dirty beam, centered on the maximum value of the dirty map. It reconstructs image from (u, v) plane samples, which begins with a dirty image $I'(x, y)$ made by the linear Fourier inversion procedure and attempts to decompose this image into a number of components, each of which contains part of the dirty beam, $S'(x, y)$.

The true image of a radio source is represented by a number of point sources in an otherwise empty sky. In this algorithm, the positions and intensities of these point sources are found by a simple interactive approach. Initially, the peak in the dirty image, $I'(\mathbf{x})$, is found and subtracted from $I'(\mathbf{x})$ the dirty beam multiplied by the peak intensity with a loop gain < 1, at the position of the peak. This is followed by the selection of the next strongest peak in the residual image and so on till the residual image is at the noise level. However, this simple algorithm is subject to instability problems leading sometimes to wrong results. One wishes to determine the set of numbers $A_i(x_i, y_i)$ such that:

$$I'(x, y) = \sum_i A_i S'(x - x_i, y - y_i) + I_R(x, y), \qquad (8.34)$$

where $I_R(x, y)$ is the residual brightness distribution after the decomposition. The solution is considered satisfactory if $I_R(x, y)$ is of the order of the expected noise.

The algorithm searches the dirty image for the pixel with largest absolute value I_{max} and subtracts a dirty beam pattern centered on this point with amplitude $G_l I_{max}$, where G_l is called the loop gain. The residual map is searched for the next largest value and the second beam stage is subtracted and so on. The iteration is stopped when the maximum residual is consistent with the noise level. The iteration consists of a residual image that contains noise and low level source calculations plus a set

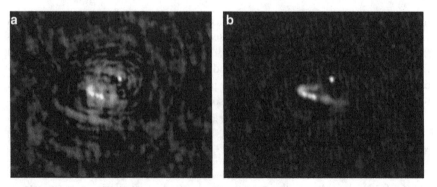

Fig. 8.8 (a) Dirty image and (b) CLEANed image of the radio source, $0916 + 6348$ at 244 MHz (Sirothia et al. 2009). This raw image was obtained by using GMRT synthesis array (courtesy: Ishwara Chandra)

of amplitudes and positions of the components removed. These components can be considered as an array of delta function and convolved with clean beams and added to produce a CLEANed image. The clean beam is usually chosen as a truncated elliptical Gaussian about the same size of the main lobe of the dirty beam. CLEAN results in a map with the same resolution as the original dirty map without sidelobes.

Figure 8.8 depicts a reconstructed image of a source, $0916 + 6348$ (Sirothia et al. 2009) using CLEAN algorithm. Another important aspect is to have polarization mapping. However, it may not be feasible to measure all four correlations (see Sect. 1.3.1) by all interferometric pairs due to instrumental problems (Weiler 1973), which in turn, provides different spatial coverage and therefore different synthesized beams for different polarization components. Owing to different convolving beams for the four components, it is possible to encounter a situation when some sources in the observed map exhibit greater than 100% polarization. In intensity imaging, these spurious sources are removed by CLEAN algorithm.

8.3.2 Bayesian Statistical Inference

The name 'Bayesian' comes from the Bayes' theorem, derived from the work of Bayes (1764), in the inference process. This inference is statistical inference where observations are used to infer the probability that a hypothesis may be true. Bayes formula provides the mathematical tool that combines prior knowledge with current data to produce a posterior distribution. Bayes' theorem about conditional probabilities states that,

$$\mathcal{P}(B|A)\mathcal{P}(A) = \mathcal{P}(A|B)\mathcal{P}(B)$$
$$= \mathcal{P}(AB), \tag{8.35}$$

which follows product rule, $\mathcal{P}(B)$ is the prior probability of B, A and B the outcomes of random experiments, B represents a hypothesis, referred to as a null hypothesis, which is inferred before new evidence, A, became available, and $\mathcal{P}(B|A)$ the probability of B given that the hypothesis A has occurred and for imaging, $\mathcal{P}(A|B)$ is the likelihood function[3] of the data given B, and $\mathcal{P}(A)$ the marginal or unconditional probability of A, a constant which normalizes $\mathcal{P}(B|A)$ to a sum of unity, and provides the probability of the data.

The term $\mathcal{P}(A)$ may be expanded by using the law of total probability,

$$\mathcal{P}(A) = \sum_{j=1}^{n} \mathcal{P}(A|B_j)\mathcal{P}(B_j), \tag{8.36}$$

with the events B_j being mutually exclusive and exhausting all possibilities and including the event B as one of the B_j.

This theorem (8.36) is a consequence of the definitions of joint probabilities denoted by $\mathcal{P}(AB)$, conditional probabilities, $\mathcal{P}(A|B)$, and marginal probabilities, $\mathcal{P}(B)$. On rearranging (8.35) Bayesian approaches may be expressed as,[4]

$$\mathcal{P}(B|A) = \frac{\mathcal{P}(A|B)\mathcal{P}(B)}{\mathcal{P}(A)} = \mathcal{L}(B|A)\mathcal{P}(B), \tag{8.37}$$

with $\mathcal{L}(B|A)$ as the likelihood of B given fixed A; in this case, the relationship $\mathcal{P}(A|B) = \mathcal{L}(B|A)$.

8.3.3 Maximum Entropy Method (MEM)

The Maximum Entropy Method (MEM) is commonly employed in astronomical synthesis imaging, where the resolution depends on the signal-to-noise (S/N) ratio, which must be specified. Therefore, resolution is image dependent and varies across

---

[3] A likelihood function, $\mathcal{L}(\theta)$, is the probability density for the occurrence of a sample configuration, $x_1, x_2, \cdot x_N$, given that the probability density $f(x_1, \cdot x_N|\theta)$ with parameter, θ is known (Harris and Stocker 1998),

$$\mathcal{L}(\theta) = f(x_1, x_2, \cdot, x_N|\theta).$$

[4] In terms of probability density function, Bayes' formula takes the form,

$$g(\lambda|x) = \frac{f(x|\lambda)g(\lambda)}{\int_0^\infty f(x|\lambda)g(\lambda)d\lambda},$$

where $f(x|\lambda)$ is the probability model, or likelihood function, for the observed data x given the unknown parameter, $g(\lambda)$ the prior distribution model for λ, and $g(\lambda|x)$ the posterior distribution model for λ given that the data x have been observed.

the map. This procedure governs the estimation of probability distributions when limited information is available. In addition, it treats all the polarization component images simultaneously (unlike CLEAN which deconvolves different polarization component images independently) and guarantees essential conditions on the image. MEM is also biased, since the ensemble average of the estimated noise is non-zero. Such a bias is smaller than the noise for pixels with a $S/N \gg 1$. It may yield super-resolution, which can be trusted to an order of magnitude in solid angle. MEM is employed in a variety of other fields like medical imaging, crystallography as well.

The principle of maximum entropy[5] was first expounded by Jaynes (1957), where he suggested that the thermodynamic entropy may be seen as a particular application of a general tool of inference and information theory[6](Shannon 1948). This concept was introduced for obtaining high resolution spectra by Burg (1967). Later, it was adopted in image processing by Frieden (1972). The main difficulty in the image processing is that for a given set of measurements, one does not invariably find a unique image.

The basic philosophy of MEM is to use the entropy of the image as a quality criterion and to obtain a physically acceptable image which is consistent with the measurements. Several authors have discussed the justification for MEM and the word 'entropy' in image processing (Frieden 1972; Gull and Daniell 1978; Jaynes 1982; Nityananda and Narayan 1982; Cornwell and Evans 1985). The entropy can be thought of as a means of providing *apriori* knowledge about the image. The minimum *apriori* knowledge one can have is that the image (considering total intensity rather than polarization) is positive with a compressed range in pixel values. Therefore, the MEM must reconstruct an image which is at least positive. For a given measurement, this method should provide a unique image irrespective of the user and the starting guess of the image. In order to define the entropy of an image, these constraints permit the usage of a few functions such as, $f(I) = \ln I$ (for Burg's form of entropy); $f(I) = -I \ln I$ (for Shannon's form of entropy). Besides these two functions there are a few more like $I \ln(I/I_0)$, $(I/I_0)^{1/2}$ in which I is the image intensity and I_0 the prior model of the image, etc., which may provide good measure of entropy from the view point of image reconstruction (Nityananda and Narayan 1982; Shevgaonkar 1986). All these functions used for explaining entropy share the following common properties (Shevgaonkar 1987):

- The entropy function $f(I)$ provides a non-linear restoration filter, i.e., $\partial f/\partial I \neq$ constant.
- The function is explicitly definable for positive values (positive constraint).
- The function possesses a single maximum over the acceptable range of the intensity, i.e., $(\partial^2 f/\partial I^2) < 0$; the uniqueness condition.

---

[5] In physics, the word entropy has important physical implications as the amount of disorder of a system.

[6] Information theory is the branch of mathematics dealing with the efficient and accurate storage, transmission, and representation of information.

The MEM, according to Jaynes (1982), governs the estimation of probability distribution when limited information is available. If one possesses some linear constraints upon a probability distribution $\mathcal{P}(x)$ in the form of K expected values:

$$\int_{-\infty}^{\infty} \mathcal{P}(x) f_k(x) dx = \langle f_k \rangle; \qquad k = 1, 2, \cdots K, \tag{8.38}$$

and that one wishes to assign a form to the probability distribution, which is optimal and agrees with the constraints. Here $\mathcal{P}(x)$ is the probability and $\mathcal{P} \ln \mathcal{P}$ is defined as 0 if $\mathcal{P} = 0$. The image selected is that which fits the data, to within the noise level, and also has maximum entropy. While enforcing positivity and conserving the total flux in the frame, smoothness is estimated by the entropy, S, that is of the form,

$$S(\mathcal{P}) = -\int_{-\infty}^{\infty} \mathcal{P}(x) \ln \left[\frac{\mathcal{P}(x)}{\mathcal{Q}(x)} \right] dx, \tag{8.39}$$

where $\mathcal{Q}(x)$ represents a prior probability function.

The entropy is dependent upon the meaning of brightness, which governs the probability of emission of waves or photons (D'Addario 1976). From the information theory (Shannon 1948), the entropy, which is a measure of ignorance about the system is defined in continuous case as,

$$S_{ph} = \int_{-\infty}^{\infty} I(\mathbf{x}) \ln I(\mathbf{x}) d\mathbf{x}, \tag{8.40a}$$

$$S_{waves} = \int_{-\infty}^{\infty} \ln I(\mathbf{x}) d\mathbf{x}, \tag{8.40b}$$

in which $\mathbf{x}$ is the orthogonal image coordinate, and in the discrete (or quanta) case,

$$S_{quanta} = -\sum_{j=1}^{n} I_j \ln I_j, \tag{8.41}$$

where, S_{quanta} is the configurational entropy; S_{quanta} is prominent amongst astronomers, while S_{waves} is used by electronics technocrats.

The maximum entropy method solves the multi-dimensional constraints minimization problem. It uses only those measured data and derives a brightness distribution which is the most random, i.e., has the maximum entropy S of any brightness distribution consistent with the interferometric data. Maintaining an adequate fit to the data, it reconstructs the final image that fits the data within the noise level. Assuming the intensity is uncorrelated from one pixel to other, the entropy of an image is equal to the sum of the entropies of individual pixels. Mathematically, the entropy is written as,

$$S = \int_{-\infty}^{\infty} f[I(\mathbf{x})] d\mathbf{x}. \tag{8.42}$$

If the measurements are carried out in the Fourier plane, consistency with the measured data requires (Narayan and Nityananda 1986),

$$\sum_N \frac{|\rho - \hat{\rho}|^2}{\sigma^2} \leq N, \tag{8.43}$$

where ρ is the measured Fourier coefficient (or visibility coefficient), $\hat{\rho}$ the true visibility coefficient, σ the root-mean-square of the statistical random noise of the measurement, an N the total number of measured data points.

On combining (8.42) and (8.43) to define an objective function F (Ables 1974; Wernecke and D'Addario 1976) which when maximized produces an image having maximum entropy,

$$F = S - \frac{\lambda}{N} \sum_N \frac{|\rho - \hat{\rho}|^2}{\sigma^2}, \tag{8.44}$$

in which λ is the Lagrange multiplier which essentially defines the relative weights given to the two terms in (8.44). A proper value of λ can be obtained by trial and error method.

Monnier et al. (2001) have reconstructed the dust shells around two evolved stars, IK Tauri and VY CMa using (u, v) coverage from the contemporaneous observations at Keck-I and IOTA. Figure 8.9 depicts the MEM reconstructions of the dust shells around these stars. Their results clearly indicate that without adequate spatial resolution, it is improbable to cleanly separate out the contributions of the star from the hot inner radius of the shell (left panel in Fig. 8.8). They opined that image reconstructions from the interferometer data are not unique and yield results which depend heavily on the biases of a given reconstruction algorithm. By including the long baselines (>20 m) data from the IOTA interferometer (right panel), the fraction of the flux arising from the central star can be included in the image reconstruction process by using the MEM prior. One can see for a dust shell such as IK Tau, that additional IOTA data are critical in accurately interpreting the physical meaning of interferometer data. The thick dashed lines show the expected dust shell inner radius from the data obtained at the Infrared Spatial Interferometer.

8.3.4 Self-calibration Method

A data correcting method, called the 'self-calibration method' (Readhead and Wilkinson 1978; Readhead et al. 1980; Pearson and Readhead 1984; Cornwell and Fomalont 1989) is based on the error free nature of the closure phase introduced by Jennison (1958). Indeed the closure phase lies in the heart of self-calibration techniques. Such a technique, widely used at radio wavelengths, is essentially an iterative technique which is a combination of the image deconvolution method CLEAN and the closure quantities. The method begins with a model distribution

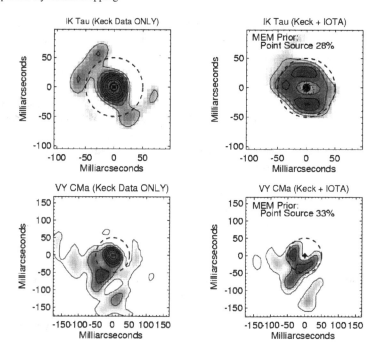

Fig. 8.9 MEM image reconstructions of the dust shell around IK Tauri and VY CMa (courtesy: J. D. Monnier); the *left panels* show the reconstructions of them from data obtained with a single Keck telescopes using aperture masking (baselines up to 9 m at 2.2 μm, and the *right panels* show the dust shell reconstructions when the fractional amounts of star and dust shell emission is constrained to be consistent with both the Keck and IOTA data (Monnier et al. 2001)

which predicts some of the visibility coefficients, and the rest are computed through closure relations. This method is generally used to improve the maps that are obtained from synthesis telescopes.

Self-calibration method takes advantage of the fact that many of the systematic errors in visibility measurements may be ascribed to individual array elements and has become a routine technique to correct the synthesis observations. In the real observations, as stated earlier that the measured visibilities, $\widehat{\mathcal{V}}(\mathbf{u})$, are contaminated by noise, as well as by errors in the antenna gains as function of time. The gains are derived for correct direction on celestial sphere as well; antenna gains may be found using least squares fit to visibility data. These errors reduce the quality and dynamic range of the final image. The error in the gain solution for a point source is,

$$\sigma_g = \frac{1}{\sqrt{n-2}} \frac{\sigma_{\mathcal{V}}}{S}, \qquad \text{for phase,} \tag{8.45a}$$

$$\sigma_g = \frac{1}{\sqrt{n-3}} \frac{\sigma_{\mathcal{V}}}{S}, \qquad \text{for amplitude and phase,} \tag{8.45b}$$

where $\sigma_{\mathcal{V}}$ is the noise per visibility sample, n the number of antennae, and S the strength of the unresolved point source.

The basic philosophy of the self-calibration method is to obtain a model of the sky intensity distribution. The method should be astronomically plausible, for example, possible constraints are positivity of the brightness and confinement of the structure. This method is essentially a combination of CLEAN and the closure phases. Since the $n(n-1)/2$ baselines are affected by n sources of error, for instance, antenna or IF gains, with a given a rough estimate of the true source visibility (by means of the first iteration CLEAN map), one may solve for the unknown antenna gains. This procedure involves adjusting the estimates of the n complex gain errors to minimize the mean square difference between the measured visibilities and true visibilities, the sum of the squares of residuals:

$$\chi^2 = \sum_{i<j} w_{ij} \left| \mathcal{V}_{ij} - G_i \bar{\mathcal{V}}_{ij} \right|^2 , \tag{8.46}$$

in which w_{ij} are the weighting factors decided from S/N ratio considerations; this should be set to the inverse of the variance of the noise, $G_{ij} (= g_i g_j^*)$ the gain factors associated with individual antennae, g_i and g_j the complex gain factors for the element, i and j, respectively, $\mathcal{V}_{ij}$ the measured complex visibility, and $\bar{\mathcal{V}}_{ij}$ the true visibilities.

This (8.46) is a non-linear least squares problem, which can be re-written as,

$$\chi^2 = \sum_{i<j} w_{ij} \left| \bar{\mathcal{V}}_{ij} \right|^2 \left| \chi_{ij} - g_1 g_j^* \right|^2 , \qquad \text{with} \tag{8.47}$$

$$\chi_{ij} = \frac{\mathcal{V}_{ij}}{\bar{\mathcal{V}}_{ij}}. \tag{8.48}$$

This step turns the object into a pseudo point source. The method can be solved by iterative approach (Schwab 1980). Deconvolution method CLEAN (Cornwell and Wilkinson 1981) or MEM (Sanroma and Estallela 1984) is used to derive a new map at every step (Pearson and Readhead 1984). In case of producing images with accurate visibility amplitude and poor or absence of phases, the technique of self-calibration method has proved to be very powerful in removing such errors as the atmospheric and instrumental errors. However, such a method requires a sufficiently bright source and introduces more degrees of freedom into the imaging so that the results may not become optimum. The other notable drawbacks are (1) it is sensitive to the choice of the model to some extent and (2) the method faces difficulties for extended object if CLEAN is used. For large field-of-view (FOV), closure quantities may be considered as an alternative to the Self-calibration method along with redundancy techniques (Wieringa 1991; Ramesh et al. 1999). Figure 8.10 depicts the radioheliogram of the solar corona obtained with the Gauribidanur radio heliograph (near Bangalore in India) using the latter scheme. The radio map is overlaid on the white light image of the corona obtained with the Large Angle Spectroscopic Coronagraph (LASCO) onboard the orbiting Solar and Heliospheric Observatory (SoHO).

Fig. 8.10 Radioheliogram of the 'halo' coronal mass ejection from the solar atmosphere observed with the Gauribidanur radioheliograph on 10 April 2006 at 115 MHz (courtesy: R. Ramesh)

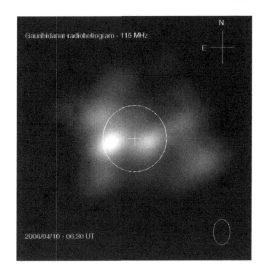

Cornwell and Wilkinson (1981) introduced a modification by explicitly solving for the telescope-specific error as part of the reconstructing step. The measured Fourier phases are fitted using a combination of intrinsic phases plus telescope phase errors. When the errors are expressed as antenna based, i.e., variations in the complex gains of individual antenna, it is possible to calibrate the antenna gains at any instant using a model for the sky brightness and the observed visibilities at a given instant. If the field of observation contains one dominating internal point source, which can be used as an internal phase reference, the visibility phase at other spatial frequencies is derived. In this method, the complex gains of individual antennae are corrected such that the visibilities computed by Fourier transforming the model sky brightness reproduce the observed visibilities within the noise level. A preliminary sky brightness obtained from the calibrated visibilities can be used as initial model. The solution to the antenna brightness obtained from the visibilities calibrated by the improved antenna gains as model sky brightness.

A hybrid map can be made with the measured amplitudes together with model phase distributions. Since the measured amplitudes differ from the single point source model, the hybrid map diffuses from the model map. Adding some new feature to the original model map, an improved model map is obtained in the next iteration. With clever selection of features to be added to the model in each iteration, the procedure converges.

Chapter 9
Astronomy with Diluted Aperture Interferometry

9.1 Astronomical Measurements

A star is a massive luminous ball of heated gases (mostly ionized) in space that produces tremendous amount of light and other forms of energy. Accurate determination and prediction of its physical properties, e.g., luminosity, temperature, mass, elemental abundance, etc., is a fundamental goal of astrophysics. Of these, stellar mass is the crucial parameter, which provides a benchmark for stellar evolution calculations. The relationship between stellar mass, luminosity, and radius over a range of metallicities[1] from the existing solar values to the very low metallicities representative of the early Universe essential to be known. The morphological details, such as granulations, oblateness of giant stars, and the image features i.e., spots and flares on their surfaces can be detected by means of Long Baseline Optical Interferometers (LBOI). High angular information is also useful to (1) study star-formation, (2) resolve binaries, and (3) detection and characterization of exoplanets.

A noted fundamental problem that can be addressed with LBOIs fitted with complete adaptive optics systems is the origin and evolution of galaxies. They would be able to provide imaging and morphological information on the faint extragalactic sources such as, galactic centers in the early Universe, deep fields, and host galaxies. Measurement of such objects may be made feasible by the instruments with a fairly complete u, v-coverage and large field of view. The derivations of motions and parallaxes of galactic centers seem to be feasible with phase reference techniques (Quirrenbach 2009). With interferometric polarization measurement, faint structures close to non-obscured central sources can also be studied in detail. Most of the astronomical objects posses spherical symmetry and their radiation follow Planck's law (see 3.17). The received light may get polarized if any asymmetry is introduced, for instance, by the presence of circumstellar disk around young stars,

---

[1] Metallicity of an astronomical object is the proportion of its matter made up of heavy (chemical) elements besides hydrogen and helium. In general, it is given in terms of the relative amount of iron and hydrogen present, as determined by analyzing absorption lines in a stellar spectrum, relative to the solar value. The ratio of the amount of iron to the amount of hydrogen in the object, [Fe/H], is calculated from the logarithmic formula given by, $[Fe/H] = \log(Fe/H)_\star - \log(Fe/H)_\odot$.

S.K. Saha, *Aperture Synthesis*, Astronomy and Astrophysics Library,
DOI 10.1007/978-1-4419-5710-8_9, © Springer Science+Business Media, LLC 2011

patchy distribution of dust in star-forming regions, etc. The polarization signatures of circumstellar disks around Young Stellar Objects (YSO) depend on location and wavelength. Spectrally and spatially resolved polarimetric interferometry allows one to resolve local polarized stellar features such as (1) abundance inhomogeneities and stellar rotation, (2) size, and number density distribution of dust grains, (3) scattering and mass-loss phenomena, and (4) magnetic field. These parameters, in turn, act as inputs to planet formation models (Elias II et al. 2008). Measurements of spatial coherence can be performed with such an instrument simultaneously with measurements of coherence between orthogonal polarization states. Other objects that may be targeted are main sequence stars, red- and super-giant stars, luminous blue variables, Be stars, eclipsing binaries such as Algol-type binary stars, AM Herculis binary stars, and Seyfert galaxies (Elias II et al. 2008).

The capabilities of the proposed large arrays offer a revolution in the study of compact astronomical sources from the solar neighborhood to cosmological distances. The aim ranges from detecting other planetary systems to imaging the black hole driven central engines of quasars and active galaxies; gamma-ray bursters may be the other candidate. Another important scientific objective is the recording of spectra to derive velocity and to determine black hole masses as a function of redshift.

9.1.1 Limiting Magnitude

Magnitude scale[2] is used to discuss the wavelength dependent flux density from a celestial source. It is a basic observable quantity for a star or other celestial quantities, and is defined as the logarithm of the ratio of brightnesses. The magnitude scale of such an object is a measure of its brightness as seen by a terrestrial observer (the apparent brightness). In general, the observed magnitudes of two stars at any given wavelength, denoted by m_1 and m_2 and their corresponding flux densities F_1 and F_2, are related by the following expression,

$$m_1 - m_2 = -2.5 \log \frac{F_1}{F_2}. \tag{9.1}$$

in which the subscripts 1 and 2 refer to two different spectral regions.

---

[2] The stellar magnitude system was introduced by the Greek astronomer Hipparchus as early as second Century BCE, who had divided the visible stars with six classes, according to their apparent brightness. He produced a catalog of over 1,000 stars, ranking them by 'magnitudes' one through six, from the brightest (first magnitude) to the dimmest (still visible to the unaided eye, which are of 6th magnitude). This type of classification was placed on a precise basis by N. R. Pogson. Pogson redefined the magnitude scale, a difference of five magnitudes was exactly a factor of 100 in light flux. The light flux ratio for a one magnitude difference is, therefore $\sqrt[5]{100} = 2.512$. Subsequently, the Pogson ratio became the standard method of assigning magnitudes.

The (9.1) relates the magnitudes and brightnesses of two object. The derived magnitude is known as apparent magnitude. The brightness of any celestial object is often derived in terms of observed flux density, $F_v (= W \, m^{-2})$. The 0.03 visual magnitude star, α Lyrae (Vega), a spectral type A0V star with surface temperature, $T = 9,602 \pm 180°K$ (Kinman and Castelli 2002), is presently taken to be a 'standard' star. The flux of such a star at 0.63 μm is 2.5×10^{-12} W cm^{-2} per μm bandwidth (Johnson 1966) and the photon energy, $h\nu$, is derived as 3.5×10^{-19} joules.

Another term, called absolute magnitude, is also used, which is measured in terms of its intrinsic brightness. It is defined as the apparent magnitude that a star has when placed at a distance of 10 parsecs[3] (pc) from the Earth. The magnitude difference is given by,

$$m - M = -2.5 \log \frac{F(r)}{F(10)} = 5 \log \frac{r}{10 \, pc}. \tag{9.2}$$

If the distance is expressed in parsecs, (9.2) may be expressed as,

$$m - M = 5 \log r - 5. \tag{9.3}$$

The apparent magnitude m of a star is related to its distance, r and M given in (9.3). The distance r is expressed as,

$$r = 10^{0.2(m-M+5)} \quad \text{pc}, \tag{9.4a}$$
$$= \beta 10^{0.2(m-M+5)} \quad \text{m}, \tag{9.4b}$$

with $\beta (= 3.086 \times 10^{16})$ as the number of meters in a parsec.

The color of a star is defined by a ratio of brightness in two different wavelengths. As determined by the measurements at two different spectral regions give information about the star's temperature. Astronomers have defined several filter systems comprising of a number of standardized filters, which are being employed to measure a star's brightness in some portion of the spectrum. The peaks for transmission for these filters are in the ultraviolet (U), blue (B), visible (V), red (R), and infrared (I), respectively. The UBV system, also referred as Johnson system, is a photometric system for classifying stars according to their colors (Johnson and Morgan 1953). In order to measure the color index one observes the magnitude of an object successively through two different filters. For example, the ($B - V$) provides the color index, which is the difference between the magnitudes in the blue and visual regions of the spectrum and the ($U - B$) color index is the analogous difference between the ultraviolet and blue regions of the spectrum and so on. The UBV system provides

---

[3] Parsec (parallax second; pc) is defined by the distance to a star that exhibits a parallax angle of one arcsecond employing a baseline of 1 AU (the average distance between the Earth and the Sun). For instance, at a distance of 100 pc, the parallax is 10 milliarcseconds (mas).

only the V magnitude and the color indices $(U - B)$ and $(B - V)$. The zero point of the color indices, $(B-V)$ and $(U-B)$, are defined by six stars of spectral type $A0 \, V$. These stars are α Lyrae, γ UMa, 109 Virgo, α CrB, γ Ophiuchi, and HR 3314. The Sun has a $(B - V)$ index of 0.656 ± 0.005, while the average index of these stars is defined to be zero, i.e., $(B - V) = (U - B) = 0$ (Johnson and Morgan 1953). The energy flux above the Earth's atmosphere due to a star of $B - V = 0 = m_B - m_V$ emits flux, $F_0 = 10^8$ photons s^{-1} m^{-2} Hz^{-1} at 550 nm.

For extended source, for example sky background, one gives the specific intensity, I_ν, that has the units of J s^{-1} m^{-2} Hz^{-1} sr^{-1}. To note, extinction causes the apparent reduction in the brightness of a celestial object. This is due to the absorption and scattering of electromagnetic radiation by the interstellar dust particles. The light intensity gets reduced by traverse through the terrestrial atmosphere as well due to scattering and absorption by air molecules and aerosols. The amount of light loss depends on the position of the star above the horizon, the wavelength of observation, and the prevailing atmospheric conditions. This quantity is described by the term extinction. The effects due to both interstellar, as well as atmospheric extinctions should be taken into account for recovering stellar intrinsic properties. The atmospheric extinction needs to be corrected when calibrating instrumental magnitudes. The measured signal, the photon fluxes, from the star, $F_\star$ and from the sky background, F_{sky} at the focal plane of a telescope (within the selected spectral region, $\Delta\lambda$) photons s^{-1}:

$$F_\star = F_0 \eta_{tr} \eta_d \, D^2 \Delta\lambda 10^{-0.4} k \frac{\pi(1 - \epsilon)}{4}, \tag{9.5}$$

$$F_{sky} = I_\nu \theta^2 \eta_{tr} \eta_d \, D^2 \Delta\lambda 10^{-0.4} k \frac{\pi(1 - \epsilon)}{4}, \tag{9.6}$$

where D is the diameter of the telescope, η_{tr} the unitless transmittance of the atmosphere, telescope, instrument (filters, grating, and other optics), η_d the quantum efficiency of the detector, k the extinction coefficient, θ the seeing in radians, $\Delta\lambda$ the bandpass, ϵ the obscuration ratio, i.e., the ratio of the size of the secondary mirror to the size of the primary mirror, I_ν the specific intensity of radiation that is defined as the amount of energy passing a unit area of surface, perpendicular to the direction of propagation, per unit time, per unit solid angle, per unit frequency interval; I_ν, is measured in photons s^{-1} m^{-2} nm^{-1} arcsec^{-2}.

To note, the range of wavelength sensitivity of the human eye ranges from 400 to 625 nm (a range falls under visual or V-band). These limits represent the wavelengths at which the sensitivity of a dark-adapted eye falls to a value of about 0.01 relative to the wavelength of maximum sensitivity (Hughes 1983). Within the area of stellar image, one measures light from both the sky background and the star.

9.1.2 Stellar Luminosity

Stars have an overall spectrum showing evidence of emission and absorption processes in lines as well as in the continuum. The overall profile of the spectrum is governed largely by the black body (3.17). The brightness of a celestial object can be derived in terms of observed flux density, $F_\nu (= \text{W m}^{-2})$, the energy received per unit time per unit telescope area per unit frequency. The flux, F is defined as the total amount of radiation crossing a unit area of surface in all directions and in unit frequency interval.

Figure 9.1 depicts the relation between the intensity, I_ν and the energy entering the solid angle element, $d\Omega$. The total flux density is given by,

$$F = I \int_{\theta=0}^{\pi/2} \int_{\phi=0}^{2\pi} \cos\theta \sin\theta \, d\theta \, d\phi$$
$$= \pi I, \quad \text{for isotropic radiation,} \tag{9.7}$$

the flux, $I = \int_0^\infty I_\nu d\nu$, the total intensity, I_ν the specific intensity of radiation over the entire solid angle element, $d\Omega = \sin\theta \, d\theta \, d\phi$, ϕ the azimuthal angle, and θ the polar angle between the solid angle $d\Omega$ and the normal to the surface.

The luminosity, an intrinsic constant independent of distance, is defined as the amount of energy a star radiates per unit time in the form of electromagnetic radiation expressed in watts, or in terms of solar luminosity, $\mathcal{L}_\odot (= 3.839 \times 10^{26} \text{ W})$. Unlike the observed apparent brightness that is related to distance with an inverse square relationship, luminosity is an intrinsic constant independent of distance. Consider a star of absolute magnitude, $M_\star$ at distance d parsecs from the Earth, hence its luminosity, $\mathcal{L}$ is given by,

$$\mathcal{L} = \mathcal{L}_\odot 10^{0.4(M_{bol\odot} - M_\star)} \quad \text{W,} \tag{9.8}$$

where $M_{bol\odot}$ are the absolute bolometric magnitude of the Sun.

A star radiates isotropically, therefore its radiation at a distance r is distributed evenly on a spherical surface whose area is $4\pi r^2$. If the flux density of the radiation passing through the surface is F, the luminosity is written as,

$$\mathcal{L} = 4\pi r^2 F. \tag{9.9}$$

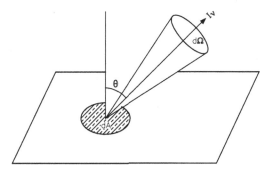

Fig. 9.1 Radiation from a small aperture depicting the relation between the specific intensity, I_ν and the energy entering the solid angle element, $d\Omega$

Stellar luminosity should be made over the whole electromagnetic spectrum; the most accurate measurements of total stellar irradiance may be made from the spacecraft. The stellar luminosity can be used to determine the distance of the star.

9.1.3 Hertzsprung–Russell (HR) Diagram

Stars are classified on the basis of their observed absorption line spectrum, particularly with regard to the appearance or disappearance of certain discrete features. The stars are arranged in a sequence O, B, A, F, G, K, M of decreasing surface temperature. Table 9.1 shows typical values for stars along the main-sequence (Habets and Heinze 1981). These are further subdivided into subclasses denoted by numbers such as, $0, 1, \ldots, 9$; occasionally decimals are used, for example B0.5. There are R, N, S-type stars as well. The R, N-type stars, also known as carbon (C) stars, are giant stars in which there is an excess of carbon in their atmosphere. The S-type stars show zirconium oxide lines in addition of titanium oxide.

Table 9.1 Main-sequence stellar parameters

Type of stars	Spectral features	Mass $(M_\odot)$	Temperature $(^\circ K)$	Luminosity (bolometric) $\mathcal{L}_\odot$
O	He II, He I, N III, Si IV, C III, O III	≥ 16	$\geq 33{,}000$	$\geq 30{,}000$
B	He I, H, C III, C II, Si III, O II	2.1–16	10,000–30,000	25–30,000
A	H I, Ca II & H, Fe I, Fe II, Mg II, Si II	1.4–2.1	7,500–10,000	5–25
F	Ca II H & K, CH, Fe I, Fe II, Cr II, Ca I	1.04–1.4	6,000–7,500	1.5–5
G	CH, CN, Ca II, Fe I, Hδ, Ca I	0.8–1.04	5,200–6,000	0.6–1.5
K	CH, TiO, CN, MgH, Cr I, Fe I, Ti I	0.45–0.8	3,700–5,200	0.08–0.6
M	TiO, CN, LaO, VO	≤ 0.45	$\leq 3{,}700$	≤ 0.08
C	C_2, CN, CH, CO		$\leq 3{,}000$	
S	ZrO, YO, LaO, CO, Ba		$\leq 3{,}000$	

The relationship between absolute magnitude, luminosity, spectral classification, and surface temperature of stars can be depicted in a diagram, called Hertzsprung–Russell (HR) diagram, which was created in 1910 by Hertzsprung and Russell. Traditionally the HR diagram plots the spectral type against the absolute magnitude – a plot of color vs. magnitude is called a color-magnitude diagram. This diagram summarizes the stages of stellar evolution. The location of a star in such a diagram depends on the rate at which it is generating energy via nuclear fusion in its core and the structure of the star itself. If the spectral class of a star is known, the H-R diagram (see Fig. 9.2) may be used to estimate its intrinsic luminosity. For main sequence stars with masses $M_\star$ in the range $2 \leq M_\star \leq 20 M_\odot$, the mass-luminosity relationship is given as $\sim \mathcal{L}_\star \propto M_\star^{3.6}$, where $\mathcal{L}_\star$ is the luminosity of the star and $M_\odot (= 1.9889 \times 10^{30}$ kg) the solar mass. For less massive stars, with $0.5 \leq M_\star \leq 0.2$ $M_\odot$, the relation changes to $\mathcal{L}_\star \propto M_\star^{2.6}$ (Salaris and Cassisi 2005). A higher stellar mass implies a shorter lifetime.

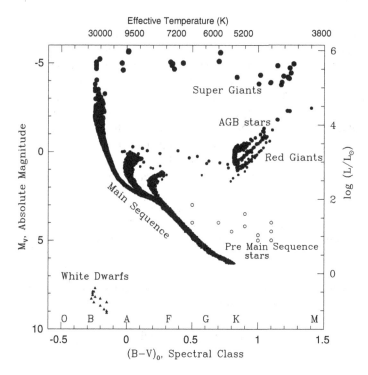

Fig. 9.2 Hertzsprung–Russell (H-R) diagram or color-magnitude diagram for population I stars (metal-rich, young stars). Various stellar evolutionary stages are marked. The locii of various evolutionary stages for stars in open clusters of ages 100 Myr, 500 Myr, and 1 Gyr are shown. The locii of pre-main sequence stars are taken from young clusters and those of white dwarf from old clusters (courtesy: A. Subramaniam)

Most of the stars are concentrated in the region along a band, called main-sequence, which stretches from upper left corner to the lower right corner. Stars located on this band are known as main-sequence stars or dwarf stars; the hotter the stars, it is brighter. The coolest dwarfs are the red dwarfs.[4] Stars along the main-sequence seem to follow mass-luminosity relations. There are other prominent sequences such as giants, supergiants (above the giant sequence) sequences (see Fig. 9.2), which lie above the main-sequence. The stars lie in those sequences, have the similar color or spectrum as the dwarfs in the main-sequence. The gap between the main-sequence and the giant sequence is referred to as Hertzsprung gap. The Asymptotic Giant Branch (AGB) rises from the horizontal branch (where the absolute magnitude is about zero) and approaches the bright end of the red giant branch. The very small stars falling in the lower left corner are called white dwarfs.

9.1.4 Derivation of Effective Temperatures

Stars have an overall spectrum of continuum emission, whose radiation pattern is governed largely by Planck's law (see 3.17). The key parameter of this law is the effective temperature. Depending on the type of observational data used, astronomers define various kinds of temperature (see Saha 2007). Among others, the brightness temperature (see Sect. 3.2.1.5) is an important aspect. The notable other temperature is the effective temperature (T_e), which can be used to estimate stellar angular diameter indirectly (Binney and Merrifield 1998), if the distance to the star is known. In the following, the effective temperature is elucidated.

The effective temperature may be defined by the luminosity and stellar radius. Measurement of accurate radius leads to the direct means of deriving the effective temperature and the surface temperature indicating the amount of radiant energy that it radiates per unit of surface area. These parameters are related to the absolute bolometric magnitude, M_{bol}, of a star, which is a measure of its total energy emitted per second, or luminosity $\mathcal{L}(= 4\pi r^2 F)$, with F as the total flux at a distance $r = 10$ pc. The bolometric magnitude, m_{bol}, is defined as the brightness of an object integrated over all frequencies, so that it takes into account the total amount of energy radiated. The intensity of light emitted by any celestial object varies strongly with wavelength. If a star is a strong infrared or ultraviolet emitter, its bolometric magnitude differs vastly from its visual magnitude, $m_v(= m_{bol} + BC)$, in which BC is the bolometric correction, that depends on the spectral energy distribution of

---

[4] Some dwarfs are dull red in color (brightest in infrared), which are often referred to as brown dwarfs. These are sub-stellar objects having a mass below that is required to maintain hydrogen-burning nuclear fusion reactions in their cores. These dwarfs may have masses between 10 to 84 Jupiter mass. Their surface temperatures are known to be around $1,000°$ K and their luminosities are of the order of 10^{-5} $\mathcal{L}_\odot$, which indicates the absolute magnitude are going to be around +17. Their chemistry in the cool and dense atmospheres plays an important role in the spectral and temperature classification of these objects, and in understanding their spectral signatures.

the star involving effective temperature, gravity, and chemical composition, and is independent of distance of the star. Let $F_\odot$ be the flux for the Sun, hence

$$M_{bol} - M_{bol\odot} = -2.5 \log \frac{F}{F_\odot}$$

$$= -2.5 \log \frac{L/4\pi r^2}{L_\odot/4\pi r_\odot^2} = -2.5 \log \frac{L}{L_\odot}. \tag{9.10}$$

With a known spectral type and luminosity class for a star, one computes its absolute magnitude.

If the star is assumed to radiate like a blackbody, from Stefan-Boltzmann's law, one may write the relation between the luminosity and temperature of a star as,

$$L = 4\pi R_\star^2 F = 4\pi\sigma R_\star^2 T^4, \tag{9.11}$$

where $R_\star$ is the radius of the star.

The (9.11) reveals that the luminosity, radius, and temperature of a star are interdependent quantities. They are related to the absolute bolometric magnitude as well. In terms of radii and temperatures, (9.10) is written as,

$$M_{bol} - M_{bol\odot} = -2.5 \log \frac{R_\star^2 T_\star^4}{R_\odot^2 T_\odot^4}$$

$$= -5 \log \frac{R_\star}{R_\odot} - 10 \log \frac{T_\star}{T_\odot}. \tag{9.12}$$

where $R_\odot (= 6.96 \times 10^8$ m) is the radius of the Sun and $T_\odot (= 5780°$ K) the solar effective temperature; the surface brightness, $B_\odot$ of the Sun can be derived from the luminosity that is related to the total radiation received at the mean distance of Earth, $L_\odot/4\pi r_\odot^2$, in which $r_\odot$ is the distance between the Sun and the Earth.

Combining photometry with the measurement of a limb-darkened stellar diameter yields the stellar emergent flux, F or surface brightness, which is found from the relation,

$$F = \frac{4F_\nu}{\theta_{LD}^2}, \tag{9.13}$$

where F_ν is the measured absolute monochromatic flux received from the star at frequency ν and θ_{LD} the limb-darkened diameter of the source.

The stellar effective temperature T_e is defined in terms of the emergent flux by Stefan–Boltzmann's law. Integrating over all frequencies,

$$F = \int_0^\infty F_\nu d\nu = \sigma T_e^4. \tag{9.14}$$

where $\sigma (= 5.67 \times 10^{-8} \text{W m}^{-2} \text{K}^{-4})$ is Stefan–Boltzmann's constant.

The most direct way of measuring effective temperatures is the combination of bolometric fluxes with angular diameters although bolometric corrections do depend on models to an extent. These measurements are of particular importance to effective temperatures since they are relatively poorly determined especially for the coolest spectral types. The effective temperature is not a thermodynamic temperature, which is defined by a stars surface brightness. Considering a star of physical radius, R, the star's total luminosity is given by its radiant emittance times its surface area (see 9.11). At a distance r, the incident bolometric flux (total radiant flux per unit collecting area) is expressed as,

$$F = \frac{\mathcal{L}}{4\pi r^2} = \left(\frac{R}{r}\right)^2 \sigma T_e^4$$

$$= \left(\frac{\theta_{LD}}{2}\right)^2 \sigma T_e^4. \tag{9.15}$$

So, the effective temperature, T_e is recast as,

$$T_e = \left(\frac{4F}{\theta_{LD}^2 \sigma}\right)^{1/4}. \tag{9.16}$$

An accurate determination of the angular diameter of a single star may yield to its effective temperature,

$$\theta_{LD} = \sqrt{\frac{4F}{\sigma T_e^4}}$$

$$\approx 8.17 \times 10^{-0.2(m_v + BC)} \left(\frac{T_e}{5800° \, K}\right)^{-2} \quad \text{mas,} \tag{9.17}$$

where m_v is the star's apparent visual magnitude and BC the bolometric correction.

It appears from (9.17) that one needs to measure a star's bolometric flux and effective temperature for an operational angular diameter estimate.

9.1.5 Stellar Spectra

A host of physical insights can be obtained from stellar spectra, such as temperature, kinematics, chemical composition, pressure, density and surface gravity. Spectra are useful to determine the position of the central fringe, obtained by LBOIs, at null optical path difference (OPD) as well. A stellar spectrum contains both continuum emission and a overlying several absorption lines. The interior layers of a star's atmospheres are very hot and optically thick. They can be considered as a perfect blackbody. The continuum emission originates from these interiors of the star's atmosphere, through free-free and bound-free atomic transitions. The energy distribution of the continuum emission is close to a black body. The outer layers are cooler and they produce the absorption lines, by absorbing specific wavelengths corresponding to various atomic transitions. These lines are valuable in inferring the physical conditions (effective temperature, density) and the interior make-up of the star. Some stars also have very extended optically thin outer layers, which are far from thermo-dynamical equilibrium (for example, Corona, circumstellar winds), these regions produce, emission lines. Many of these lines are due to collisionally excited quadrupole transitions. These forbidden line ratios are very accurate diagnostic tools to study the density and temperatures of the line forming regions.

Due to the thermal motion of the gas, each emitter can be red or blue shifted, and the net effect is a broadening of the line that is known as a Doppler profile, and the width of which is proportional to the square root of the temperature of the gas. This permits the Doppler-broadened line to be used to measure the temperature of the emitting gas. The strength of a spectral line is the amount of absorption measured in units of equivalent width (EW). Such a line appears as a curve with a shape defined by the velocity field and various line forming mechanism. The equivalent width, W_ν, is the width of a rectangle centered on a spectral line that on a plot of intensity against wavelength, which has the same area as the line; it takes out the same amount of flux as does the line integrated over the entire profile. Mathematically for normalization, one writes,

$$W_\nu = \int_{\text{line}} \frac{I_c - I_\nu}{I_c} d\nu, \qquad (9.18)$$

where I_c is the intensity at the continuum, which needs to be interpolated from its value outside the line and I_ν the intensity at frequency ν within the line.

The equivalent width may be expressed in wavelength units by the relation, $|\Delta\lambda| = c|\Delta\nu|/\nu^2$, in which $\lambda = c/\nu$ and c is the velocity of light. Thus,

$$W_\lambda = \frac{c}{\nu^2} W_\nu = \frac{\lambda^2}{c} W_\nu. \qquad (9.19)$$

The number of atoms in the atmosphere along the line of sight that are in a state to absorb a photon determine the equivalent width. The number of absorbing atoms per gram of stellar material can be determined from the measurement of such widths. The width of a line depends on various factors, viz., (1) natural broadening arising from the intrinsic uncertainty in energy of the states involved in the transition, (2) thermal or Doppler-broadening due to motions of atoms, and (3) collisional broadening resulting from collisions between atoms.

9.2 Stellar Parameters

The stellar parameters such as mass, $M_\star$, radius, $R_\star$, and absolute luminosity, $\mathcal{L}$ are directly related to the stellar atmospheric parameters, for example, effective temperature, T_e, surface gravity, g. Mathematically (Straizys and Kuriliene 1981), one may write,

$$
\begin{aligned}
\log g &= \log(M_\star/M_\odot) + 4\log T_e + 0.4 M_{bol} - 12.50 \\
&= \log(M_\star/M_\odot) - 2\log(R_\star/R_\odot) + 4.44,
\end{aligned} \tag{9.20}
$$

where $\log(R_\star/R_\odot) = 8.47 - 2\log T_e - 0.2 M_{bol}$.

These are based on the assumption of the parameters for the Sun, $T_e = 5780° K$, $\log g = 4.44$, and $M_{bol} = +4.72$. for the Sun. The long baseline interferometry has made a dent in measuring the diameters and asymmetries of stars, and has answered a wide range of current astrophysical interests (Saha and Morel 2000; Quirrenbach 2001; Saha 2002; and references therein). In what follows, some of these results obtained with such an instrument are enumerated in brief.

9.2.1 Determining Stellar Distance

With a precise distance measurement to a star, its effective temperature can be defined by the luminosity and stellar radius. The relative orbits of a binary system can also provide dynamical masses of stars for those transitioning to the giant phases (Boden et al. 2005), high mass main sequence and Wolf-Rayet (WR) stars (Kraus et al. 2007; North et al. 2007a; b). Light from a distant object falls off as the square of the distance. The distance indicators include, parallax, Cepheids variables, most luminous supergiants, globular clusters, and H II regions, and supernovae. Hubble's constant and red shifts can also be considered as distance indicator.

The measurement of parallax is an essential parameter. It is the angle subtended by the star at two diametrically opposite points on the Earth's orbit, with the orbital radius as the baseline. Effectively it is measured as the angular shift in the position of a nearby stars measured relative to far-off (and hence 'fixed') background star. Measurements of the parallax over a half year period as the Earth moves in its orbit was the first reliable way of determining the distances to the closest stars, for example 61 Cygni (distance 3 pc) and Barnard's star. At a distance of 1.8 pc, Barnard's star exhibits the large proper motion that is the change in a star's position as a result of its true motion through space, measured in arcsec yr^{-1}. To note, the measurements of proper motion can confirm stars as members of cluster (known distance) that may elucidate the dynamics of the Galaxy. Given the small angular shifts, parallax measurements have been carried out for stars within a distance of 1,000 pc (by the Hipparcos – High Precision Parallax Collecting Satellite – mission).

At larger distance, the periods of Cepheids and other pulsating stars,[5] and moving star clusters are used for estimating the distance. The absolute luminosity of Cepheids is proportional to their period of variability. These variable stars can be calibrated as standard candles for calculating distance. Cepheids have instabilities in their envelopes that cause them to radially plusate and causes changes in temperature and luminosity over timescales of 1–70 days. The pulsations (time between maxima and minima) of type I Cepheids are very regular, for example, δ Cephei[6] has a period of 5.366341 days. They have a direct relationship between period and the luminosity. Their light curves depict a fairly rapid brightening, followed by a sluggish fall off, while type II light curves show a characteristic bump on the decline side and they have an amplitude range of 0.3–1.2 magnitudes. Precise parallax distances of Cepheids would help to establish a period/absolute magnitude relationship in order to calibrate distances of galaxies, thus reducing the uncertainty on the value of the Hubble parameter, H, which is a value that is time dependent.

Direct measurements of Cepheid angular sizes combined with photometric and spectroscopic measurement of the physical radius of the star provides access to their distances. One way of estimating the diameters and other properties of certain kinds of pulsating stars including Cepheids is the usage of Baade-Wesselink method. This method measures flux and color at two different times, t_1 and t_2, in the pulsation cycle to find the ratio of the stars radii, $R_\star(t_2)/R_\star(t_1)$, at these times. The high resolution spectra provides the radial velocity, $v_r(t)$, curve over the pulsation period of the star. When integrated, this in turn, delivers the difference in the radius between t_1 and t_2. Given the difference, $R_\star(t_2)-R_\star(t_1)$, and ratio of the radius, $R_\star(t_2)/R_\star(t_1)$, it is possible to solve the radii. With interferometric data, the Cepheid period luminosity data can be improved considerably. Such an improvement may propagate to the extragalactic distance scale determined through Cepheids. Measurements of these stars with such a technique can be used to obtain improved calibration of the period-luminosity relation (Kervella 2006). The Grand Interféromètre à deux Télescopes (GI2T) has been employed to determine the mean angular diameter and accurate distance estimate of δ Cephei by Mourard et al. (1997). Based on their measurement of limb-darkened diameter, $\theta_{LD} = 1.60 \pm 0.12$ mas and adopting a mean linear radius, R = 42.7 ± 1 R$_\odot$, these authors derived a distance of 240 ± 24 pc, which corresponds to a parallax of 4.2 ± 0.4 mas. Later, Armstrong et al. (2001), with the NPOI instrument measured the limb-darkened diameter of the said star, $\theta_{LD} = 1.52 \pm 0.014$ mas. By adopting R = 41.5 ± 5.1 R$_\odot$, they derived its linear

---

[5] Pulsating stars expand and contract periodically and have instabilities in their envelopes, which cause them to pulsate in size, temperature, and luminosity over timescales of a few days. They show periodic changes in brightness accompanied by shifts in their spectral lines. The main characteristics of these stars are (1) regular periodicity in their light and radial velocity curves, (2) amplitudes of light variation are of the order of $\Delta m_v = 1.0 \pm 0.5$, (3) radial velocity curve is a mirror image of the light curve, (4) the star's spectral type changes with phase, and (5) there is a tight correlation between the absolute luminosity and the period.

[6] δ Cephei ($\alpha 22^h 29^m 10.2^s$, $\delta + 58°24^m 55^s$) is a prototype supergiant F-type star with a radius of about 70 M$_\odot$. It has a small blue companion at a distance of 12,000 AU.

distance of 254 ± 30 pc (parallax 3.94 ± 0.47 mas). Interferometric measurements (Mérand et al. 2005) using the Center for High Angular Resolution Astronomy (CHARA) array, the projection factor, in the Baade-Wesselink method, of this star is found to be 1.27 ± 0.06.

Another promising technique of determining distance involves the use of type Ia Supernovae (SNe Ia) as standard candles. These supernovae (SN) are among the most spectacular events, which have right characteristics for standard candles. They reach the same brightness as an entire galaxy and are extremely bright enough to be detected. Their homogeneity in light curve make them excellent distant indicators (Hoeflich 2005).

9.2.2 Evolution of Stars

The process of making a star begins inside giant clouds of gas and dust in the InterStellar Medium (ISM) lying primarily within the plane of the Galaxy, whose density, mass and temperature allow the formation of molecules, containing H, He, H_2, H_2O, OH, CO, H_2CO, dust of silicates, iron, ices, etc. These clouds are supported by their gravity against the thermal pressure, turbulent gas motions, and magnetic fields within. They are approximately in a state of virial[7] equilibrium, which occurs when gas pressure equals gravity. The outward pressure exactly balances the inward gravitational pull, a condition that is known as hydrostatic equilibrium. Assuming a cylindrical volume element at a distance r from the center of a star, the equation of hydrostatic equilibrium is written as,

$$\frac{dP}{dr} = -\rho g = -\frac{GM_r\rho}{r^2}, \tag{9.21}$$

in which $g(= M_r G/r^2)$ is the gravitational acceleration, $G(= 6.672 \times 10^{-11}$ N.m^2/kg^2) the gravitational constant, ρ the gas density at radius r, P the pressure, and M_r the mass contained within the radius r.

The mass continuity equation is expressed as,

$$\frac{dM_r}{dr} = 4\pi r^2\rho. \tag{9.22}$$

---

[7] The virial theorem states that, for a stable, self-gravitating, spherical distribution of equal mass objects such as stars, galaxies, etc., the time-averaged value of the kinetic energy of the objects is equal to $-1/2$ times the time-averaged value of the gravitational potential energy, i.e., $\langle T \rangle = -\langle U \rangle/2$, where $\langle T \rangle$ is the time average of the total kinetic energy, and $\langle U \rangle$ the time average of the total potential energy.

The expression for radiative equilibrium,

$$\frac{dT}{dr} = \frac{dT}{dP}\frac{dP}{dr}$$
$$= -\frac{3kL_r\rho}{16\pi acr^2 T^3}, \tag{9.23}$$

in which $a = B(0) = B(T_0)$, B is the brightness distribution, k the absorption co-efficient, which is a function of temperature, pressure, and chemical composition, L_r the amount of energy passing through the surface r per unit time, and T the absolute temperature.

For the convective equilibrium, according to Schwarzchild criterion, a fluid layer in hydrostatic equilibrium in a gravitational field would become unstable if the rate of change of pressure with density exceeds the corresponding adiabatic derivative,

$$\frac{d \log T}{d \log P} = \frac{\gamma - 1}{\gamma}\frac{T}{P}, \tag{9.24}$$

$\gamma(= C_P/C_v)$, C_P and C_v being the specific heats at constant pressure and constant volume respectively.

Combining with (9.21), one writes,

$$\frac{dT}{dr} = -\frac{\gamma - 1}{\gamma}\frac{T}{P}\frac{GM_r\rho}{r^2}. \tag{9.25}$$

The gravity pulls gas and dust towards a common center (the core) where the temperature and pressure increase as the density increases. The gravitational binding energy of such clouds is balanced by the kinetic energy of its constituent molecules. The molecular clouds are observed to have turbulent velocities imposed on all scales within the cloud. These turbulent velocities compress the gas in shocks generating filaments and clumpy structures within the cloud over a wide range of sizes and densities, and this process is known as turbulent fragmentation. When a fragment of such a cloud reaches a critical mass, called Jeans mass, M_J, it becomes gravitation-ally unstable and may again fragment to form a single or multiple star system. Jeans mass can be expressed as,

$$M_J = C\left(\frac{k_B T_k}{\bar{\mu}G}\right)^{3/2}\frac{1}{\sqrt{\rho}}, \tag{9.26}$$

in which C is the constant and T_k the kinetic temperature, $\mu(= m/m_P)$ the mean molecular weight (the average mass per particle in units of m_P), and m the mean particle mass.

As the star accretes more gas and dust, it tries to maintain equilibrium. On reaching equilibrium, accretion stops, failing which the star collapses. The present observational upper limit to stellar mass is $\sim$ 150 M$_\odot$, from observation of the Arches cluster (Figer 2005). It should be noted that this is well in excess of the theoretical limit of 95 M$_\odot$, assuming that undamped stellar pulsations are allowed, and that ionizing flux inhibits accretion beyond this mass. It is possible that damping of stellar pulsations and/or collisions of protostellar clumps may be responsible for the higher observed upper limit to the stellar mass.

As the stellar formation progresses, the rotating cloud collapses to form a central source with a Keplerian (accretion) disk. Considering dynamics of contraction, the equation of motion for a shell of material at a distance r from the center is given by,

$$\frac{d^2 r}{dt^2} = -\frac{GM_r}{r^2} - 4\pi r^2 \frac{\partial P}{\partial M_r}. \tag{9.27}$$

The rate at which energy is produced depends on the distance to the center. The increase in the luminosity as one passes through the shell from inside, moving outwards is equal to the energy produced within the shell. The energy conservation equation is given by,

$$\frac{dL_r}{dr} = 4\pi r^2 \rho \varepsilon, \tag{9.28}$$

where ε is the coefficient of energy production, defined as the amount of energy released in the star per unit time and mass.

The (9.21–9.23 and 9.28) are the fundamental equations of stellar structure. Coinciding with the onset of accretion there is a steady outflow of material in powerful winds that emanate from the poles of the young protostar (which is a star in its first stage of development). When the density and temperature are high enough, the nuclear fusion commences, and the outward pressure of the resultant radiation slows down the collapse. Material comprising the cloud continues to fall onto the protostar that undergoes continuous contraction. At this stage, bipolar flows are produced, probably due to the angular momentum of the infalling material. Observed edge-on, such a protostar may be mistaken as a binary star. This early phase in the life of a star is known as the T Tauri phase. During this phase, the star has (1) vigorous activity (flares, eruptions), (2) strong stellar winds, and (3) variable and irregular light curves. Stars in this phase are usually surrounded by massive, opaque, protoplanetary disk, which gradually accretes onto to the stellar surface. A fraction of the material accreted onto the star is ejected perpendicular to the plane of the disk in a highly collimated stellar jet. Table 9.2 summarizes the use of optical/IR interferometry in understanding the various stages of stellar evolution.

Table 9.2 Possibilities of optical/IR interferometric studies on astronomical sources

Objects	Possible detections	Interferometric mode
Molecular clouds	Resolving multiple cores	IR
Young stellar objects: T-Tauri, Herbig Ae/Be, FU Orionis	Resolving disk-debris, detection of exoplanets	IR
Main-sequence stars	Diameters (UD and LD), resolving cores of star-clusters, gravity-darkening coefficient	Optical
Variable stars: Cepheids/Mira	Diameter, distance	Optical/IR
Giant/Super giant phase, Horizontal Branch (HB), AGB, Post-AGB phase	Diameter, spots, gravity-darkening coefficient	Optical
Planetary nebulae	Resolving circumstellar shells	Optical
Binary/multiple systems	Deriving sizes, masses, orbital parameters	Optical
Novae/supernovae	Appearance/mapping of shell	Optical/IR
Galactic center	Probing Sgr A* region	IR
Galaxies	Star-forming regions	Optical/IR
AGN	Structure of core	Optical/IR

9.2.3 Resolving Young Stellar Objects (YSO)

During the initial collapse, a Pre-Main Sequence (PMS) star generates energy through gravitational contraction. Such a star is contracting towards the main-sequence, whose evolutionary phase lasts for tens of millions of years, as dictated by the Kelvin-Helmholtz timescale. Throughout this period, the protostar progressively increases in density as it shrinks in size. The infalling material begins to spin up as the radius of the protostar decreases. Since all of the material in the envelope surrounding the protostar rotates in the same direction, the matter falls into orbits of various sizes forming a circumstellar disk. Most matter flows onto the protostar through the disk, but some remains in orbit. As the surrounding envelope of dust disperses, the accretion process stops, and the central ball of gas is now considered to be a PMS star. These protostars and PMS stars are often grouped as Young Stellar Objects (YSO).

The understanding of stellar structure is uncertain in the case of PMS stars, particularly for low-mass systems (Palla and Stahler 2001, Hillenbrand and

White 2004). Long Baseline Optical/IR Interferometry is ideally suited to probe directly the innermost regions of the circumstellar environment around young stars, the physics of mass accretion, and their evolution.

9.2.3.1 Disks Around YSOs

Most of the young stars have a disk-like distribution of matter, which is believed to be the location of planet-formation. The Young Stellar Objects (YSO) play an important role in understanding the star and planet-formation. Also, they offer a rich circumstellar environment, namely accretion disks, stellar companions, optical jets, and stellar winds. These objects display a variety of phenomena like IR excesses, luminosity variations, and jets. Most of these phenomena occur within the region of 1 AU. In general, YSOs are obscured in the optical, but directly visible in the IR and have different effective optical and IR extinctions.

LBOIs operating at near and mid-IR wavelengths have spatially resolved the circumstellar emission of several YSOs of various types such as T-Tauri and Herbig Ae/Be objects providing new insights into stellar formation and star-forming regions. FU Orionis objects, located in active star-forming regions, are believed to be T-Tauri stars surrounded by a disk, whose unusual properties may arise from the presence of a rapidly accreting (10^{-4} $M_\odot$/yr) circumstellar disk (Hartmann and Kenyon 1996); the accretion disk model explains the broad spectral energy distributions (SED) of these objects. They exhibit large IR excesses, double peaked line profile, apparent spectral types varying with wavelength, broad, blue-shifted Balmer line absorption, and are associated with strong mass outflows. Malbet et al. (1998) have resolved a YSO, FU Orionis, using the long baseline Palomar testbed interferometer (PTI) in the near IR wave band with a projected resolution better than 2 AU at the 450 pc distance. They have resolved this object (see Fig. 9.3). This object has shown the distinct signature of a disk hotspot (Malbet et al. 2005).

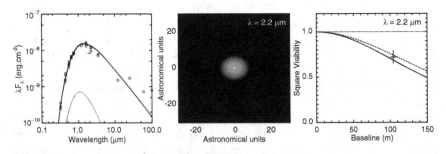

Fig. 9.3 Accretion disk model of FU Orionis (Malbet et al. 1998). *Left panel*: spectral energy distribution from literature data (*circles*), of the accretion disk model (*dashed line*), the star (*dotted line*) and the whole system (*solid line*), *middle panel*: the synthetic image of the accretion disk at 2.2 μm, and *right panel*: the predicted visibility curves of the accretion disk model for the major and minor axes (respectively, *solid and dashed lines*; courtesy: F. Malbet)

Herbig Ae/Be objects are intermediate-mass pre-main sequence stars (Natta et al. 2000) of spectral type A and B, with strong emission lines (especially Hα and the calcium H and K lines). Located close to the zero-age main-sequence, these objects are associated with circumstellar material, in the form of either proto-planetary disk, or remnant envelope, or both; many of the best structures are located in the nearby Orion nebula. The spectral type of these stars are earlier than F0, and Balmer emission-lines can be envisaged in the stellar spectrum. The excess IR radiation in comparison with normal stars are due to circumstellar dust. These stars may show significant brightness variability, which can be due to clumps (proto-planets and planetesimals) in the circumstellar disk.

Millan-Gabet et al. (1999, 2001) have resolved circumstellar structure of Herbig Ae/Be stars in near-IR. Figure 9.4 depicts the examples of H-band visibility data and models for 2 sources. The lower right panel of this figure illustrates the observed lack of visibility variation with baseline rotation, consistent with circumstellar emission from dust which is distributed either in a spherical envelope or in a disk viewed almost face-on. Figure 9.5 summarizes the existing set of measurements of near-IR sizes of Herbig Ae/Be and T Tauri objects. The data used for this plot by

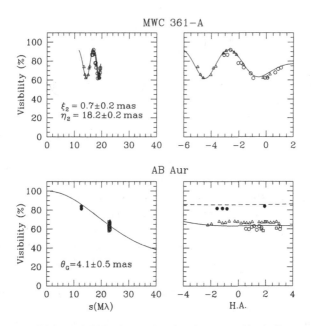

Fig. 9.4 Examples of H-band visibility data and models for two Herbig Ae/Be stars (courtesy: R. Millan-Gabet). The data and models are plotted as a function of baseline (*left panels*), and hour angle (*right panels*), which determines the baseline position angles; different symbols correspond to data obtained on different nights (Millan-Gabet et al. 2001). *Solid symbols* and *dashed lines* are for 21 m baseline data and models, and *open symbols* and *solid lines* are for 38 m baseline data and models. The *upper panels* show the data and best fit model for the binary detection in MWC 361-A, and displays the companion offsets in right ascension (ξ_2) and declination (η_2). The *lower panels* show the data and best fit Gaussian model for AB Aurigae, and displays the angular FWHM (θ_G)

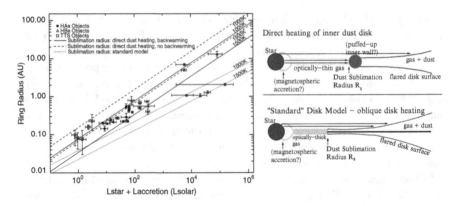

Fig. 9.5 Measurements of near-IR sizes of Herbig Ae/Be and T Tauri objects. The sizes are plotted as a function of the central luminosity (stellar plus accretion shock); and comparison with subli-mation radii for dust directly heated by the central luminosity (*solid and dashed lines*) and for the oblique heating implied by the classical models (*dotted line*). A schematic representation of the key features of inner disk structure in these two classes of models is also shown (Millan-Gabet et al. 2007; courtesy: F. Malbet)

Millan-Gabet et al. (2007) are from Keck-aperture masking (Danchi et al. 2001; Tuthill et al. 2001), IOTA (Millan-Gabet et al. 2001), PTI (Eisner et al. 2004), and KI (Monnier et al. 2005) for the former objects, while for the latter objects, the data are from PTI (Akeson et al. 2005a), and KI (Akeson et al. 2005b; Eisner et al. 2005). Monnier et al. (2001) have reported the results of decomposing the circumstellar dust and stellar signatures from the evolved stars (see Fig. 8.9) as well.

Spatially resolved studies involve the absolute calibration of PMS tracks, the mass accretion process, continuum emission variability, and stellar magnetic activ-ity. There are ∼50 known PMS stars with $m_v < 8$ within 50 pc. Their ages lie within the range 8–50 Myr (Zuckerman and Song 2004). The evolution of these objects may be studied by resolving them to calibrate evolutionary tracks or by observing hot and cold spots associated with mass accretion and magnetic activity (de Witt et al. 2008). Hot spots deliver direct information regarding the accretion of material onto the stellar surface, while cool spots are the product of the slowly decaying rapid rotation of young stars. Imaging these spots may explain the anomalous photometry observed in young stars (Stauffer 2003).

VLTI/AMBER spectro-interferometric instrument (see Sect. 6.1.3.5) was em-ployed to observe the emitting region of a young stellar object, MWC 297. It has been spatially resolved (Malbet et al. 2007a) in the continuum with a visibility of ∼50 mas, as well as in the Br γ emission line, where it decreased to a value of ∼33 mas (see Fig. 9.6). This YSO, MWC 297, is an embedded Herbig Be star exhibiting strong hydrogen emission lines and a strong near-IR continuum excess. The central star has a large projected rotational velocity of 350 km/s. A possible interpretation given by these authors was that the gas emitting the Br γ emission line is located in a region slightly larger than the one from which the dust continuum

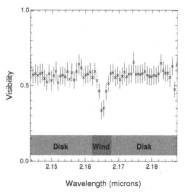
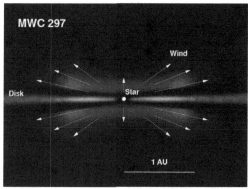

Fig. 9.6 *Left panel*: spectral dependence of the visibility as measured with VLTI/AMBER for MWC 297 around the Br γ line. *Right panel* edge-on view of the model including an equatorial optically thick disk and an outflowing wind. The wind geometry has been computed to both fit the visibility drop in the Br γ line and reproduce the object spectrum. The apparent size of the wind is larger than the apparent size of the disk (Malbet et al. 2007a; courtesy: F. Malbet)

emission arises. With the same set up, a few Herbig Ae/Be stars, carried out by Kraus et al. (2008a), showed evidence for at least two distinct line-formation mechanisms, giving evidence for mass in-fall in one object and outflow for the rest. They were able to measure the spatial distribution of the hydrogen and CO gas emission in the inner disks around six such stars. The inner region of the disk of material surrounding another YSO, MWC 147 (2,600 ly away having $m_v = 8.8$), a Herbig Ae/Be object with a few solar masses in the constellation, Monoceros, was also observed (Kraus et al. 2008b). This object is surrounded by disk of gas and dust and is increasing in mass by accreting material from outside. The observation shows that the temperature changes with radius are much steeper than predicted by the models. This indicates that most of the near-IR emission emerges from hot material located very close to the star. The young stellar object, MWC 419 that lies at a distance of 650 pc in the constellation Cassiopeia, is a B8 spectral type object having 330 solar luminosity. It is a photometrically variable emission-line star, typical of the Herbig Ae/Be class, including illumination of a reflection nebula (Herbig 1960). Narrow band Hα imaging studies show unipolar large scale structures around this object (Marston and McCollum 2008). MWC 419 was spatially resolved by the Palomar Testbed Interferometer (PTI; see Sect. 7.3.4) in the K-band with a uniform-disk angular diameter of 3.34 ± 0.16 mas (Wilkin and Akeson 2003). Recently, Ragland et al. (2009) measured the temperature of dust at various regions throughout the inner disk of MWC 419 using Keck Interferometer (see Sect. 7.3.5) through simultaneous K and L-band near-IR (3.5–4.1 μm) wavelengths. The L-band observations gave the capability in probing the density and temperature of planet-forming regions. The temperature differences throughout the disk may indicate that the dust has different chemical compositions and physical properties, which can affect how planets form. The multiwavelength observations distinguish between models and gaining insight into the 3-dimensional geometry of the inner disk. The measured disk size at and around Br γ reveals that the emitting hydrogen gas is located in the inner regions of the dust disk. However, in order to address the intriguing inner disk geometry, sufficient u, v-coverage is required. Recently, Eisner et al. (2009) have reported spatially resolved near-IR (K-band) spectroscopic observations of 15 young stars. Using a grism spectrometer at the Keck interferometer, they have obtained an angular resolution of a few milliarcseconds. Their data constrain the relative spatial and temperature distributions of dust and gas in the inner disks around these stars. The angular size of the near-IR emission was found to increase with wavelength, indicating hot, presumably gaseous material within the dust sublimation radius. They have detected hot hydrogen gas through Br γ emission line. Their observations suggest the presence of water vapor and carbon monoxide gas in the inner disks of several objects. In all cases, this gaseous emission is more compact than the dust continuum emission.

Another highly luminous Herbig Ae/Be star, LkHα 101, thought to be a transitional object which is on (or nearly on) the main-sequence, surrounded by a hot circumstellar disk, has also been observed by means of interferometry. It was identified as the source of illumination of the irregular reflection nebula NGC 1579 by Herbig (1956). The near-IR images of this object have been obtained by Tuthill et al. (2001) utilizing interferometry on the Keck 1 telescope. They have resolved the structure of the inner accretion disk, which show a face-on circular disk with a central cavity. These authors opined that the asymmetric brightness seen in the circumstellar ring of emission might be due to flaring of the disk observed at a slight inclination.

9.2.3.2   Resolving Debris Disks

A debris disk is a ring-shaped circumstellar disk, probably analog to Kuiper belt encircling the solar system, which contains Pluto and other small icy bodies, of gas-poor dust and small grains (1–100 μm in size) in orbit around a star. The dust comes from small bodies colliding and grinding themselves down to tiny debris. Radiation from the host star can cause these particles to spiral inward. Several evolved and young stars harbor debris disks (Wyatt et al. 2003; Rhee et al. 2007); one debris disk in orbit around a neutron star has also been found (Wang et al. 2006). Debris disks are optically thin around main-sequence stars. The presence of dusty disks around stars with ages of $\sim$10 Myr serves as a marker for the existence of planetesimals that are confined to narrow belts (Setiawan et al. 2008). Although these disks were first detected by the InfraRed Astronomical Satellite (IRAS) in 1983, their innermost parts are poorly known due to the high contrast and small angular separation with their parent stars. This inner part is important to understand, as it directly probes the location of planets (if any), the formation, evolution, and dynamics of planetary system.

Debris disks can constitute a phase in the formation of a planetary system following the proto-planetary disk phase. They were inferred from an infrared excess flux between 25 and 100 μm many times brighter than expected from the stellar photosphere. They usually absorb excess radiation from the star and radiate away as infrared energy. Resolved images of these systems help to constrain their physical and geometrical properties. Adrila et al. (2004) resolved debris disk around the G 2V star, HD 107146, with the Hubble Space Telescope's (HST) advanced camera for surveys (ACS) coronagraph multiwavelength imaging. Observations on a relatively young star, α PsA (Fomalhaut A, HR 8728; RA $22^h 57^m 39.1^s$, δ $-29°37'20''$) in the southern constellation Piscis Austrinus, were carried out by HST, as well as with ground-based telescopes equipped with adaptive optics systems. At a distance of 25.07 ly, with an apparent magnitude of $m_v = 1.16$, this A3 V spectral type star has a surface temperature of $\sim$8,720°K. α PsA is about 200 million years old. Kalas et al. (2005) have found a large belt of cold dust in a toroidal shape with a very sharp inner edge orbiting far from the star. They have estimated that the belt contains about 50–100 Earth masses, is about 133–158 AU away from its parent star and that its geometric center of the ellipse is offset from the star by as large as about 15 AU.

With a hope of detecting the signature of hot dust around α PsA, Absil et al. (2009) analyzed the archival data obtained using the VLTI/VINCI instrument (see Fig. 9.7). They have studied the fringe visibility at projected baseline lengths ranging

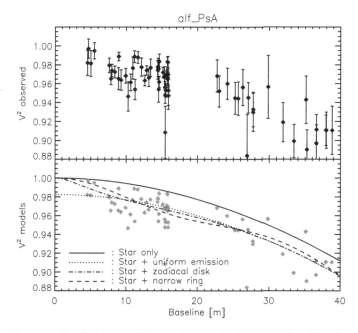

Fig. 9.7 *Top panel*: short baseline visibilities collected with VLTI/VINCI on Fomalhaut. *Bottom panel*: Expected square visibilities for various disk models. The narrow ring is located at the sublimation radius of blackbody dust grains, assuming a sublimation temperature of 1,700°K, while the zodiacal disk follows the Kelsall et al. (1998) model. The data collected at projected baselines between 0 and 40 m are superimposed for comparison (*light gray diamonds*). This analysis shows that various disk models fit almost equally well the VLTI/VINCI data set, so that the disk morphology cannot be directly constrained (Absil et al. 2009; courtesy: O. Absil)

from 4 to 140 m in the K band and found a significant deficit of visibility at short baselines with respect to the expected visibility of the sole stellar photosphere. This is interpreted as the signature of resolved circumstellar emission, producing a relative flux of 0.88% ± 0.12% with respect to the stellar photosphere. These authors opined that the thermal emission from innermost hot dusty grains located within 6 AU from Fomalhaut causes the detected excess. Their study also provides a revised limb-darkened diameter for α PsA, which is 2.223 ± 0.022 mas (see Fig. 9.8), taking into account the effect of the resolved circumstellar emission.

With the FLUOR (Fiber-Linked Unit for Optical Recombination) instrument at the CHARA interferometer (see Sect. 7.3.7), Absil et al. (2008) reported the detection of circumstellar emission around the 2.99 visual magnitude (m_v), main-sequence A 0V-type star, ζ Aquilae (RA $19^h05^m24.5^s$; δ + $13°51'48''$). Their observations suggest the presence of circumstellar emission around this star. However, the morphology of the emission source cannot be directly constrained because of the sparse spatial frequency sampling of the interferometric data. With adaptive optics observations and radial velocity measurements, they found that the

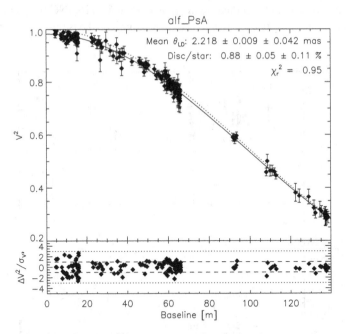

Fig. 9.8 Result of the fit of a star-disk model to squared visibilities obtained on α PsA with the VLTI/VINCI interferometer. The *solid line* represents the best fit star-disk model, while the *dotted line* represents the best fit with a single star for comparison. In the latter case, a systematic offset between the model and the observations is noticed at short baselines, which reveals the presence of resolved circumstellar emission around Fomalhaut. The amplitude of this visibility deficit at short baselines is directly proportional to the flux ratio between the disk and the star. The star-disk model provides a good fit to the full data set and gives simultaneous estimations of the limb-darkened diameter of the star and of the star-disk flux ratio (Absil et al. 2009; courtesy: O. Absil)

presence of a low-mass companion with a K-band contrast of four magnitudes, a probable mass of about 0.6 $M_\odot$, is a likely origin for the excess emission, which orbits between about 5.5 AU and 8 AU from its host star (assuming a purely circular orbit). Another viable interpretation for the near-IR excess, according to them, is the presence of hot dust within 10 AU from ζ Aquilae, producing a total thermal emission equal to 1.69% $\pm$ 0.31% of the photospheric flux in K-band. The hot debris disks around a few more stars are also found by detecting near-IR excesses in α Lyrae (Absil et al. 2006), τ Ceti (Di Folco et al. 2007), and β Leonis (Akeson et al. 2009) using the said instrument. In all cases (with possible exception of ζ Aquilae), the presence of hot dust seems to be the most plausible explanation to the detected near-infrared excess.

9.2.4 Diameter across Stellar Evolution

A protostar continues to collapse while the center core heats up to $\sim 10^7$ degree kelvin, and hydrogen begins to fuse in its core by fusion reactions. The collapse is halted by the pressure of the heated gas and radiation which counteract gravity. An equilibrium is reached between opposing forces. After a few million years or so, much of the dust and gas in this disk dissipates, leaving a PMS star. Subsequently it develops into a main-sequence star, which generates energy at the core using an exothermic nuclear fusion process, where protons (^{1}H) are being converted into the atoms of helium (He) nuclei. There are 4 p-p chains, the most common form of which is the pp-I chain given as (Hansen and Kawaler 1994),

$$^1H \ + \ ^1H \rightarrow \ ^2D + e^+ + \nu_e \tag{9.29a}$$

$$^2D \ + \ ^1H \rightarrow \ ^3He + \gamma, \tag{9.29b}$$

$$^3He \ + \ ^3He \rightarrow \ ^4He + 2 \ ^1H. \tag{9.29c}$$

The pp II chain is given by,

$$^3He \ + \ ^4He \rightarrow \ ^7Be + \gamma, \tag{9.30a}$$

$$^7Be \ + \ e^- \rightarrow \ ^7Li + \nu_e + \gamma, \tag{9.30b}$$

$$^7Li \ + \ ^1H \rightarrow \ ^4He + \ ^4He. \tag{9.30c}$$

The pp-III chain is,

$$^7Be \ + \ ^1H \rightarrow \ ^8B + \gamma, \tag{9.31a}$$

$$^8B \ \rightarrow \ ^8Be + e^+ + \nu_e, \tag{9.31b}$$

$$^8Be \ \rightarrow \ ^4He + \ ^4He. \tag{9.31c}$$

Apart from the above three pp-chain reactions, a fourth chain is also predicted,

$$^3\text{He} + \text{p} \rightarrow \alpha + e^+ + \nu_e, \tag{9.32}$$

however, this term has never been observed due to its rarity. The pp-I chain dominates upto $10 < T_6 < 14$, the pp-II chain dominates in the temperature range $14 < T_6 < 23$, and the pp-III chain dominates when $T_6 > 25$. Here, $T_6 = T/10^6$ degree Kelvin. The carbon (CNO) cycle starts at much higher temperature and uses carbon (C), nitrogen (N) and oxygen (O) as catalysts for the production of He. During this stage, a star is located along the main-sequence at a position determined by its mass.

For a star of above 1.5 $M_\odot$, the CNO cycle becomes dominant since its reaction rates increases more rapidly with temperature. The abundance of helium in the stellar interior increase as a consequence of the reactions. At the time of leaving the main-sequence, the star possesses (1) an isothermal He-core, (2) an H-burning shell surrounding such a core, and (3) an envelope. As the He produced by the H-burning shell comes down to the He core, the core grows in mass and heats up until, at around 100 million degree kelvin (K), He starts thermonuclear fusion to form C and O. After it grows to its maximum size, it begins to contract. The contraction of the core liberates gravitational energy causing the central temperature, as well as the surrounding H-burning shell temperature, which in turn, increases in the rate of nuclear reactions in the shell. As the nuclear reactions continue in H-burning shell, the hydrogen burnt becomes part of the He-core and the shell moves outward in respect of mass; the temperature in the shell is maintained at H-burning level by the contraction of core. For a very low mass star, the envelope is convective. As a result, the energy produced in the shell pushes the stellar envelope outward, against the pull of gravity with the result of enhanced luminosity. The opacity impedes the radiative flow of the generated energy towards the surface.

Stars remain near their initial position on the main-sequence until a significant amount of hydrogen in the core has been exhausted. Subsequently, it begins to evolve into a more luminous star. Stars with initial masses between ~2 $M_\odot$ and 8 $M_\odot$ on the main-sequence develop through the red giant phase until they ignite central helium burning. The luminosity becomes constant at the initial phase, but with the drop of the surface temperature, their envelopes become convective and the luminosity rises. The mass-loss[8] increases when the star swells up to the size and

---

[8] Mass-loss is a crucial physical process for the life, which occurs due to (1) gravitational attraction of a binary companion; this effect is prominent when the companion has high gravity such as a white dwarf, a neutron star, or a black hole, (2) coronal mass ejection-type events, particularly in the Wolf–Rayet (WR) stars that have a stellar wind with a high velocity and high mass loss rate of order 10^5 $M_\odot$ per year; their upper layers are weak, often events like solar flares and coronal mass ejections blast some of the material from these layers, and (3) ascension to red giant or red supergiant status; stars entering into red giant phase loose mass rapidly because gravitational hold on their upper layers is weakened, they may be shed into space by violent events such as the beginning of a helium flash in the core. All stars exhibit mass-loss with rates ranging from ~10^{-14} to 10^{-4} $M_\odot$ per year, depending on spectral type, luminosity class, rotation rate,

low gravity of a red giant. The mass of these stars is not high enough to start further nuclear reactions and therefore its core ends up as carbon-oxygen white dwarf. The helium fusion continue in the core until the core fuel supply is exhausted. This double-shell burning phase is referred to as the Asymptotic Giant Branch (AGB) stage. As the central temperature of contracting core rises to a value of the order of about 10^8, the helium can be transformed into carbon in the triple-alpha reactions. On the HR diagram, such a star moves up to the right of the main-sequence, called red giant region.

A giant star has a radius and luminosity higher than a main-sequence star of the same surface temperature. Typically, these type of stars have diameters between 5 and 25 $R_\odot$ and luminosity between 10 and 1,000 $\mathcal{L}_\odot$. Their surface layers are expanded and cooled following the exhaustion of its core hydrogen reserves. The energy generated by helium fusion causes the core to expand, causing the pressure in the surrounding hydrogen-burning shell to decrease. This reduces its energy-generation rate. The luminosity of the star decreases, its outer envelope contracts again, following which the star leaves the red giant branch. The surface gravity of such a star is also much lower than for a dwarf star since the radius of a giant star is much larger than a dwarf. Given the lower gravity, gas pressures and densities are much lower in giant stars than in dwarfs.

9.2.4.1 Uniform Disk Diameter

In order to convert an angular diameter into a physical diameter, the distance must be known. From observations of its apparent magnitude, the star's probable distance is derived, and hence its probable diameter. The (9.3) provides a conservative estimate of a star's expected diameter, since it has neglected the interstellar absorption term. On adding absorption term, A to this equation,

$$m - M = 5 \log r - 5 + A, \tag{9.33}$$

where m denotes an apparent magnitude of the star, M the absolute magnitude of the star, and r the distance (pc) of the star, the apparent distance of the star can be derived; if the distance of the star is known, the effective temperature of a star can also be derived from its luminosity and radius.

Precise measurement of radii and relative temperature of stars were carried out by the astronomers in early days in eclipsing binary system (Andersen 1991 and references therein). However, these binaries are very rare since the orbits of the stars

---

companion proximity, and evolutionary stage; mass-loss is apparently related to the brightness variations. Mass-loss determines evolution of pre-main sequence, time on the main sequence, morphology of the horizontal branch, the formation and characteristics of planetary nebulae, impacts the pre-explosion characteristics of type-II supernovae (SN II), and their interaction of their remnant shells. Also, it helps to understand the evolutionary connections between Be stars, WR stars, hypergiants (Lamers 2008), and other massive, hot stars.

should be edge-on to the solar system. In a span of less than a decade (1965–1974), Narrabri intensity interferometer had produced accurate measurements of the diameters of 32 stars brighter than 2.5 m_v, in the spectral range O5 to F8 resulting in the effective temperature scale for early-type stars; some as small as ∼0.41 ± 0.03 mas (Hanbury Brown 1974). All these stars were limited to the declinations South of 20°N in order to avoid excessive atmospheric extinction at low elevations, exception being the star, α Lyrae at 39°N. These measurements still form the basis of the temperature scale for stars hotter than the Sun.

Single aperture telescopes employing speckle interferometry have measured diameters of the nearest and the brightest giants and supergiants with a formal accuracy of about 10%. On the contrary, the LBOI observations provide a better accuracy. More than 50 stars have been measured (Labeyrie 1985 and references therein; Di Benedetto and Rabbia 1987; Dyck et al. 1993; Perrin et al. 1999, 2004; Kervella et al. 2001) with such a method. The angular diameters of some of these stars, observed with the small interferometer, are presented in Table 9.3.

Davis and Tango (1986), reported a direct experimental comparison between the measurements of the diameter of α Orionis (M2 Iab) at the same wavelength made by the Narrabri intensity interferometer and by an amplitude interferometer comprising of two 15 cm apertures separated by a 11.4 m baseline. This experiment showed that equally accurate results could be obtained by the latter instrument. Subsequently, with Sydney University Stellar Interferometer (SUSI), Davis et al. (1999b) measured precise diameter (3.474±0.091 mas) of δ CMa (F8 Ia) that was the faintest star measured with Narrabri Intensity Interferometer by Hanbury Brown et al. (1974a) and had the largest uncertainty (±14%) of all the angular diameters determined with the latter instrument. With PTI, measurements of diameters and

Table 9.3 Stars resolved with the interferometers in optical wavelengths (550 nm; Labeyrie 1985)

Star	Spectral type	Angular size (mas)	$\dfrac{R}{R_\odot}$	$T_e(°K)$
α Arietis	K2 III	7.6 ± 1	15 ± 5	4,300 ± 350
α Aurigae-A	G1 III	8.0 ± 1.2	12 ± 2	5,400 ± 200
α Aurigae-B	K0 III	4.8 ± 1.5	7 ± 2	5,950 ± 200
α Cassiopeiae	K0 III	5.4 ± 0.6	26 ± 8	4,700 ± 300
α Cygni	A2 Ia	2.7 ± 0,3	145 ± 45	8,200 ± 600
α Lyrae	A0 V	3.0 ± 0.2	2.6 ± 0.2	9,602 ± 180
α Persei	F5 Ib	2.9 ± 0.4	55 ± 9	7,000 ± 600
β Andromeda	M0 III	13.2 ± 1.7	33 ± 9	3,800 ± 250
β Gem	K0 III	7.8 ± 0.6	8 ± 2	4,900 ± 220
β UMi	K4 III	8.9 ± 1	30 ± 9	4,220 ± 300
γ^1 Andromeda	K3 IIb	6.8 ± 0.6	50 ± 14	4,650 ± 250
γ Draconis	K5 III	8.7 ± 0.8	45 ± 10	4,300 ± 230
δ Draconis	G9 III	3.8 ± 0.3	15 ± 5	4,530 ± 220

effective temperatures of G, K, and M-giants and supergiants have been reported (van Belle et al. 1999). With the two of the VLTI telescopes Glindemann and Paresce (2001) have measured the angular diameter of the blue star, Achernar (α Eridani), which was found to be 1.92 mas. Subsequently they (1) derived the diameters of a few red dwarf stars, and (2) determined the variable diameters of a few pulsating Cepheid stars. With the CHARA array, Boyajian et al. (2008) have measured the angular diameter of the subdwarf μ Cassiopeiae (G5 V), which is a halo population star having an effective temperature of $5297 \pm 32° K$.

9.2.4.2 Limb-Darkened Diameter

Measurement of the center-to-limb intensity variation of the Sun was carried out in the beginning of the twentieth century (Schwarzschild 1906), which had helped solar community to constrain models for the transport of energy in the solar atmosphere. The Narrabri intensity interferometer was the first instrument that measured the limb-darkening of a hot star, α CMa (A1 V), by Hanbury Brown et al. (1974b). As LBOIs have come on-line, the measurement of center-to-limb intensity variation of a stellar photosphere have become reality.

Limb-darkened measurements of several stars were carried out using the small interferometer, I2T (see Sect. 7.1.1) at the Observatoire de la Calern, France (see Table 9.4; Labeyrie 1985; Di Benedetto and Rabbia 1987). With the Cambridge Optical Aperture Synthesis Telescope (COAST) instrument, images of α Orionis reveals the presence of a circularly symmetric data with an unusual flat-topped and limb-darkening profile (Burns et al. 1997). Subsequently, Young (2004) reported the measurement of the squared visibilities and closure phases for this star at wavelengths of 782 nm and 905 nm. The raw interference fringe data recorded at COAST and IOTA were reduced by using a Maximum Entropy Method (MEM; Sect. 8.3.3) to obtain a set of estimates of these quantities for each observing waveband. Prior knowledge was incorporated in some of the MEM reconstructions by using a moderately limb-darkened symmetric disc as the default image. They have detected an asymmetry in the appearance of the stellar disk in all the wavebands. The IOTA/FLUOR instrument was employed to measure diameter of Cepheids

Table 9.4 Comparative statement of Uniform-Disk (UD) and limb-disk (LD) measurements innear-IR wavelengths (2.2 μm; Labeyrie 1985; Di Benedetto and Rabbia 1987)

Star	UD size (mas)	$T_e(°K)$	LD size (mas)	$T_e(°K)$
α Boötis	21.5±1.2	4,240±120	20.95±0.20	4,294±30
α Tauri	20.7±0.4	3,904±34	20.21±0.30	3,970±49
β Andromeda	14.4±0.5	3,711±64	14.34±0.19	3,705±45
γ Draconis	10.2±0.1.4	3,960±95	10.13±0.24	3,981±62
μ Gem	14.6±0.8	3,860±95	13.97±0.16	3,660±43

(Kervella et al. 1999), as well as to derive the effective temperature of giant stars (Perrin et al. 1998). Also, the diameter and effective temperature of α Boötis (K1.5 III) were derived (Quirrenbach et al. 1996; Lacour et al. 2008).

Employing Navy Prototype Optical Interferometer (NPOI), Hajian et al. (1998) measured the limb-darkened angular diameters of two K type giants, α Arietis (K2 III) and α Cassiopeiae (K0 III) with 20 spectral channels covering 520–850 nm. They were able to extend the spatial frequency coverage beyond the first zero of the stellar visibility function for these stars. The non-zero closure-phase measurement for these stars were also carried out, which is important to identify stars with significant departures from circularly symmetric brightness profiles; the closure-phase showed a jump of 180° at the first zero in the visibility amplitude. In order to reach beyond the first zero, wavelength bootstrapping (see Sect. 6.1.6.3) was used. Since the determination of angular diameter of a star depends on model-dependent limb-darkening corrections (see Sect. 8.1.1.2), these authors fit both uniform-disk (see Sect. 8.1.1.1) and linear limb-darkened models to the complex triple-product. The former model diameters for the respective stars are 6.34±0.06 mas and 5.25±0.06 mas, while the limb-darkened model diameters are 6.80±0.07 mas and 5.62±0.07 mas respectively.

With the VLTI/VINCI instrument, Wittkowski et al. (2004) measured the limb-darkened intensity profile of a giant star, ψ Phoenicis (M4 III, HR 555), which is a short-period, photometric variable star. The squared visibility amplitudes in the second lobe of the visibility function were obtained using two of the VLT's 8.2 m Unit telescopes (see Sect. 7.3.6). Their measurement constrains the diameter of this star, as well as its center-to-limb intensity variation. They constructed a spherical hydrostatic model atmosphere based on the available spectro-photometric data and confronted its center-to-limb prediction with the observed interferometric measurement. These authors derived the fundamental parameters for this star, such as the angular diameter (8.13±0.2 mas), linear radius (86±3 $R_\odot$), effective temperature (3,550±50° K), and luminosity ($\log \mathcal{L}/\mathcal{L}_\odot = 3.02\pm0.06$).

Although most of the interferometric limb-darkening measurements have been limited to cool stars with the exception of α CMa, several such measurements have also been made in the recent past for the hot stars, namely α Aquilae[9] (HR 7557; Ohishi et al. 2004), α Lyrae (Aufdenberg et al. 2006; Peterson et al. 2006). Aufdenberg et al. (2008) reported the limb-darkened measurements of two bright stars, α Cygni (HR 7924; A2 Ia-type, supergiant) and β Orionis (HR1713; B8 Ia-type) with CHARA/FLUOR instrument. They have measured squared visibilities, $\mathcal{V}^2$, in the first and second lobes as a function of projected baseline. The important aspect of these observations is the detection of stellar wind of the former interferometrically. They estimated the diameter of the latter ($\simeq$ 2.75 mas) as well.

The visibility amplitudes have been measured recently by Domiciano de Souza et al. (2008) on the rare F-type yellow-white (F0 Ib) supergiant ($8 - 9$ $M_\odot$),

---

[9] The first magnitude A7-V-type star, α Aquilae (Altair) at visual magnitude 0.77, is the brightest star in the constellation Aquila. It is notable for its extremely rapid rotation.

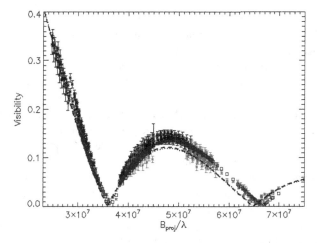

Fig. 9.9 Visibility amplitudes measured on the α Carinae in the H (gray;+) and K bands (black;+). The model visibilities (*squares*) were calculated with a linear limb-darkened (LLD) disk model fitted to the data. Theoretical visibilities from a LLD disk model with limb darkening coefficients from Claret (2000) are also shown (*dots* for H and *dashes* for K). Clearly, these theoretical LLD models do not account for the observations, especially after the first minimum (Domiciano de Souza et al. 2008; courtesy: Domiciano de Souza)

α Carinae (Canopus; RA $06^h23^m57.11^s$, δ $-52°41'44.38''$), a southern star in the constellation of Carina in the H- and K-bands using VLTI/AMBER instrument. Fifteen thousand times more luminous than the Sun, with visual magnitude $m_v =$ -0.72, its surface temperature has been estimated at $7,350 \pm 30° K$ (Desikachary and Hearnshaw 1982). Domiciano de Souza et al. (2008) have determined its linear limb-darkened angular diameter of 6.93 ± 0.15 mas by adopting a linearly limb-darkened disk model. They have also derived the stellar radius $R = 71.4 \pm 4.0 R_\odot$ from this angular diameter and *Hipparcos* distance. The relatively important height of the second visibility lobe in Fig. 9.9 is interpreted as due to the presence of surface convective cells on α Carinae.

9.2.5 Stellar Rotation

Stellar rotation is an important aspect in understanding of star-formation and evolution. Rotation causes meridional circulation. It is related to the activity, winds, mass loss, and braking mechanism, which could be either magnetic field, or internal oscillations (g-modes). It is possible to derive how quickly the star is rotating from the spectral line profile, which also reflects rotational broadening.

Rotating stars are significantly more complicated than spherically symmetric stars. The difficulty arises from interpreting data as the observed parameters depend on the unknown inclination between the observer and the rotation axis. If the star is observed from near its equatorial plane, an oblate image results, as for the α Eridani (Kervella 2007), or the α Aquilae (Monnier et al. 2007). To note, the star becomes oblate because the centrifugal force resulting from rotation creates additional outward pressure on the star. The ratio of equatorial/ polar diameter is greater than unity in such cases; the equatorial regions of a star has a greater centrifugal force compared to the pole. Such a force pushes mass away from the axis of rotation, and results in less pressure on the gas in its equatorial regions causing the gas to become less dense and cooler. However, the measurement of oblateness require precise diameter measurement.

Rapidly rotating main-sequence dwarf stars naturally take on an oblate shape, with an equatorial bulge that for stars rotating close to their braking speed may extend into a circumstellar disk, while the higher effective-gravity regions near the stellar poles become overheated, driving a stellar wind. The equatorial region is cooler than the polar region and due to the differences in limb-darkening in the polar and equatorial directions, the polar diameter may appear larger in observations limited to low spatial frequencies. With adequate resolution, both oblateness and equatorial darkening can be determined in some favorable cases. For rapid rotators, interferometry can map the ejected material, which may be found in disks (Tycner et al. 2006) and/or in polar winds (Kervella et al. 2009). The interferometric observations suggest that the very rapid stellar rotation causes enhanced mass-loss along the rotation axis rather than from the equatorial regions, resulting from the large temperature difference between pole and equator that develops in rapidly rotating stars (van Boekel et al. 2003). Such observations may provide further parameter constraints on stellar evolution models (van Belle et al. 2006).

Another physical effect is the gravity darkening effect, also known as von Zeipel law, making the surface gravity and the emitted flux decrease from the poles to the equator. The gravity darkening is a phenomenon, where a star rotates so rapidly that has a detectable oblate shape. The poles of such a star will have a higher surface gravity, and thus temperature and brightness. Thus, the poles are gravity brightened, and the equator gravity darkened. This effect was predicted theoretically by von Zeipel (1924), which states that the emergent flux, F of the total radiation at any point on the surface of a rotating star or tidally distorted star in radiative and hydrostatic equilibrium varies proportionally to the local effective gravity, g, i.e., $F \propto g^{\beta}$, where β is known as gravity-darkening coefficient. This effect has been directly measured by Domiciano de Souza et al. (2003) with VLTI for the Be star, α Eridani (see Sect. 9.2.9.2).

The α Aquilae was found to be flattened at the poles due to its high rate of rotation. This phenomenon was imaged by CHARA array combined combined with the Michigan Infra-Red Combiner (MIRC; Monnier et al. 2006a). Monnier et al. (2007) have measured its temperature contours. The atmospheric temperatures range was found to be between 7,000 and 8,000°K from equator to pole. The exact temperatures depend on gravity darkening and differential rotation models. If a star instead is observed from above its poles, it is possible to envisage a radial temperature gradient. From the optical interferometric observations,

Peterson et al. (2006) showed that α Lyrae is a distorted, rapidly rotating pole-on star. The temperature drops more than 2,400°K from pole to equator, creating an 18-fold drop in intensity at 500 nm, compared to 5-fold drop for non-rotating model. Stars with rapid and strong differential rotation may take on weird shapes, midway between a donut and a sphere (MacGregor et al. 2007).

However, the stellar rotation can be studied if the line profiles reflect the rotational broadening and the interferometer measurements resolve the profile. Differential speckle interferometric technique can be employed to obtain some rotation related information to a spatial resolution beyond the limit dictated by the interferometer baseline. However, Townsend et al. (2004) showed that the diagnosis of Doppler-broadened spectral lines might underestimate the true rotational velocity of the star due to such gravity darkening effect, and consequently the reported rotational velocities of Be stars may be systematically lower. Through a combination of K-band interferometric with the CHARA array and optical spectroscopic measurements, McAlister et al. (2005) have determined the equatorial and polar radii and temperatures, the rotational velocity and period, the inclination and position angle of the spin axis, and the gravity darkening coefficient ($\beta = 0.25 \pm 0.11$) for the rapidly rotating star, α Leonis (B7V-type). This star, the brightest in the constellation Leo, has a distorted shape exhibiting gravity darkening, a phenomenon that detected in binary stars. Its equatorial radius ($R_e = 4.16 \pm 0.08\ R_\odot$) is 32% larger than the polar radius ($R_p = 3.14 \pm 0.06\ R_\odot$). The darkening occurs since it is colder at the equator than at its poles. The temperature at its poles is found to be $15,400 \pm 1,400$°K, while the equator's temperature is $10,314 \pm 1,000$°K. Recently, Gies et al. (2008) have found that α Leonis has a companion that could be a white dwarf having a mass of 0.3 $M_\odot$ and an orbital period of 40.11 days.

9.2.6 Be Stars

Be-stars[10] are stars of spectral type B, on (or near) the main-sequence, which exhibits Balmer lines in emission and infrared excess produced by free-free and free-bound processes in an extended circumstellar disk. Weak emission lines from other atomic ions may also be present. A heterogeneous group of objects such as pre-main sequence B[e]-type stars, B[e] supergiants, proto-planetary nebulae B[e]-type stars, and symbiotic B[e]-type stars emit forbidden emission lines in their optical spectrum. In a given stellar field, approximately 20% of the B stars are in fact Be stars. This percentage can be much higher in some young clusters where more than 50% of the B stars are showing the Be phenomenon. They are very bright

---

[10] the 'e' stands for emission.

having high luminosity ($\mathcal{L}_\star \sim 10^6 \mathcal{L}_\odot$). The luminosity function may seem to contain too massive stars, leading to an artificially top-heavy Initial Mass Function[11] (IMF).

Generally, Be stars have high rotational velocities, which is of the order of $\sim 350\ \mathrm{km\ s^{-1}}$. They manage to eject gas into a circumstellar disk (Porter and Rivinius 2003). The stars of this category play an important role in the heating of galaxies. Many Be stars show variable discrete absorption components in UV resonance lines (Prinja 1989). In fact, all B-type stars with $M_V \geq -7$ with such components are Be stars and more of 80% of O-type stars are showing these components in their UV spectra (Howarth and Prinja 1989).

The formation of rotationally flattened gas shell is due to the centrifugal forces at their surfaces. Such a force makes the effective gravity and the brightness temperature decrease from pole to equator, and thus the corresponding radiative force will depend on the stellar latitude. The circumstellar envelopes of Be stars are typically assumed to be thin, axisymmetric, circumstellar disks viewed at different angles, and therefore the axial ratio is mostly due to a projection effect (Wood et al. 1997). Hummel and Vrancken (2000) presented a model for the circumstellar disks of Be stars, which has azimuthal velocities close to the Keplerian. The basic concept is that the disk is axisymmetric and begins at the stellar surface. The gas density in the disk is given by,

$$\rho(R, Z) = \rho_0 R^{-m} e^{-\frac{1}{2}\left[\frac{Z}{H(R)}\right]^2}, \tag{9.34}$$

in which $H(R)(= C_s R^{3/2}/v_k)$ is the gas scale height in units of stellar radius, $C_s(= \sqrt{\gamma R_g T_d})$ the sound speed, $\gamma = 5/3$ for mono-atomic gases, R_g the universal gas constant, $T_d = 2T_e/3$, T_e the stellar effective temperature, $v_k(= \sqrt{GM_\star/R_\star})$ the Keplerian velocity at the stellar equator, $M_\star$ the mass of the star, G the gravitational constant, $R_\star$ the radius of the star, $R = (\sqrt{x^2 + y^2})/R_\star$, $Z = z/R_\star$ the radial and vertical cylindrical coordinates in units of stellar radii, ρ_0 the base density at the stellar equator, m the radial density component. To note, these authors have assumed that the temperature is constant throughout the disk.

The disks of Be-stars exhibit large temporal variations, ranging from days to decades, reflecting large structural changes in their geometry and/or density; in some cases they disappear altogether for extended periods. This time-variability suggests

---

[11] Initial Mass Function (IMF) is a relationship that specifies the distribution of masses created by the process of star-formation. This function infers the number of stars of a given mass in a population of stars, by providing the number of stars of mass $M_\star$ per pc^3 and per unit mass. Generally, there are a few massive stars and many low mass stars. For masses $M_\star \geq 1\ M_\odot$, the number of stars formed per unit mass range $\xi(M_\star)$, is given by the power law (referred to as Salpeter IMF),

$$\xi(M_\star) = \xi_0 M_\star^{-(1+x)},$$

in which $x = 1.35$ was determined by Salpeter (1955); a star's mass determines both its lifetime and its contribution to enrich the ISM with heavy elements at the time of its death.

that the gas injection into the disk is partially the result of the fast spin and equatorial extension of the Be star (Townsend et al. 2004) and that magnetic, pulsational, wind driving, and/or other processes may also be required for mass-loss into the disk (Owocki 2005). Photometric observations of Be star disks indicate that they may evolve into ring structures before disappearing into the interstellar medium (de Witt et al. 2006). These stars loose mass along the poles through a strong stellar wind and are surrounded at the equator by a disk of matter. Their gravity and temperature are aspect angle-dependent.

In addition to photometric and spectroscopic observations, the presence of circumstellar disk around Be stars can be studied by the interferometry equipped with a polarimeter (Mackay et al. 2009). The polarization in these stars arises from Thomson scattering in the ionized circumstellar material and can reach upto 2% (Porter and Rivinius 2003). Study of Be stars overlaps with other fields in stellar physics, such as the evolution of massive stars, interacting binaries that exhibit both mass-loss and mass transfer, magnetic field evolution, asteroseismology, formation of stellar winds and mass-loss. Although γ Cassiopeiae has been a favorite target for long baseline interferometry, with further systematic monitoring of multiple emission-lines, the formation, structure, and dynamics of other Be stars can also be addressed. In what follows, some interesting results that are obtained by means of LBOIs are illustrated.

9.2.6.1 γ Cassiopeiae

The emission-line star, γ Cassiopeiae (HD 5394, HR 264), is a third brightest star in the W-shaped constellation Cassiopeia. A B0.5 IVe-type subgiant having 25,000°K surface temperature, this star was observed in 1886 by Angelo Secchi. It is an eruptive variable star, whose brightness changes irregularly between 2.20 m_v and 3.40 m_v. It rotates with enormous speed, spinning about 300 km s^{-1} at its equator. The rotation and high luminosity conspire to drive mass from the star into a surrounding disk that radiates the emissions. This object is a spectroscopic binary with an orbital period of 203.59 days and an eccentricity of 0.26. The mass of the companion is about 1 $M_\odot$, implying a white dwarf or a neutron star but this could also be a normal late-type dwarf (Harmanec et al. 2000; Miroschnichenko et al. 2002). The presence of discrete absorption components in the spectra of γ Cassiopeiae was reported by Hammerschlag-Hensberge (1979) depicting blue-shifted absorption enhancements in the resonance lines of N V, Si IV and C IV.

γ Cassiopeiae, also radiates X-rays. The origin of X-ray emission has been attributed mostly to an unseen accreting compact companion. However, combination of results obtained with far-UV observations using the International Ultraviolet Explorer (IUE) and optical spectroscopic and photometric observations Horaguchi et al. (1994) found no correlation between the X-ray, far UV and optical behavior, which suggest that the X-ray source of this star must be far from the central star. Recent evidence suggests that the X-rays may be associated with the Be star itself or in some complex interaction between the star and surrounding decretion disk. The X-ray production is known to vary on both short and long time scales with respect to various UV line and continuum diagnostics associated with a B star or with circumstellar matter close to the star (Smith and Robinson 1999; Cranmer et al. 2002). Moreover, the X-ray emissions exhibit long-term cycles that correlate with the visible wavelength light curves (Smith et al. 2004).

The first interferometric measurement of γ Cassiopeiae was obtained by Thom et al. (1986) in the continuum, as well as in the Hα emission-line with the I2T, measuring a 3.25 mas diameter for the envelope. Following this result, a lot of progress and results have been obtained by a few more interferometers. The measurements of spectrally resolved visibilities of this star with GI2T by Mourard et al. (1989) distinguish the hydrogen emission in the envelope from the continuum photospheric emission. With the central star as a reference, they determined the relative phase of the shell visibility. Combining high spatial resolution and medium spectral resolution of 1.5 Å centered on Hα, they have shown that the envelope around this star was in rotation and was approximately fitting a disk model; the star is distorted due to its fast rotation. As many as $\sim$300,000 short-exposure images were digitally processed using the correlation technique. The results were co-added to reduce the effect of atmospheric seeing and photon noise, according to the principle of speckle interferometry. Through subsequent observations with the same instrument of γ Cassiopeiae in the He I (λ 6678 Å) and Hβ emission-lines on later dates, Stee et al. (1998) found that the envelope size in the visible increases following the sequence: He I at λ 6678 (2.3 R$_\star$), 0.48 μm continuum (2.8 R$_\star$), 0.65 μm continuum (3.5 R$_\star$), Hβ ($\leq$ 8.5 R$_\star$), and Hα (18 R$_\star$). They have also derived the radiative transfer model for this star based on the spectroscopic and interferometric data. Figure 9.10 depicts the intensity maps from the model of γ Cassiopeiae of different Doppler-shifts across the Hα emission-line (Stee et al. 1995). Figure 9.11 depicts the global picture of circumstellar environment of the said star.

Using the data obtained since 1988 with GI2T, Berio et al. (1999a, b) have found the direct evidence of a slowly prograde one-armed oscillation of its equatorial disk. This prograde precession agrees with the Okazaki's model (1997) of a one-armed ($m = 1$) oscillation confined by radiative effects. They have investigated how such oscillations occur within the equatorial regions of the latitude-dependent radiative wind model developed by Stee and de Araújo (1994). The star has a highly ionized fast wind in the polar regions and a slow wind of low ionization, with a higher density near the equator. In the inner ($\leq$ 10 R$_\star$), cool, dense equatorial regions, the kinematics is dominated by the Keplerian rotation, while the one-armed oscillations developed in the region where the radiative wind is driven by optically thin lines (Stee 2005 and references therein).

Marlborough (1997) found an apparent similarity between the variations of γ Cassiopeiae during the period 1933–1942 and the variations of 1-2^m, which occur in the LBV objects on time scales of tens of years. However, Hummel (1998) opined that the specular emission-line variations occurring in γ Cassiopeiae may be produced by a circumstellar Keplerian disk, which is tilted with respect to the equatorial plane. The precessing nodal and apsidal line causes a variation in

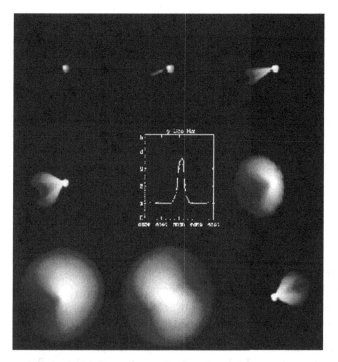

Fig. 9.10 Intensity maps from the model of γ Cassiopeiae of different Doppler-shifts across the Hα emission-line (courtesy: P. Stee). Each map is computed within a spectral band of 0.4 nm; theoretical visibilities are computed from these maps and compared to GI2T data

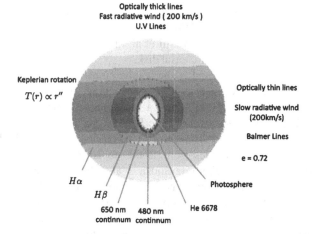

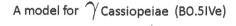

Fig. 9.11 Global picture of circumstellar environment of γ Cassiopeiae (courtesy: P. Stee)

the shapes of emission-line widths and profile. According to him, the alternating shell-phases and narrow single-peak phases might be due to an apparent variation in the disk inclination. In order to cover a large (u, v) plane by using the 'super-synthesis' effect, Quirrenbach et al. (1993a) exploited the capability of the Mark III interferometer, where it was proved that γ Cassiopeiae envelope was not spherical but rather elongated to the East from $20°$ with respect to the North–South direction. Further observations with the Navy Prototype Optical Interferometer (NPOI), Tycner et al. (2003) have resolved the Hα emitting region. Their best fit of an elliptical Gaussian model has an average angular size of the major axis of 3.67 ± 0.09 mas, an average axial ratio of 0.79 ± 0.03 and an average position angle of $32° \pm 5°$.

9.2.6.2   Other Be Stars

Apart from γ Cassiopeiae several other Be stars such as, ψ Persei (Gies et al. 2007), ζ Tauri, κ Draconis (Gies et al. 2007), α Eridani, α Arae (Meilland et al. 2007a), κ CMa (Meilland et al. 2007b), δ Centauri (Meilland et al. 2008), CI Camelopardalis, 51 Ophiuchi (Tatulli et al. 2008), etc., have also been observed; a few of these observations are described below:

1. ζ Tauri: A single-lined spectroscopic binary with a period of about 133 days and an eccentricity of ~ 0.15 (Harmanec 1984) in the constellation Taurus, ζ Tauri is a hot blue class B2 subgiant star, approximately 417 ly from the Earth, whose surface temperature is about $22,000°$ K. Owing to the intrinsic variability of both the components, its luminosity varies from magnitude, $m_v = 2.88$ to $m_v = 3.17$. Its equatorial rotation speed has been as high as 330 km s^{-1}, spinning around with a period of only one day. The rotation of this star is related to a thick disk of matter that surrounds it. The disk radiates bright emissions from hydrogen in the red and blue parts of the spectrum. Tycner et al. (2004) have measured the angular size of the major axis (3.14 ± 0.21 mas), the position angle ($-62.3° \pm 4.4°$), and the axial ratio (0.310 ± 0.072). Vakili et al. (1998) have detected its prograde one-armed oscillation in the equatorial disk.

2. α Eridani (HD 10144): Located at the southernmost point of the constellation Eridanus, α Eridani is a fast rotator ($v \sin i$ ranging from 220 to 250 km s^{-1}) with a rotational period of hours and a high magnetic field around one kG. It is a first magnitude ($m_v = 0.46$) blue B3 Ve-type star that possesses six to eight solar masses lying at a distance of 44 pc (*Hipparcos*) from the Earth. The effective temperature T_e varies from 15,000 to $20,000°$K. Observations with VLTI instrument by Domiciano de Souza et al. (2003) showed that α Eridani is highly deformed star, which may be due to rotation. Their measurements correspond to a major/minor axis ratio of 1.56 ± 0.05. This observation revealed the flattened shape of the star, which rules out the commonly adopted Roche approximation in which the rotation is supposed to be uniform and the mass centrally condensed. This might also be the evidence that Be stars are rotating close to their critical velocities. Its rapid rotation induces mainly two effects

on the stellar structure such as (1) a rotational flattening and (2) a gravity darkening effect described by von Zeipel (1924). Kanaan et al. (2008) presented a model based on the interferometric observations, as well as on spectroscopic observations of α Eridani. Such a model allowed them to investigate possible geometries of the circumstellar environment. They have computed three different kinds of models: an equatorial disk, a polar wind, and a disk and wind model. In this model, the overheated polar caps of this star (due to the von Zeipel effect) eject a fast stellar wind that radiates free-free emission in the K-band.

3. α Arae (HD 158427): Located about 74 pc away from the Earth α Arae (B3 Ve) is a blue-white variable Be star in the constellation Ara. It has a mean apparent magnitude, $m_v = 2.84$, varying brightness from magnitude $m_v = 2.79$ to $m_v = 3.13$. This star was observed by infrared interferometry, using the MIDI and AMBER instruments (see Sect. 6.1.3.5) at the VLTI, which allowed to study the kinematics of the gas envelope inner part, as well as to estimate its rotation law. It has an equatorial rotational speed of about 470 km per second, very close to its critical velocity (Meilland et al. 2007a), which may bring sufficient energy to levitate material in a strong gravitational field to provide sufficient energy and angular momentum to create the circumstellar disk. The disk around α Arae that is in Keplerian rotation (rather than uniform) is compatible with a dense equatorial matter confined in the central region whereas a polar wind may be contributing along the rotational axis of the central star. Figure 9.12 depicts the intensity map in the continuum at 2.15 μm obtained with model parameters for the α Arae central star and its circumstellar environment obtained from their work.

4. CI Camelopardalis: CI Camelopardalis (MWC 84; spectral type B4 III-V), a possible optical counterpart of the X-ray transient, XTE J0421+560, is known to be a supergiant Be star emitting a two-component wind, a cool, low-velocity wind and a hot, high velocity wind. The X-ray-radio-optical outburst and the

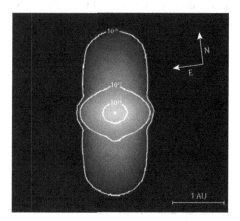

Fig. 9.12 Intensity map in the continuum at 2.15 μm obtained with model parameters of α Arae (courtesy: F. Malbet)

appearance of relativistic S-shaped SS 433[12] like jets make this star to be an interesting object. Thureau et al. (2009) have observed CI Camelopardalis in near-infrared H- and K-spectral bands by means of a few LBOIs, such as IOTA2T and its upgraded version IOTA3T and PTI, for a long period of time from soon after its outburst in 1998. Their analysis of the visibility data shows that the dust visibility has not changed over the years. The star seems to be elongated confirming the disk-shape of the circumstellar environment. The closure phase measurements also show direct evidence of asymmetries in the circumstellar environment of CI Camelopardalis.

9.2.7 Stellar Surface Structure

Interferometry plays an important role in understanding the pulsation mode of Mira variables. These variables are M-type low or intermediate-mass AGB stars and have long periods of 80–1,000 days, usually with emission-lines in their spectrum. They have cool ($T_e \leq 3000°K$) and extended ($R_\star > R_\odot$) photosphere. These stars vary in brightness by 2.5–11 magnitudes. Mira stars have masses similar to that of the Sun but have a feeble gravitational hold on the material in their outer layers. As a result they are losing gas in a steady stellar wind at the rate of about $10^{-7} - 10^{-6}$ $M_\odot$ yr^{-1}. Such material accumulates around the star as an extensive circumstellar shell. The effective temperature of the Mira variables is about 2,000°K; 95% of their radiation is in the infrared. Haniff et al. (1995) have reported that the derived linear diameters of some Cepheids are not compatible with fundamental mode of pulsation. Mira-type Long-Period Variables (LPV) pulsate more exuberantly than any other type of stars. There are several LPVs with $K < +2$ and photospheric apparent angular diameters exceeding 5 mas. Interferometric observation of Cepheid binary systems can provide improvement in the calibration, which allow to determine their masses as well. The direct determination of stellar surface fluxes and effective temperature would reveal the Cepheid atmosphere with clarity.

---

[12] The enigmatic star SS 433 is a bright compact emission line object (either a neutron star or a black hole) in the constellation Aquila having a companion (A-type star; Hillwig et al. 2004) with an orbital period of 13.1 d, a large accretion disk, and two collimated relativistic jets moving at 26% of the speed of light. The disk-axis makes an angle of 78° with the line of sight, while the jet precesses with the axis at an angle of 19° (Margon 1984) with a periodicity of about 162.15 d. Material from the normal star is falling in the direction of the compact object. The angular momentum of the material forms the disk, which radiates at wavelengths ranging from the optical to the X-ray. A multiwavelength campaign of this star had found sharp variations in intensity on time-scales of a few minutes in the X-ray, IR and radio wavelengths (Chakrabarti et al. 2005).

The pulsation constant, Q, can be determined by combining the period, angular diameter, and parallax with plausible values for the mass range of Mira stars, for instance,

$$Q = P \left(\frac{M}{M_\odot} \right)^{1/2} \left(\frac{R}{R_\odot} \right)^{-3/2}, \qquad (9.35)$$

where P is the period, M and R the mass and radius of the star respectively, $M_\odot$ and $R_\odot$ the respective solar mass and radius.

By substituting R from (9.11), one finds,

$$M \propto L^{3/2} T^{-6} P^{-2}. \qquad (9.36)$$

This may allow to know if the LPVs pulsate in the fundamental-mode or overtone-mode. Stellar analogs of the solar 5 min oscillation[13] may be detected. Astero-seismology aims to study the internal structure and dynamics of stars with a high precision; it constraints the stellar evolution theory. Two methods such as (1) photometric measurements that essentially detect stellar variations ranging from several milli-magnitudes to one or two magnitudes and (2) spectroscopic measurements that permits to detect small amplitude on solar-type oscillators, have been mainly used from the ground, following the amplitudes of the modes. The latter method requires measurements of radial velocity through the use of Doppler shift of many spectral lines. Systematic observations with sufficient spectral resolution and coverage, with a range of baseline lengths in a LBOI, and covering the full pulsation cycle may be needed for critical tests of the model atmospheres, and for the eventual determination of the pulsation mode of Mira variables.

It may be feasible to study the surface structures of supergiants stars, which are predicted from three-dimensional (3-D) and time-dependent models of large-scale

---

[13] Waves travel in all directions inside the Sun's interior, from the deep core to the surface. They may be classified as acoustic waves, Alfvén waves, shock waves etc., depending on the restoring forces and other mechanisms that come into play. The photometric oscillations observed by helioseismology reveal interior solar properties through their dependence on the density distribution within the Sun. Such oscillations are acoustic waves (or p-mode, since they have pressure as their restoring force) generated by convection. The p-mode oscillations are strong in the 2–4 MHz range frequencies, where they are often referred to as 5-min oscillations; the g-mode or gravity waves have gravity as restoring force and their oscillations are low frequency (0–0.4 MHz). At the solar surface, p-modes have amplitudes of hundreds of kilometers, however beyond the surface of the Sun they are unable to move any further, since the solar atmosphere is not dense enough for the wave propagation (Fossat 2005). These waves are trapped and their resonance is organizing eigenmodes. Optically, they are visible at the surface, and by measuring the frequencies of many such modes and using theoretical models, one may infer the Sun's internal structure and its dynamical behavior. These acoustic waves travel very fast (in an hour or so) from the Sun's center to its surface, which provide a nearly real time access to the solar core. To note, photons take hundreds of thousand years to come out from the very dense and opaque internal layers. In order to measure solar oscillation, a technique, known as helioseismology, which is a part of the NOAO program called the Global Oscillation Network Group (GONG) operating an international network of instrument stations is in place. Such observational facilities provide continuous measurements of solar oscillations, by having a station in daylight at all times.

stellar convection (Freytag et al. 2002). Multi-wavelength studies of α Orionis have found a strong variation in the asymmetry of the stellar brightness as a function of wavelength, including three bright spots interpreted as unobscured areas of elevated temperature, shining through the upper atmosphere at the shorter wavelengths (Young et al. 2000).

Although it is not feasible to image the smaller convective surface features on dwarf stars directly, predictions from 3-D stellar atmospheric models may possibly be tested from limb-darkening curves observed at different wavelengths, for instance, the brightest star in the constellation Canis Minor, α CMi having a visual apparent magnitude, m_v 0.34, a white main-sequence star of spectral type F5 IV-V (Aufdenberg et al. 2005), and α Centauri B, an orange star[14] with a spectral type K1 V with a visual magnitude m_v 1.34 (Bigot et al. 2006). For some classes of active dwarf stars with suitably rapid rotation, Doppler imaging may enable for inferring of spotted stellar surface structure (Wolter et al. 2008).

9.2.8 Stellar Atmospheres

Stars are gaseous spheres, which do not have a well-defined edge. The intensity profile across the stellar disk depends on the stellar atmospheric structure. The stellar atmosphere is the outer region of a star extending upto interstellar medium. This low-density outer region produces the stellar spectrum. The physical depths in the atmosphere where the spectral features form depend on the atmospheric conditions, such as temperature, density, level populations, optical depth etc. For example, in the case of the Sun, the lowest and coolest part is the photosphere. Light escaping from the surface of the Sun stems from this region and passes through the higher layers. Above the photosphere, lies the chromosphere, a thin spherical shell region extending upwards to heights from ten to fifteen thousand kilometers. It is an irregular layer, in which the temperature rises from $5000°K$ to $20,000°K$ at the upper boundary. In the transition region above the chromosphere, the temperature increases rapidly on a distance of only around 100 km. Beyond this region, a pearly white halo called corona extends to tens of thousands of kilometers into space; corona is extremely hot having temperature about a few million degrees kelvin. Note that a number of evolved stars, as well as B- and A-type mainsequence stars, do not have transition regions or corona.

The stellar diameter can be envisaged differently when observed in chromospheric or photospheric spectral lines. Many young, cool stars, M-type stars in

---

[14] α Centauri is in the constellation of Centaurus in the Southern Hemisphere as the outermost pointer to the Southern Cross. Two components of the system are too close to be resolved as separate stars by the naked eye and so are perceived as a single source of light with a total visual magnitude, m_v 0.27. It is a triple star system consisting of two main stars, α Centauri A (G2 V-type) and α Centauri B, at a distance of 4.36 ly from the Sun. An estimated age of these stars is about 6 billion years (England 1980).

particular, exhibit evidence of very strong chromospheric activity in the form of emission-line cores to their absorption features. The chemical composition of the stellar atmosphere can be determined from the strength of the observed spectral lines (see Sect. 9.1.5) by comparing them with theoretically generated synthetic spectra using stellar model atmosphere. In the stellar atmosphere, (1) the spectra may turn from photospheric absorption to chromospheric emission, (2) the forbidden lines from metastable atomic states as produced in regions with very low densities, for example planetary nebulae, (3) the fluorescence emission-lines[15] in complex spectra, like Fe II, occur in stars like symbiotic stars,[16] and (4) the stimulated emission-lines appear in extended low-density clouds close to hot, eruptive stars, such as the exceptionally luminous and highly variable object, η Carinae.

In some stars, the atmospheres do not have perfect symmetric structure. The most extreme cases are Ap- and Bp- stars[17] that have strong magnetic fields, in some cases reaching several kilogauss, and have peculiar chemical abundances (Shulyak et al. 2008). With polarimetric interferometry, Hanbury Brown et al. (1974c) and subsequently Rousselet-Perraut et al. (1997) attempted to study scattering and mass-loss phenomena inside hot and extended environments. Donati et al. (2008) studied polarized line profiles to tomographically reconstruct the magnetic field structures of M dwarfs.

Several supergiants have extended gaseous atmosphere, which can also be imaged in their chromospheric lines. However, they become opaque at a larger radius in absorption bands of molecules such as TiO than in the continuum. In such a situation, measuring photospheric diameter turns out to be a bane (Baschek et al. 1991). Strong variations of diameter with TiO absorption depth have been observed in the Mira variables (Labeyrie et al. 1977; Bonneau et al. 1982; Di Giacomo et al. 1991; Haniff et al. 1995); the hotspots and other asymmetries on the surface of these variables and giants have been noticed (Tuthill et al. 1997). The red giants represent the final stage of massive star evolution. Their photospheres are highly convective, contain many molecular species, and may exhibit hot/cool spots. Their diameters depend on wavelength and can vary greatly between line and continuum (Dyck and Nordgren 2002). For example, α Orionis is a good candidate to understand the process of mass loss in red supergiants.

With LBOI equipped with adaptive optics system, the photosphere and close environment of such a star can be studied. Wilson et al. (1997) have detected a

---

[15] Fluorescence lines appear when higher energy photons excite a sample, which would emit lower energy photons when the atom cascades down in the energy levels.

[16] Symbiotic stars are believed to be binaries, whose spectra indicate the presence of both cool and hot objects; one component may be an irregular long period cool M-type and the other is a hot compact star. The spectra of these stars contain strong emission lines of high ionization potential like [O III]. Both the components are associated with a nebular environment. Interaction of both components may lead to accretion phenomena and highly variable light curves.

[17] Ap- and Bp- stars are peculiar stars of types A- and B-type stars, which show over abundances of certain elements such as, magnesium (Mg), silicon (Si), chromium (Cr), strontium (Sr), and europium (Eu). The additional nomenclature, p, is used to indicate peculiar features of the spectrum. These stars have a slower rotation than normal for A- and B-type stars.

complex bright structure in the surface intensity distribution of α Orionis, which changes in the relative flux and positions of the spots over a period of eight weeks; a new circularly symmetric structure around the star with a diameter of $\geq 0.3''$ is found. Recent observations, carried out at the VLTI/AMBER facility, revealed that the gas in the star's atmosphere is moving vigorously up and down. It has a vast plume of gas, almost six times the diameter of the star, extending into space from its surface and a gigantic bubble boiling on its surface (Kervella et al. 2009; Ohnaka et al. 2009). These investigations provide important clues to explain how supergiant stars lose shed material at such a high rate. With a larger baseline interferometer, it is possible to resolve the flattened shapes of rapidly rotating stars, and periodically changing sizes of pulsating ones.

Longer period pulsating variable like Mira stars show very large amplitudes with strong wavelength dependent distortions. With SUSI instrument studded with the Optical Interferometric Polarimetric (OIP; LBOI with polarimetry) observation, Ireland et al. (2005) were able to place constraints on the distribution of circumstellar material in R Carinae and RR Scorpii. The inner radius of dust formation for both stars was found to be less than three stellar radii; much smaller than expected for 'dirty silicate' grains. A geometrically thin shell fit the data better than an outflow. Recent studies, with the ISI three telescopes, of o Ceti and M8 IIIe-type variable R Leonis by Tatebe et al. (2008) showed that the shape of the former star is rather symmetric, while the latter star appears more asymmetric.

Figure 9.13 represents two visibility curves of the oxygen-rich Mira-type variable star R Leonis, obtained by the FLUOR/IOTA combination in the K band at two different epochs, which show both the change in equivalent Uniform Disk (UD) diameter and (for the 1997 data) the diffusion by circumstellar material containing ionized gas or dust, whose signature is a visibility curve substantially different from that of a uniform disk (Perrin et al. 1999). From the data obtained with IOTA during 2005–2006 in the H-band, Lacour et al. (2009) have found $\sim$40% variation in the diameter of the S-type Mira star, χ Cygni. The object was observed at H-band where the images show 40% variation in the stellar diameter. The data analysis was carried

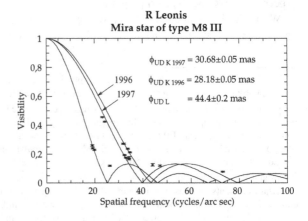

Fig. 9.13 Visibility curves of the Mira variable R Leonis (courtesy: V. Coudé du Foresto)

out by measuring squared visibilities and obtaining closure-phase. The imaging was performed by the Multi-aperture Image Reconstruction Algorithm (MIRA; Thiébaut 2008). They also observed significant changes in the limb-darkening and stellar inhomogeneities. The radius of the star, corrected for limb-darkening, has a mean value of 12.1 mas and shows a 5.1 mas amplitude pulsation. Their model reveals presence and displacement of a warm molecular layer as well.

Long-period (radially or non-radially) pulsating variables of the Mira-type not only show substantial distortions from circular symmetry, but different variations in different spectral features, for example the molecular atmosphere seen in TiO bands, as opposed to the photospheric continuum (Haniff et al. 1992; Ragland et al. 2006). Further, the stellar diameters undergo huge cyclic variations on the order of 50% (Burns et al. 1998), and the combination with infrared and radio data reveals intricate relationships between the photosphere, molecular layer, dust shell, and SiO maser emission (Ohnaka et al. 2007; Wittkowski et al. 2007). Although the pulsational amplitudes in Cepheid variables are smaller, the changing stellar size is now seen by interferometers (Kervella et al. et al. 2004).

9.2.9 Circumstellar Shells

Strong stellar winds from cool red giants give rise to circumstellar shells that are less dense than the photospheres of such giants, but dense by ISM standards (Green et al. 2004). In their inner regions, these shells are kept warm by the star's luminosity, with temperature around $1,000°K$. As the circumstellar matter surrounding objects ejecting mass in the form of dust grains and molecules, moves outwards, it turns out to be less dense and also cools. It might get widespread, what is referred to as thin interstellar cirrus; space-borne IRAS had discovered it in 1983. All matter from stellar winds from stellar winds becomes thinly dispersed.

Post-AGB phase (or proto-planetary nebula phase), the stars lie for a short period between the AGB phase and the Planetary Nebula (PN) stage. The rising central temperature enhances the rate of energy production during helium burning, causing large increase in luminosity with accompanying mass ejection, what is referred to as helium flash. For low mass stars below 3 solar mass, the degeneracy is removed with high enough temperature and the stars settles down in a helium burning phase. When the envelope mass reaches a critically small value, the large mass-loss rates decrease to much lower values, the post-AGB phase commences. The central star contracts at constant luminosity and hence begins to increase in effective temperature. An unstable giant star throws off a significant fraction of its mass. This is more violent than mass-loss via stellar winds. When the star has become hot enough to ionize its circumstellar shell, it may become observable as a PN. Most stars that are formed with a mass of less than 8 $M_\odot$ evolve through the AGB-post-AGB-PN sequence to end as white dwarfs (Pottasch 1992; Kwok 1993; Zijlstra 1995).

When carbon burning takes place in the core of the stars, they also have helium and hydrogen burning regions in two outer shells. The final stage of these stars is the gradual shedding of all of the outer H and He layers, which gives rise to phenomenon displaying gaseous nebulae, what is referred to as Planetary Nebula (PN), surrounding the hot stellar core in their center. This core, known as white dwarf, is comprised of carbon and oxygen with a surface temperature of about 100,000°K.

The resulting planetary nebula is the interaction of the ejected shell of gas and the ultraviolet light from the hot stellar remnant, which heats the gas. The nucleus photoionizes and excites the ejected gaseous shell, giving rise to a nebula with a rich emission-line spectra that contains both allowed and forbidden transitions. The nebulae drift from the C–O–Mg core, which cools down to become a white dwarf, at a speed of 10–30 km s^{-1}. These objects represent the relatively short-lived phase, lasting for about 10,000 years, formed as a result of the stars in their AGB phase losing their outer shells before reaching the white dwarf stage (Pottasch 1984).

High resolution imagery may depict the spatial distribution of circumstellar matter particularly in compact young planetary nebula or newly formed stars. Observationally, they are found to be emitter of strong infrared emission originating from a dense and very extended circumstellar shell of cool gas and dust. The central stars are usually obscured by the circumstellar dust, and in many cases, they are optically invisible. In order to explore the immediate environment of young stellar objects, speckle interferometry with large telescope is used. Some of the speckle interferometric measurements have shown many previously unknown complex structures around these objects. In many cases, the images are very complex reflecting the complicated multicomponent structure of the circumstellar material. The accretion disk, gaseous streams, jets, and scattering envelopes were found in an interacting binary star, β Lyrae (HD 174638). GI2T was employed to determine subtle structures such as jets in this star, as well as the clumpiness in the wind of P Cygni (30 CMa). It is confirmed from the interferometric observations of the former star that the bulk of the Hα and the He I 6678 emission is associated with bipolar jet-like structure perpendicular to the orbital plane creating a light scattering halo above its poles (Harmanec et al. 1996). Figure 9.14 depicts the jet-like structure of β Lyrae.

Using high spatial resolution data at Hα and He I (667.8 nm) emission-lines, Vakili et al. (1997) resolved the extended envelope of the variable B1-type ($m_v = 5.0$) hypergiant Luminous Blue Variable[18] (LBV) star, P Cygni, located in the constellation Cygnus (RA (in the epoch J2000.0, J is the Julian year) $= 20^h17^m47^s$; Dec. (in the epoch J2000.0)$= 38°01'59''$), with the GI2T. Detection of a localized asymmetry at 0.8 mas to the south of the same star's photosphere has also been reported (Vakili et al. 1998). In order to determine precisely the clumps mass and nature, continuous monitoring of the clumps formation and evolution from the base

---

[18] Luminous Blue Variable (LBV) stars are extremely energetic and massive early type supergiants. They are unstable and variable, which exhibit occasional episodes of substantial mass-loss superimposed on slow brightness changes. They exhaust their nuclear fuel very fast, therefore their life-time is very short. These stars are formed in the regions of intense star-formation in the galaxies.

Fig. 9.14 Global scheme of the β Lyrae binary system observed with the CHARA interferometer. The *dashed line* at the center corresponds to the baseline orientation. The system is represented at the binary phase 0.80 but was also observed at 3 other phases, respectively 0.5, 0.9 and 0.2. The extensions are given in milliarcseconds (courtesy: D. Bonneau)

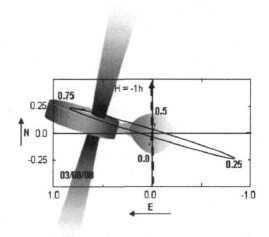

Fig. 9.15 Diffraction-limited K'-band image of Frosty-Leo reconstructed with the bispectrum speckle interferometry. The image shows the bipolar nebula, the clumpy intensity distribution resolved in both lobes, a dust lane between the lobes, and the central star (Murakawa et al. 2008a; courtesy: G. Weigelt)

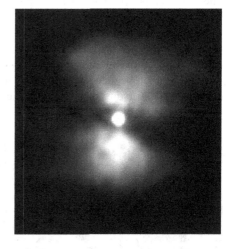

of the wind to larger scales is required. The high resolution instrument, such as LBOI, is ideally suited for monitoring the spatial activity from the innermost regions to the outer nebula.

In order to derive various physical properties of an oxygen-rich proto-planetary nebula, Frosty-Leo (IRAS 09371+1212), Murakawa et al. (2008a) have performed bispectrum speckle interferometric (K'-band) and adaptive optics imaging polarimetric (H- and K-band) observations. Reconstructed with the bispectrum speckle interferometry, these authors could resolve a point-symmetric bipolar lobes (see Fig. 9.15), an equatorial dust-lane, and complex clumpy structures in the lobes, but did not find any evidence of a binary companion of the central star as reported earlier (Roddier et al. 1995). The polarization signatures such as (1) a polarization disk extending $\sim 5''$ along the dust lane and (2) an elongated region with low polarizations along a position angle of $-45°$ are seen. Using 2-D Monte Carlo radiative transfer code (Ohnaka et al. 2006), these authors derived the dust disk mass (2.85×10^3 M$_\odot$), the inner radius of the disk (1,000 AU), the mass-loss rate (8.97×10^3 M$_\odot$ yr^{-1}), and the total envelope mass including the disk (4.23 M$_\odot$).

Another object CW Leo, a carbon star IRC+10216, which is a long-period variable star evolving along the AGB phase, also showed a resolved central peak surrounded by patchy circumstellar matter (Osterbart et al. 1996). This is a peculiar object with the central star being embedded in a thick dust envelope. A separation of $0.13''$–$0.21''$ between bright clouds was noticed, implying a stochastic behavior of the mass outflow in pulsating carbon stars. Weigelt et al. (1998) found that five individual clouds were resolved within a $0.21''$ radius of the central object in their high resolution K$'$ band observation. Monitoring of the components A, B, C, and D has revealed that the dust shell is evolving and the separations between the different components have steadily been increasing (see Fig. 9.16). Weigelt et al. (2002) reported that the separation between the two brightest components A and B was increased by $\sim 36\%$ from 1995 to 1999 and by 2001, it had increased to $\sim 80\%$. The latter component was found to be fading, while the faint components C and D became brighter. Based on the 2-D radiative transfer model of IRC+10216, these authors observed that the relative motion of components A and B is not consistent with the outflow of gas and dust at the terminal wind velocity of 15 km s^{-1}. The observed the relative motion of components A and B with a de-projected velocity of 19 km s^{-1} might be, according to these authors, caused by a displacement of dust density peak due to dust evaporation in the optically thicker and hotter environment.

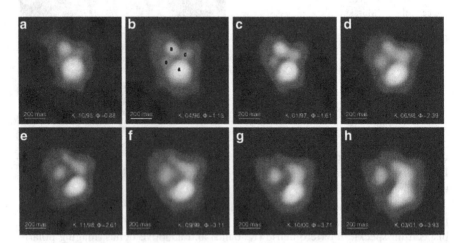

Fig. 9.16 Bispectrum speckle reconstructed images of IRC+10216 in K$'$-band for eight epochs from 1995 to 2001, which show the dynamic evolution of the inner dust shell (Weigelt et al. 2002). Courtesy: G. Weigelt

9.2.10 Binary Systems

A binary star is a stellar system consisting of two close stars moving around each other in space and gravitationally bound together. In most of the cases, the two members are of unequal brightness. The brighter star, in general, is called the 'primary', while the fainter is called the 'companion' or secondary. The position of the companion of a binary system with respect to the primary is specified by two coordinates, namely (1) the angular separation, ρ and (2) the position angle, θ. Binary systems are classified into four types based on the techniques used for their discoveries. These are:

1. Visual binaries: These binary systems are gravitationally bound to each other, whoose components can be individually resolved through a telescope, since they are close to the observer. Their relative positions can be plotted from long-term observations to determine their orbits.
2. Astrometric binaries: They are relatively nearby stars; the presence of the faint component is deduced by observing a wobble in the proper motion of the bright component.
3. Spectroscopic binaries: In these systems the separation between the stars is usually very small, and the orbital period ranges from a fraction of an hour to several years (Batten and Fletcher 1990); in some spectroscopic binaries, spectral lines from both stars are visible and the lines are alternately double and single.
4. Eclipsing binaries: Eclipsing binary system is a direct method to measure the radius of a star, both the primary and the secondary, from the time for the light curve to reach and rise from minimum. The light curve is characterized by periods of practically constant light, with periodic drops in intensity; when two stars have their orbital planes close and along the observer's line of sight, the two stars take turns blocking wholly or partially each other from the sight during each orbital period, thus causing dips in the light curve.

Modern interferometric techniques allow accurate astrometric study of close binary stars. Many such binaries have been resolved by means of single aperture interferometry (Saha 1999, 2002 and references therein). Determining precise stellar masses of these systems addresses the queries related to star-formation, stellar evolution, and population synthesis. The procedure in obtaining masses of stars requires information about the period, shape, orientation, and physical shape of the orbit. On the other hand, the measurements of stellar density provided by asteroseismology combined with precise radius determination by interferometry offer a method to determine mass for a single star. A long-term benefit of interferometric imaging is to obtain better calibration of the main-sequence mass-luminosity relationship, as well as to derive accurate orbital elements and distances. Also, combining the angular diameter of the orbit with the physical scale set by spectroscopy yields the orbital parallax. But to get a good estimation of the masses speckle orbits must be defined very accurately.

The evolution of close binaries differs from that of single stars. The transfer of mass between the two stars in binaries may change the evolution, chemical composition, as well as alter the orbits of both components. The tidal pull of one star on another may lead to direct mass exchange. The tidal distortion reflects in the sinusoidal variations in their light curve and wide diffused absorptions lines in their spectra. The rotation and revolution in these systems are synchronized due to tidal distortions, and rotational speeds. The components of such a contact binary rotate very rapidly ($v \sin i \sim 100 - 150$ km s^{-1}). The radiative transfer concerning the effects of irradiation on the line formation in the expanding atmospheres of the component, which is distorted mainly by physical effects, viz., (1) rotation of the component and (2) the tidal effect, can be modeled as well. For instance, W UMa variables consist of two solar type components sharing a common outer envelope. These variables are tidally distorted stars in contact binaries. Aperture synthesis using LBOI may provide the means to image the small spatial scales of interacting binaries in order to study tidal distortions and hot spot activity.

Spectroscopic observations for these stars are difficult to interpret due to shorter period, faint components, and rotational broadening of spectral lines. These observations may yield mass-ratio, and absolute dimensions in conjunction with the photometric analysis of these eclipsing binaries. Resolving the components of spectroscopic binaries, by means of LBOI, would determine all of the system parameters, including the respective members' masses (Davis et al. 2005). However, component mass determinations for non-eclipsing spectroscopic binaries are possible with measurement of the orbital inclination.

Spectroscopic and interferometric measurements are required to derive precise stellar masses since they depend on $\sin^3 i$. A small variation on the inclination, i, implies a large variation on the radial velocities. Although speckle interferometry has been successful in resolving detached spectroscopic binaries, most of the orbital calculations carried out with such a method are not precise to provide masses better than 10% (Pourbaix 2000). In this respect, LBOI has the ability to resolve short period binaries and often results in orbits with precisions in the dynamical masses precisely as high as 1% (Hummel et al. 2001; Boden et al. 2005). Since an interferometer measures the visual orbit of the components in a binary system, it (combined with radial velocity data) allows one to solve for all the orbital parameters of the system, including orbital inclination and hence component masses.

Eclipsing binary systems are complex objects, because of the interaction between the components, physical processes like mutual reflection, tidal interaction, mass loss, and mass transfer. Information on circumstellar envelopes such as, the diameter of inner envelope, color, symmetry, radial profile etc., can also be obtained. The spectroscopic and photometric studies of circumstellar environment in these stars revealed that their structures are very complicated in many cases, for example, β Lyrae (see Fig. 9.14). The dusty environment of a single-line spectroscopic binary, υ Sagittarii (HD 181615; $P = 137.9$ d), has been observed with the VLTI/MIDI by Netolický et al. (2009). This system, with an A-type low-mass supergiant, is the brightest member of the extremely hydrogen-deficient binary stars. These binary stars are a rather rare class of evolved binary systems, which are in a second phase of mass-transfer in which the primary has ended the core helium-burning phase. With the distance of 513 ± 189 pc, its absolute magnitude, M_v, is estimated to be -4.8 ± 1

from the distribution of the interstellar reddening, polarization, and interstellar lines of the surrounding stars. The hotter secondary is surrounded by a disk with colors of a B8-B9 star. Their observations provide evidence of a thin, flat circumbinary disk around this binary, whose inner rim lies close to the radius of sublimation temperature. Recently, using Keck interferometer, Boden et al. (2009) have reported evidence for resolved warm dust in the double-lined Pre-Main Sequence (PMS) binary system DQ Tauri in the near-IR wave band, which suggests the IR excess from this system is distributed on the physical scale of the binary orbit (0.1–0.2 AU). DQ Tauri as an eccentric, short-period (15.8 d), double-lined spectroscopic binary composed of similar late K-stars. The mid-IR excess is due to emission from a circumbinary disk.

Another eclipsing binary system, ϵ Aurigae, with a period of 27.1 years, has been observed recently; the eclipse lasts nearly 2 years. At a distance of 625 pc from the Earth (Perryman et al. 1997), this system is the the fifth brightest star in the northern constellation Auriga. It consists of a visible F-type post-AGB star having low mass, with a gigantic disk of gas that enshrouds a single B5V-type main-sequence star (Hoard et al. 2010). The thickness of this body covers about 50% of the F-star, reducing the brightness by 0.7–0.8 magnitude. Using the CHARA/MIRC instrument, Kloppenborg et al. (2010) have reported the interferometric near-IR images of this object in H-band ($\lambda = 1.50 - 1.74$ μm) during its eclipse, which show the eclipsing body moving in front of the F-type star; the body is an opaque disk and appears tilted. Figure 9.17 shows the transit of the opaque and flat disk at multiple epochs during the most recent eclipse ingress. The synthesized image is created by using bispectrum maximum entropy method. The mass of the F-type star is estimated to be 3.63 ± 0.7 $M_{\odot}$, while the disk mass is dynamically negligible. There is a need, as suggested by these authors, for the frequent photometric and spectroscopic measurements to create a longitudinal profile of the disk. Combined with geometry derived from interferometric observations, the composition, density scale height, and temperature structure of the disk may be determined.

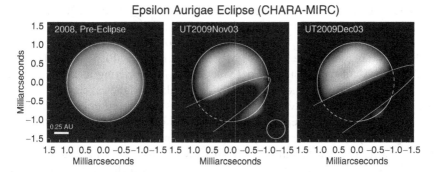

Fig. 9.17 The mysterious dust disk in orbit around the bright visible star ϵ Aurigae was imaged during eclipse by Kloppenborg et al. (2010). The angular resolution of CHARA using the MIRC instrument at H band is shown as a *circle* (courtesy: J. Monnier)

Algol-type binaries are mass-transferring binary stars; the dynamics of the mass-transfer process depends on the rate of radiative cooling. A proto-type Algol exhibits phase-locked and sporadic light curves (Kim 1989) and phase-locked polarization curves produced by Thompson scattering in the mass-transfer stream (Wilson and Liou 1993; Elias II et al. 2008). Resolving interacting binaries of this kind is a challenging task. However, the recent advancement in LBOIs has enhanced their capabilities to discern the components in these systems. These instruments are able to determine their physical parameters directly and can provide information on mass transfer and binary evolution.

Algol (β Persei; HR 936) is a proto-type eclipsing binary system with spherical or slightly ellipsoidal components in the constellation Perseus. It varies regularly in magnitude from 2.3 to 3.5 over a period of a few days. Careful observations indicate that Algol is a multiple (trinary) system comprising of (1) Algol A (primary; β Persei-A), a blue spectral class B8 V-type main sequence star, (2) Algol B (sub-giant; β Persei-B), a red-yellow spectral class K2 IV-type star, and (3) Algol C (β Persei-C), A5-class that orbits the close binary pair, Algol AB. The average separation between Algol AB system and the Algol C is about 2.69 AUs, which makes an orbit of 1.86 years (Labeyrie et al. 1974). Algol A and B form a close binary system, the eclipsing binary, that are separated by 10.4 million km. This eclipsing system is semi-detached with the sub-giant filling its Roche-lobe and transferring the material at a modest rate to its more massive companion star. The Roche-lobes are gravitational equipotential surfaces surrounding the components of a binary system, which touch each other only at the inner Lagrangian point of the system. While matter inside a Roche-lobe is confined only to the corresponding star, mass transfer to the binary component, or to the environment of the binary occurs if the star overflows its Roche-lobe. Recently, Zavala et al. (2010) have produced images resolving all three components in the Algol triple system using Navy Prototype Optical Interferometer (NPOI). They are able to separate the tertiary component from the binary and simultaneously resolved the eclipsing binary pair. These authors have presented the astrometric orbits of the Algol A-B and AB-C systems as well. An important result of this observation is finding the magnitude difference of these three components. At 550 nm, the differences are: (1) Δm_v 2.70 $\pm$ 0.3 for the components A-B, (2) Δm_v 2.80 $\pm$ 0.2 for A-C, and Δm_v 2.90 $\pm$ 0.1 for AB-C. The respective masses of these individual components, A, B, and C are found to be 3.7 $\pm$ 0.2 $M_\odot$, 0.8 $\pm$ 0.1 $M_\odot$, and 1.5 $\pm$ 0.21$M_\odot$.

Another Algol-type system is β Lyrae (HD 174638), lying at a distance of 270$\pm$39 pc, whose components are tidally-distorted; its brightness changes continuously. It has been extensively studied since it was discovered to be an eclipsing system over a couple of centuries (Goodricke and Englefield 1785); it remains a baffling binary systems. According to the current model, it consists of a $\sim$3 $M_\odot$ B6-8II star (second star), which has filled its Roche-lobe and is losing mass at a rate of 2×10^5 $M_\odot$ per year to a $\sim$13 $M_\odot$ early B-type star (first star). It is thought that the gainer is completely embedded in and hidden by a thick accretion disk. The orbit of the system is circular and is close to edge-on. It has a period of 12.94 days, which is increasing at a rate of $\sim$19 s per year due to the high mass transfer rate of the system. This star exhibits a large spike in the position angle produced by the eclipse of the stream/disk impact region. The spectrum of β Lyrae contains at least six spectroscopic line components (Harmanec 2002). The major components are (1) strong absorption lines from the second star, (2) shell absorption components seen in the hydrogen, helium, and many metallic lines, which are believed to arise from accretion disk, and (3) strong lines of hydrogen and helium witnessed in emission which vary with orbital phase and with time. Schmitt et al. (2008) observed its relative motion between the continuum and Hα emission with the NPOI at limited spatial and spectral resolution. They have detected a differential phase signal in the channels containing He I emission lines at 587.6 nm and 706.5 nm, with orbital behavior different from the Hα, indicating that it originates from a different part of the interacting system. Recently, Zhao et al. (2008) resolved images of β Lyrae at six different epochs using CHARA array interferometer and the MIRC combiner in the H-band, resolving the mass-losing star and the accretion disk surrounding the mass-gaining star.

Using the VLTI/VINCI instrument, Verhost et al. (2007) observed SS Leporis (HR 2148; RA $6^h0^m59.1^s$, $\delta - 16°28'39''$), an Algol-type semi-detached binary having visual magnitude $m_v = 5.2$ (period $= 260$ days) in the constellation Lepus. This binary system consists of an A0 V-type star as its primary and a M4 III-type giant star as secondary, which fills it Roche-lobe, and therefore loses mass to the primary star. These authors have resolved spatially the giant star and the circumstellar environment. The limb-darkened diameter of the star is found to be 3.11 ± 0.32 mas, which probably was a regular main-sequence star but at present shows an increased size and luminosity due to relatively high accretion rate. The mass transfer appears to be non-conservative, since they detected an optically thick circumbinary dust disk.

Another group of close binary stars, referred to as AM Herculis binary stars, that exhibit rapidly changing degrees of circular and linear polarization, often as large as a few tens of percent peak-to-peak, can also be observed with LBOI using optical interferometric polarimeter. Such observations would enable to map the mass flowing along the magnetic field lines. These evolved binaries consist of a super-strong magnetic white dwarf primary and a closely-orbiting late type main-sequence dwarf M-type star or K-type star that is filling its Roche-lobe. These are cataclysmic variables (CV) type systems (Hilditch 2001), in which the white dwarf accreter has a strong (10–100 mega-gauss) magnetic field that captures infalling material from the red dwarf and forces into an accretion stream or funnel. The materials directly heads to the poles of the white dwarf, as opposed to other CV-type systems where the mass accretes onto a disk. The orbital periods are short, which are of the order a few hours.

Finding the orbital elements of a binary system is of paramount importance in the study of binary stars since it is the only way to obtain the masses of the individual stars in that system. Due to the mutual gravitational attraction both the stars move around the center of mass of the system, but the motion of the secondary with respect to the primary would be a Keplerian elliptic orbit.[19] This is the true orbit, the plane of which is not generally coincident with the sky plane (the tangent plane to the celestial sphere) at the position of the primary, and its plane is the true orbital plane. By Kepler's third law,

$$4\pi^2 a^3 = G(m_1 + m_2)P^2, \tag{9.37}$$

where a is the semi-major axis of the orbit, G the gravitational constant, P the period and m_1 and m_2 the masses of the stars.

The orientation of the orbits for an eclipsing system is such that a large percentage of each star is eclipsed during the primary and secondary eclipses. Since the stellar dimensions are about ~ 1 mas (the primary is about three times the solar diameter, while the secondary is about 3.5 solar diameters), the stellar shapes can be determined by means of interferometric techniques. Such measurements for binary systems may lead to accurate determination of apsidal motion,[20] the test of which is a check on the theory of stellar structure. In a binary system, the precession of the line of apsides is due to rotational and mutual tidal distortion, and is sensitive to the central condensation of the stars.

The orbit obtained by observations would be a projection of the true orbit in a plane perpendicular to the line of sight. This observed orbit, known as 'apparent orbit', is the projection of the true orbit on the plane perpendicular to the line of sight and is also an ellipse. The observations of the position of the secondary with respect to primary at different times give the apparent orbit and this in turn provides the information of the true orbital elements. The apparent orbit of a binary system can be deduced from the interferometric measurements.

The most common binary orbit periods (as estimated from their separations and typical distances) lie between 10 and 30 years. Thus at the present stage, a large number of binary systems have completed one or more revolutions under speckle study and speckle data alone can be sufficient to construct the orbits. Figure 9.18 depicts the orbit of HR 2134, a system of two stars situated in Gemini constellation. Its orbital parameters is calculated using speckle data and other interferometric data. The algorithm is based on standard least square technique with iterative improvement of the orbital parameters (Saha 2007). The normal equations

---

[19] Kepler's three laws are defined as: (1) the orbit of every planet is elliptical in shape, with the Sun at a focus, (2) a line connecting a planet and the Sun sweeps out equal areas in equal intervals of time, and (3) the square of the period of any planet is proportional to the cube of the semi-major axis of its orbit. An orbit is characterized by the orbital elements, such as, (1) semi-major axis, a, defining the size of the orbit, (2) eccentricity, e, that measures the deviation of an orbit from a perfect circle; a circular orbit has $e = 0$ and an elliptical orbit has $0 < e < 1$, (3) inclination, i, which is the angle of the object's orbit as seen by the observer on Earth, (4) longitude of the ascending node, Ω, that defines the angle between line of nodes and the zero point of longitude in the reference plane, (5) argument of perigee (perihelion – the point in an orbit when the planet is closest to the Sun), ω, defining the low point, perigee, of the orbit with respect to the Earth's surface, and (6) true (mean) anomaly at epoch, ν, that defines the position of the orbiting body along the ellipse at a specific time.

[20] Apsidal motion is the rotation of the line of apsides (major axis of an elliptical orbit) in the plane of the orbit.

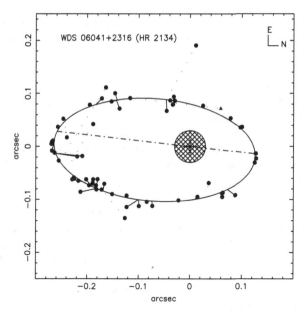

Fig. 9.18 Orbit of HR 2134

are solved using Cracovian matrix elimination technique[21] (Kopal 1959). This algorithm does not require any previous knowledge of the period and the eccentricity of the binary systems.

The first spectro-interferometric study of a double-lined spectroscopic binary, α Virgo was made possible with the Narrabri intensity interferometer (Hanbury Brown 1974); nine multiple systems were also observed with the said instrument. From the data obtained at the COAST, aperture-synthesis maps of the double-lined spectroscopic binary α Aurigae (Baldwin et al. 1996) depict the milli-arcsecond orbital motion of the system over a 15 day interval. Employing NPOI, Hummel (1998) have determined the orbital parameters of two spectroscopic binaries, ζ Ursae Majoris (Mizar A), η Pegasi (Matar) and derived masses and luminosities based on the data obtained with NPOI; published radial velocities and *Hipparcos* trigonometrical parallaxes were used for the analysis. In addition to these two interferometers where three or more telescopes are employed to resolve binary systems, IOTA interferometer with three telescopes (Traub et al. 2003) was also employed to resolve binary systems (Monnier et al. 2004); two such system, λ Virgo

---

[21] Cracovian matrix, introduced by Banachiewicz (1955), adopts a column-row convention for designating individual elements as against the normal row-column convention of matrix analysis. The Cracovian products of two matrices A and B is equal to the matrix product $A^T B$, in which A^T is the transpose of A. This theory is used to solve astronomical, geodesic, mechanical and mathematical problems.

and a Wolf-Rayet (WR) were resolved. The underlying binary in the prototypical colliding-wind source, WR 140 (WC7 + O4/O5) was found to have a separation of ~13 mas with a position angle of 152°.

Recently, Raghavan et al. (2009) have reported an updated spectroscopic orbit and a new visual orbit for the double-lined spectroscopic binary σ^2 Coronae Borealis (CrB; HD 146361) based on radial velocity measurements at the Oak Ridge Observatory in Harvard and interferometric visibility measurements at the CHARA array. This binary, a RS CVn binary with circularized orbit with a period of 1.14 days, is a central component of a quintuple system, along with σ^1 CrB (HD 146362) and a M-dwarf binary, σ CrB–C. σ^2 CrB is composed of two approximately equal mass stars having a projected angular separation of about 1.1 mas in the sky. The CHARA instrument is able to resolve the visual orbit for this pair, hence enables the authors to determine the masses of the components.

Studies of WR–O-type binary systems are important for understanding stellar evolution (North et al. 2007a), as well as for studying massive stars and their interacting winds (Millour et al. 2007). The proximity of the O-type star helps stellar winds from these stars to interact. The WR and O-type binary system, γ^2 Velorum (WR 11; $m_v = 1.8$), was observed earlier by Hanbury Brown et al. (1970) with the Narrabri intensity interferometer, which provided an angular semi-major axis of the orbit of 4.3±0.5 mas and an angular size for the largest component of 0.44±0.05 mas (17±4 $R_\odot$ at a distance estimated to be 350±50 pc). Angular size in C III–IV emission at 465 nm shows stellar envelope five times bigger than its disk in the continuum. With a goal to understand the wind from a WR star in an interacting binary on the double-line spectroscopic system, Millour et al. (2007) had observed γ^2 Velorum using VLTI/AMBER instrument (see Sect. 6.1.3.5). Their analysis infers that the binary system lies at a distance of 368 pc with +38 pc or −13 pc, which is significantly larger than the *Hipparcos* value of 258 pc with +41 pc or −31 pc.

The visual orbit for the spectroscopic binary ι Pegasi with interferometric visibility data has also been derived using PTI (Pan et al. 1998; Boden et al. 1999); its visual orbit with separation of 1 mas in RA, having a circular orbit with a radii of 9.4 mas was also determined. With VLTI/AMBER instrument in the H- and K-band, Meilland et al. (2008) have detected an oscillation in the visibility curve plotted as a function of the spatial frequency, which is a signature of a companion star around δ Centauri (HD 105435, B2 IVne). It is a variable star with a brightness variation between 2.51 and 2.65 m_v. Jilinsky et al. (2006) have measured a radial velocity of this star of 3.8±2.8 km per second.

9.2.11 Multiple Systems

The multiple star systems consist of more than two stars, which are gravitationally bound, and generally move around each other in stable orbits. Studies of the orbital parameters, distribution of angular momentum, and system coplanarity of these multiple systems are of interest for understanding their formation and evolution (Sterzik and Tokovinin 2002). The coplanarity of angular momentum vectors between the inner and outer orbits in hierarchical systems may reflect the initial conditions during core collapse and fragmentation. When two such stars have a relatively close orbit, their gravitational interaction may have a significant impact on their evolution; in some cases tidal distortions take place.

Studies of multiple stars are also an important aspect that can reveal mysteries. For instance, the R 136a was thought to be a very massive star with a mass of ∼2500 $M_\odot$ (Cassinelli et al. 1981). Speckle imagery revealed that R 136a was a dense cluster of stars (Weigelt and Baier 1985). R 64 (Schertl et al. 1996), HD 97950, and the central object of giant H II region starburst cluster NGC 3603 (Hofmann et al. 1995) have been observed as well.

The intriguing massive (∼100 $M_\odot$) LBV object, η Carinae, NGC 3372, is exceptionally luminous (∼4 × 10^6 $\mathcal{L}_\odot$). It is highly variable southern object in the Galaxy located in the constellation Carina (α 10 h 45.1 m, δ 59°41′) surrounded by a large, bright nebula, known as the Eta Carinae nebula (NGC 3372). A variety of observations suggest that the central source of this object is a binary system, very close to each other, with a mass of about sixty solar mass each. η Carinae is suffering from a high mass-loss rate. From the spectroscopic studies of the Homunculus nebula showed that its wind is latitude-dependent. This object is believed to be on the verge of exploding as supernova. With VLTI telescopes, Glindemann and Paresce (2001) have measured of the core of this object. Subsequently, van Boekel et al. (2003) resolved the optically thick, aspheric wind region with NIR interferometry using the VLTI/VINCI instrument. Their observations suggest that the very rapid stellar rotation causes enhanced mass-loss along the rotation axis resulting from the difference in temperature between the pole and the equator, which develops in rapidly rotating stars. Weigelt et al. (2007) have detected the primary star's dense wind from the near-IR (K band) spectro-interferometric observations of this object with the VLTI/AMBER using three UTs with baselines from 42 to 89 m. At two different epochs using both medium and high spectral resolutions in the spectral regions around He 1 and Br γ emission lines, these authors have resolved the η Carinae's optically thick regions. The results are consistent with a fast-rotating, luminous hot stars with enhanced high velocity mass-loss at the poles. They have also developed a model that shows that the asymmetries measured within the wings of the Br γ line with differential and closure phases are consistent with the geometry expected for an aspherical, latitude-dependent stellar wind (see Fig. 9.19).

The close companions of θ^1 Orionis A and θ^1 Orionis B (Petr et al. 1998), subsequently, an additional faint companion of the latter and a close companion of θ^1 Orionis C with a separation of ∼33 mas were detected (Weigelt et al. 1999) in the IR band. These Trapezium system, θ^1 Orionis ABCD, are massive O-type and early B-type stars and are located in the Orion star-forming region. Both the θ^1 Orionis A and θ^1 Orionis B stars are the eclipsing binary systems. The θ^1 Orionis C is the dominant and most luminous star having ∼40 $M_\odot$ with a temperature of about 40,000°K. The intense radiation of this star is ionizing the whole Orion nebula. Appears to be a single star, both with conventional telescope and HST, θ^1 Orionis C

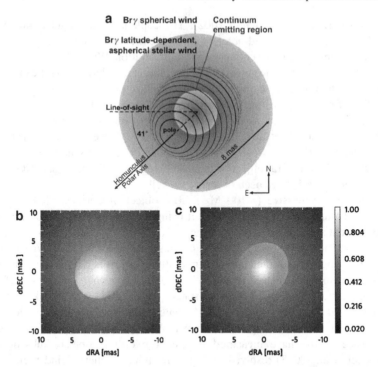

Fig. 9.19 Illustrations (**a**) of the components of the geometric model for an optically thick, latitude-dependent wind (for the weak aspherical wind component; the lines of latitudes are to illustrate the 3-D-orientation of the ellipsoid), (**b**) and (**c**) for two representative wavelengths, the total brightness distribution of the model including the aspherical wind component and the contributions from the two spherical constituents (Weigelt et al. 2007; courtesy: F. Malbet)

is one of the two O-type stars, which has a strong magnetic field (Donati et al. 2002). Recently, Kraus et al. (2009) discovered the existence of a close companion with the VLTI/AMBER. This was the sharpest ever image of the young double star θ^1 Orionis C, which clearly separates the two young, massive stars of this system. The angular distance between the two stars was found to be about 20 mas. These authors have been able to derive the orbit of this binary system using position measurements obtained over several years (see Fig. 9.20). Using Kepler's third law, the masses of the two stars were derived to be 38 and 9 $M_\odot$. The measurements allow a trigonometric determination of the distance to the said system as well. Their measurements shows that the two massive stars are on a very eccentric orbit with a period of 11 years.

An interesting result on a pre-main sequence quadrupole system, HD 98800, was obtained by Prato et al. (2001), using speckle interferometry and adaptive optics at the Keck telescopes. This system, 10 Myr old post-T-Tauri stars – two spectroscopic binaries A and B in orbit about one another having separation of $\sim$0.8″ – is located in the TW Hydrae association. The B component is a double-lined spectroscopic binary associated with infrared excess. Although it harbors a dust disk, the

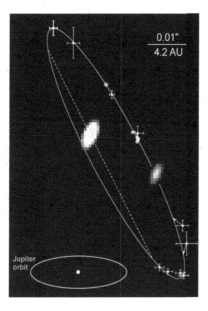

Fig. 9.20 Aperture-synthesis image of the high-eccentricity binary system, θ^1 Orionis C (brightest Orion Trapezium star), which was imaged with VLTI. The orbit of the binary system was derived from the interferometric monitoring campaign, covering the period of 1997 and 2008 using the 6 m telescope, Special Astrophysical Observatory, Russia, 3.6 m telescope, ESO, Chile, IOTA-3 interferometer, and VLTI/AMBER instrument, revealing the orbital motion of the companion (Kraus et al. 2009). The size of the orbit of Jupiter around the Sun is shown for comparison. Courtesy: G. Weigelt

system is apparently non-accreting. The authors opined that the mid-IR excess flux is emanating from this system and that dust disk around the pair is the source of the flux having an inner gap extending to ~2 AU and a height of ~1.7 AU. The orbital solution of HD 98800B system suggests that both components are under 0.7 $M_\odot$ (Boden et al. 2005).

With the aperture-masking method at 3.9 m Anglo-Australian telescope, Robertson et al. (1999) reported the primary component of a visual binary star, β Centauri (HR 5267), the second brightest star in the constellation Centaurus, is a double-lined spectroscopic binary with two β Cephei-type[22] variable giants, separated by 0.015″. Observations with Sydney University Stellar Interferometer (SUSI) by Davis et al. (2005) confirmed the binary nature of the primary component having roughly equal masses. These authors have enabled to determine the orbital parameters of the system as well. The magnitude difference of two components at

---

[22] Variables of β Cephei-type, are pulsating O8-B6 I-V stars with periods of light and radial velocity variations in the range of 0.1–0.6 days. They consist of massive, non-supergiant stars lying just above the main-sequence in the HR-diagram (see Fig. 9.2), whose low-order pressure and gravity mode pulsations result in light, radial and/or line-profile variations (Stankov and Handler 2005).

the visible wavelength of 442 nm is found to be 0.15 ± 0.02. They are able to derive the masses of both primary and secondary as well.

The orbital solution for another spectroscopic pair in the multiple system, σ Scorpii (HR 6084), has also been determined from measurements with the afore-mentioned interferometer by North et al. (2007b). The primary component of the system, σ Scorpii A, is also of β Cephei-type. This system is classified as a binary with B2 IV + B9.5 V components separated by $20''$ on the sky and a visual magnitude, m_v difference of 5.31. With the speckle interferometry, lunar occultations and spectroscopy have shown that the former component (B2 IV) comprises of three stars, a spectroscopic pair and a B7 tertiary. The tertiary is approximately 2.2 mag fainter and separated by $0.4''$ from the spectroscopic pair. In combination with the double-lined radial velocity measurements, the distance, spatial scale, mass and age of the spectroscopic pair were quantified by these authors; the age of the system is estimated to be 10 Myr.

9.3 Exploding Stars

Both novae and supernovae (SN) have complex nature of shells viz., multiple, secondary and asymmetric. The latter objects are considered to be the laboratory for advanced radiation hydrodynamics, combustion theory and nuclear and atomic physics as they are the major contributors to the chemical enrichment of the interstellar matter with heavy elements. The high resolution mapping using LBOI may depict the core structure of supernovae, early expansion and the interaction zones between gas clouds with different velocities. However, for imaging the environment of neutron stars[23] of 10 km size, a space-born very large array approaching 200,000 km would be required. Another interesting science target is to get spatial information of gamma ray bursts for low redshift cases; early warning systems may permit to obtain some spatial and spectral information at the peak of the optical flux.

9.3.1 Novae

A nova occurs in an interacting binary system consisting of Roche-lobe filling secondary on a main-sequence, when a surface thermo-nuclear explosion fueled by hydrogen on a white dwarf primary takes place. The secondary loses hydrogen-rich material through the inner Lagrangian point onto an accretion disk surrounding the primary. The matter accreted by the white dwarf reaches a critical amount, the temperature at the base of the envelope rises sufficiently to ignite thermonuclear

---

[23] Neutron stars that are formed from supernova explosions, collapsed under gravity to an extent that they consist almost entirely of neutrons. Such objects are about 10 km across and have a density of 10^{17} kg m^{-3}.

processes leading to a runaway reaction. This thermonuclear runaway results in the explosive ejection of the accreted material causing a nova outburst, which is characterized by a rapid and unpredictable rise in brightness ranging from 7 to greater than 14 magnitudes in the optical within a few days. The envelope is ejected with velocities more than 300 km per second.

The eruptive event is followed by a steady decline back to the pre-nova magnitude over a few months, which suggests that the event causing the nova does not destroy the original star. According to the features of light variations, novae are sub-divided into dwarf and classical novae. The classical novae were also sub-divided into NA (fast), NB (slow), NC (very slow), and NR (recurrent) categories. The primary characteristic of the nova outburst is the optical light curve. A rapid rise to maximum is followed by an early decline, whose rate defines the speed class of the nova, with the very fast classical novae declining at greater than 0.2 magnitude per day (Warner 2008). The speed class is, in turn, related to both the ejection velocity and peak absolute magnitude. Their spectral types at minimum are between A to F with corresponding absolute magnitudes ranging from $M_v = -0.5$ to 1.5.

In the constellation of Cygnus, a tremendous explosion occured on February 19, 1992, which is referred to as Nova Cygni 1992. It has attained maximum brightness three days after its eruption at a magnitude of $m_v = 4.4$. The Mark III interferometer was pressed into service to observe this phenomenon 10 days after maximum light. Combining the diameter of 5.1 mas measured in a 10 nm wide filter centered on the Hα line with an average observed expansion velocity of ~1,100 km per second, Quirrenbach et al. (1993b) derived its distance, which was ~2.5 kpc. Due to the lack of sufficient u, v-coverage, the data did not permit detailed modeling of the structure and evolution of the nova envelope. The HST image taken on May 31, 1993 reveals a remarkably circular yet slightly lumpy ring-like structure. A bar-like structure across the middle of the ring is also noticed, which may mark the edge-on plane of the orbits of a binary system triggering the nova. Alternately, the bar is produced by twin jets of gas ejected from the star and spanning the distance between the shell and the star.

Using VLTI/AMBER, as well as VLTI/MIDI, Chesneau et al. (2008) witnessed the appearance of a shell of dusty gas around a star, Nova Scorpii 2007a (V 1280 Scorpii). Twenty three days after its discovery, the source was found to be very compact having a diameter of 1 mas, a few days later, after the detection of the major dust formation event, the source measured to 13 mas. These authors opined that the latter size could be the diameter of the dust shell in expansion, while the size previously measured was an upper limit of the erupting source. They were able to trace its evolution for more than 100 days, starting from the onset of its formation. The measurement of the angular expansion rate, together with the knowledge of the expansion velocity, enables to derive the distance of the object. In this case, it is found to be 5,500 ly.

The recurrent novae, where multiple outbursts have been observed, are two types. For example, (1) U Sco, V394 CrA, in this an evolved main-sequence or sub-giant secondary star is transferring material onto a white dwarf; rapid optical declines and extremely high ejection velocities are the main characteristics of this class and

(2) RS Ophiuchi, T CrB, in which the outburst occurs inside the outer layers of the red giant star; both these novae have giant secondaries and long orbital periods.

RS Ophiuchi (HD 162214; RA $17^h 50^m 13.2^s$, δ $-06° 42'28''$), having apparent magnitude $m_v = 9.6 - 13.5$ (quiet phase), lies several thousand light years away in the constellation Ophiuchus. It is a wind accreting binary system with a white dwarf, close to 1.4 $M_\odot$, widely known as 'Chandrasekhar limit', in orbit with a red giant star (M2 III; Evans et al. 2008 and references therein); the system is closest known type-SN Ia progenitor system. The intense gravitational field of the former continuously pulls a stream of gas from the outer layers of the red giant. This process sets off a thermonuclear explosion on the surface of the white dwarf. RS Ophiuchi had several recorded outbursts (1898, 1933, 1958, 1967, 1985, 2006). The most recent explosion began on February 12, 2006, where the apparent magnitude reached $m_v = 4.5$. This event triggered off an intensive multi-wavelength observational campaign, from the X-ray to radio wavelengths (Bode 2010 and references therein). RS Ophiuchi was detected at radio emission in frequencies below 1.4 GHz; radio emission was detected at 0.61 GHz on day 20 with a flux density of ∼48 mJy and at 0.325 GHz on day 38 with a flux density of ∼44 mJy (Kantharia et al. 2007). This object was found to be 3 mas in size, measured by means of the interferometers such as IOTA, PTI, and Keck at multiple epochs by Monnier et al. (2006b). Chesneau et al. (2007) have also measured the extension of the expanding milli-arcsecond scale emission with VLTI/AMBER instrument five days after the discovery using three telescopes and the medium spectral resolution in the K-band continuum, the Br γ 2.17 μm line and the He I 2.06 μm line. Both the continuum and the line emissions are flattened, sharing apparently the same geometry, at different scale (see Fig. 9.21). Two radial velocity fields were also detected by these authors in the Br γ line: a slower one expanding ring-like structure and a faster structure extended in the East–West direction.

9.3.2 Supernovae

A supernova (SN) is a cataclysmic stellar explosion. More energetic than a nova, a typical supernova is characterized by a sudden and dramatic rise in brightness by several magnitudes that often briefly outshines the rest of its galaxy for several days or a few weeks. Massive stars (>8 $M_\odot$) form a dense iron core that eventually collapses. The temperature of such a core can reach upto 6×10^8 degree kelvin to initiate the thermonuclear fusion of C and proceeds towards more heavier elements, so that an iron core is developed at the center with a temperature of about ∼10^9 degrees Kelvin. Following several steps, the burning of silicon (Si) produces nickel (Ni) and iron (Fe) by the following reactions:

$$^{28}Si + {}^{28}Si \rightarrow {}^{56}Ni + \gamma$$
$$^{56}Ni \rightarrow {}^{56}Fe + 2e^+ + 2\nu. \tag{9.38}$$

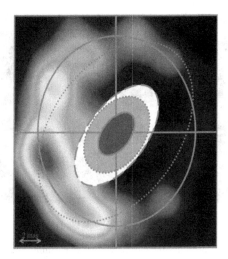

Fig. 9.21 Sketch of the fitted elliptical extension in the near-IR for RS Ophiuchi nova at t = 5.5 d compared with the radio structure observed at t = 13.8 d (*thick extended ring*). The continuum ellipse is delimited by the *solid line*, the ellipse that corresponds to the core of Br γ by the *dotted line* and the one corresponding to the core of He I by the *dashed line*. The outer *small dotted line* delimits the Br γ ellipse scaled at t = 13.8 d (Chesneau et al. 2007; courtesy: F. Malbet)

The collapsing matter releases its gravitational binding energy in the form of neutrino emission. The occurrence of burning stages subsequent to H-burning can be halted if the degeneracy of the core is not lifted at the end of the burning stage, and that this is a function of the initial mass of the star. This results in He-white dwarfs, C–O-white dwarfs (the end product of He burning) and even C–Si-white dwarfs (e.g., GD 349; Dupuis et al. 2000). New shell-burning and red giant phases occur before every new stage of core-burning. The result of these ever-expanding outer layers is a supergiant (luminosity class I) whose radius can reach 1,000 R$_{\odot}$.

Unable to sustain thermonuclear fusion, the Fe-core contracts rapidly to produce the heat and energy required to stabilize the system. With further contraction, the central temperature rises to about 5×10^9 degrees, where the energy of the photons becomes large enough to destroy certain nuclei. The photo-dissociation of Fe converts into He, which gives rise to an instability causing the contraction of the core to high density. The collapse is then halted by quantum effects. According to the helium flash model, a white dwarf in a binary system accretes matter through Roche-lobe overflow from an evolving companion star (see Fig. 9.22). If sufficient amount of matter is transferred to push white dwarf over the Chandrasekhar limit for electron-degeneracy support, the white dwarf begins to collapse under gravity (Mazzali et al. 2007). Unlike massive stars with iron core, a dwarf having C–O core may undergo further nuclear reactions. Another model suggests that the supernova can be an explosion of a rotating configuration formed from the merging of two low-mass white dwarfs on a dynamical scale, following the loss of angular momentum due to gravitational radiation (Hoeflich 2005).

Fig. 9.22 Artist's concept of a possible scenario for a progenitor system of a SN Ia. A white dwarf having a mass between 0.6 and 1.2 $M_\odot$ accretes material from a close companion by Roche-lobe overflow and approaches the Chandrasekhar mass limit. The companion star may be a main sequence star or red giant, or a helium star or another white dwarf. Depending on this, the accreted material may be either H, He or C–O-rich. If H or He is accreted, nuclear burning on the surface converts it to a C–O mixture at an equal ratio in all cases. Despite the different evolutionary pathways, the net result is the same, i.e., the explosion of a C–O-white dwarf with a mass close to the Chandrasekhar mass limit and with very similar supernova properties. Some small fraction of SNe Ia may also be the result of merging of two white dwarfs on a dynamical time scale (Hynes 2002; courtesy: P. Hoeflich)

The outer layers of the star crash onto the core and bounce off, driving a shock wave into the surrounding interstellar medium (Schawinski et al. 2008). The explosion expels much of the material at a very high speed, which sweeps up an expanding shell of gas and dust called a supernova remnant, leaving behind a neutron 'star or a black hole[24] (if the mass of the core is more than 3 $M_\odot$; Adams and Laughlin 1997; Willson 2000; Giacobbe 2005). The collapse of the star liberates heat that ignites nuclear reactions and blow the remnant apart in a thermonuclear deflagration, and subsequently turns into a detonation wiping out the signatures of the deflagration phase and causes a radially stratified chemical layering. The expanding shock waves from supernova explosions can trigger the formation of new stars.

---

[24] Black holes are the end point of gravitational collapse of massive celestial objects, which are conceived as singularities in space time. The space time metric defining the vacuum exterior of a classical black hole, and the black hole itself, is characterized by parameters such as, the mass of the black hole M_{BH}, the rotation (spin) J, and charge q. For $J = q = 0$, one obtains a Schwarzschild black hole, and for $q = 0$, one obtains a Kerr black hole. Black holes may be broadly classified into two categories, the stellar mass ($M_{BH} > 20 \, M_\odot$) and Super Massive ($M_{BH} \geq 10^6 M_\odot$) Black Holes (SMBH). The birth history of the former is theoretically known with almost absolute certainty; they are the endpoint of the gravitational collapse of massive stars, while the latter may form through the monolithic collapse of early proto-spheroid gaseous mass originated at the time of galaxy formation or a number of stellar/intermediate mass ($M_{BH} \sim 10^{3-4} M_\odot$) black holes may merge to form it; they are ubiquitously found at the centers of large galaxies.

Observationally, the supernovae are classified according to the lines of different chemical elements that appear in their spectra, for example, if a spectrum contains a hydrogen line, it is classified type II, otherwise it is type I. There are subdivisions according to the presence or absence of other lines and the shape of the light curve of the supernova. These are: type Ia has Si II line at 615.0 nm, type Ib contains He I line at 587.6 nm, and type Ic possesses weak or no helium lines. Their light curves exhibit sharp maxima, followed by gradual fadeout. To note, a Wolf-Rayet (WR) star is interpreted as a central He-burning core that has lost the important part of its H-rich envelope. This stage of the most massive stars lasts about $\leq 5 \times 10^5$ years before they explode as type-Ib/Ic supernovae (Meynet and Maeder 2005). The light curves of type II have less sharp peaks at maxima, and then die away more sharply than the type I. The type II supernovae are classified based on the shape of their light curves into type IIP (Plateau) and type IIL. The former reaches a plateau in their light curve while the latter has a linear decrease in their light curve, in which it is linear in magnitude against time, or exponential in luminosity against time. The type II supernovae are thought to occur in population I-type stars in the spiral arms of galaxies. On the contrary, type I supernovae are probably population II stars (metal-poor, old stars) and occur typically in elliptical galaxies.

The supernova, SN 1987A, in the Large Magellanic Cloud (LMC) that is 50 kpc away, provided confirmation of the basic picture of core collapse (Arnett et al. 1989). Although, the spectra of such an object provides the spectral energy distribution and profiles of spectral lines for a supernovae, which can be related to its angular size, optical interferometry, speckle interferometry in particular, gave an opportunity for the direct measurement of the spatial distribution and evolution of the optical emission of this object. Soon after the explosion of the supernova SN 1987A, various observers monitored routinely the expansion of the shell in different wavelengths by means of speckle imaging (Nisenson et al. 1987; Papaliolios et al. 1989; Wood et al. 1989). Nulsen et al. (1990) have derived the velocity of the expansion as well and found that the size of this object was strongly wavelength dependent at the early epoch − pre-nebular phase indicating stratification in its envelope. HST measurements of debris velocity agree with the speckle interferometric results (Wang et al. 2002). Papaliolios et al. (1989) reported the asymmetry of the shell, which was also confirmed by HST image. A bright source at $0.06''$ away from this supernova with a magnitude difference of 2.7 at Hα had been detected 30 and 38 days after the explosion by Nisenson et al. (1987) and 50 days after by Meikle et al. (1987). Nisenson and Papaliolios (1999) have detected a faint second spot, $\Delta m_v \sim 4.2$, on the opposite side of SN 1987A with separation, $\rho = 160$ mas.

9.4 Extragalactic Sources

A galaxy is a gravitationally bound system of stars, neutral and ionized gas, dust, molecular clouds, and dark matter. Typical galaxies contain billions of stars, which orbit a common center of gravity. Most galaxies contain a large number of multiple star systems and star clusters, as well as various types of nebulae. At the center of many galaxies, there is a compact nucleus. The luminosities of the brightest galaxies may correspond to 10^{12} $\mathcal{L}_\odot$; a giant galaxy may have a mass of about 10^{13} $M_\odot$ and a radius of 30 kiloparsecs (kpc). Most galaxies are, in general, separated from one another by distances on the order of millions of light years. The space between galaxies, known as intergalactic space, is filled with a tenuous plasma with an average density less than one atom per cubic meter. There are more than a hundred billion galaxies in the Universe. They are rarely isolated and found as galaxy pairs, small groups, large clusters, and superclusters. Galaxies form in dark matter halos, which on larger timescales merge, and therefore become massive structures. These structures can be in small galaxy groups or in clusters of galaxies bound together by gravity. These clusters are composed of hundreds to thousands of galaxies, The bulk of their mass contain hot gas and dark matter.

Various shapes of galaxies were discovered. Hubble (1936) classified them into elliptical, lenticular,[25] spiral, and irregular galaxies. Based on the visual morphological type, these galaxies are ordered in a sequence, what is referred to as the Hubble sequence, from early to late types. However, it may miss certain important characteristics of galaxies such as star-formation rate in starburst galaxies and activity in the core of active galaxies. The galaxies are arranged in a tuning fork sequence, the base of which represents elliptical galaxies of various types, while the spiral galaxies are arranged in two branches, the upper one represents normal spirals, and the lower one represents barred spirals.

The Hubble sequence classified the elliptical galaxies on the basis of their ellipticity, ranging from $E0$, which is almost spherical, up to $E7$, which is highly elongated. The density of stars in the elliptical galaxies falls off in a regular fashion as one goes outwards. The $S0$ type galaxies are placed in between the elliptical and spiral galaxies. These galaxies have relatively little interstellar matter, and consequently have a low portion of open clusters and a reduced rate of star-formation. They have more evolved stars that orbit the common gravitational center in all possible orbits.

Spiral galaxies are relatively bright objects and have three basic components such as (1) the stellar disk containing the spiral arms that are extended outward from the bulge, (2) the halo, and (3) the nucleus or central bulge composed of older stars. Some have large scale two-armed spiral pattern, while in others the spiral structure is made up of a many short filamentary arms. In addition, there is a thin disk of gas and other interstellar matter. These galaxies are divided into normal and barred spirals. The Hubble sequence listed the spiral galaxies as type S, followed by a letter (a, b, c), which indicates the degree of tightness of the spiral arms and the size of the central bulge. There are two sequences of spirals, normal Sa, Sb, and Sc and barred SBa, SBb, and SBc. The spiral arms in spiral galaxies have approximate logarithmic shape. These arms also rotate around the center, but with constant angular velocity. Most of the interstellar gas in such galaxies is in the form of neutral hydrogen.

---

[25] Lenticular galaxies ($S0$-type) have properties of both elliptical and spiral galaxies. These galaxies possess ill-defined spiral arms with an elliptical halo of stars.

   Irregular galaxies cannot be classified in the above sequence. There are two major Hubble types of irregular galaxies, Irr I-type are rich in gas and are natural extension of spirals with further reducing bulge and opening out of spiral arms. Irr II-type are dusty, and probably originate through galaxy mergers. In addition, a few percent of the galaxies do not follow the standard pattern fully, hence are referred as peculiar galaxies. Many of them have bridges, tails, and counterarms of various sizes and shapes. Such peculiarities may have resulted from the interactions of two or more galaxies. The galaxies with close companions experience tidal friction, which decreases their orbital radii and leads to their gradually forming a single system in equilibrium, what is known as dynamical friction. They are expected to merge in a few galactic crossing times. Giant luminous galaxies at the cores of dense clusters are supposed to have formed by the merger of smaller neighbors. If the galaxies are centrally concentrated and have similar mass, merging occurs nonrapidly than disruption. On the contrary, if the masses are dissimilar, the interaction between them is likely to cause considerable disruption to the less massive companion.

   Every large galaxy, including the Galaxy (Milky Way), harbors a nuclear SMBH (Kormendy and Richardson 1995), or several of them as a consequence of galaxy mergers (Milosavljevic and Merritt (2001). The SMBHs have masses between a million and several billion solar masses. Due to the deep gravitational potential, they accrete matter from their surroundings. Since the galaxies are rotating, the infalling matter would preserve its angular momentum (rotational motion) and can form a rapidly spinning accretion disk around the central SMBH. The extraction of gravitational energy from a SMBH accretion is assumed to power the energy generation mechanism of Active Galactic Nuclei (AGN; Frank et al. 2002).

   The Hubble sequence of galaxies is applicable to massive galaxies as one sees them to-day. There is strong evidence that galaxies evolve. The bottom up scenario currently gaining acceptance suggests galaxies were born as smaller systems and grew by intermerginal look-back times corresponding to redshifts 7–3. The galaxy downsizing hypothesis suggests that more massive galaxies evolved faster than less massive ones. Merger process enhanced the star-formation; galaxies grouped and clustered while individual galaxies merged and evolved. The SMBH grew during merger phase through coalescing stellar or intermediate mass black holes and further through accretion.

9.4.1 Active Galactic Nuclei

The Active Galactic Nuclei (AGN) are one of the most energetic phenomena in the Universe. In some galaxies, the core generates huge energy that are orders of magnitudes higher than for normal galaxies. The study of the physical processes, viz., temperature, density, and velocity of gas in the active regions of AGN is an important aspect in observational astronomy. The activity is generally detected by the presence of broad-emission lines in the nuclear spectrum of the galaxy and/or intense emission in the γ ray, X-ray or radio region. The galaxy hosting an AGN is known as 'active galaxy'. These galaxies are more luminous long lived sources having luminosities ranging between $10^{33} - 10^{39}$ W. The spectrum of radiation emitted by an AGN is markedly different from that of an ordinary galaxy, which implies that the radiation from the former is dominated by non-stellar processes (gravitation) as compared to the stellar process (nuclear fusion) operating in normal galaxies. A widely supported hypothesis suggests that AGNs are powered by a SMBH at the center of the active galaxy.

AGN may also posses an obscuring torus of gas and dust, with a radius of a few parsecs surrounding the central SMBH, which obscures the Broad-Line Region (BLR) from some directions. The BLR are the in the innermost regions, a few light years away from the center and contain gas with density of $\approx 10^9 - 10^{12}$ atoms cm^{-3} and moving with speeds of a few thousands of kilometers per seconds. They are responsible for the broad emission lines in the spectrum of an AGN. The Narrow Line Region (NLR) clouds are far away from the BLR region having a low density of $\approx 10^2 - 10^6$ atoms cm^{-3} and move with speeds of a few hundreds of kilometers per second. Thus they produce the narrow emission lines in the optical spectrum of an AGN. The optical imaging by emission-lines on sub-arcsecond scales can reveal the structure of the NLR. The scale of narrow-line regions is well resolved by the diffraction-limit of a moderate-sized telescope. The time variability of AGNs ranging from minutes to decades is an important phenomenon that constrains the models of active regions. In addition, AGN may also have (1) an accretion disk and corona in the immediate vicinity of the SMBH and (2) a relativistic jet emerging out of the nucleus. The strength, size, and extent of the various components vary from one AGN to another.

Role of powerful AGN feedback through winds and ionization of the interstellar media is now seen as an integral part of the process of galaxy formation. Some of the most recent X-ray surveys revealed unexpected populations of AGN in the distant Universe, and suggest that there may have been more than one major epoch of black hole mass accretion assembly in the history of the Universe (Gandhi 2005). The sizes of accretion disk are thought to be of the order of light-days for typical SMBHs of mass 10^6 M$_\odot$. However, even at the distance of the nearest AGN, such sizes are too small to be resolved by the current generation of telescopes, since the resolution required is close to 1 mas (Gallimore et al. 1977). LBOIs with very large telescopes may be able to resolve the disks for the very nearest AGN.

Active galaxies emitting high energy radiation in the form of X-rays are classified either as (1) Seyfert galaxies, traditionally AGN that are fainter than $M_B \sim -21.5 + 5 \log H_0$ in the optical, where M_B is the absolute B-magnitude and H_0 the Hubble constant[26] in units of 100 km s^{-1} Mpc^{-1}, or (2) QUASARs (QUASi-stellAR radio source).

---

[26] The Hubble constant is a measure of expansion rate of the Universe, measured by the ratio of the speed of recession of a galaxies to their distance from the observer. The Hubble constant can be used to determine the distances to galaxies from the observed recession velocity, when peculiar velocities of galaxies are negligible and the acceleration/deceleration of the Universe is not important.

Seyferts are classified into two types such as, Seyfert 1, in which the spectra depict both narrow and broad emission-lines and Seyfert 2, where the spectra show narrow lines. Present models suggest that both these Seyferts are intrinsically the same, a black hole surrounded by a small and rapidly orbiting disk plus an extended region. Seyfert 2 galaxies, where the broad component is believed to be obscured by dust, produce polarization at arcsecond scales, resolvable by HST; the broad component can be observed in polarized light. Seyfert 1 galaxies, however, have reflection and scattering regions closer to the nucleus. Intermediate types, namely, Seyfert 1.5, Seyfert 1.8, Seyfert 1.9 are also identified based on their optical spectrum by Osterbrock (1981).

The active galaxies are broadly divided into two classes in terms of their radio power such as (1) radio-quiet AGN and (2) radio-loud AGN. The latter type AGNs are associated with large scale relativistic jets and radio lobes and are found in elliptical galaxies, while the former type AGNs have very weak radio emitting regions and are predominantly found in spiral galaxies. Radio-quiet AGNs include LINERs (Low-Ionization Nuclear Emission line Region), whose optical spectra are quite distinct from those of both H II regions and classical AGNs, Seyfert galaxies, and radio-quiet quasars. Quasars are the distant luminous objects, which display a very high redshift. Some of them display rapid changes in luminosity as well. They can be observed in radio, IR, visible, UV, X-ray and gamma ray bands. The central black hole is drawing in material from the inner parts of the galaxy and producing a jet of extremely energetic particles. Accretion on to such a black hole transforms gravitational potential energy into radiation and outflows, emitting nearly constant energy from the optical to X-ray wavelengths. Most of the quasars are radio-quiet. Radio-loud AGNs include radio-loud quasars, radio galaxies, and Blazars that are very compact energy source associated with a SMBH at the center of a host galaxy and are having a relativistic jet.

Many quasars do not emit radio radiation, these objects are designated as QSO (quasi-stellar object). They are closely related to the active galaxies such as more luminous AGN, Seyfert galaxies or BL Lac objects. QSOs are powered by accretion of material onto supermassive black holes in the nuclei of distant galaxies. They are found to vary in luminosity on a variety of time scales such as a few months, weeks, days, or hours, indicating that their enormous energy output originates in a very compact source. Such objects exhibit properties common to active galaxies, for example, radiation is nonthermal and some are observed to have jets and lobes like those of radio galaxies. QSOs may be gravitationally lensed by the object located along the line of sight. Gravitational lensing, an effect predicted by Einstein's general relativity theory (Einstein 1916), occurs when the gravitational field from a massive object, such as galaxy cluster (the lens) warps space and deflects light from a distant object (the source) behind it. Depending on the cluster mass distribution a host of interesting effects are produced, such as magnification, shape distortions, giant arcs, and multiple images of the same source. The resulting magnification is time variable as the projected separation between the source and lens varies.

There are three classes of gravitational lensing (1) strong lensing in which Einstein rings[27] (Chwolson 1924), arcs, and multiple images are formed, (2) weak lensing, where the distortions of background objects are much smaller, and (3) microlensing in which distortion in shape is invisible, but the amount of light received from a background object changes in time. Foy et al. (1985) had resolved the gravitational image of the multiple QSO PG1115+08 by means of speckle interferometry; one of the bright components, discovered to be double (Hege et al. 1981), was found to be elongated that might be, according to them, due to a fifth component of the QSO.

A large number of bright radio sources in the sky are very powerful non-thermal synchrotron[28] sources. The spectrum of these radio sources is opposite to that of thermal emission – usually the flux increases sharply with increasing wavelength. The typical examples are SN remnants, radio galaxies, pulsars etc.

Extragalactic radio sources emit a continuum of radio wave lengths and lie beyond the confines of the galaxy. The characteristic feature of a strong radio galaxy is a double (fairly) symmetrical, roughly ellipsoidal structures, in which two large regions of radio emission are situated in a line on diametrically opposite sides of an optical galaxy. The parent galaxy is, in general, a giant elliptical, sometimes with evidence of recent interaction, for example Cygnus A. The typical spectrum of the observed radio waves decreases as a power of increasing frequency, which is conventionally interpreted, by analogy with situation known to hold for the Galaxy (Milky Way) in terms of radiation by cosmic-ray electrons, with a decreasing power-law distribution of energies. These radio sources exhibit a wide variety of polarization. For instance, the non-thermal emission from the Galaxy and from some of the localized objects show a low degree of linear polarization. To note, some radio sources like flare stars, the Sun, the Jupiter etc. exhibit a high degree of circular polarization but very little or no linear polarization. Since, the emission is polarized due to the presence of the magnetic field within the emitting source, it is possible to estimate the strength and the orientation of the magnetic field from the polarization measurements. It was found from aperture-synthesis studies of fine structure of double-lobed radio galaxies that many such sources possess radio jets that point from the nuclei of the parent galaxies to the radio lobes. It is believed that the nucleus of an active galaxy supplies the basic energy that powers the radio emission.

Active galactic nuclei, in the form of Seyfert galaxies, quasars, and other violent sources containing accretion disks have very small angular sizes. The high energy

---

[27] An Einstein ring is a special case of gravitational lensing, caused by the perfect alignment of two galaxies one behind the other. The angular radius of this ring-like structure is called Einstein radius, θ_E, which is given by,

$$\theta_E = \sqrt{\frac{4GM}{c^2} \frac{d_{LS}}{d_L d_S}},$$

in radians. Here, G is the gravitational constant, M the mass of the lens, c the speed of light, d_L the distance to the lens, d_S the distance to the source, and d_{LS} is the distance between the lens and the source.

[28] Synchrotron radiation is generated by the acceleration of ultra-relativistic charged particles around magnetic field lines; at non-relativistic velocities, this results in cyclotron radiation. The radiation typically includes radio waves, infrared light, visible light, ultraviolet light, and x-rays.

nucleus consists of a central source, called a continuum region, with a size of less than one-tenth of a parsec, but surrounded by a clumpy region called the BroadLine Region with a size on the order of one parsec. Interferometry is a very good technique to get more useful information. For example, NGC 1068, an archetype Seyfert 2 galaxy, is one of the brightest and nearest Seyfert 2 galaxies. Located at a distance of 14.4 Mpc, it harbors an active galactic nucleus. Its nucleus has been studied in the entire spectrum, from X-rays and UV to radio wavelengths, including the optical, near- and mid-IR wavelength ranges. It was classified as a Seyfert 2 due to the narrow emission-lines it emits. However, Antonucci and Miller (1985) have discovered broad, polarized emission-lines, suggesting the presence of a Seyfert 1 nucleus, hidden by a geometrically and optically thick dusty torus surrounding it.

Observations of NGC 1068 corroborated by theoretical modeling like radiative transfer calculations have made significant contributions on its structure. Single aperture bispectrum speckle interferometry at K′- and H′-bands resolved structure consisting of a compact core and an extended Northern and South-Eastern component (see Fig. 9.23; Weigelt et al. 2004). The K′-band FWHM diameter of

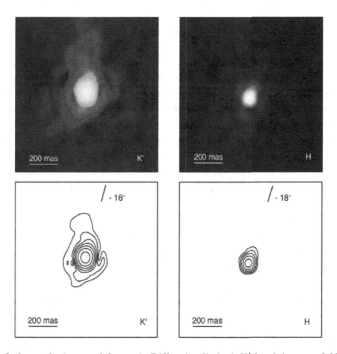

Fig. 9.23 *Left panels (top and bottom)*: Diffraction-limited K′-band image of NGC 1068, reconstructed by bispectrum speckle interferometry showing the compact core with its tail-shaped, North-Western extension; the first diffraction ring around the compact component is visible. *Right panels (top and bottom)*: Diffraction-limited H-band image, which also show North-Western extension (Weigelt et al. 2004). Courtesy: G. Weigelt

this core was found to be $\approx 18 \times 39$ mas, while it is $\approx 18 \times 45$ mas at H$'$-band. The dust sublimation radius for NGC 1068 is estimated to $\sim 0.5 - 1$ pc. The VLTI/MIDI has also succeeded in resolving structures in the dusty torus of NGC 1068 at 8 μm (Jaffe et al. 2004) on scales of 30 mas. Its nucleus was also observed using the said instrument at 10 μm by Poncelet et al. (2006). The visibilities measured across the N-band indicate that the core of NGC 1068 is well resolved in the mid-IR.

In addition, several other AGNs such as Circinus (an inclined spiral galaxy harboring a Seyfert 2 active nucleus, as well as a circumnuclear starburst; Tristram et al. 2007), Centaurus A (NGC 5128), Mrk 1239 (a narrow-line Seyfert 1 galaxy located at a distance of about 80 Mpc), NGC 3783 (Beckert et al. 2008), and MCG-05-23-016 (a lenticular Seyfert 2 galaxy at a distance of 35 Mpc) have also been observed. Among these, the radio galaxy, Centaurus A, is the closest active galaxy undergoing late stages of a merger event with a spiral galaxy in the constellation Centaurus. Interferometric results obtained by Meisenheimer et al. (2007) reveal the existence of two-components in the inner parsec of Centaurus A, a resolved component, the disk, which is extended along a position angle of $\sim 110°$ and an unresolved core (< 10 mas), which presumably represents the base of the radio jet. For QSO HE2217-2818, the VLTI/AMBER instrument could discern details like starburst knots in their host galaxies and resolve emissions cones and reflection nebulae.

With Keck interferometer, Swain et al. 2003) showed that near-IR emission of the nucleus of NGC 4151, 12.26 Mpc far from Earth, is very compact (≤ 0.1 pc). NGC 4151, a 10.8 m_v spiral galaxy in Canes Venatici, is identified as the archetype type 1 Seyfert galaxy. The composite spectrum of this galaxy shows the wide variety of emission lines present, from the Lyman limit at 912 Å to the mid-infrared. The observed variability in high resolution observations in optical region suggests that the sizes of the emission region in NGC 4151 is of the order of 0.2 mas. From the interferometric data, using the afore-mentioned interferometer, of the innermost dust in this active nuclei, Pott et al. (2010) observed that the size of the torus does not vary along with the changing luminosity of the central source. According to them, dust destruction and formation time scales around Seyfert nuclei are significantly longer than a few years.

Observations, in order to explore the inner region, of four type-1 AGNs, namely NGC 4151, Mrk 231, NGC 4051, and the QSO IRAS 13349+2438 at $z = 0.108$, have also been made in the near-IR (K-band) wavelengths at the Keck interferometer by Kishimoto et al. (2009). They have detected high visibilities ($\mathcal{V}^2 \sim 0.8 - 0.9$) for all these objects, although a marginal decrease of $\mathcal{V}^2$ with increasing baselines is noticed in the case of the first object. These observations partially resolved the dust sublimation region. The decrease and absolute $\mathcal{V}^2$ are well fitted with a ring having a radius of 0.45 ± 0.04 mas (0.039 ± 0.003 pc). The effective radius of these four AGNs, obtained from ring model, is comparable to the light traveling distance for the time lag between the K-band flux variation and the UV/optical variation.

9.4.2 Star-Formation in Galaxies

As stated earlier in Sect. 9.2.2, the star-formation takes place in dense molecular clouds within galaxies. The parts of molecular clouds collapse, which is associated with the segregation of different chemical species occurring, into a ball of plasma to form a star. Its process involves an enormous range of physical scales, from whole molecular cloud complexes of the size of several hundred parsecs to individual stars of the size of a few solar radii. Stellar chemical evolution or nucleo-synthesis[29] further influences evolutionary process.

A single nebula may give birth to many stars. The evolution of young stars is from a cluster of protostars deep in a molecular clouds core, to a cluster of T Tauri stars whose hot surface and stellar winds heat the surrounding gas to form H II (ionized atomic hydrogen) region, in which star-formation is taking place. The H II regions are composed primarily of hydrogen and have the temperatures around 10,000°K. Generally, less than 10% of the available gases of this region is converted into stars, with the remainder of the gases dispersed by radiation pressure, supernova explosions, and strong stellar winds from the most massive stars, leaving behind open clusters. The massive star-formation takes place inside dense cluster environments, via coalescence of lower mass (<10 $M_\odot$) stars. The evolutionary time scales of high-mass stars are very short. These stars begin to burn hydrogen while still accreting material from the surrounding proto-stellar cloud. Following the formation of the stellar core, they vigorously affect their environment and have a fundamental influence over the galaxy evolution. They generate most of the UV radiation in galaxies, powering the far-IR luminosities through the heating of dust. The evolution of bubbles of hot plasma is formed in the InterStellar Medium (ISM) by supernova explosions or mass-loss from massive stars. In regions of OB associations, these bubbles collide with one another and merge to form large structure (Dopita and Sutherland 2003). Speckle interferometric observations of R 64, HD 32228, the dense stellar core of the OB association LH 9 in the LMC, revealed 25 stellar components within a $6.4'' \times 6.4''$ field of view (Schertl et al. 1996). The ultra-compact H II region, K3-50 A, is highly obscured in short wavelengths, but bright at long wavelengths. Centered on near-IR and mid-IR wavelengths, Hofmann et al. (2004) observed the nebula using speckle techniques. The reconstructed image resolved the central K'-band emission of K3-50 A into several point-like sources (see Fig. 9.24). The brightest K'-band source, located at the tip of the cone-shaped nebulosity, dominates the near-IR emission. Some of the other point-like sources are considered to be massive stars.

Later, the cluster breaks out, the gas is blown away, and the stars evolve. To note, the star clusters are groups of stars that are gravitationally bound and are formed at the same time, from the same cloud of interstellar gas. They provide information about stellar evolution, because they are similar in age and chemical composition. There are two types of star cluster such as open clusters (or galactic clusters) and

---

[29] Nucleosynthesis is a process of creating new atomic nuclei from the pre-existing nucleons.

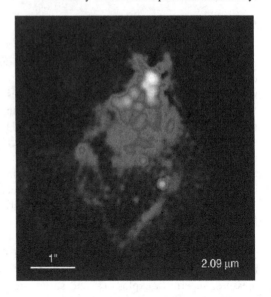

Fig. 9.24 K′-band image of
K3-50 A, reconstructed with
the bispectrum speckle
interferometry (Hofmann
et al. 2004; courtesy:
G. Weigelt)

globular clusters. The open clusters are confined to the galactic plane and are found
within spiral arms. These clusters are loosely bound collection of 100–1,000 hot and
young population I stars within a region up to about a few parsecs. On the contrary,
the globular clusters possess 100,000–1,000,000 population II stars in a region of
about 50 parsecs. They populate the halo or bulge of the Galaxy and other galaxies
with a significant concentration toward the galactic center. Since all the stars in a
cluster formed at the same time, they are all the same age.

One of the important features of galaxies is the wide range in young stellar
content and star-formation activity (Kennicutt Jr. et al. 1998 and references therein).
The star-formation history is an indicator of galaxy evolution. Individual young
stars are unresolved in most of the galaxies by any present day available telescope.
Most of the information on the star-formation properties of galaxies arrive from the
integrated light measurements in the UV (1,250–2,500Å), far-IR (in the range of
10–300 μm), or nebular recombination lines. The UV stellar continuum depends on
the amount of dust within the galaxy and its distribution with respect to star-forming
regions. The far-IR and sub-mm spectral energy distributions of luminous star-
forming galaxies are determined by the reradiation by the dust of energy absorbed
in the visible and UV region of the spectrum. The total star-formation rate (SFR)
can be estimated by the following relation (Kennicutt 1998b),

$$\text{SFR}(\text{M}_\odot \ yr^{-1}) = 4.5 \times 10^{-44} L_{FIR}(erg \ s^{-1}) \quad \text{(starbursts)}, \qquad (9.39)$$

where L_{FIR} refers to the IR luminosity integrated over the full-, mid-, and far-IR
spectrum (8–1,000 μm).

The far-IR luminosity depends on the contribution of young stars to heating of the dust and on the optical depth of the dust in the star-forming regions. These spectra of galaxies contain both a warm (60 μm) and cooler (≥ 100 μm) components associated with dust around young star-forming regions. The radio continuum flux is also a good indicator of star-formation, which results predominantly from the non-thermal supernova remnant flux. However, it is model dependent and is critically influenced by the density of the interstellar gas.

The nebular lines effectively re-emit the integrated stellar luminosity of galaxies shortward of the Lyman limit. They provide a direct probe of the young massive stellar population. Most applications of this method have been based on measurements of the Hα line, but other recombination lines, including Hβ, Brα, and Brγ, have also been used. The flux in a hydrogen line is proportional to the number of photons produced by these stars, which is, in turn, proportional to their birth-rate (Kennicutt 1998a, b). The luminosities of forbidden lines are not directly coupled to the ionizing luminosity, and their excitation is sensitive to abundance and the ionization state of the gas. The strongest emission feature in the blue is the [O II], λ 3727Å, forbidden-line doublet which depends on the star-formation rate, as well as on the degree of enrichment of the interstellar gas; knowledge of the metallicity of gas is required to determine the star-formation rate.

The large-scale star-formation takes place in the extended disks of spiral and irregular galaxies, as well as in the compact, dense gas disks in the centers of galaxies. The giant molecular clouds, which occupy spiral arms in the galaxy, are the sites of both low mass and high mass star-formation, while the low mass stars form through out the galactic disk. To note, the low mass star-formation is common throughout the galaxy, although it can be studied within a few hundred parsecs due to its relative faintness compared to massive star-formation. High-resolution observations at a wide range of wavelengths are critical for probing the physical phenomena associated with the birth of low-mass stars.

The strong trends in disk star-formation rates arise from the relationships between the global star-formation rate and other physical properties of galaxies, such as their gas contents or dynamical structure (Kennicutt Jr. 1998a, b and references therein). The global properties of a galaxy such as the mass, bar structure, spiral arm structure, influence its star-formation rate. In the disk galaxies, for a self-regulated star-formation, there exists a connection between the local star-formation rate in the disk and the local disk properties. The star-formation rate is generally derived from Hα surface photometry.

The external environmental influences can also have effects on the star-formation rate, for example galaxy interacting with another galaxy, although its enhancement is variable. The average enhancement is a factor of 2–3, while in extreme case, it may of the order of 10–100; the intense bursts of star-formation are believed to be driven by mergers between gas-rich galaxies (Tacconi et al. 2008). Larger enhancements may be envisaged in the circumnuclear regions of strongly interacting and merging systems (Kennicutt et al. 1998). A cluster environment may also alter the star-formation properties of galaxies. Many spiral galaxies located in rich clusters exhibit significant atomic gas deficiencies (Cayatte et al. 1994).

Moss and Whittle (1993) have found a 37–46% lower H_α detection rate among Sb, Sc, and irregular galaxies in the clusters, albeit a 50% higher detection rate among Sa–Sab galaxies.

The circumnuclear regions of many spiral galaxies host luminous star-forming regions, for example starburst regions, which refer to a region of space with a violent high-mass star-formation, compared to the usual star-formation rate. Galaxies are often observed to have a burst of star-formation after a collision or close encounter between two galaxies. They are often associated with merging or interacting galaxies. Scaled-down versions of starbursts are found in the local group of galaxies, like 30 Doradus in the LMC. To note, the spectra of massive WR stars are often observed in extragalactic H II regions. For example, recently López-Sánchez and Esteban (2010) have studied the broad stellar features originated by winds of WR stars in a sample of starburst galaxies, which confirm the presence of a substantial population of WR stars.

In order to study the complex process of star-formation in external galaxies that can be very different from those found in the Galaxy, observations at different spatial scales are required to be carried out. The study of the formation of massive young clusters in mergers and merger remnants may allow the formation of similar-mass globular clusters in the early Universe. The nearby dwarf galaxies such as LMC, SMC (Small Magellanic Cloud), permit to enlighten the effect of metallicity on the star-formation process. Although large telescopes can identify the cores, but long baseline interferometry is required to study their structure. Recently, Swinbank et al. (2010) have made direct measurements of the size and brightness of regions of star-birth in a very distant sub-mm galaxy, SMM J2135-0102, at redshift $z = 2.3259$, which is gravitationally magnified by a massive foreground galaxy cluster lens, with the 12 m Atacama Pathfinder Experiment (APEX) telescope. This magnification, when combined with high-resolution sub-mm imaging, resolved the star-forming regions at a linear scale of 100 pc. According to these authors the luminosity densities of these star-forming regions are comparable to the dense cores of giant molecular clouds in the local Universe. They have opined that this galaxy is producing stars at a rate much faster than in more recent galaxies.

9.5   Infrared Astronomy

Introduction of the new generation infrared (IR) detectors resulted in exciting discoveries in IR astronomy. These are in the form of accurate measurements of both luminosity and temperature-sensitive indices in IR spectral regions, detection of disks of material and planets orbiting other stars (exoplanets). In addition, many objects such as cool stars, IR galaxies, clouds of particles around stars, nebulae, interstellar molecules, and brown dwarfs, which are too cool and faint to be detected in optical light, can also be detected. With increased sensitivity due to large aperture telescopes on ground and in space, it has become possible to probe deeper into the distant Universe and reach the earliest galaxies formed when the Universe was less than a million years old, and study the first stars formed in the Universe. The ultraviolet light emitted by such distant objects is redshifted into the infrared band.[30]

In the near-IR wavelength region, generally large red giants stars and low mass red dwarfs dominate. In this region, the interstellar dust becomes transparent, allowing to observe regions hidden by dust in the visible image. In the mid-IR image, one may envisage proto-planetary disks, planets, comets, and asteroids. Planets absorb light from the Sun and heat up, which is re-radiated by them as infrared light. The temperature range of the planets in the solar system is about 53–573°K. Objects having temperatures in this range emit most of their light in mid-IR. For instance, the Earth radiates strongly at about 10μm. Asteroids emit most of their light in this region as well, such data may help to determine their diameter and surface composition. Dust, which is composed of silicate ranging in size from sub-micron grains to large rocks, lies in the plane of the solar system is bright at around 10 μm. The dust around stars that have ejected material shines in the mid-IR region; the proto-planetary disks, also shines in this region of IR wavelengths. Another component of dust is graphite or more exotic molecules and crystals of carbon.

Further down the spectrum that falls between the wavelength range of 25–40 to 200–350 μm, called Far-IR spectral region, the temperature range is between the 10.6–18.5 to 92.5–140°K. In this spectral range, the cold clouds of gas and dust in the Galaxy, as well as in nearby galaxies become discernible. Owing to the thick concentration of stars embedded in dense clouds of dust, which heat up the dust, cause the center of the Galaxy to glow brightly in the Far-IR. Observations with this range of IR wavelength can detect protostars. Knowledge of the coolest stellar component in galaxies would set constraints on both empirical and evolutionary synthesis models, which would lead to new insight into the relation between systems with different star-formation histories.

The first infrared survey of the sky was carried out in the mid-1960s at the Mt. Wilson Observatory. The survey covered about 75% of the sky and found about 20,000 IR sources. The surface temperature of these stars are of the order of 1000–2000°K. The advancement in IR detector technology made it possible to detect IR emissions from the centers of many galaxies. The longer wavelength of IR light, and reduced effect of atmospheric turbulence renders adaptive optics system easier in this region. Today, many ground-based infrared telescopes are employing adaptive optics systems to create sharp images.

---

[30] The Universe is expanding as a result of the Big Bang, the explosion that marked the onset of the Universe, and the most of the galaxies are moving away; the farther they are, faster they move. This recession of galaxies away from the Earth-based observer has an effect on the light emitted from these galaxies. As a result of Doppler effect, at large redshifts, all the ultraviolet and much of the optical light from distant sources may shift into the IR by the time it reaches the telescope.

9.5.1 Astronomy with IR Interferometry

Although the IR interferometer, SOIRDÉTÉ, yielded interesting results on α Orionis (Gay and Mekarnia 1988), it did not work well. In contrast, the Infrared Spatial Interferometer (ISI) has delivered good results. Its scientific programs range from the early to the late stages of stellar evolution as well as to the precision astrometry. These stars return some fraction of their mass to the circumstellar environment, by cooling off, this material condenses into dust grains. Such grains absorb visible light and re-radiate this energy in the infrared region of the spectrum. The high resolution mapping in this region can bring out the spatial structure and temporal evolution of dust shell around long period variables.

The ISI has been employed to measure the diameter of several stars, as well as to study the spatial distribution of dust around late-type stars. The diameter of Mira was found to be 47.8 ± 0.5 mas, at pulsation phase 0.9, substantially larger than previously expected, with a systematic variation of $\pm 13\%$ over its ~ 330 day period. From the data obtained with this instrument at 11.15 μm, Danchi et al. (1994) showed that the radius of dust formation depends on the spectral type of the stars. Some of these stars such as the supergiants α Orionis, α Scorpii, and α Herculis, have dust shells far from their photospheres; the mean distance from the photosphere is ~ 38 stellar radii. Tatebe et al. (2007) have observed the asymmetry of the surface of α Orionis at mid-IR wavelengths; a 15% reduction in size at 11.15 μm of this star has also been noticed during the period of 1993–2009 by Townes et al. (2009). The mean distance between the inner radius of the dust shells and the photospheres of other stars like carbon star IRC+10216, the Mira variables, R Leonis, o Ceti, the supergiants VX Sag, VY CMa, and the symbiotic star R Aquarii is found to be within 1.8–3.5 stellar radii. Recently, Wishnow et al. (2010) resolved dust shells surrounding the star W Hydrae (M7e) at a wavelength of 11.15 μm using the aforementioned instrument. It is an AGB star that is known to be surrounded by dense gas and dust. These authors used two different models for the star and dust shells to fit the data, one for recent data and another for the combined data of several years. Their model shows the presence of two dust shells with diameters of approximately 100 mas and 250 mas.

Continuous monitoring of some of the evolved stars would determine the changes in the visibilities due to variations in stellar luminosity and movements and changes in the dust shells. Observations have been made of NML Cygni, as well as of changes in the dust shell around Mira and IK Tauri (Hale et al. 1997). Closure-phase measurements at 11.15 μm carried out by the ISI, showed temporal variations and asymmetries in the surrounding dust, with a difference of about 15% in intensity between two sides of the star, IK Tauri (Weiner et al. 2006).

With the ISI instrument, it was possible to determine the precise occurrence of inorganic and organic molecules in the circumstellar environment, as well as the clues about their formation. Heavy mass-loss on the AGB promotes circumstellar formation of a variety of polyatomic molecules such as ammonia (NH_3), silane (SiH_4), and ethylane. The location of molecular formation may be determined by measuring the angular size of the absorption region in the line. It can be derived from a comparative study of the visibilities made on the line frequency in which some of the dust radiation has been absorbed, and at a frequency where the molecular line or absorption is absent. Monnier et al. (2000) were able to determine the mid-IR molecular absorption features of NH_3 and SiH_4 of IRC+10216 and VY CMA. A variety of dust condensations that include a large scattering plume, a bow shaped dust feature around the latter have been found; a bright knot of emission $1''$ away from the star is also reported.

From the high resolution near-IR adaptive optics imaging and polarimetry, mid-IR imaging, analysis of HST images, and theoretical modeling of the Herbig Ae/Be star, R Mon, Close et al. (1997) resolved a faint source, $0.69''$ away from R Mon and identified it as a T Tauri star. Located at a distance of $\sim$800 pc in the constellation Monoceros, R Mon (B0-type) is an intermediate-mass pre-main sequence star, which illuminates the fan-shaped reflection nebula, NGC 2261. Also known as variable nebula, its appearance could change on the timescales of months. NGC 2261 contained a complex set of filaments. Optical polarimetric observations found that this nebula was strongly polarized with a centrosymmetric pattern centered on R Mon. These authors modeled the system as a YSO surrounded by an optically and geometrically thick accretion disk that is embedded in an extended dusty envelope. The jet from R Mon, according to them, has cleared a conical cavity in the envelope, through which light escapes and illuminates the NGC 2261. Their images also revealed a complex of twisted filaments with a double-helical structure along the Eastern edge of the parabolic shell. Speckle interferometric observations by Leinert et al. (2001) resolved the halo around the central object into two components with sizes of $0.4''$ and $0.3''$. From the reconstructed images with bispectrum speckle interferometry in the H-band, Weigelt et al. (2002) reported that the primary R Mon-A star was having a bright arc-shaped structure (see Fig. 9.25), pointing away from R Mon in north-western direction,

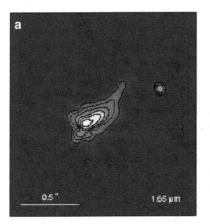
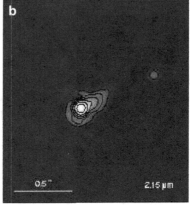

Fig. 9.25 (a, b) Central parts of H- and K-band speckle images of R Mon (Weigelt et al. 2002; courtesy: G. Weigelt)

while the secondary star, R Mon-B appeared to be point source. They interpreted this feature as the surface of a dense structure close to the thick circumstellar disk or torus around R Mon. They have also found several twisted filaments of helical shape similar to the twisted filaments in the outer parts of the nebula. The near-IR polarimetric data obtained at the VLT and Subaru telescopes by Murakawa et al. (2008b) suggests the presence of multiple grain populations in the R Mon nebula. A variety of interesting astronomical sources can be resolved by infrared interferometry using a kilometer baseline array. In what follows, a few frontline areas are enumerated.

9.5.1.1 The Galactic Center

The galactic center is the rotational center of the Galaxy (Reid 1993; Eisenhauer et al. 2003), which is obscured (in visible wavelengths) by a prodigious amount of absorbing gas and dust. Most of the information of the Galaxy, thus far, comes from observations of radio and infrared radiation, since their long wavelengths have the ability to pass through the dust along the plane. To note, gas emits X-rays when heated to extremely hot temperatures, while dust, which is cooler than stars or gas, emits a continuum of light mostly in the infrared and longer wavelengths. The galactic center was first detected by Jansky (1932). Observations with Very Large Array (VLA) have produced images representing the wide variety of phenomena occurring at the galactic center, for example radio filaments, which are strongly magnetized. These filaments are coherent and oriented perpendicular to the plane of the Galaxy. The radio image obtained at the VLA shows a complex structure of ionized gas, known as mini-spiral, covering about 4 arcminutes at the center of the Galaxy (Yusef-Zadeh et al. 1998). It is composed of the Northern arm, Eastern arm and Western arc, which is surrounded by a thick ring of molecular material called the circumnuclear disk (CND).

Near-IR images of the galactic center mostly probe the stellar content showing a large amount of stars. Early near-IR images of this could resolve groups of stars. Approaching near the center, the stellar density increases to rise towards a sharp central peak; the galactic gas disk has a central hole and the central bulge is bar-shaped. Inside the central hole is a dense nuclear gas disk, whose radius is about 1.5 kpc in neutral hydrogen. However, most of its mass is molecular and is concentrated within 300 kpc of the nucleus. The mass of this molecular gas is about 10^8 $M_\odot$, which is about 5% of the total molecular mass of the Galaxy. The central 10 pc are dominated by the complex radio continuum source Sagittarius A (Sgr A). Recent observations of the galactic center, using near-IR speckle techniques and adaptive optics on large telescopes, can resolve the individual stars within the clusters. Stellar populations in galaxies in near-IR region provides the peak of the spectral energy distribution for old populations. Bedding et al. (1997) have observed the Sgr A window at the center of the Galaxy. They have produced an IR luminosity function and color-magnitude diagram for 70 stars down to $m_v \simeq 19.5$ mag. Figure 9.26 depicts an adaptive optics (AO) corrected image of the Sgr window of the Galaxy.

Fig. 9.26 The ADONIS K′ image of the Sgr window in the bulge of the Galaxy. The image is 8″ × 8″ (courtesy: T. R. Bedding)

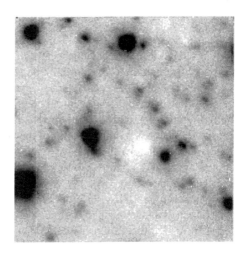

Within the core (about 5 arcsecs in size) of the galactic center, a prominent feature is the point-like radio source called Sgr A*. Studies of radio properties of this source reveal that it is a compact object approximately one astronomical unit in size. However, with the diffraction-limited speckle interferometry and adaptive optics system at the 10 m Keck-I telescope, Ghez et al. (1998) have observed a 6″ × 6″(0.23 × 0.23 pc) region to monitor the proper motion of 90 stars. The collected speckle images were processed using shift-and-add algorithm to produce a PSF with a diffraction limited core and seeing halo. Fitting orbits to the stellar motions showed that the stellar surface density and the velocity dispersion were consistent with the position of Sgr A*. The observations are also consistent with Sgr A* being located close to or at center of the Galaxy. Their measurements allow an independent estimate of the central density, which is at least 10^{12} $M_\odot$ pc^{-3}. Subsequent observations of Sgr A* covering an area of 76″ × 76″ by Ghez et al. (2005) at the 10 m Keck-II telescope equipped with adaptive optics system, have found significant IR intensity variations in $\mathcal{L}$ (3.8 μm) band; the magnitudes varied from 12.6 to 14.5. A decrease in flux density of a factor of 2 over an interval of 8 min. has also been noticed.

The galactic center is interesting for exploring the interactions between a SuperMassive Black Hole (SMBH; Reynolds 2008) and its stellar environment. By tracking the orbits of stars close to such a black hole, one may obtain information on the gravitational potential in which they move. The measurement of distance to the galactic center and its mass are of paramount importance. A precision determination of the distance to Sgr A*, in combination with its proper motion may lead to a precision measurement of the Galaxy's local rotation speed. High resolution measurements of this object have found that it is located at a distance of 8.33 ± 0.35 kpc from the Earth having a mass of ~4 × $10^6 M_\odot$ (Ghez et al. 2008; Gillessen et al. 2009). The resulting enclosed mass combined with the minute size of Sgr A* constraint provided by the radio emission, suggests that the stars must be swiftly circling around a SMBH.

There are different classes of stars observed at the center of the Galaxy. Three stellar superclusters: the Central Parsec, Quintuplet, and Arches clusters have also been discovered at the galactic center. The late-type (K and M-type) stars are identified through the CO absorption bands in their stellar spectra. The early-type (O, A and B-type) stars, characterized by the He I and H I emission lines in their spectra, contribute most of the ultraviolet radiation that ionizes the surrounding gas along the mini-spiral. The combined speckle and adaptive optics data allowed to trace two thirds of a complete orbit of the star, S 2, close to Sgr A* (Schödel et al. 2002). The data show that the star is on a bound, in a highly elliptical Keplerian orbit around Sgr A*, with an orbital period of 15.4 years and a peri-center distance of 17 light hours.

By means of speckle interferometry at the Keck I telescope, Tanner et al. (2002) have obtained image of a cool source, IRS 21 an enigmatic object, near the galactic center. Mid-IR observations revealed that it is a self-luminous source rather than heated clump of gas. According to the authors, the best model for this source and for similar sources is that these massive stars experience bow shocks as they move through the Northern arm of the Galaxy. Tanner et al. (2005) found bow-shock-like morphologies in a near-IR adaptive optics imaging experiment. Such bow-shocks can be created by stellar winds of evolved massive stars, moving through the interstellar medium of the mini-spiral.

The large-aperture Long-Baseline Optical Interferometers (LBOI) can be employed to resolve the innermost environment of stars at the galactic center. This offers a possibility to study the distribution and dynamics of stars surrounding a SMBH. By combining the light from two main telescopes of VLTI in K and N-band with the instruments, AMBER and MIDI beam combiners, Pott et al. (2006) could resolve the enigmatic dust-forming and spectrally featureless source IRS 3.

9.5.1.2 Exoplanets

Detection of exoplanets is the first step towards the detection of life beyond the Earth. These planets are thought to form in a proto-planetary disk (Armitage 2007). Although more than four hundreds exoplanets (several of these stars are known to have multiple planet systems) have been found (http://exoplanet.eu) since the first confirmed discovery of a planet, 51 Pegasi-b[31] orbiting the G-type star 51 Pegasi (located 15.4 pc from Earth in the constellation Pegasus) by Mayor and Queloz (2009), their observations can be difficult. The distances involved, as well as the faintness of any such planet with respect to its parent star, make it hard to make direct observations. Many of the planets discovered so far are either in highly eccentric orbit or have very small (0.1–0.02 AU) distance from parent star, or possess high surface temperature up to 2,000° K and mass comparable to Jupiter. They are known as 'hot Jupiters'.

---

[31] Capital letters after a star's name are reserved for stellar companions.

   The various methods such as (1) radial velocity, (2) transits, (3) astrometry, (4) direct imaging, (5) microlensing, and (6) timing detections, have been employed for the detection of exoplanets. However, most of these objects have been observed indirectly by tracking the radial velocity of stars, by detecting tiny, periodic wavelength shifts in the spectra of the parent star (Rivera et al. 2005; Udry et al. 2007; Mayor et al. 2009). The radial velocity measurement using a very high resolution spectrograph capable of detecting spectral line shift, may reveal a Jupiter-like planet. Every star with planets is affected by their mass. As planets orbit their parent star, they periodically pull it back and forth. Since the planet's gravitational pull should be large enough to move its parent star significantly, the potential for detection of planets with this method increases with planet mass and decreases with orbital radius. The radial velocity reflex motion of a star due to a planetary companion is proportional to $m_p P^{1/3}$, in which m_p is the mass of the planet and P the orbital period. This method is sensitive to the inclination of the planet's orbit with respect to the observer. It is also possible to measure the radial component of the star's velocity; its companion's mass is known only to within a factor of $\sin i$, in which i is the inclination angle between the orbital plane and a reference plane. However, stellar rotation and intrinsic variability are the major sources of noise for radial velocity measurements (Lisauer 2002). Recently, Setiawan et al. (2008) detected a giant planet, TW Hydrae-b, whose mass is estimated to be 9.8 ± 3.3 M$_J$ around TW Hydrae, a young star with an age of 810 Myr surrounded by a circumstellar disk. It orbits the star with a period of 3.56 d at 0.04 AU, inside the inner rim of the disk.

   The photometric transit method detects the drop in the luminosity of the star as the planet passes between the observer and the star, and detects the planet from the shape of the light curve. The transit signal is proportional to the occulted area of the stellar disk by the planet, and thus depends on the radii of the planet and its host star. This method provides a direct measure of the size of exoplanets, however, it may be difficult to draw any conclusion if the large planets orbit very close to a small star. The transit geometry also determines the inclination of the planet orbital plane. Combining with the radial velocity measurement, it can deduce the mass of the planet. With highly accurate transit measurements, the chemical elements in an exoplanet's atmosphere can also be identified by observing the polarization of light from the star harboring a planet as it passes through the atmosphere of the exoplanet. It is pertinent to note that the transit planet candidates in stellar light curves always require an extensive follow-up programme. High-resolution imaging and photometric observations are required to confirm the signal on the target star; medium- and high-spectral resolution radial-velocity observations to derive the mass of the planet. Several exoplanets have been characterized using photometric transit method (Torres et al. 2008; Charbonneau et al. 2009). To note, for a basic characterization of exoplanets, it is essential to derive their fundamental parameters, such as radius, mass, density, and orbit. The planetary orbit determines the dynamical evolution of the planet and the energy it received from its host star, while the other parameters determine the overall nature of a planet and allow insights into their interior structure. Recently, the CoRoT (COnvection ROtation and planetary Transits) satellite[32] has discovered a small exoplanet, known as CoRoT-7b, around the star, TYC 4799-1733-1 (now known as CoRoT-7), located towards the constellation of Monoceros at a distance of about 500 ly. This planet is 2.5 million km away from its host star. With the High Accuracy Radial velocity Planet Searcher (HARPS; Rupprecht et al. 2004) spectrograph measurements at the 3.6 m telescope at La Silla on the star CoRoT-7, Queloz et al. (2009) have measured its mass as 4.8 ± 0.8 $M_\oplus$, where $M_\oplus$ is the Earth's mass. Unlike the exoplanet, Gliese 581-e,[33] whose geometry of the orbit is undefined making its real mass unknown, in the case of CoRoT-7b, as the planet is transiting (Léger et al. 2009), the geometry is well defined, allowing the measurement of its mass accurately; its radius as 1.68 ± 0.09 $R_\oplus$, where $R_\oplus$ is the Earth's radius.

Characterization of exoplanets can be made by the interferometric techniques, such as (1) direct detection by means of interferometric nulling (see Sect. 5.1.3) or differential visibility measurements and (2) astrometry (see Sect. 5.1.2). Another technique, called differential astrometry that measures the relative angular separation of two or more objects in the sky by estimating the differential piston of the incoming starlight, can also be used. To note, the differential piston originates from a differential geometric delay from above the atmosphere, and is uncorrupted by the atmospheric turbulence. Astrometry with interferometer can determine masses and orbits of exoplanets. Astrometric surveys of young and old planetary systems may provide insight into the mechanisms of planet formation, orbital migration and evolution, orbital resonances, and interaction between planets (Quirrenbach 2009). Both the radial velocity and astrometry measurements provide estimates of the planet's mass, eccentricity, and average distance from its host star. Combining data from these methods, the inclination angle of the planetary system can be constrained, which may reduce the uncertainty in the planet's estimated mass. The motion observed in astrometry is the movement of the star in the celestial sphere. Since these motions are very small, such precision necessitates observation with either a large telescope equipped with adaptive optics system or a space-based telescope. The astrometric signature of a Jupiter-Sun system has an amplitude of 0.5 mas from a distance of 10 pc (Colavita 1999). The amplitude of such a signature is

$$\Theta = \frac{M_\star}{m} \frac{r}{l}, \tag{9.40}$$

in which m and $M_\star$ are the respective masses of the planet and star, r the orbital radius, and l the distance of the system from the Earth.

---

[32] The COnvection ROtation and planetary Transits (CoRoT) is a space mission led by the French Space Agency (CNES) in conjunction with ESA and other international partners: Austria, Belgium, Brazil, Germany, and Spain. Launched in December, 2006, it consists of an afocal telescope with 27 cm pupil and has two main scientific objectives: (1) search for exoplanets with short orbital periods and (2) perform asteroseismology by measuring solar-like oscillations in bright stars.

[33] Gliese 581-e is the fourth exoplanet found around Gliese 581, a red dwarf star approximately 20.5 ly away from Earth in the constellation of Libra (Mayor et al. 2009).

The (9.40) can be recast in terms of the period, T, of the system, i.e.,

$$\Theta = \frac{M_\star}{m^{2/3}} \frac{T^{2/3}}{l}. \tag{9.41}$$

Thus astrometry is sensitive to planets having large orbital radii and long periods.

Model-independent measurement of the angular diameter of the transiting exoplanet host star, HD 189733, was made possible by the CHARA array (Baines et al. 2007). The linear radius of the host star was found to be 0.779 ± 0.052 $R_\odot$ and of the planet, HD 189733b is 1.19 ± 0.08 R_J, where R_J is the Jupiter radius. However, the imaging of an exoplanet is beyond the capabilities of any ground-based telescope that does not employ interferometry or adaptive optics system. Large contrast and short separations between the planet and its host star are the main drawbacks to obtain an image of exoplanet. With the coronagraph in the high resolution camera on HST's advanced camera, Kalas et al. (2008) have directly imaged a planet orbiting α PsA, that is responsible for the sharp inner edge of the debris disk, and whose presence was already predicted by theoretical models. To note, planets form within disks and gas orbiting newly born stars (see Sect. 9.2.3.2). The planet, α PsA-b (most often referred to as Fomalhaut-b), is brighter than expected for an object of three Jupiter-masses and is about 18 AU closer to the star than the inner edge of the debris disk. Three planets around HR 8799 are also discovered around the same time. These were the only exoplanets whose orbital motion was confirmed by means of direct imaging so far. Such a method may expose more intricate structure in the disk such as gaps and clumps.

The detection of the gravitational lensing effect caused by large planets is a promising technique (Beaulieu et al. 2006). Microlensing events occur on timescales of months. If the source image passes close to the planetary companion of the lens star, a further perturbation of the magnification occurs. Several planet candidates having masses in the range of 0.02–3.5 M_J have been detected using such a method (Bond et al. 2004; Beaulieu et al. 2006; Gould et al. 2006; Goudi et al. 2008). Microlensing is sensitive to (1) low-mass planets, the timescale for the Earth-sized planetary events is about 1.5 h, (2) planets that have a projected distance within a factor of 1.6 of the Einstein ring, and (3) free-floating planets since the microlensing event from a planet occurs regardless of whether it orbits a host star or not. Unlike Doppler method that is sensitive to late-type stars, the planetary microlensing events are independent of the type of host star. Also, it can detect multiple planets in a single event. However, the drawbacks of such a method are: (1) impossible to carry out any follow up observation since the rare alignment of both the lensed object and the lensing system never occurs again, (2) both sets of these objects need to be large and very distant in order to have any signature of planets, and (3) difficult to assign types and luminosities to both the lensed object and the lenser; the planetary mass depends on both of these parameters.

Detection of planetary companions to stars is also possible by searching for timing variations of periodic phenomena associated with the star. Tiny anomalies in the timing of observed maximum of the periodic phenomenon caused by differences in the light travel time as the star orbits the barycenter of the star-planet system can be used to track changes in its motion caused by the presence of planets. Wolszczan and Frail (1992) discovered the planets around neutron stars using the time variations provided by a milli-second pulsar, PSR 1257+12. Eclipsing binary stars can also be used to search for timing variations in the eclipses caused by planets either around one companion, or circumbinary planets (Lee et al. 2009; Qian et al. 2010a, b).

9.5.2 Astrobiology

Astrobiology, study of life as a planetary phenomenon, aims to understand the fundamental nature of life such as origins, evolution, distribution, and future of life on the Earth and the possibility of life elsewhere in the Universe. An interdisciplinary subject of this kind which comprises of astronomy, biology, chemistry, and geology requires to get acquainted with the fundamental concepts of life and habitable environments. To note, the circumstellar habitable zone is defined as the region around a star within which an Earth-like planet can sustain liquid water or terrain convenient for life on its surface for a significant period of time. The main requirement for life is an energy source. The presence of water, a condition necessary for life, over long time periods is required; the availability of the basic organic compounds and nutrients should also be present (Mix et al. 2006).

Knowledge of the chemical composition of any planetary atmosphere provides information on the likelihood of finding carbon-based life. Ozone (O_3), in the presence of liquid water, is taken as a biomarker representative of the Earth. Methane plays the same role in history of early Earth as well. Lovelock (1965) has suggested that the simultaneous presence on Earth of a highly oxidized gas, like O_2, and highly reduced gases, like CH_4 and N_2O is the result of the biochemical activity. However, finding spectral features that are specific to biological activity on an exoplanet would be difficult. An alternative life indicator would be ozone (O_3), detectable as an absorption feature at 9.6 μm. On Earth, ozone is photochemically produced from O_2 and, as a component of the stratosphere, is not masked by other gases. Finding ozone would, therefore, indicate a significant quantity of O_2 that should have been produced by photosynthesis (Léger et al. 1993). Moreover, for a star-like the Sun, detecting ozone can be done 1,000 times faster than detecting O_2 at 0.76 μm: estimates made by Angel and Woolf (1997) show that the requirements for planet detection in the visible with an 8 m telescope are not achievable with current technology. Most of these aspects may be observed by a large space-based interferometer such as Luciola concept (Labeyrie et al. 2009). Such an interferometer fitted with coronagraphic channels may produce images and low-resolution spectra of habitable planets near their parent star. In the 10–20 μm IR, habitable planets having relative luminosity 10^{-6} with respect to the parent star can be imaged (Boccaletti et al. 2000).

Appendix A
Transfer Function of an Optical System

A.1 Linear System

A linear system is characterized by its response to a delta function, known as impulse response, for example, the term, $\mathcal{L}[\delta(\mathbf{x})]$, is the impulse response of an optical system. Invariant impulse response is an important condition that can be imposed upon. In a time- and/ or shift-invariant linear system, the eigen functions are exponentials e^{iux}. Such a system responds to an harmonic input by an harmonic output at the same frequency u and these responses specify the properties of the system. Many physical processes may be approximated as being linear shift-invariant systems. An optical element described by an operator $\mathcal{L}$, that maps an input function onto an output function, yields an output g_1 for any input g_0,

$$g_1(\mathbf{x}_1) = \mathcal{L}\left[g_0(\mathbf{x}_0)\right]. \tag{A.1}$$

The Fourier transform of the convolution of a single aperture function with the Dirac delta function, δ, array is equal to the product of the individual transform. The input is decomposed as a sum of impulse function,

$$g_0(\mathbf{x}_0) = \int_{-\infty}^{\infty} g_0(\boldsymbol{\xi})\delta(\mathbf{x}_0 - \boldsymbol{\xi})d\boldsymbol{\xi}, \tag{A.2}$$

where the coefficient $g_0(\boldsymbol{\xi})$ is the weighting factors of the decomposition.

Suppose for a rectangular slit,

$$g_0(\boldsymbol{\xi}) = \begin{cases} 1 & \text{for, } |\xi| \leq a/2; |\eta| \leq b/2, \\ 0 & \text{otherwise}; \end{cases} \tag{A.3}$$

i.e., for an optical system, this term $g_0(\boldsymbol{\xi})$ denotes the transmittance pattern, $\boldsymbol{\xi} = (\xi, \eta)$ the 2-D position vector, and therefore,

$$g_1(\mathbf{x}_1) = \mathcal{L}\left[\int_{-\infty}^{\infty} g_0(\boldsymbol{\xi})\delta(\mathbf{x}_0 - \boldsymbol{\xi})d\boldsymbol{\xi}\right]$$

$$= \int_{-\infty}^{\infty} g_0(\boldsymbol{\xi})\mathcal{L}\left[\delta(\mathbf{x}_0 - \boldsymbol{\xi})\right]d\boldsymbol{\xi}. \tag{A.4}$$

A system is linear if the principle of superposition applies. This principle states that the system response to the sum of two inputs is equal to the responses to the each of the two inputs as expressed in (A.4). In order to ensure the shift-invariance, the isoplanatism condition imposes that the supports of the functions g_0 and g_1 is restricted to the isoplanatic patch. In the presence of atmospheric turbulence, the time-invariance condition for the impulse response is needed. If these conditions are met, the impulse response is constant, and

$$\mathcal{L}\left[\delta(\mathbf{x}_0 - \boldsymbol{\xi})\right] = h(\mathbf{x}_1 - \boldsymbol{\xi}), \tag{A.5}$$

hence,

$$g_1(\mathbf{x}_1) = \int_{-\infty}^{\infty} g_0(\boldsymbol{\xi})h(\mathbf{x}_1 - \boldsymbol{\xi})d\boldsymbol{\xi} \tag{A.6a}$$

$$= g_0(\mathbf{x}) \star h(\mathbf{x}), \tag{A.6b}$$

where $*$ denotes 2-D convolution.

The (A.6) states that if the image formation system is space-invariant, the point spread function (PSF) depends on the differences of the corresponding coordinates. The output of a shift-invariant system is the convolution of the input and the impulse response of the system. In this case, a displacement of the object would result in a displacement of the image but not a change of image configuration. The convolution of two functions results in a function which is broader than both, the implication of which is that all optical systems decrease the resolution of an image. In a perfect image formation system, the image of a point source is not blurry. A linear and space-invariant image formation system is represented in the spatial frequency domain and taking the Fourier transform (FT) of both sides of (A.6), one expresses,

$$\widehat{g}_1(\mathbf{u}) = \widehat{g}_0(\mathbf{u})\widehat{h}(\mathbf{u}), \tag{A.7}$$

in which $\widehat{h}(\mathbf{u})$ is the transfer function of the system,

A cosinusoidal input to a real time invariant linear system produces a cosinusoidal output. The (A.7) states that each of the elementary sinusoids is filtered by the system with a complex gain $\widehat{h}$; the output is the sum of the filtered sinusoids. The properties of $\mathcal{L}$ are determined either by the impulse response $h(\mathbf{x})$ or by the transfer function $\widehat{h}(\mathbf{u})$.

A.2 Measures of Coherence (Table A1)

Table A.1 Definitions of various measures of coherence

Symbols	Definitions	Names	Coherence
$\mathcal{J}(\mathbf{r}_1, \mathbf{r}_2)$	$\langle U(\mathbf{r}_1, t)U^*(\mathbf{r}_1, t)\rangle$ $= \Gamma(\mathbf{r}_1, \mathbf{r}_2, 0)$	Mutual intensity	Spatial quasi-monochromatic
$\Gamma(\mathbf{r}_1, \mathbf{r}_1, \tau)$	$\langle U(\mathbf{r}_1, t + \tau)U^*(\mathbf{r}_1, t)\rangle$	Self-coherence function	Temporal
$\gamma(\mathbf{r}_1, \mathbf{r}_1, \tau)$	$\dfrac{\Gamma(\mathbf{r}_1, \mathbf{r}_1, \tau)}{\Gamma(\mathbf{r}_1, \mathbf{r}_1, 0)}$	Complex degree of self-coherence	Temporal
$\Gamma(\mathbf{r}_1, \mathbf{r}_2, \tau)$	$\langle U(\mathbf{r}_1, t + \tau)U^*(\mathbf{r}_2, t)\rangle$	Mutual coherence function	Spatial & temporal
$\gamma(\mathbf{r}_1, \mathbf{r}_2, \tau)$	$\dfrac{\Gamma(\mathbf{r}_1, \mathbf{r}_2, \tau)}{\sqrt{\Gamma(\mathbf{r}_1, \mathbf{r}_1, 0)\Gamma(\mathbf{r}_2, \mathbf{r}_2, 0)}}$	Complex degree of coherence	Spatial & temporal
$\mu(\mathbf{r}_1, \mathbf{r}_2)$	$\dfrac{\mathcal{J}(\mathbf{r}_1, \mathbf{r}_2)}{\sqrt{\mathcal{J}(\mathbf{r}_1, \mathbf{r}_1)\mathcal{J}(\mathbf{r}_2, \mathbf{r}_2)}}$ $= \gamma(\mathbf{r}_1, \mathbf{r}_2, 0)$	Complex coherence factor	Spatial quasi-monochromatic

Appendix B
Fourier Optics

B.1 Fourier Transform

A field with general time dependence may be thought of as a linear superposition of fields varying harmonically with time at different frequencies. This relationship is known as Fourier transform (FT; J. B. J. Fourier, 1768–1830). The Fourier transform is used extensively to analyze the output from optical, as well as radio telescopes. It is based on the discovery that it is possible to take any periodic function of time $f(t)$ and transform it to a function of frequency $\widehat{f}(v)$, in which the notation $\widehat{}$ denotes the Fourier transform (FT) of a particular physical quantity, the frequency v, and the time t are Fourier dual coordinates. The linear addition of various components of a quantity is carried out by summing up the phasor vectors. Adding two components with identical phasors but with opposite signs of the angular frequency, $\omega = 2\pi v$, provides the real quantity.

In the analysis of time signal, the one-dimensional Fourier transform plays an important role, which can be evaluated as a sum of exponential frequency terms, each weighted by the corresponding monochromatic response $\widehat{f}(v)$. A waveform of a given signal, $f(t)$, on a time axis is related to the spectrum (on a frequency axis) by FT and is represented as a superposition of harmonic oscillations $e^{i2\pi vt}$. In the case of electrical filters, transfer functions depend on frequency, while in the case of spectroscopy where the scanning function varies with frequency, the transfer function depends on time. The Fourier equation associated with the harmonic function of frequency (Bracewell 1965),

$$\widehat{f}(v) = \int_{-\infty}^{\infty} f(t)e^{-i2\pi vt}\,dt, \tag{B.1}$$

$$f(t) = \int_{-\infty}^{\infty} \widehat{f}(v)e^{i2\pi vt}\,dv. \tag{B.2}$$

in which $f(t)$ relates to the spectral distribution function $\widehat{f}(v)$ by the dual transformation theorem.

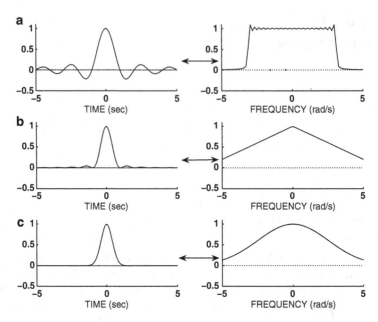

Fig. B.1 1-D Fourier transform of (**a**) $\operatorname{sinc} t$, (**b**) $\operatorname{sinc}^2 t$, and (**c**) $e^{-\pi t^2}$

The spectral distribution function, $\widehat{f}(\nu)$, is the Fourier transform of $f(t)$. The reciprocal relation between such pairs of functions can be expressed as, $f(t) \leftrightarrow \widehat{f}(\nu)$. A set of examples of Fourier transform pairs may be found (see Fig. B.1) in:

$$e^{-\pi t^2} \rightleftharpoons e^{-\pi \nu^2}, \tag{B.3a}$$

$$\delta(t) \rightleftharpoons 1, \tag{B.3b}$$

$$\cos(\pi t) \rightleftharpoons \frac{1}{2}\delta\left(\nu - \frac{1}{2}\right) + \frac{1}{2}\delta\left(\nu + \frac{1}{2}\right), \tag{B.3c}$$

$$\sin(\pi t) \rightleftharpoons \frac{i}{2}\delta\left(\nu - \frac{1}{2}\right) - \frac{i}{2}\delta\left(\nu + \frac{1}{2}\right), \tag{B.3d}$$

$$\operatorname{sinc} t \rightleftharpoons \Pi(\nu). \tag{B.3e}$$

The Fourier components are the elementary sinusoidal (or cosinusoidal) waves which contribute to a complex waveform. Any complex wave may be built up or broken down from a number of simple waves of suitable amplitude, frequency and phase. The conditions for the existence of these integrals are:

1. $f(t)$ is absolutely integrable, i.e., $|\int_{-\infty}^{\infty} f(t)dt| < \infty$ and
2. $f(t)$ has no infinite discontinuities, but has a finite number of discontinuities and/or extrema in any finite interval.

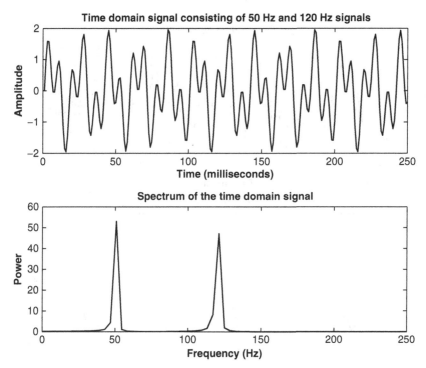

Fig. B.2 Fourier transform and power of two added sinusoidal waves

The Fourier transform of the addition of two sinusoidal waves,

$$x(t) = \sin 2\pi f_1 t + \sin 2\pi f_2 t,$$

in which $f_1 = 50$ Hz and $f_2 = 120$ Hz is shown in Fig. B.2. The Fourier transform of $x(t)$, $[\hat{x}(f)]$ is two delta functions peaked at f_1 and f_2. The power, $\left[|\hat{x}(f)|^2\right]$, would again be two delta functions positioned at 50 Hz and 120 Hz. In the second example (see Fig. B.3), the random Gaussian noise has been added (normally distributed with mean zero variance $\sigma^2 = 1$ and standard deviation $\sigma = 1$) to $x(t)$, so that

$$x_1(t) = x(t) + \text{noise}.$$

The Fourier transform and the power of $x(t)$ would again be two delta functions at 50 Hz and 120 Hz. However, this spectrum is not as clean as the FT of $x(t)$ because of the presence of noise.

The Fourier transform is important in dealing with the problems involving linear time and space-invariant systems. The eigen functions e^{ivt} are respondent to an harmonic input by an harmonic output at the same frequency v. These responses specify the properties of the system. The transform of a real function $f(t)$ has $\hat{f}(-v) = \hat{f}^*(v)$, in which $*$ stands for complex conjugate. The other related theorems are given below.

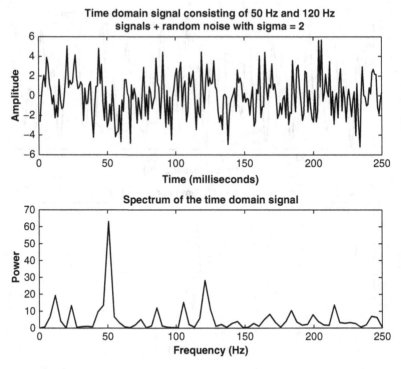

Fig. B.3 Fourier transform and power of two sinusoidal waves added with random noise

1. Uniqueness theorem: The input produces a unique output. The Fourier transform of the function $f(t)$, is denoted symbolically as,

$$\mathcal{F}[f(t)] = \widehat{f}(v). \tag{B.4}$$

2. Linearity theorem: If $h(t) = af(t) + bg(t)$, the transform of sum of two functions is simply the sum of their individual transforms, i.e.,

$$\widehat{h}(v) = \mathcal{F}[af(t) + bg(t)] = a\widehat{f}(v) + b\widehat{g}(v). \tag{B.5}$$

where a, b are complex numbers.

3. Similarity theorem:

$$\mathcal{F}[f(a\,t)] = \frac{1}{a}\widehat{f}\left(\frac{v}{a}\right). \tag{B.6}$$

4. Shift theorem: A shift in the time at which the input starts is seen to cause a shift in the frequency at which the output starts; the shape of the input is unchanged by the shift,

$$\mathcal{F}[f(t - a)] = \widehat{f}(v)e^{-i2\pi av}. \tag{B.7}$$

5. Parseval theorem: The Parseval or Power theorem is generally interpreted as a statement of conservation of energy. For instance, if $\mathcal{F}[f(t)] = \widehat{f}(\nu)$,

$$\int_{-\infty}^{\infty} \left|\widehat{f}(\nu)\right|^2 d\nu = \int_{-\infty}^{\infty} |f(t)|^2 dt. \tag{B.8}$$

6. Area: If $\mathcal{F}[f(t)] = \widehat{f}(\nu)$,

$$\widehat{f}(0) = \int_{-\infty}^{\infty} f(t) dt. \tag{B.9}$$

7. Derivative theorem: The transform of the temporal derivative of a function is written as,

$$\frac{df(t)}{dt} \leftrightarrow i\nu \widehat{f}(\nu). \tag{B.10}$$

and therefore the transform of the nth temporal derivative of a function is expressed as,

$$\frac{d^n f(t)}{dt^n} \leftrightarrow (i\nu)^n \widehat{f}(\nu). \tag{B.11}$$

8. Dirac function: Dirac impulse function, $\delta(t)$ has value at a single point otherwise it is zero. The Fourier transform of such a function is given by,

$$\widehat{f}(\nu) = \int_{-\infty}^{\infty} \delta(t) e^{i2\pi\nu t} dt = 1. \tag{B.12}$$

A unit impulse has a flat spectrum of amplitude unity. When such an impulse occurs at time T other than zero, the spectral distribution posses all frequencies with unit amplitude, but with a frequency-dependent phase, $2\pi\nu t$.

9. Sampling function: Signals and images are continuous in reality, therefore they can be sampled at regular intervals, called point sampling. Sampling consisting of integrating the intensity over a small region is referred as area sampling. A band-limited signal $f(t)$ with bandwidth $\Delta\nu$ can be uniquely represented by a time series obtained by periodically sampling $f(t)$ at a frequency ν_s (the sampling frequency), which is greater than a critical frequency (Shannon 1949). The signal can be sampled with exactly equal to or greater than the critical frequency $2\Delta\nu$. The maximum measurable frequency, the Nyquist limit[1] is that half the sampling frequency. The spectrum of signals sampled at a frequency $<2\Delta\nu$ (under sampled) is distorted. Hence, the time series obtained is not a true representative of the band-limited signal. Such an effect is called aliasing. In order to eliminate or reduce aliasing, one needs to (1) increase the sampling rate by reducing T, or/and (2) reduce the high frequency component contained in the signal by low-pass filtering it before sampling. Electrical signals are sampled with a specific sampling frequency, e.g. in the analog-to-digital (A/D)

---

[1] Nyquist limit is the highest frequency of a signal that can be coded with at a given sampling rate in order to reconstruct the signal.

converter, as well as in the sample and hold unit of pulsed Doppler instruments. The sampling theorem has been established by Cauchy.

For a band-limited function, $f(t)$, one notes that $\widehat{f}(v) = 0$ for $v \geq \sigma$, in which σ is the highest frequency component for this signal. The sampled version $f_s(v)$ of $f(t)$ is written as,

$$f_s(t) = comb\,(t/T)f(t),\qquad\qquad (\text{B}.13)$$

with $comb\,(t) = \sum_{n=-\infty}^{\infty}\delta(t-n)$, as the $comb$ function,[2] $\delta(t)$ the delta function, and T is the sampling interval.

According to the convolution theorem, the spectrum of the resultant signal is,

$$\widehat{f}_s(v) = T\,comb\,(vT) \star \widehat{f}(v) = \sum_{n=-\infty}^{\infty}\delta\left(v - \frac{n}{T}\right)\star \widehat{f}(v).\qquad (\text{B}.14)$$

The square step with unit amplitude from $-T/2$ to $T/2$ and zero elsewhere, transforms into the sinc function.

10. Gaussian function: The Fourier transform of the Gaussian function (see Fig. B.4) gives a Gaussian. i.e.,

$$f(t) = e^{-(t/T)^2} \leftrightarrow \frac{1}{\pi^{1/2}T}e^{-(\pi vT)^2} = \widehat{f}(v).\qquad (\text{B}.15)$$

To note, if the characteristic time T is short, its spectrum turns out to be wide, and vice versa. Similarly, when a sinusoidal function is switched on and off at times, $-T/2$ and $T/2$, its spectrum has a value over a bandwidth, Δv, which is of the order $1/T$.

11. Modulated wave function: The spectrum of the wave $g(t)\cos(2\pi v_1 t)$, in which $g(t)$ is the modulating function, is derived as,

$$\begin{aligned}\widehat{f}(v) &= \int_{-\infty}^{\infty} g(t)\cos(2\pi v_1 t)e^{i2\pi vt}\,dt\\ &= \int_{-\infty}^{\infty} g(t)\frac{e^{i2\pi(v+v_1)t} + e^{i2\pi(v-v_1)t}}{2}\,dt\\ &= \frac{1}{2}\widehat{g}(v+v_1) + \frac{1}{2}\widehat{g}(v-v_1),\end{aligned}\qquad (\text{B}.16)$$

where $\widehat{g}$ is the transform of the modulation function, $g(t)$.

12. Gate function: It is defined as,

$$\Pi(t/\tau) = \begin{cases} 1 & \text{if } |t| < \tau/2, \\ 0 & \text{if } |t| > \tau/2. \end{cases}\qquad (\text{B}.17)$$

---

[2] The comb function is a sequence of infinite Dirac delta functions uniformly spaced at intervals of time (or space). It is often called as Shah function.

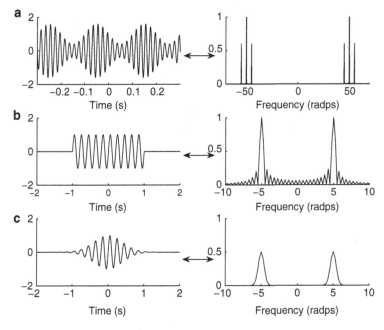

Fig. B.4 Fourier transform of sine wave modulated by (**a**) low frequency signal, (**b**) rectangular function, and (**c**) Gaussian function, $e^{-\pi t^2}$

Its Fourier transform is given by,

$$
\widehat{f}(\omega) = \int_{-\infty}^{\infty} \Pi(t/\tau)e^{-i\omega t}\,dt = \int_{-\tau/2}^{\tau/2} e^{-i\omega t}\,dt
$$
$$
= \tau \left[\frac{\sin(\omega\tau/2)}{(\omega\tau/2)} \right]. \tag{B.18}
$$

Figure B.5 displays the Fourier transform and power spectrum of a gate function.

13. Discrete Fourier transform: The discrete Fourier transform (DFT) reveals periodicities in input data as well as the relative strengths of any periodic components. It is widely employed in signal processing and related fields to analyze the frequencies contained in a sampled signal, solve partial differential equations, as well as to perform other operations such as convolutions. The DFT is used, in general when the number of samples N is finite. It can be computed efficiently with a Fast Fourier Transform[3] (FFT) algorithm. If $f(\tau)$

---

[3] Fast Fourier Transform (FFT) algorithm optimizes the computation of the DFT, which reduces the number of computations needed for N points from $2N^2$ to $2N \log N$.

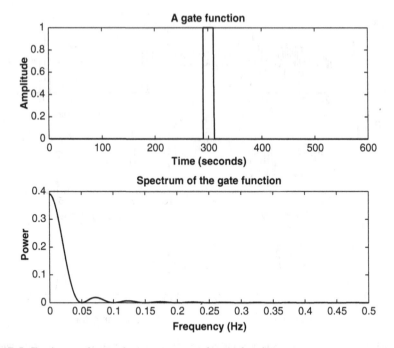

Fig. B.5 Fourier transform and power spectrum of a gate function

and $f(v)$ consist of sequence of N samples, the respective direct and inverse discrete Fourier transform (DFT) of a signal are defined as,

$$\widehat{f}(v) = \frac{1}{N} \sum_{\tau=0}^{N-1} f(\tau)e^{-i2\pi\tau v/N}, \tag{B.19}$$

$$f(\tau) = \sum_{v=0}^{N-1} \widehat{f}(v)e^{i2\pi\tau v/N}. \tag{B.20}$$

The change of notation emphasizes that the variables are discrete. The DFT assumes that the data, $f(\tau)$, is periodic outside the sampled range, and returns a transform, which is periodic as well,

$$\widehat{f}(v + N) = \frac{1}{N} \sum_{\tau=0}^{N-1} f(\tau)e^{-i2\pi\tau(v+N)/N},$$

$$= \frac{1}{N} \sum_{\tau=0}^{N-1} f(\tau)e^{-i2\pi\tau v/N}e^{-i2\pi\tau},$$

$$= \widehat{f}(v). \tag{B.21}$$

A DFT requires $N \times N$ complex multiplications for computations. All the basic theorems that describe below for the Fourier transform have a counterpart, for example, the shift theorem can be written as,

$$D[f(t - T)] = e^{-i2\pi T v/N} \widehat{f}(v), \tag{B.22}$$

in which D stands for the DFT operator.

B.1.1 Convolution and Cross-Correlation

A convolution is an integral that expresses the amount of overlap of one function $g(t)$ as it is shifted over another function $f(t)$. It describes the action of an observing instrument over a physical quantity, as the instrument takes a weighted mean of the physical quantity over a narrow range of some variable. If the form of the weighting function does not change appreciably with the central value of the variable, the observed quantity is the convolution of the distribution of the object function with the weight function, rather than the object function itself. Convolution theorem plays a vital role in formulating the effect of blurring by telescope beam, particularly in an aperture synthesis map. The convolution of two functions $f(t)$ and $g(t)$, which is demonstrated in Fig. B.6 is defined by,

$$h(t) = \int_{-\infty}^{+\infty} f(\tau)g(t - \tau)d\tau = \int_{-\infty}^{+\infty} g(\tau)f(t - \tau)d\tau = f(t) \star g(t), \tag{B.23}$$

where $\star$ denotes the convolution parameter.

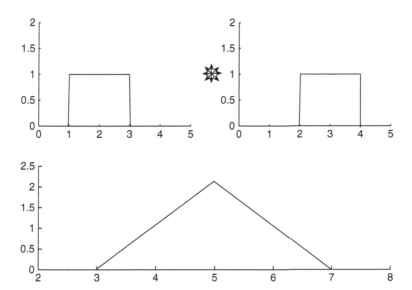

Fig. B.6 1-D convolution of two functions

In Fourier space the effect becomes a multiplication, point by point, of the transform of $\widehat{f}(v)$ with the transfer function, $\widehat{g}(v)$,

$$\widehat{h}(v) = \widehat{f}(v)\widehat{g}(v). \tag{B.24}$$

Cross-correlation analysis is essentially a means for assessing and specifying mutual relationships of two signals in order to determine the degree of similarity between them. This method is needed to analyze the outputs of two telescopes which are used in interferometry to view a single object. Cross correlation of two complex functions $f(t)$ and $g(t)$ of a real variable t, denoted $f(t) \otimes g(t)$ is defined by,

$$f(t) \otimes g(t) \equiv f^*(-t) \star g(t), \tag{B.25}$$

in which $\otimes$ stands for correlation.

The cross-correlation of a pair of functions bears a close relation to convolution, but differs with a shift of the function $g(t)$, i.e.,

$$h(t) = \int_{-\infty}^{\infty} f(\tau)g(\tau - t)d\tau = f(t) \otimes g(t). \tag{B.26}$$

For a complex cross-correlation function, $\gamma_{cc}(t)$ is defined as,

$$\gamma_{cc} = \int_{-\infty}^{\infty} f(\tau)g^*(\tau - t)d\tau = f(t) \otimes g^*(t). \tag{B.27}$$

It is necessary for some purposes to normalize correlation functions by dividing (B.27) by the central value of the correlation, i.e., by zero shift, $t = 0$. Thus it provides,

$$\gamma_{cc} = \frac{\displaystyle\int_{-\infty}^{\infty} f(\tau)g^*(\tau - t)d\tau}{\displaystyle\int_{-\infty}^{\infty} f(\tau)g^*(\tau)d\tau}, \tag{B.28}$$

where the normalized value is given by, $\gamma_{cc}(0) = 1$.

This is the form of the Wiener–Khintchine theorem, which allows the determination of the spectrum by way of the autocorrelation of the generating function. It is a mathematical tool used in signal processing[4] for analyzing functions or series of values. The square of the magnitude of the Fourier transform is called the power spectrum and can be obtained by taking the Fourier transform of the

---

[4] Signal processing refers to various techniques for improving the accuracy and reliability of communications by means of analysis and manipulation of signals. All communication circuits contain some noise regardless of the signals if they are analog, in which the signal varies continuously according to the information, or digital where the signal varies according to a series of discrete values containing information.

cross-correlation between a signal and its complex conjugate. By switching to time and frequency, one may write,

$$\mathcal{F}[f(t) \otimes f(t)] = \widehat{f}(v)\widehat{f}^*(v) = \left|\widehat{f}(v)\right|^2, \tag{B.29}$$

where $|\widehat{f}(v)|^2$ is referred as the power spectrum in terms of temporal frequency and $\mathcal{F}$ represents the Fourier operator.

Autocorrelation is the correlation of a signal shape, or waveform, resembles a delayed version of itself. It is one of the most important theorems that is employed as basic data treatment tools. The auto-correlation of continuous real function $f(t)$ is,

$$\Re(t) \equiv \lim_{T \to \infty} \frac{1}{2T} \int_{-T}^{T} f(\tau)f(t + \tau)d\tau. \tag{B.30}$$

If $f(t)$ is a complex function, its autocorrelation is

$$\Im(t) \equiv f(t) \star f(t) = \int_{-\infty}^{+\infty} f(t + \tau)f^*(\tau)d\tau, \tag{B.31}$$

where $f^*(t)$ is the complex conjugate of $f(t)$.

B.1.2 Hankel Transform

The Hankel transform, developed by the mathematician H. Hankel, is an integral transform equivalent to a two-dimensional Fourier transform with a radially symmetric integral kernel. It is also known as the Fourier–Bessel transform. If a two-dimensional signal has circular symmetry, its Fourier transform shrinks to a cyclical transform. This technique is useful for optical systems with circular symmetry in the plane perpendicular to the optical axis (Papoulis 1968; Gaskill 1978). The Fourier transform pair in two dimensional (2-D) spatial domain may be written as,

$$\widehat{f}(\mathbf{u}) = \int_{-\infty}^{\infty} f(\mathbf{x})e^{-i2\pi \mathbf{u} \cdot \mathbf{x}}d\mathbf{x}; \tag{B.32a}$$

$$f(\mathbf{x}) = \int_{-\infty}^{\infty} \widehat{f}(\mathbf{u})e^{i2\pi \mathbf{u} \cdot \mathbf{x}}d\mathbf{u}, \tag{B.32b}$$

in which $\mathbf{x} = (x, y)$ is the 2-D position vector and the dimensionless variable is the 2-D spatial vector $\mathbf{u} = (u, v) = (x/\lambda, y/\lambda)$.

If the function $f(x, y)$ is separable, i.e.,

$$f(x, y) = g(x)h(x), \tag{B.33}$$

then,

$$\widehat{f}(u, v) = \widehat{g}(u)\widehat{h}(v). \tag{B.34}$$

In order to transform the rectangular coordinates (x, y) into polar coordinates (ρ, θ), if the functions are separable, one may write,

$$f(x, y) = g(\rho)h(\theta), \tag{B.35}$$

in which

$$x = \rho\cos\theta, \qquad y = \rho\sin\theta, \qquad \rho = \sqrt{x^2 + y^2}, \qquad x + iy = \rho e^{i\theta},$$

with $i = \sqrt{-1}$.

Given a function $g(\rho)$ and a real constant w, the Hankel transform can be expressed as,

$$\widehat{g}(w) = \int_0^\infty g(\rho)J_0(w\rho)\rho d\rho, \tag{B.36a}$$

$$g(\rho) = \int_0^\infty \widehat{g}(w)J_0(w\rho)wdw, \tag{B.36b}$$

where

$$u = w\cos\psi, \qquad v = w\sin\psi, \qquad w = \sqrt{u^2 + v^2};$$
$$ux + vy = w\rho\cos(\theta - \psi), \qquad dxdy = \rho d\rho d\theta,$$

and

$$J_0(\rho) = \frac{1}{2\pi} \int_{-\pi}^{\pi} e^{\rho\cos(\theta-\alpha)} d\theta. \tag{B.37}$$

is the zeroth-order Bessel function, and (B.36b) is the inverse Hankel transform of $\widehat{g}(w)$.

Figure B.7 depicts the Hankel transform of $g(\rho)$. To note, the forward and inverse transforms (B.36a and B.36b) are identical. As $\cos\theta$ is periodic, the integral is independent of α. The (B.37) defines as the Hankel transform. This transform is

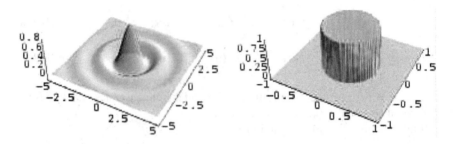

Fig. B.7 Hankel transform of $g(\rho)$

useful for analyzing circular apertures. The function $\widehat{g}(w)$ can be expressed in terms of the 2-D Fourier transform $\widehat{f}(u, v)$ of the function $g(\rho)$. If the function is independent of θ or $h(\theta) = 1$, $\widehat{f}(u, v)$ has circular symmetry and is given by,

$$\widehat{f}(u, v) = 2\pi\widehat{g}(w), \tag{B.38}$$

therefore (B.37) can be recast in as,

$$\widehat{f}(u, v) = \int_0^\infty \rho g(\rho) \int_{-\pi}^{\pi} e^{-iw\rho\cos(\theta-\psi)} d\theta d\rho$$

$$= 2\pi \int_0^\infty \rho g(\rho) J_0(w\rho) d\rho. \tag{B.39}$$

The (B.38) shows that $\widehat{g}(w)$ and $g(\rho)$ are linked by the Hankel transform of zero order,

$$g(\rho) = 2\pi \int_0^\infty w\widehat{g}(w) J_0(w\rho) dw. \tag{B.40}$$

The convolution theorem of two functions $g_1(\rho)$ and $g_2(\rho)$ with Hankel transform $\widehat{g}_1(w)$ and $\widehat{g}_2(w)$, one may write,

$$g_1(\rho) \star g_2(\rho) \leftrightarrow 2\pi\widehat{g}_1(w)\widehat{g}_2(w). \tag{B.41}$$

Another transformation related to the one-dimensional Fourier transform, known as Hartley transform, which transforms real function into a real function and shares some properties of the FT, is defined for a real function $f(x)$ as,

$$Hart| f(x)| = \int_{-\infty}^{\infty} f(x) \left[\cos(2\pi ux) + \sin(2\pi ux)\right] dx. \tag{B.42}$$

The Fourier transform of $f(x)$ can be retrieved from the Hartley transform by even-odd decomposition.

Appendix C
Spatial Frequency Response

C.1 Transfer Function

It has been shown by Booker and Clemmow (1950) that the Fourier transform of the antenna power pattern is proportional to the complex autocorrelation function of the aperture distribution, i.e.,

$$\overline{P}(x_0) \propto \int_{\infty}^{\infty} E(x - x_0)E^{\star}(x_0)dx_0, \qquad (C.1)$$

where $\overline{P}(x_0)$ is the Fourier transform of antenna power pattern $P_n(\phi)$, $E(x_0)$ the amplitude and phase of the current distribution over the aperture, $x = x/\lambda$ the distance, $x_0 = x/\lambda_0$ the displacement, $a = a/\lambda$ the aperture or antenna width, and $s = s/\lambda$ the spacing between the antennas.

The autocorrelation function involves displacement x_0, multiplication and integration. The aperture distribution of a single telescope is shown in the top panel in Fig. C.1 and its displacement by x_0 in the middle panel. The autocorrelation function shown in the bottom panel is proportional to the area of overlap. It is apparent that the autocorrelation function is zero for values of x_0 greater than the aperture width a since $\overline{P}(x0) = 0$ for $|x_0| > a$.

The angular resolution of a radio telescope can be improved, for example, increasing the aperture a. However, this may not be economically and mechanically feasible. A common method is to use two antennas spaced a distance s apart. If each antenna has an uniform aperture distribution of width a, the resulting autocorrelation function is as shown in Fig. C.2. It is apparent that by making observations with spacings out to s, it is possible to obtain higher spatial-frequency components in the pattern to a cutoff $x_0 = s + a$.

Figure C.3 shows the autocorrelation function of a phase switched interferometer. A comparison with the Fig. C.2 shows that the response to the zero and other nearby lower spatial frequency components is not present here. In other words, the phase switched interferometer responds comparatively only to the higher spatial frequency components.

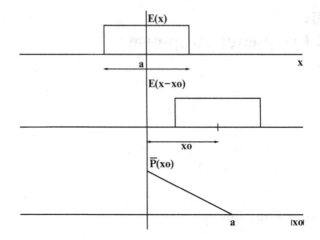

Fig. C.1 Transfer function of a single telescope

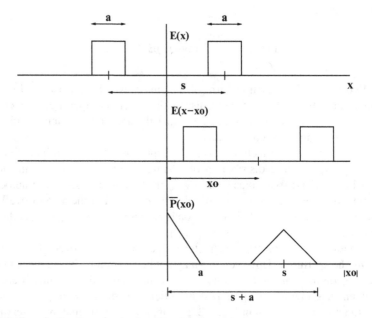

Fig. C.2 Transfer function of a simple adding interferometer

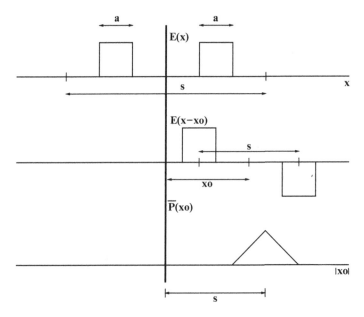

Fig. C.3 Transfer function of a phase-switched or cross-correlation interferometer

Appendix D
Zernike Representation of Atmospheric Turbulence

Zernike polynomials are a set of orthogonal polynomials over unit circle. In optics, the aberrations are often represented by such polynomials. It is convenient to express the turbulent wavefront in the circular aperture telescope in terms of them; the central obscuration may be treated as a special case. A minor problem with this representation is that non-rotationally symmetric aberrations like coma and astigmatism (these aberrations generate from subsets of a larger centered bundle of spherically aberrated rays) are decomposed into two components, one along the x-axis and the other along the y-axis. These, however, may be combined in a single aberration with a certain orientation that depends on the magnitude of the two components.

Zernike polynomials are similar to the eigen functions of Kolmogorov phase statistics, hence most of the energy of the wave front aberrations can be described by the first few Zernike polynomials. One may derive every single mode like tip-tilt, astigmatism, coma, etc.; however, the coefficients of these aberrations are random function changing in time. One proceeds with the calculations of the residual aberration after correcting a specified number of modes with an adaptive optics system. Noll (1976) had used a normalized set of Zernike polynomials for the application of Kolmogorov turbulence.

Zernike polynomial, $Z_n^m(\rho)$, is defined in polar coordinates on a circle of unit radius, $R < 1$. They are characterized by the radial order, n, as well as by the azimuthal order, m. The arbitrary wavefront, $\psi(\rho)$, in which $\rho = \rho, \theta$ over the unit circle may be represented as an infinite sum of Zernike polynomials,

$$\psi(\rho) = \sum_{j=1}^{\infty} a_j Z_j(\rho), \tag{D.1}$$

where $Z_j(\rho)$ is the Zernike polynomial of order j, and a_j the coefficients of expansion that is given by,

$$a_j = \int_{aperture} \psi(\rho) Z_j(\rho) W(\rho) d\rho. \tag{D.2}$$

The aperture weight function, $W(\rho)$,

$$W(\rho) = \begin{cases} 1/\pi & \rho \le 1, \\ \\ 0 & \text{otherwise,} \end{cases} \qquad \text{(D.3)}$$

is added so that the integral can be taken over all space. The Zernike polynomials are usually written in polar form ρ and θ, as given by the following formulae,

$$Z_{j\,=\,even} = \sqrt{n+1}\,R_n^m(\rho)\sqrt{2}\cos(m\theta), \qquad \text{for } m \ne 0, \qquad \text{(D.4a)}$$

$$Z_{j\,=\,odd} = \sqrt{n+1}\,R_n^m(\rho)\sqrt{2}\sin(m\theta), \qquad \text{for } m \ne 0, \qquad \text{(D.4b)}$$

$$Z_j = \sqrt{n+1}\,R_n^0(\rho), \qquad \text{for } m = 0, \qquad \text{(D.4c)}$$

with

$$R_n^m(\rho) = \sum_{s=0}^{(n-m)/2} \frac{(-1)^s(n-s)!\,\rho^{n-2s}}{s!\left[\dfrac{n+m}{2}-s\right]!\left[\dfrac{n-m}{2}-s\right]!}, \qquad \text{(D.5)}$$

where the indices n and m are the radial degree and the azimuthal frequency respectively and s the mode-ordering number.

The variance of any Zernike coefficient is determined by the total power in its spectrum,

$$\langle \sigma_{a_j} \rangle^2 = \int_0^\infty A_c(f)df. \qquad \text{(D.6)}$$

The general expression for the covariance of the expansion coefficients derived by Noll (1976) for $n, n' \ne 0$, as

$$\langle a_j a_{j'} \rangle = \begin{cases} \dfrac{0.046}{\pi}\left(\dfrac{R}{r_0}\right)^{5/3}\sqrt{(n+1)(n'+1)} \\ \quad \times (-1)^{(n+n'-2n)/2}\delta_{mm'} \\ \quad \times \displaystyle\int_0^\infty \dfrac{J_{n+1}(2\pi\kappa)J_{n'+1}(2\pi\kappa)}{\kappa^2}\kappa^{-8/3}d\kappa, & (j-j') \text{ even,} \\ \\ 0 & (j-j') \text{ odd.} \end{cases} \qquad \text{(D.7)}$$

When $j = j'$, the (D.7) reduces to,

$$\langle \sigma_{a_j} \rangle^2 = \frac{0.046}{\pi}\left(\frac{R}{r_0}\right)^{5/3}(n+1)\int_0^\infty \frac{J_{n+1}^2(2\pi\kappa)}{\kappa}\kappa^{-8/3}d\kappa. \qquad \text{(D.8)}$$

If the phase obeys Kolmogorov statistics, one can determine the covariance of the Zernike coefficients corresponding to the atmospheric phase aberrations (Noll 1976) had used a normalized set of Zernike polynomials for the application of Kolmogorov

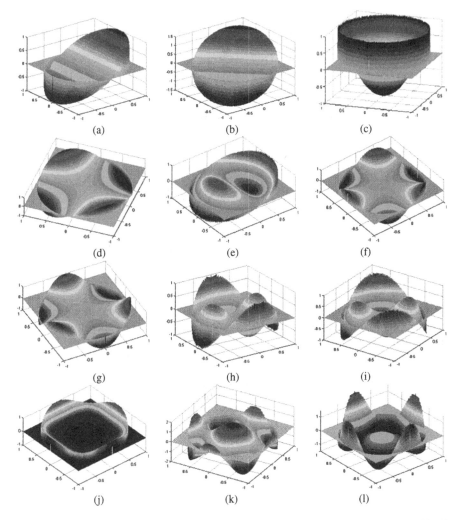

Fig. D.1 Graphical representation of Zernike modes: (**a**) tilt about y-axis, (**b**) x and y combined tilt, (**c**) defocus, (**d**) astigmatism at $0°$, (**e**) x-axis coma, (**f**) triangular astigmatism (or trefoil) on x-axis, (**g**) trefoil on y-axis, (**h**) 5th order astigmatism at $90°$, (**i**) 5th order astigmatism at $45°$, (**j**) tetrafoil x-axis, (**k**) 3rd order spherical, and (**l**) 3rd order defocus; these aberrations are generated at the laboratory (Courtesy: V. Chinnappan)

statistics. A Zernike representation of the Wiener spectrum of the phase fluctuations due to Kolmogorov turbulence (4.45) can be obtained by evaluating the covariance of the expansion coefficients in (D.1). The mean square residual aberration is expressed as the variance of the difference between the uncorrected phase and of the removed modes. If the aberration due to first J, Zernike polynomials is expressed as,

$$\psi_c(\rho) = \sum_{j=1}^{J} a_j Z_j(\rho). \tag{D.9}$$

The mean square residual error is expressed as,

$$\Delta = \int_{aperture} \langle [\psi(\rho) - \psi_c(\rho)]^2 \rangle W(\rho) d\rho. \tag{D.10}$$

By inserting (D.9) into (D.10), one finds,

$$\Delta_J = \langle \psi(\rho) \rangle^2 - \sum_{j=1}^{J} \langle |a_j|^2 \rangle, \tag{D.11}$$

in which $\langle a_j \rangle = 0$ and $\langle \psi(\rho) \rangle^2$ is the variance of the phase fluctuations, which is infinite for the Kolmogorov spectrum.

Figure D.1 depicts Zernike polynomials in terms of higher order optical errors such as tilt, astigmatism, defocus, coma, etc. The analysis in Zernike polynomials shows that this infinity contains in the piston mode of the spectrum. Removing the piston term provides a finite value for the variance of the residual aberrations. When no correction is applied, then

$$\Delta_1 = 1.0299 \left(\frac{D}{r_0} \right)^{5/3}. \tag{D.12}$$

Appendix E
Celestial Coordinate System (Table E1)

Table E.1 Four celestial coordinate systems

Coordinate system	Principal axis	Principal reference circle	Coord. (units)	Secondary reference circle	Coord. (units)
Alt-azimuth	Zenith-Nadir	Celestial horizon	Azimuth $0° - 360°$	Vertical circle	Altitude $0° \pm 90°$
Ecliptic	N-S ecliptic pole	Ecliptic	Celestial longitude $0° - 360°$	–	Celestial latitude $0° \pm 90°$
Equatorial	N-S celestial pole	Celestial equator	RA $0^h - 24^h$	Hour circle	Dec $0° \pm 90°$
Galactic	N-S galactic pole	Plane of geometry	Galactic longitude $0° - 360°$	–	Galactic latitude $0° \pm 90°$

References

Ables J. G., 1974, Astron. Astrophys. Suppl., **15**, 383.
Absil O. et al., 2006, Astron. Astrophys., **452**, 237.
Absil O. et al., 2008, Astron. Astrophys., **487**, 1041.
Absil O. et al., 2009, Astrophys. J., **704**, 150.
Acernese F. et al., 2006, Class. Quant. Grav., **23**, S63.
Adams F. C., Laughlin G., 1997, Rev. Mod. Phys., **69**, 337.
Adrila D. R., Golimowski D. A., Krist J. E., Clampin M., Williams J. P., Blakeslee J. P., Ford H. C., Hartig G. F., Illingworth G. D., 2004, Astrophys. J., **617**, 147.
Aharonian F. A. et al., 2005, Science, **307**, 1938.
Aime C., Vakili F., 2006, IAU Colloq., **200**, Direct Imaging of Exoplanets, Cambridge University Press, UK.
Akeson R. L. et al., 2005a, Astrophys. J., **622**, 440.
Akeson R. L. et al., 2005b, Astrophys. J., **635**, 1173.
Akeson R. L. et al., 2009, Astrophys. J., **691**.
Alexander G., 2003, Rep. Prog. Phys., **66**, 481.
Andersen J., 1991, Astron. Astrophys. Rev., **3**, 91.
Anderson, J. A., 1920, Astrophys. J., **51**, 263.
Ando M., 2005, Class. Quant. Grav., **22**, S881.
Angel J. R. P., Cheng A., Woolf N. J., 1986, Nature, **322**, 341.
Angel J. R., Woolf N. J., 1997, Astrophy. J., **475**, 373.
Antonucci R. R. J., Miller J. S., 1985, Astrophys. J., **297**, 621.
Armitage P. J., 2007, Astrophys. J., **665**, 1381.
Armstrong J. et al., 1998, Astrophys. J., **496**, 550.
Armstrong J. et al., 2001, Astron. J., **121**, 476.
Arnett W. D., Bahcall J. N., Kirshner R. P., Woosley S. E., 1989, Ann. Rev. Astron. Astrophys., **27**, 629.
Arsenault R. et al., 2002, SPIE, **4839**, 174.
Assus P., Choplin H., Corteggiani J. P., Cuot E., Gay J., Journet A., Merlin G., Rabbia Y., 1979, J. Opt., **10**, 345.
Aufdenberg J. P., Ludwig H. G., Kervella P., 2005, Astrophys. J., **633**, 424.
Aufdenberg J. P. et al., 2006, Astrophys. J., **645**, 664.
Aufdenberg J. P. et al., 2008, in The Power of Optical/IR Interferometry: Recent Scientific Results & 2nd Generation Instrumentation, Eds. A. Richichi, F. Delplancke, F. Paresce, A. Chelli, Springer, Berlin, 71.
Ayers G. R., Dainty J. C., 1988, Opt. Lett., **13**, 457.
Baade W., Minkowski R., 1952, Astrophys. J., **119**, 215.
Babcock H. W, 1953, Pub. Astron. Soc. Pac., **65**, 229.
Bahcall J. N. et al., 1991, The Decade of Discovery in Astronomy and Astrophysics, National Academy Press, Washington DC.

Baines E. K., van Belle G. T., ten Brummelaar T. A., McAlister H. A., Swain M., Turner N. H., Sturmann L., Sturmann J., 2007, Astrophys. J., **661**, L195.

Balasubramanian K., 2006, in IAU Colloq., 200, Direct Imaging of Exoplanets, Eds. C. Aime, F. Vakili, Cambridge University Press, UK, 406.

Baldwin J. et al., 1996, Astron. Astrophys., **306**, L13.

Baldwin J., Boysen R., Cox R. C., Haniff C., Rogers J., Warner P., Wilson D., Mackay C., 1994, SPIE, **2200**, 118.

Baldwin J., Boysen R., Haniff C., Lawson P., Mackay C., Rogers J., St-Jacques D., Warner P., Wilson D., Young J., 1998, SPIE, **3350**, 736.

Baldwin J., Haniff C., Mackay C., Warner P., 1986, Nature, **320**, 595.

Banachiewicz T., 1955, Vistas Astron., **1**, 200.

Barger A. J. et al., 2001, Astron. J., **122**, 2177.

Barnes J. A., Chi A. R., Cutler L. S., Winkler G. M. R., 1971, IEEE Trans. Instr. Meas., **IM-20**, 105.

Baschek B., Scholz M., Wehrse R., 1991, Astron. Astrophys., **246**, 374.

Basden A. G., Buscher D. F., 2005, Mon. Not. R. Astron. Soc., **357**, 656.

Batten A. H., Fletcher J. M., 1990, Bull. Astron. Soc. Ind., **18**, 285.

Bayes T., 1764, Phil. Trans. R. Soc. Lond., **53**, 370.

Baym G., 1998, Acta. Phys. Polonica, B., **29**, 1839.

Beaulieu J. -P. et al., 2006, Nature, **439**, 437.

Beckers J. M., 1986, SPIE, **628**, 255.

Beckers J. M., 1988, in Difffraction-Limited Imaging with Very Large Telescopes, Eds., D. M. Alloin, J. M. Mariotti, Kluwer, Dordrecht, 355.

Beckers J. M., 1998, Proceedings of the ESO Conference on Very Large Telescope and their Instrumentation, Ed., M -H Ulrich, 693.

Beckers J. M., 2001, Expt. Astron., **12**, 1.

Beckert T., Driebe T., Hönig S. F., Weigelt G., 2008, Astron. Astrophys., **486**, L17.

Beaulieu J. -P. et al., 2006, Nature, **439**, 437.

Bedding T. R., Minniti D., Courbin F., Sams B., 1997, Astron. Astrophys., **326**, 936.

Beiging J. H. et al., 2006, Astrophys. J., **639**, 1053.

Bender P. L. et al., 1998, Doc. MPQ 233, Garching, Germany.

Benson J., Mozurkewich D., Jefferies S., 1998, SPIE, **3350**, 493.

Berger J. -P. et al., 2003, SPIE, **4838**, 1099.

Berio P., Mourard D., Bonneau D., Chesneau O., Stee P., Thureau N., Vakili F., 1999a, J. Opt. Soc. Am. A., **16**, 872.

Berio P., Stee P., Vakili F., Mourard D., Bonneau D., Chesneau O., Thureau N., Le Mignant D., Hirata R., 1999b, Astron. Astrophys., **345**, 203.

Berkefeld T., 2008, Private communication.

Bernlöhr K. et al., 2003, Astropart. Phys., **20**, 111.

Bigot L., Kervella P., Thévenin F., Ségransan D., 2006, Astron. Astrophys., **446**, 635.

Binney J., Merrifield M., 1998, Galactic Astronomy, Princeton University Press, Princeton, NJ.

Blazit A., 1986, SPIE, **702**, 259.

Blazit A., Bonneau D., Koechlin L., Labeyrie A., 1977, Astrophys. J., **214**, L79.

Boal D. H., Gelbke C. -K., Jennings B. K., 1990, Rev. Mod. Phys., **62**, 553.

Boccaletti A., Moutou C., Labeyrie A., Kohler D., Vakili F., 1998, Astron. Astrophys., **340**, 629.

Boccaletti A., Riaud P., Moutou C., Labeyrie A., 2000, Icarus, **145**, 628.

Bode M. F., 2010, Astron. Nachr., **331**, 160.

Boden A. et al., 1999, Astrophys. J., **515**, 356.

Boden A. et al., 2005, Astrophys. J., **635**, 442.

Boden A., Akeson R. L., Sargent A. I., Carpenter J. M., Ciardi D. R., Bary J. S., Skrutskie M. F., 2009, Astrophys. J., **696**, L111.

Boden A., Torres G., Hummel C., 2005, Astrophys. J., **627**, 464.

Bond I. A. et al., 2004, Astrophys. J., **606**, L155.

Bonneau D., Foy R., Blazit A., Labeyrie A., 1982, Astron. Astrophys., **106**, 235.

Booker H. G., Clemmow P. C., 1950, Proc. Instn. Elec. Engrs. Lond., Ser. 3, **97**, 11.

Booth R. S., 1985, 15th. Advanced Course, Swiss Society of Astrophysics and Astronony, Eds. A. Benz, M. Huber, M. Mayor, Geneva Observatory, Sauverny, Switzerland, 65.

Bordé P., Traub W., 2007, C. R. Physique, **8**, 349.

Born M., 1926, Zur Quant. der Stovorgnge, Zeitschrift fr Phys., **37**, 863.

Born M., Wolf E., 1984, Principles of Optics, Pergamon Press, NY, USA.

Bosc I., 1988, Proceedings of the ESO-NOAO Conference, Ed. F. Merkle, ESO, FRG, 735.

Boyajian T. S. et al., 2008, Astrophys. J., **683**, 424.

Bracewell R. N., 1965, The Fourier Transform and Its Applications, McGraw-Hill, NY.

Bracewell R. N., 1978, Nature, **274**, 780.

Bridle A. H., Laing R. A., Scheuer P. A. G., Turner S., 1994, The Physics of Active Galaxies, Eds. G. V. Bicknell, M. Dopita, P. Quinn, ASP Conf. Ser. **54**, ASP, San Francisco, 187.

Brouwer D., 1951, Phys. Today, **4**, 6.

Brouwer D., Hori G. I., 1962, Physics and Astronomy of the Moon, Ed. Z. Kopal, Academic Press, New York, Ch. 1.

Bruns D., Barnett T., Sandler D., 1997, SPIE, **2871**, 890.

Buckley J., 2009, Private communication.

Burg J. P., 1967, 37th Annual Society of Exploration Geophysicists, Meeting, Oklahoma City.

Burke B. F., Graham-Smith F., 2002, An Introduction to Radio Astronomy, Cambridge University Press, UK.

Burns D. et al., 1997, Mon. Not. R. Astron. Soc., **290**, L11.

Burns D., Baldwin J. E., Boysen R. C., Haniff C. A., Lawson P. R., Mackay C. D., Rogers J., Scott T. R., St.-Jacques D., Warner P. J., Wilson D. M. A., Young J. S., 1998, Mon. Not. R. Astron. Soc., **297**, 462.

Buscher D. F., Armstrong J. T., Hummel C. A., Quirrenbach A., Mozurkewich D., Johnston K. J., Denison C. S., Colavita M. M., Shao M., 1995, Appl. Opt., **34**, 1081.

Buscher D. F., Bakker E. J., Coleman T. A., Creech-Eakman M. J., Haniff C. A., Jurgenson C. A., Klinglesmith III D. A., Parameswariah C. B., Young J. S., 2006, SPIE, **6307**, 11.

Cameron P. B. et al., 2005, Nature, **434**, 1112.

Cannon J. M., Walter F., Skillman F. D., van Zee L., 2005, Astrophys. J., **621**, L21.

Carilli C., Rawlings S., 2004, New Astron. Rev., **48**, 1039.

Carleton N. et al., 1994, SPIE, **2200**, 152.

Carpenter K. G. et al., 2009, Astrophys. Space Sci. **320**, 217.

Cassaing F., Fleury B., Coudrain C., Madec P. -Y., Di Folco E., Glindeman A., Lévéque S., 2000, SPIE, **4006**, 17.

Cassinelli J. P., Mathis J. C., Savage B. D., 1981, Science, **212**, 1497.

Catanzarite J., Shao M., Tanner A., Unwin S., Jeffrey Yu., 2006, Pub. Astron. Soc. Pac., **118**, 1319.

Cayatte V., Kotanyi C., Balkowski C., van Gorkom J. H., 1994. Astron. J., **107**, 1003.

Chakrabarti S. K., Anandarao B. G., Pal S., Mondal S., Nandi A., Bhattacharyya A., Mandal S., Ram Sagar, Pandey J. C., Pati A., Saha S. K., 2005, Mon. Not. R. Astron. Soc., **362**, 957.

Charbonneau D. et al., 2009, Nature, **462**, 891.

Chesneau O. et al., 2007, Astron. Astroph., **464**, 119.

Chesneau O. et al., 2008, Astron. Astroph., **487**, 223.

Chou C. -W., Hume D. B., Koelemeij J. C. J., Wineland D. J., Rosenband T., 2010, Phys. Rev. Lett., **104**, 070802.

Christiansen W. N., Högbom J. A., 1985, Radio Telescopes, Cambridge University Press, Cambridge.

Chwolson O., 1924, Astron. Nachr., **221**, 329.

Clampin M., Croker J., Paresce F., Rafal M., 1988, Rev. Sci. Instru., **59**, 1269.

Claret A., 2000, Astron. Astrophys., **363**, 1081.

Close L. M., Roddier F., Hora J. L., Graves J. E., Northcott M. J., Roddier C., Hoffman W. F., Doyal A., Fazio G. G., Deutsch L. K., 1997, Astrophys. J., **489**, 210.

Colavita M., 1999, Pub. Astron. Soc. Pac., **111**, 111.

Colavita M. et al., 1999, Astrophys. J., **510**, 505.

Colavita M. et al., 2008, SPIE, **7013**, 70130A.

Colavita M. M., Shao M., Staelin D. H., 1987, Appl. Opt., **26**, 4106.

Collett E., 1993, Polarized Light: Fundamentals and Applications, Marcel Dekkar, Inc. N. Y.

Conan R., Ziad A., Borgnino J., Martin F., Tokovinin A., 2000, SPIE, **4006**, 963.

Cornils R. et al., 2003, Astropart. Phys., **20**, 129.

Cornwell T. J., 1994, Synthesis Imaging in Radio Astronomy, Eds. R. A. Perley, F. R., Schwab, A. H. Bridle, Astronomical Society of the Pacific, San Francisco, **277**.

Cornwell T. J., Evans K. F., 1985, Astron. Astrophys., **143**, 77.

Cornwell T. J., Fomalont E. B., 1989, in Synthesis Imaging in Radio Astronomy, Eds. R. A. Perley, F. R. Schwab, A. H. Bridle, ASP, San Francisco.

Cornwell T. J., Wilkinson, P. N., 1981, Mon. Not. R. Astron. Soc., **196**, 1067.

Coudé du Foresto V., Perrin G., Boccas M., 1995, Astron. Astrophys., **293**, 278.

Coudé du Foresto V. C. et al., 1998, SPIE, **3350**, 856.

Coudé du Foresto V. C. et al., 2001, Compt. Rend. Acad. Sci. Paris, **IV 2**, 45.

Coudé du Foresto V., Ridgway S. T., 1992, Proc. ESO-NOAO, Conference, Eds. J. M. Beckers, F. Merkle, 731.

Coulman C. E., Vernin J., Coqueugniot Y., Caccia J. L., 1988, Appl. Opt., **27**, 155.

Cranmer S., Smith M., Robinson R., 2002, Astrophys. J., **537**, 433.

Crane P. C., Napier P. J., 1994, Synthesis Imaging in Radio Astronomy, Eds. R. A. Perley, F. R., Schwab, A. H. Bridle, Astronomical Society of the Pacific, San Francisco, 139.

D'Addario L. R., 1976, SPIE, Image analysis and Evaluation, 221.

Danchi W. C., Bester M., Degiacomi C., Greenhill I., Townes C., 1994, Astron. J., **107**, 1469.

Danchi W. C., Tuthill P. G., Monnier J. D., 2001, Astrophys. J., **562**, 440.

D'Arcio L. et al., 2003, SPIE, **4852**, 184.

Davies J. M., Cotton E. S., 1957, J. Sol. Energy Sci. Eng., **1**, 16.

Davis J. et al., 2005, Mon. Not. R. Astron. Soc., **356**, 1362.

Davis J., Lawson P. R., Booth A. J., Tango W. J., Thorvaldson E. D., 1995, Mon. Not. R. Astron. Soc., **273**, L53.

Davis J., Tango W. J., 1986, Nature, **323**, 234.

Davis J., Tango W. J., 1996, Pub. Astron. Soc. Pac., **108**, 456.

Davis J., Tango W. J., Booth A. J., ten Brummelaar T. A., Minard R. A., Owens S. M., 1999a, Mon. Not. R. Astron. Soc., **303**, 773.

Davis J., Tango W. J., Booth A. J., Thorvaldson E. D., Giovannis J., 1999b, Mon. Not. R. Astron. Soc., **303**, 783.

Davis J., Tango W. J., Booth A. J., Minard R. A., ten Brummelaar T. A., Shobbrook R. R., 1992, Proc. ESO-NOAO Conf., Eds. J. M. Beckers, F. Merkle, 741.

Deil C. et al., 2008, in The Universe at Sub-second timescales, Eds. D. Phelan, O. Ryan, A. Shearer, AIP Conf. Proc., **984**, 140.

Dejonghe J., Arnold L., Lardière O., Berger J. -P., Cazalé C., Dutertre S., Kohler D., Vernet D., 1998, SPIE, **3352**, 603.

Derie F., Ferrai M., Brunetto E., Duchateau M., Amestica R., Aniol P., 2000, SPIE, **4006**, 99.

Desikachary K., Hearnshaw J. B., 1982, Mon. Not. R. Astron. Soc., **201**, 707.

de Witt W. J. et al., 2006, Astron. Astrophys. **456**, 1027.

de Witt W. J. et al., 2008, in The Universe at Sub-second timescales, Eds. D. Phelan, O. Ryan, A. Shearer, AIP Conf. Proc., **984**, 268.

Di Benedetto G. P., Conti G., 1983, Astrophys. J., **268**, 309.

Di Benedetto G. P., Rabbia Y., 1987, Astron. Astrophys., **188**, 114.

Di Folco E. et al., 2007, Astron. Astrophys., **475**, 243.

Di Giacomo A., Richichi A., Lisi F., Calamai G., 1991, Astron. Astrophys., **249**, 397.

Domiciano de Souza A., Bendjoya P., Vakili F., Millour F., Petrov, R. G., 2008, Astron. Astrophys., **489**, L5.

Domiciano de Souza A., Kervella P., Jankov S., Abe L., Vakili F., di Folco E., Paresce F., 2003, Astron. Astrophys., **407**, L47.

Donati J. -F. et al., 2002, Mon. Not. R. Astron. Soc., **333**, 55.

Donati J. -F. et al., 2008, Mon. Not. R. Astron. Soc., **390**, 545.

Dopita M. A., Sutherland R. S., 2003, Astrophysics of the Diffuse Universe, Springer, Berlin, Heidelberg, New York.

Downes D., 1988, Diffraction-limited Imaging with Very Large Telescopes, Eds. D. M. Alloin, J. -M. Mariotti, Kluwer Academic Publishers, Dordrecht (Hingham, MA), 53.

Dravins D., LeBohec S., 2008, SPIE, **6986**.

Dupuis J., Chayer P., Vennes S., Christian D. J., Kruk J. W., 2000, Astrophys. J., **537**, 977.

Dyck H., Benson J., Ridgway S., 1993, Pub. Astron. Soc. Pac., **105**, 610.

Dyck H. M., Nordgren T. E., 2002, Astron. J., **124**, 541.

Eidson J. C., 2006, Measurement, Control, and Communication using IEEE 1588, Springer, Berlin, Heidelberg, New York.

Einstein A., 1905a, Ann. der Phys., **17**, 891.

Einstein A., 1905b, Ann. der Phys., **17**, 132.

Einstein A., 1916, Sitzungsber. Preuss. Akad. Wiss., 688.

Eisenhauer F., Schödel R., Genzel R., Ott T., Tecza M., Abuter R., Eckart A., Alexander T., 2003, Astrophys. J., **597**, L121.

Eisenhauer F. et al., 2008, ESO Astrophysics Symposium, Springer, Berlin, 431.

Eisner J. A., Graham J. R., Akeson R. L., Najita J., 2009, Astrophys. J., **692**, 309.

Eisner J. A., Hillenbrand L. A., White R. J., Akeson R. L., Sargent A. I., 2005, Astrophys. J., **623**, 952.

Eisner J. A., Lane B. F., Hillenbrand L. A., Akeson R. L., Sargent A. I., 2004, Astrophys. J., **613**, 1049.

Elias II N. M., 2001, Astrophys. J., **549**, 647.

Elias II N. M., 2004, Astrophys. J., **611**, 1175.

Elias II N. M., Jones C. E., Schmitt H. R., Jorgensen A. M., Ireland M. J., Perraut K., 2008, Olbin communication.

Elias II N. M., Koch R. H., Pfeiffer R. J. 2008, Astron. Astrophys., **489**, 911.

England M. N., 1980, Mon. Not. R. Astron. Soc., **191**, 23.

Esposito S., Riccardi A., 2001, Astron. Astrophys., **369**, L9.

Essen L., 1963, J. Inst. Elect. Engrs., **9**, 247.

Evans A., Bode M. F., O'Brien T. J., Darnley M. J., 2008, in ASPC 401, Astron. Soc. Pac.

Ewan H. I., Purcell E. M., 1951, Nature, **168**, 356.

Figer D., 2005, Nature, **434**, 192.

Fizeau H., 1867, Compt. Rend. Acad. Sci. Paris, 932.

Fomalont E. B., Wright M. C. H., 1974, in Galactic and Extra-galactic Radio Astronomy, Eds. G. L. Verschuur, K. I. Kellerman, 256.

Fontana P. R., 1983, J. Appl. Phys., **54**, 473.

Fossat E., 2005, 21st Century Astrophysics, Eds. S. K. Saha, V. K. Rastogi, Anita Publications, New Delhi, 163.

Fowler R. H., 1929, Statistical Mechanics, Cambridge University Press, Cambridge.

Foy R., Bonneau D., Blazit A., 1985, Astron. Astrophys., **149**, L13.

Foy R., Bonneau D., Blazit A., 1985, Astron. Astrophys., **149**, L13.

Foy R., Labeyrie A., 1985, Astron. Astrophys., **152**, L29.

Frank J., King A. R., Raine D. J., 2002, Accretion Power in Astrophysics, Cambridge University Press, Cambridge.

Freire P., Gupta Y., Ransom S., Ishwara-Chandra C., 2004, Astrophys. J., **606**, L53.

Freytag B., Steffen M., Dorch B., 2002, Astron. Nachr., **323**, 213.

Fridlund C. V. M. et al., 2004, SPIE, **5491**, 227.

Fried D. L., 1965, J. Opt. Soc. Am., **55**, 1427.

Fried D. L., 1966, J. Opt. Soc. Am., **56**, 1372.

Frieden B. R., 1972, J. Opt. Soc. Am., **62**, 511.

Gallimore J. F., Baum S. A., O'Dea C. P., 1997, Nature, **388**, 852.

Gamo H., 1963, J. Appl. Phys., **34**, 875.

Gandhi P., 2005, 21st Century Astrophysics, Eds. S. K. Saha, V. K. Rastogi, Anita Publications, New Delhi, 90.

Garnier D., Coburn D., Dainty J. C., 2005, SPIE, **5891**.

Gaskill J. D., 1978, Linear Systems, Fourier Transforms, and Optics, Wiley, NY.

Gay J., Mekarnia D., 1988, Proceedings of the ESO-NOAO Conference, Ed. F. Merkle, ESO, FRG, 811.

Gendreau et al., 2004, SPIE, **5488**, 394.

Gerchberg R. W., Saxton W. O., 1972, Optik, **35**, 237.

Ghez A. M., Klein B. L., Morris M., Becklin E. E., 1998, Astrophys. J., **509**, 678.

Ghez A. M. et al., 2005, Astrophys. J., **635**, 1087.

Ghez A. M. et al., 2008, Astrophys. J., **689**, 1044.

Giacobbe F. W., 2005, Electron. J. Theoret. Phys., **2**, 30.

Gies D. R. et al., 2007, Astrophys. J., **654**, 527.

Gies D. R. et al., 2008, Astrophys. J., **682**, L117.

Gillessen S., Eisenhauer F., Trippe S., Alexander T., Genzel R., Martins F., Ott T., 2009, Astrophys. J., **692**, 1075.

Glauber R. J., 1963, Phys. Rev., **130**, 2529.

Gledhill T. M., 2005, Mon. Not. R. Astron. Soc., **356**, 883.

Glindemann A. et al., 2001, Proceedings of the 36th Liege International Astrophysics Colloquium, Eds. J. Surdej, J. D. Swings, D. Caro, A. Detal, 27.

Glindemann A., Paresce F., 2001, http://www.eso.org/outreach.

Glindemann A. et al., 2003, SPIE, **4838**, 89.

Golimowski D. A., Clampin M., Durrance S. T., Barkhouser R. H., 1992, Appl. Opt., **31**, 4405.

Goodricke J., Englefield H. C., 1785, Phil. Trans. Ser., **75**, 153.

Goodman J. W., 1985, Statistical Optics, Wiley, NY.

Gopal-Krishna, Wiita P. J., 2005, 21st Century Astrophysics, Eds., S. K. Saha, V. K. Rastogi, Anita Publications, New Delhi, 108.

Goudi B. S. et al. 2008, Science; astro-ph/arXiv:0802.1920.

Gould A. et al. 2006, **644**, L37.

Green S. F., Jones M. H., Burnell S. J., 2004, An Introduction to the Sun and Stars, Cambridge University Press, UK.

Greenwood D. P., 1977, J. Opt. Soc. Am., **67**, 390.

Grindlay J. E., Helmken H. F., Hanbury Brown R., Davis J., Allen L. R., 1975a, Astrophys. J., **197**, L9.

Grindlay J. E. et al., 1975b, Astrophys. J., **201**, 82.

Gull S. P., Daniell G., 1978, Nature, **272**, 686.

Guyon O., 2006, in IAU Colloq., 200, Direct Imaging of Exoplanets, Eds. C. Aime, F. Vakili, Cambridge University Press, Cambridge, 559.

Habets G. M. H. J., Heinze J. R. W., 1981, Astron. Astrophys. Suppl., **46**, 193.

Haguenauer P., Sevei M., Schanen-Duport I., Rousselet-Perraut K., Berger J., Duchéne Y., Lacolle M., Kern P., Melbet F., Benech P., 2000, SPIE, **4006**, 1107.

Hajian A. et al., 1998, Astrophys. J., **496**, 484.

Hale D., Bester M., Danchi W., Fitelson W., Hoss S., Lipman E., Monnier J., Tuthill P., Townes C. H., 2000, Astrohys. J., **537**, 998.

Hale D. et al., 1997, Astrophys. J., **490**, 826.

Hale D., Weiner J., Townes C., 2004, SPIE, **5491**.

Hammerschlag-Hensberge G, 1979, Int. Astron. Union Circ., 3391.

Hanbury Brown R., 1974, The Intensity Interferometry- its Applications to Astronomy, Taylor & Francis, London.

Hanbury Brown R., 1991, Boffin, Adam Hilger, Bristol.

Hanbury Brown R., Davis J., Allen L. R., 1967, Mon. Not. R. Astron. Soc., **137**, 375.

Hanbury Brown R., Davis J., Allen L. R., 1974a, Mon. Not. R. Astron. Soc., **167**, 121.

Hanbury Brown R., Davis J., Allen L. R., 1974c, Mon. Not. R. Astron. Soc., **168**, 93.

Hanbury Brown R., Davis J., Herbison-Evans D., Allen L. R., 1970, Mon. Not. R. Astron. Soc., **148**, 103.

Hanbury Brown R., Davis J., Lake R. J. W., Thompson R. J., 1974b, Mon. Not. R. Astron. Soc., **167**, 475.

Hanbury Brown R., Jennison R. C., Das Gupta M. K., 1952, Nature, **170**, 1061.

Hanbury Brown R., Twiss R., 1954, Phil. Mag., **45**, 663.

Hanbury Brown R., Twiss R., 1956a, Nature, **177**, 27.

Hanbury Brown R., Twiss R., 1956b, Nature, **178**, 1046.

Hanbury Brown R., Twiss R., 1957a, Nature, **179**, 1128.

Hanbury Brown R., Twiss R., 1957b, Proc. R. Soc. A., **242**, 300.

Hanbury Brown R., Twiss R., 1958, Proc. R. Soc. A., **248**, 222.

Haniff C. A., 2007, New Astron. Rev., **51**, 583.

Haniff C. A., Buscher D. F., Christou J. C., Ridgway S. T., 1989, Mon. Not. R. Astron. Soc., **241**, 694.

Haniff C. A., Ghez A. M., Gorham P. W., Kulkarni S. R., Matthews K., Neugebauer G., 1992, Astron. J., **103**, 1662.

Haniff C., Scholz M., Tuthill P., 1995, Mon. Not. R. Astron. Soc., **276**, 640.

Hansen C. J., Kawaler S. O., 1994, Stellar Interiors, Physical Principles, Structure and Evolution, Springer-Verlag, Berlin, Heidelberg, New York.

Hardy J. W., 1998, Adaptive Optics for Astronomical Telescopes, Oxford University Press, New York.

Harmanec P., 1984, Bull. Astron. Inst. Czech., **35**, 164.

Harmanec P., 2002, Astron. Nachrichten, **323**, 87.

Harmanec P. et al., 1996, Astron. Astrophys., **312**, 879.

Harris D. E., Carilli C. L., Perley R. A., 1994, Nature, **367**, 713.

Harris J. W., Stocker H., 1998, Handbook of Mathematics and Computational Science, Springer-Verlag, NY.

Hartmann L., Kenyon S. J., 1996, Ann. Rev. Astron. Astrophys., **34**, 207.

Heap S. R., Lindler D. J., Woodgate B., 2006, New Astron. Rev., **50**, 294.

Hecht J., 2000, Laser Focus World, 107.

Hege E. K., Hubbard E. N., Strittmatter P. A., Worden S. P., 1981, Astrophys. J., **248**, 1.

Heisenberg W., 1927, Zeitschrift für Physik., **43**, 172.

Hellwig H., Allan D. W., Wtt F. L., 1975, Proceedings of the Atomic Masses and Fundamental Constant, Eds. J. H. Sanders, A. H. Wapstra, Plenum Press, NY, USA, 305.

Herbig G. H., 1956, Pub. Astron. Soc. Pac., **68**, 531.

Herbig G. H., 1960, Astrophys. J. Suppl., **4**, 337.

Hestroffer D., 1997, Astron. Astrophys., **327**, 199.

Hey J. S., 1946, Nature, **157**, 47.

Hilditch R. W., 2001, An Introduction to Close Binary Stars, Cambridge University Press, UK.

Hill J. M., 2000, SPIE, **4004**, 36.

Hillenbrand L., White R., 2004, Astrophys. J., **604**, 741.

Hillwig T. C., Gies D. R., Huang W., McSwain M. V., Stark M. A., van der Meer A., Kaper L., 2004, Astrophys. J., **615**, 422.

Hinz P., Angel R., Hoffmann W., Mccarthy D., Mcguire P., Cheselka M., Hora J., Woolf N., 1998, Nature, **395**, 251.

Hinz P., Hoffmann W., Hora J., 2001, Astrophys. J., **561**, L131.

Hipparcos Catalogue, 1997, ESA, SP-1200.

Hoard D. W., Howell S. B., Stencel R. E., 2010, Astrophys. J., **714**, 549.

Hoeflich P., 2005, 21st Century Astrophysics, Eds. S. K. Saha, V. K. Rastogi, Anita Publications, New Delhi, 57.

Hofmann K. -H., Balega Y. Y., Weigelt G., 2004, Astron. Astrophys., **417**, 981.

Hofmann K. -H., Seggewiss W., Weigelt G., 1995, Astron. Astrophys. **300**, 403.

Högbom J., 1974, Astron. Astrophys. Suppl., **15**, 417.

Hogg D. E., MacDonald G. H., Conway R. G., Wade C. M., 1969, Astron. J., **74**, 1206.

Holder J. et al., 2006, Astropart. Phys., **25**, 391.

Horaguchi T. et al., 1994, Pac. Astron. Soc. J., **46**, 9.

Howarth I. D., Prinja R. K., 1989, Astrophys. J. Sup., **69**, 527.

Howe D. A., 1976, NBS tech. Note 679.

Hubble E. P., 1936, The Realm of the Nebulae, Yale University Press.

Hufnagel R. E., 1974, in Proceedings of the Optical Propagaion through Turbulence, Optical Society of America, Washington, D. C. WAI 1.

Hughes D. W., 1983, Quart. J. R. Astron. Soc., **24**, 246.

Hughes S. A., 2006, AIPC, **873**, 13.

Hummel C. A. et al., 2001, Astron. J., **121**, 1623.

Hummel C., Mozurkevich D., Armstrong J., Hajian A., Elias N., Hutter D., 1998, Astron. J., **116**, 2536.

Hummel W., 1998, Astron. Astrophys., **330**, 243.

Hummel W., Vrancken M., 2000, Astron. Astrophys., **359**, 1075.

Hutter D., 1999, in Principles of Long Baseline Stellar Interferometry, Ed. P. R. Lawson, JPL Publications, Pasadena, 165.

Hyland D., 2005, SPIE, **5905**, 330.

Hynes R., 2002, Proc. Astron. Soc. Pac. Conf. Eds. Gänsicke et al., **261**, 676.

Ireland M. J., Tuthill P. G., Davis J., Tango W., 2005, Mon. Not. R. Astron. Soc., **361**, 337.

Ishimaru A., 1978, Wave Propagation and Scattering in Random Media, Academic Press, N. Y.

Jackson J. D., 1999, Classical Electrodynamics, Wiley, London.

Jaffe W. et al., 2004, Nature, **429**, 47.

Janesick J. R., 2001, Scientific Charge Coupled Devices, SPIE Press, Bellingham Washington.

Jansky K. G., 1932. Proc. Inst. Radio Engrs., **20**, 1920.

Jaynes E. T., 1957, Phys. Rev., **106**, 620.

Jaynes E. T., 1982, Proc. IEEE,, **70**, 939.

Jefferies S. M., Christou J. C., 1993, Astrophys. J., **415**, 862.

Jennison R. C., 1958, Mon. Not. R. Astron. Soc., **118**, 276.

Jennison R. C., 1966, Introduction to Radio Astronomy, Newnes, London.

Jennison R. C., Das Gupta M. K., 1953, Nature, **172**, 996.

Jennison R. C., Das Gupta M. K., 1956, Phil. Mag. Ser., **8**, 65.

Jennison R. C., Latham V., 1959, Mon. Not. R. Astron. Soc., **174**, 1959.

Jerram P., Pool P. J., Bell R., Burt D. J., Bowring S., Spencer S., Hazelwood M., Moody I., Catlett N., Heyes P. S., 2001, SPIE, **4306**, 178.

Jilinsky E., Daflon S., Cunha K., de La Reza R., 2006, Astron. Astrophys., **448**, 1001.

Johnson H. L., 1966, Ann. Rev. Astron. Astrophys., **4**, 201.

Johnson H. L., Morgan W. W., 1953, Astrophys. J., **117**, 313.

Jones R. C., 1941, J. Opt. Soc. Am., **31**, 488.

Kalas P., Graham J. R., Clampin M., 2005, Nature, **435**, 1067.

Kalas P. et al., 2008, Science, **322**, 1345.

Kanaan S. et al., 2008, Astron. Astrophys. **486**, 785.

Kantharia N., Ananthakrishnan S., Nityananda R., Hota A., 2005, Astron. Astrophys., **435**, 483.

Kantharia N., Anupama G. C., Prabhu T. P., Ramya S., Bode M. F., Eyres S. P. S., O'Brien T. J., 2007, Astrophys. J., **667**, L171.

Kelsall T. et al., 1998, Astrophys. J., **508**, 44.

Kennicutt Jr. R. C., 1998a, Ann. Rev. Astron. Astrophys., **36**, 189.

Kennicutt Jr. R. C., 1998b, Astrophys. J., **498**, 541.

Kennicutt Jr. R. C., Schweizer F., Barnes J. E., 1998, in Galaxies: Interactions and Induced Star formation, Saas-Fee Advanced Course 26, Eds. D. Friedli, L. Martinet, D. Pfenniger, Swiss Society for Astrophys. & Astron., XIV, Springer, Berlin, 404.

Kern P., Berger J. -P., Haguenauer P., Malbet F., Perraut K., 2005, Astrophysique, **2**, 111.

Kern P. et al., 2008, SPIE, **7013**, 701315.

Kervella P., 2006, Mem. Soc. Astron. Italiana, **77**, 227.

Kervella P., 2007, New Astron. Rev., **51**, 706.

Kervella P., Coudé du Foresto V., Glindemann A., 2000, SPIE, **4006**, 31.

Kervella P., Coudé du Foresto V., Perrin G., Schöller M., Traub W., Lacasse M., 2001, Astron. Astrophys., **367**, 876.

Kervella P., Fouqué P., Storm J., Gieren W. P., Bersier D., Mourard D., Nardetto N., Coudé du Foresto V., 2004, Astrophys. J., **604**, L113.

Kervella P., Traub W. A., Lacasse M. G., 1999, ASP Conf. Series on Working on the Fringe, Eds. S. Unwin, R. Stachnik, Dana Point, USA, **194**.

Kervella P., Verhoelst T., Ridgway S. T., Perrin G., Lacour S., Cami J., Haubois X., 2009, astro-ph/arXiv:0907.1843.

Kim H. -L. 1989, Astrophys. J., **342**, 1061.

Kinman T., Castelli F., 2002, Astron. Astrophys., **391**, 1039.

Kishimoto M., Hónig S. F., Antonucci R., Kotani T., Barvainis R., Tristram K. R. W., Weigelt G., 2009, Astron. Astrophys., **507**, L57.

Klein I., Guelman M., Lipson S. G., 2007, Appl. Opt., **46**, 4237.

Kloppenborg B. et al., 2010, Nature, **464**, 870.

Knox K., Thompson B., 1974, Astrophys. J., **193**, L45.

Koechlin L., Lawson P. R., Mourard D., Blazit A., Bonneau D., Morand F., Stee P., Tallon-Bosc I., Vakili F., 1996, Appl. Opt., **35**, 3002.

Kolmogorov A. N., 1941a, Compt. rend. de l'Acadmie des Sciences de l'U.R.S.S., **32**, 16.

Kolmogorov A. N., 1941b, Compt. rend. de l'Acadmie des Sciences de l'U.R.S.S., **30**, 301.

Kopal Z., 1959, Close Binary Systems, International Astrophysics Series, **5**, Chapman & Hall Ltd, London.

Kormendy J., Richardson D., 1995, Ann. Rev. Astron. Astrophys., **33**, 581.

Kovalevsky J., 1995, Modern Astrometry, Springer, New York.

Kovalevsky J., 2004, Fundamentals of Astrometry, Cambridge University Press, Cambridge.

Kraus S. et al., 2007, Astron. Astrophys., **466**, 649.

Kraus S. et al., 2008a, Astron. Astrophys., **489**, 1157.

Kraus S., Preibisch Th., Ohnaka K., 2008b, Astrophys. J., **676**, 490.

Kraus S. et al., 2009, Astron. Astrophys., **497**, 195.

Krauss J. D., 1966, Radio Astronomy, McGraw Hill, NY.

Kubo H. et al., 2004, New Astron. Rev., **48**, 323.

Kuchner M. J., Traub W. A., 2002, Astrophys. J., **570**, 900.

Kwok S., 1993, Ann. Rev. Astron. Astrophys., **31**, 63.

Labeyrie A., 1970, Astron. Astrophys., **6**, 85.

Labeyrie A., 1975, Astrophys. J., **196**, L71.

Labeyrie A., 1979, Astron. Astrophys., **77**, L1.

Labeyrie A., 1985, 15th Advanced Course, Swiss Society of Astrophysics and Astron, Eds. A. Benz, M. Huber, M. Mayor, 170.

Labeyrie A., 1995, Astron. Astrophys., **298**, 544.

Labeyrie A., 1996, Astron. Astrophys. Suppl., **118**, 517.

Labeyrie A., 1998, Proceedings NATO-ASI, Planets outside the solar system, 5–15 May, 1998, Cargése, France.

Labeyrie A., 1999a, ASP Conference, **194**, Eds. S. Unwin, R. Stachnik, 350.

Labeyrie A., 1999b, Science, **285**, 1864.

Labeyrie A., 2005, 21st Century Astrophysics, Eds. S. K. Saha, V. K. Rastogi, Anita Publications, New Delhi, 228.

Labeyrie A., 2009, Private communication.

Labeyrie A., Bonneau D., Stachnik R. V., Gezari D. Y., 1974, Astrophys. J., **194**, L147.

Labeyrie A., Koechlin L., Bonneau D., Blazit A., Foy R., 1977, Astrophys. J., **218**, L75.

Labeyrie A., Le Coroller H., Dejonghec J., 2008, SPIE, **7013**.

Labeyrie A. et al., 2009, Expt. Astron., **23**, 463.

Labeyrie A., Lipson S. G., Nisenson P., 2006, An Introduction to Optical Stellar Interferometry, Cambridge University Press, UK.

Labeyrie A., Schumacher G., Dugué M., Thom C., Bourlon P., Foy F., Bonneau D., Foy R., 1986, Astron. Astrophys., **162**, 359.

Lacour S. et al., 2008, Astron. Astrophys., **485**, 561.

Lacour S. et al., 2009, Astrophys. J., **707**, 632.

Lamers J. G. L. M., 2008, ASP Conference Series, Eds. A. de Koter, L. J. Smith, L. B. F. M. Waters, **288**, 443.

Lawson P. R., 1994, Pub. Astron. Soc. Pac., **106**, 917.

Lawson P. R., 1999, in Principles of Long Baseline Stellar Interferometry, Ed. P. R. Lawson, JPL Publications, Pasadena, 113.

Lawson P. R., Lay O. P., Johnston K. J., Beichman C. A. (Eds.), 2007, Terrestrial Planet Finder Interferometer Science Working Group Report, JPL Publication 07-1, Pasadena, CA, USA.

Lawson P. R. et al., 2008, SPIE, **7013**.

Leahy J. P., Perley R. A., 1991, Astron. J., **102**, 537.

LeBohec S., 2009, Private communication.

LeBohec S., Barbieri C., de Witt W., Dravins D., Feautrier P., Foellmi C., Glindemann A., Hall J., Holder J., Holmes R., Kervella P., Kieda1 D., Le Coarer E., Lipson S., Morel S., Nunez P., Ofir A., Ribak E., Saha S., Schoeller M., Zhilyaev B., Zinnecker H., 2008, SPIE, **7013**, 7013-2E.

LeBohec S., Holder J., 2006, Astrophys. J., **649**, 399.

LeBouquin J. -B. et al., 2004, Astron. Astrophys., **424**, 719.

LeBouquin J. -B. et al., 2006, Astron. Astrophys., **450**, 1259.

Le Coroller H., Dejonghel J., Arpesella C., Vernet D., Labeyrie A., 2004, Astron. Astrophys., **426**, 721.

Lee J., Bigelow B., Walker D., Doel A., Bingham R., 2000, Pub. Astron. Soc. Pac., **112**, 97.

Lee J., Kim C., Koch R., Lee C., Kim H., Park J., 2009, Astron. J., **137**, 3181.

Léger A., 2009, Private communication.

Léger A., Pirre M., Marceau F. J., 1993, Astron. Astrophys., **277**, 309.

Léger A. et al., 2009, Astron. Astrophys., **506**, 287.

Lefèvre H. C., 1980, Electron, **16**, L778.

Leinert C. et al., 2003, The Messenger, **112**, 13.

Leinert C., Haas M., Abraham P., Richichi A., 2001, Astron. Astrophys., **375**, 927.

Leisawitz D., 2004, Adv. Space Res., **34**, 631.

Levy L., 1968, Appl. Opt., **1**, 164.

Lindengren L., Perryman M. A. C., 1996, Astron. Astrophys. Suppl., **116**, 579.

Linfield B. P., Colavita M. M., Lane B. F., 2001, Astrophys. J., **554**, 505.

Lipman E., Bester M., Danchi W., Townes C. H., 1998, SPIE, **3350**, 933.

Lisauer J. J., 2002, Nature, **418**, 355.

Lloyd-Hart M., 2000, Pub. Astron. Soc. Pac., **112**, 264.

Lohmann A., Weigelt G., Wirnitzer B., 1983, Appl. Opt., **22**, 4028.

Lopez B. et al., 2006, SPIE, **6268**, 62680.

López-Sánchez A. R., Esteban C., 2010, astro-ph/arXiv:1004.0051v1.

Lorenz E., 2004, New Astron. Rev., **48**, 339.

Lovell B., 1985, The Jodrell Bank Telescopes, Oxford University Press, UK.

Lovelock J. E., 1965, Nature, **207**, 568.

Lowman A. E., Trauger J. T., Gordon B., Green J. J., Moody D., Nissner A. F., Shi F., 2004, SPIE, **5487**, 1246.

Lück H. et al., 2006, Class. Quant. Grav., **23**, S71.

Lyot B., 1939, Mon. Not. R. Astron. Soc., **99**, 580.

Lynds C., Worden S., Harvey J., 1976, Astrophys. J., **207**, 174.

MacGregor K. B., Jackson S., Skumanich A., Metcalfe T. S., 2007, Astrophys. J., **663**, 560.

Machida Y. et al., 1998, SPIE, **3350**, 202.

Mackay F. E., Elias II N. M., Jones C. E., 2009, Astrophys. J., **704**, 591.

Malbet F. et al., 1998, Astrophys. J., **507**, L149.

Malbet F. et al., 1999, Astron. Astrophys. Suppl., **138**, 135.

Malbet F. et al., 2005, Astron. Astrophys., **437**, 672.

Malbet F. et al., 2007a, Astron. Astrophys., **464**, 43.

Malbet F. et al., 2007b, The Messenger, **127**, 33.

Malbet F. et al., 2008, Science with the VLT in the ELT era. Proceedings of the Astrophysics and Space Science, 343.

Mandel L., 1963, Prog. Opt., **II**, 181.

Mandel L., Wolf E., 1976, J. Opt. Soc. Am., **66**, 529.

Mandel L., Wolf E., 1995, Optical Coherence and Quantum Optics, Cambridge University Press, UK.

Margon B., 1984, Ann. Rev. Astron. Astrophys., **22**, 507.

Mario Gai et al., 2004, SPIE, **5491**, 528.

Mariotti J. -M., 1988, Diffraction-limited Imaging with Very Large Telescopes, Eds. D. M. Alloin, J. -M. Mariotti, Kluwer Academic Publishers, Dordrecht (Hingham, MA), 3.

Markowitz W., 1960, Stars and Stellar Systems 1, University of Chicago Press, IL, USA.

Marlborough J. M., 1997, Astron. Astrophys., **317**, L17.

Marston A. P., McCollum B., 2008, Astron. Astrophys., **477**, 193.

Mawet D., Riaud P., 2005, Proc. IAU Colloq., **1**, 361. Cambridge University Press.

Mayor M., Queloz D., 1995, Nature, **378**, 355.

Mayor M. et al., 2009, Astron. Astrophys., **507**, 487.

Mazzali P. A., Röpke F. K., Benetti S., Hillebrandt W., 2007, Science, **315**, 825.

McAlister H., Bagnuolo W., ten Brummelaar T., Hartkopf W. I., Shure M., Sturmann L., Turner N., Ridgway S., 1998, SPIE, **3350**, 947.

McAlister H. et al., 2005, Astrophys. J., **628**, 439.

McCarthy D. D., Seidelmann P. K., 2009, TIMEFrom Earth Rotation to Atomic Physics, Wiley-VCH Verlag GmbH & Co, Weinheim.

McChelland D. E. et al., 2006, Class. Quant. Grav., **23**, S41.

Meikle W. P. S., Matcher S. J., Morgan B. L., 1987, Nature, **329**, 608.

Meilland A., Millour F., Stee P., Spang A., Petrov R., Bonneau D., Perraut K., Massi F., 2008, Astron. Astrophys., **488**, L67.

Meilland A., Stee P., Vannier M., Millour F., Domiciano de Souza A., Malbet F., Martayan C., Paresce F., Petrov R., Richichi A., Spang A., 2007a, Astron. Astrophys., **464**, 59.

Meilland A. et al., 2007b, Astron. Astrophys., **464**, 74.

Meisenheimer K. et al., 2007, Astron. Astrophys., **471**, 453.

Mérand A. et al., 2005, Astron. Astrophys., **438**, L9.

Mercier C., Subramanian P., Kerdraon A., Pick M., Ananthakrishnan S., Janardhan P., 2006, Astron. Astrophys., **447**, 1189.

Meynet G., Maeder A., 2005, Astron. Astrophys., **429**, 581.

Michelson, A. A., 1890, Phil. Mag., **30**, 1.

Michelson, A. A., 1891, Nature, **45**, 160.

Michelson, A. A., Pease, F. G., 1921, Astrophys. J., **53**, 249.

Millan-Gabet R., Malbet F., Akeson R., Leinert C., Monnier J., Waters R., 2007, Protostars and Planets, Eds. V. B. Reipurth, D. Jewitt, K. Keil, University of Arizona Press, Tucson, 951.

Millan-Gabet R., Schloerb P., Traub W., 2001, Astrophys. J., **546**, 358.

Millan-Gabet R., Schloerb P. F., Traub W. A., Carleton N. P., 1999, Pub. Astron. Soc. Pac., **111**, 238.

Millour et al., 2007, Astron. Astrophys., **464**, 107.

Mills B. Y., 1952, Nature, **170**, 1063.

Mills B. Y., Little A. G., 1953, Austr. J. Phys., **6**, 272.

Milosavljevic M., Merritt D., 2001, Astrophys. J., **563**, 34.

Miroschnichenko A. S., Bjorkman K. S., Krugov V. D., 2002, Pub. Astron. Soc. Pac., **114**, 1226.

Mitton S., Ryle M., 1969, Mon. Not. R. Astron. Soc., **146**, 221.

Mix L. J. et al., 2006, Astrobiology, **6**, 735.

Mohan D., Mishra S. K., Saha S. K., 2005, Ind. J. Pure Appl. Phys., **43**, 399.

Monnier J. D., 2003, Rep. Prog. Phys., **66**, 789.

Monnier J. D. et al., 2001, AAS Meeting, **198**, 63.02.

Monnier J. D., Fitelson W., Danchi W. C., Townes C. H., 2000, Astrophys. J. Suppl. Ser., **129**, 421.

Monnier J. D. et al., 2004, Astrophys. J., **602**, L57.

Monnier J. D. et al., 2005, Astrophys. J., **624**, 832.

Monnier J. D. et al., 2006a, SPIE, **62681P**.

Monnier J. D. et al., 2006b, Astrophys. J., **647**, L12.

Monnier J. D. et al., 2007, Science, **317**, 342.

Montilla I., Sellos J., Pereira S. F., Braat J. J. M., 2005, Modern Opt., **53**, 437.

Moran E. C., 2007, ASPC, **373**, 425.

Morel S., Saha S. K., 2005, 21st Century Astrophysics, Eds. S. K. Saha, V. K. Rastogi, Anita Publications, New Delhi, 237.

Morel S., Traub W., Bregman J., Mah R., Wilson C., 2000, SPIE, **4006**, 506.

Moscadelli L., 2005, Proc. IAU, **1**, 190, Cambridge University Press, Cambridge.

Moss C., Whittle M., 1993, Astrophys. J., **407**, L17.

Mourard D., 2009, Private communication.

Mourard D. et al., 2000, SPIE, **4006**, 434.

Mourard D. et al., 2001, C. R. Acad. Sci. Paris., t2, S. IV, 35.

Mourard D. et al., 2009, Astron. Astrophys., **508**, 1073.

Mourard D., Bonneau D., Koechlin L., Labeyrie A., Morand F., Stee P., Tallon-Bosc I., Vakili F., 1997, Astron. Astrophys., **317**, 789.

Mourard D., Bosc I., Labeyrie A., Koechlin L., Saha S., 1989, Nature, **342**, 520.

Mourard D., Tallon-Bosc I., Blazit A., Bonneau D., Merlin G., Morand F., Vakili F., Labeyrie A., 1994, Astron. Astrophys., **283**, 705.

Mozurkewich D., 1994, SPIE, **2200**, 76.

Mozurkewich D., 1999, in Principles of Long Baseline Stellar Interferometry, Ed. P. R. Lawson, JPL Publications, Pasadena, CA, USA, 231.

Mozurkewich D. et al., 2003, Astron. J., **126**, 2502.

Mungall A. G., Audoin C., Lesage P., 1974, IEEE Trans. Instr. Meas., **IM-23**, 501.

Murakawa K., Ohnaka K., Driebe T., Hofmann K. -H., Oya S., Schertl D., Weigelt G., 2008a, Astron. Astrophys., **489**, 195.

Murakawa K., Preibisch T., Kraus S., Ageorges N., Hofmann K. -H., Ishii M., Oya S., Rosen A., Schertl D., Weigelt G., 2008b, Astron. Astrophys., **488**, L75.

Nakajima T., 1994, Astrophys. J., **425**, 348.

Naoko O., 2003, Astron. Herald, **96**, 537.

Narayan R., Nityananda R., 1986, Ann. Rev. Astron. Astrophys., **24**, 127.

Natta A., Grinin V., Mannings V., 2000, in Protostars and Planets IV, Eds. V. Mannings, A. P. Boss, S. S. Russell, Arizona University Press, Tucson, 559.

Netolický M., Bonneau D., Chesneau O., Harmanec P., Koubský P., Mourard D., Stee P., 2009, Astron. Astrophys., **499**, 827.

Nisenson P., Papaliolios C., 1999, Astrophys. J., **518**, L29.

Nisenson P., Papaliolios C., Karovska M., Noyes R., 1987, Astrophys. J., **320**, L15.

Nishikawa J., Abe L., Murakami N., Kotani T., 2008, Astron. Astrophys., **489**, 1389.

Nityananda R., Narayan R., 1982, J. Astrophys. Astron., **3**, 419.

Noll R. J., 1976, Opt. Soc. Am. J., **66**, 207.

North J. R., Davis J., Tuthill P. G., Tango W. J., Robertson J. G., 2007a, Mon. Not. R. Astron. Soc., **380**, 1276.

North J. R., Tuthill P., Tango W., Davis J., 2007b, Mon. Not. R. Astron. Soc., **377**, 415.

Northcott M. J., Ayers G. R., Dainty J. C., 1988, J. Opt. Soc. Am. A, **5**, 986.

Nulsen P., Wood P., Gillingham P., Bessel M., Dopita M., McCowage C., 1990, Astrophys. J., **358**, 266.

Ofir A., Ribak E. N., 2006a, Mon. Not. R. Astron. Soc., **368**, 1652.

Ofir A., Ribak E. N., 2006b, SPIE, **6268**, 1181.

Ofir A., Ribak E. N., 2006c, Mon. Not. R. Astron. Soc., **368**, 1646.

Ohishi N., Nordgren T. E., Hutter D. J., 2004, Astrophys. J., **612**, 463.

Ohnaka K., Driebe1 T., Weigelt G., Wittkowski M., 2007, Astron. Astrophys., **66**, 1099.

Ohnaka K. et al., 2006, Astron. Astrophys., **445**, 1015.

Ohnaka K. et al., 2009, astro-ph/arXiv:0906.4792.

Okazaki A. T., 1997, Astron. Astrophys., **318**, 548.

Ollivier M. et al., 2001, Compt. Rend. Acad. Sci. Paris, **IV 2**, 149.

Osterbart R., Balega Y. Y., Weigelt G., Langer N., 1996, Proceedings of IAU Symposium 180, on Planetary Nabulae, Eds. H. J. Habing, G. L. M. Lamers, 362.

Osterbrock D. E., 1981, Astrophys. J., **249**, 462.

Otte A. N. et al., 2007, Proceedings of 30th ICRC, OG.2.4, Merida, Mexico.

Owocki S., 2005, The Nature and Evolution of Disks Around Hot Stars, Eds. R. Ignace, K. G. Gayley, ASP, San Francisco, 101.

Palla F., Stahler S., 2001, Astrophys. J., **553**, 299.

Pan X. -P. et al., 1998, SPIE, **3350**, 467.

Papaliolios C., Karovska M., Koechlin L., Nisenson P., Standley C., Heathcote S., 1989, Nature, **338**, 565.

Papaliolios C., Nisenson P., Ebstein, S., 1985, App. Opt., **24**, 287.

Papoulis A., 1968, Systems and Transforms with Applications in Optics, McGraw-Hill, NY.

Papoulis A., 1984, Probability, Random Variables, and Stochastic Processes, McGraw-Hill, NY.

Pearson T. J., Readhead A. C. S., 1984, Ann. Rev. Astron. Astrophys., **22**, 97.

Pease F. G., 1931, Ergebnisse der Exakten Naturwissenschaften, **10**, 84.

Pedretti E., Labeyrie A., 1999, Astron. Astrophys. Suppl., **137**, 543.

Pedretti E., Labeyrie A., Arnold L., Thureau N., Lardiére O., Boccaletti A., Riaud P., 2000, Astron. Astrophys. Suppl., **147**, 285.

Penny A. J. et al., 1998, SPIE, **3350**, 666.

Penzias A. A., Wilson R. W., 1965, Astrophys. J., **142**, 419.

Perrin G., 1997, Astron. Astrophys. Suppl., **121**, 553.

Perrin G. et al., 1998, Astron. Astrophys., **331**, 619.

Perrin G. et al., 2004, Astron. Astrophys., **418**, 675.

Perrin G., Coudé du Foresto V., Ridgway S. T., Mariotti J. -M., Traub W. A., Carleton N. P., Lacasse M. G., 1998, Astron. Astrophys., **331**, 619.

Perrin G., Coudé du Foresto V., Ridgway S., Mennesson B., Ruilier C., Marrioti J. -M., Traub W., Lacasse M., 1999, Astron. Astrophys., **345**, 221.

Perryman M. A. C., 1998, Nature, **340**, 111.

Perryman M. A. C. et al., 1997, Astron. Astrophys., **323**, L49.

Peterson D. M. et al., 2006, Nature, **440**, 896.

Petr M. G., Du Foresto V., Beckwith S. V. W., Richichi A., McCaughrean M. J., 1998, Astrophys. J., **500**, 825.

Petrov R. et al., 2003, Astrophys. Space Sci., **286**, 57.

Petrov R. et al., 2007, Astron. Astrophys., **464**, 1.

Pinna E. et al., 2008, SPIE, **7015**, 701559-1.

Planck M., 1901, Ann. d. Physik, **4**, 553.

Poncelet A., Perrin G., Sol H., 2006, Astron. Astrophys., **450**, 483.

Porter J. M., Rivinius Th., 2003, Pub. Astron. Soc. Pac., **115**, 1153.

Pott J. -U., Eckart A., Glindemann A., Ghez A., Schödel R., 2006, J. Phys. Conf. Ser., **54**, 273.

Pott J. -U., Malkan M. A., Elitzur M., Ghez M. A., Herbst T. M., Schödel R., Woillez J., 2010, Astrophys. J., **715**, 736.

Pottasch S. R., 1984, Planetary Nebulae, D. Reidel, Dordrecht.

Pottasch S. R., 1992, Astron. Astrophys. Rev., **4**, 215.

Pourbaix D., 2000, Astron. Astrophys. Suppl., **145**, 215.

Prato L., et al., 2001, Astrophys. J., **549**, 590.

Prinja R. K., 1989, Mon. Not. R. Astron. Soc., **241**, 721.

Qian S. et al., 2010a, Mon. Not. R. Astron. Soc., **401**, L34.

Qian S. et al., 2010b, Astrophys. J., **708**, L66.

Queloz D. et al., 2009, Astron. Astrophys., **506**, 303.

Quirrenbach A., 2001, Ann. Rev. Astron. Astrophys., **39**, 353.

Quirrenbach A., 2009, in Relativity in Fundamental Astronomy, Proceedings of IAU Symposium, No. 261, Eds. S. A. Klioner, P. K. Seidelman, M. H. Soffel, 277.

Quirrenbach A., Elias N. M., Mozurkewich D., Armstrong J. T., Buscher D. F., Hummel C. A., 1993b, Astron. J., **106**, 1118.

Quirrenbach A., Hummel C. A., Buscher D. F., Armstrong J. T., Mozurkewich D., Elias N. M., 1993a, Astrophys. J., **416**, L25.

Quirrenbach A. et al., 1996, Astron. Astrophys., **312**, 160.

Quirrenbach A. et al., 1998, SPIE, **3350**, 807.

Rabbia Y., Mekarnia D., Gay J., 1990, SPIE, **1341**, 172.

Ragazzoni R., 1996, J. Mod. Opt., **43**, 289.

Ragazzoni R., Diolaiti E., Vernet E., Farinato J., Marchetti E., Arcidiacono C., 2005, Pub. Astron. Soc. Pac., **117**, 860.

Raghavan D. et al., 2009, Astrophys. J., **690**, 394.

Ragland S. et al., 2006, Astrophys. J., **652**, 650.

Ragland S. et al., 2009, Astrophys. J., **703**, 22.

Ramesh R., Subramanian K. R., Sastry Ch. V., 1999, Astron. Astrophys. Suppl., **139**, 179.

Readhead A. C. S., Nakajima T. S., Pearson T. J., Neugebauer G., Oke J. B., Sargent W. L. W., 1988, Astron. J., **95**, 1278.

Readhead A. C. S., Wilkinson P. N., 1978, Astrophys. J., **223**, 25.

Readhead A. C. S., Walker R. C., Pearson T. J., Cohen M. H., 1980, Nature, **285**, 137.

Reber G., 1940, Astrophys. J., **91**, 621.

Racine R., 2005, Pub. Astron. Soc. Pac., **117**, 401.

Reid M. J., 1993, Ann. Rev. Astron. Astrophys., **31**, 345.

Reid M. J., Brunthaler A., 2004, Astrophys. J., **616**, 883.

Reinheimer T., Weigelt G., 1987, Astron. Astrophys., **176**, L17.

Reynolds C., 2008, Nature, **455**, 39.

Reynolds O., 1883, Phil. Trans. R. Soc., **174**, 935.

Riaud P., Boccaletti A., Baudrand J., Rouan D., 2003, Pub. Astron. Soc. Pac., **115**, 712.

Rhee J. H., Song I., Zuckerman B., McEwain M., 2007, Astrophys. J., **660**, 1556.

Rivera E. J. et al., 2005, Astrophys. J., **634**, 625.

Robbe S., Sorrente B., Cassaing F., Rabbia Y., Rousset G., 1997, Astron. Astrophys. Suppl. **125**, 367.

Robbe-Dubois S. et al., 2007, Astron. Astrophys., **464**, 13.

Robertson J. G., Bedding T. R., Aerts C., Waelkens C., Marson R. G., Barton J. R., 1999, Mon. Not. R. Astron. Soc., **302**, 245.

Robertson N. A., 2000, Class. Quant. Grav., **17**, 19.

Roddier F., 1981, Prog. Opt., **XIX**, 281.

Roddier F., 1988, Phys. Rep., **170**, 97.

Roddier F., 1988b, App. Opt., 27, 1223.

Roddier F., 1999, Adaptive Optics in Astronomy, Ed. F. Roddier, Cambridge University Press, Cambridge.

Roddier F., Roddier C., Graves J. E., Northcott M. J., 1995, Astrophys. J., **443**, 249.

Roggemann M. C., Welsh B. M., Fugate R. Q., 1997, Rev. Mod. Phys., **69**, 437.

Rosenberg I., 1970, Mon. Not. R. Astron. Soc., **151**, 109.

Röttgering H. J. A., 2003, New Astron. Rev., **47**, 405.

Rouan D., Riaud P., Boccaletti A., Clénet Y., Labeyrie A., 2000, Pub. Astron. Soc. Pac., **112**, 1479.

Rousselet-Perraut K., Vakili F., Mourard D., 1996, Opt. Eng., **35**, 2943.

Rousselet-Perraut K. et al., 1997, Astron. Astrophys., **123**, 173.

Roy S., Rao A. P., 2004, Mon. Not. R. Astron. Soc., **349**, L25.

Rupprecht G. et al., 2004, SPIE, **5492**, 148.

Ryle M. A., 1952, Proc. R. Soc. A, **211**, 351.

Saha S. K., 1999, Bull. Astron. Soc. Ind., **27**, 443.

Saha S. K., 2002, Rev. Mod. Phys., **74**, 551.

Saha S. K., 2007, Diffraction-limited Imaging with Large and Moderate Telescopes, World-Scientific, New Jersey.

Saha S. K., Maitra D., 2001, Ind. J. Phys., **75B**, 391.

Saha S. K., Morel S., 2000, Bull. Astron. Soc. Ind., **28**, 175.

Saha S. K., Rajamohan R., Vivekananda Rao P., Som Sunder G., Swaminathan R., Lokanadham B., 1997, Bull. Astron. Soc. Ind., **25**, 563.

Saha S. K., Sridharan R., Sankarasubramanian K., 1999b, Triple correlation of Binary stars, Presented at the ASI conference, Bangalore.

Saha S. K., Sudheendra G., Umesh Chandra A., Chinnappan V., 1999a, Exp. Astr., **9**, 39.

Saha S. K., Venkatakrishnan P., 1997, Bull. Astron. Soc. Ind., **25**, 329.

Saha S. K., Venkatakrishnan P., Jayarajan A. P., Jayavel N., 1987, Curr. Sci., **56**, 985.

Saha S. K., Yeswanth L., 2004, Asian J. Phys., **13**, 227.

Sahlmann J. et al., 2008, SPIE, **7013**.

Salaris M., Cassisi S., 2005, Evolution of stars and stellar populations, Wiley, London.

Salpeter E., 1955, Astrophys. J., **121**, 161.

Sandler D., Cueller G., Lefébvre M., Barrette T., Arnold R., Johnson P., Rego A., Smith G., Taylor G., Spivey B., 1994, J. Opt. Soc., Am. A., **11**, 858.

Sanroma M., Estellela R., 1984, Astron. Astrophys., **133**, 299.

Sato K. et al., 1998, SPIE, **3350**, 212.

Sato T., Wadaka S., Yamamoto J., Ishii J., 1978, Appl. Opt., **17**, 2047.

Schawinski K. et al., 2008, Science, **321**, 223.

Schertl D., Hofmann K. -H., Seggewiss W., Weigelt G., 1996, Astron. Astrophys., **302**, 327.

Schmitt H. R. et al., 2008, Astrophys. J., **691**, 984.

Schrödinger E., 1926, Phys. Rev., **28**, 1049.

Schwab F., 1980, SPIE, **231**, 18.

Schwarz U. J., 1978, Astron. Astrophys., **65**, 345.

Schwarzschild K., Villiger W., 1896, Astron. Nachr., **139**, 353.

Schwarzschild K., 1906, Göttingen Nachr., **195**, 41.

Schödel R. et al., 2002, Nature, **419**, 694.

Scully M. O., Zubairy M. S., 1997, Quantum Optics, Cambridge University Press, Cambridge.

Seifert F., Kwee P., Heurs M., Willke B., Danzmann K., 2006, Opt. Lett., **31**, 2000.

Serabyn E., 2000, SPIE, **4006**, 328.

Serabyn E., Bloemhof E. E., Gappinger R. O., Haguenauer P., Mennesson B., Troy M., Wallace J. K., 2006, in IAU Colloq., **200**, Direct Imaging of Exoplanets, Eds. C. Aime, F. Vakili, Cambridge University Press, Cambridge, 477.

Serbowski K., 1947, Planets, Stars, and Nabulae Studied with Photopolarimetry, Ed. T. Gehrels, University of Arizona Press, Tuscon, 135.

Serkowski K., Mathewson D. S., Ford V. L., 1975, Astrophys. J., **196**, 261.

Setiawan J., Henning Th., Launhardt R., Müller A., Weise P., Kürster M., 2008, Nature, **451**, 38.

Shaddock D. A., 2009, Pub. Astron. Soc. Austr., **26**, 128.

Shaklan S. B., Roddier F., 1987, Appl. Opt., **26**, 2159.

Shaklan S. B., Roddier F., 1988, Appl. Opt., **27**, 2334.

Shannon C. E., 1948, Bell Syst. Tech. J., **27**, 329.

Shannon C. E., 1949, Proc. IRE, **37**, 10.

Shao M., Colavita M. M., 1987, Proceedings of ESO-NOAO Workshop, Ed. J. W. Goad, 115.

Shao M., Colavita M. M., 1992, Astron. Astrophys., **262**, 353.

Shao M., Colavita M. M., Hines B. E., Hershey J. L., Hughes J. A., Hutter D. J., Kaplan G. H., Johnston K. J., Mozurkewich D., Simon R. H., Pan X. -P., 1990, Astron. J., **100**, 1701.

Shao M. et al., 1988a, Astron. Astrophys., **193**, 357.

Shao M. et al., 1988b, Astrophys. J., **327**, 905.

Shao M., Staelin D. H., 1977, J. Opt. Soc. Am., **67**, 81.

Shao M., Staelin D. H., 1980, App. Opt., **19**, 1519.

Shevgaonkar R. K., 1986, J. Astrophys. Astron., **7**, 275.

Shevgaonkar R. K., 1987, in Image Processing in Astronomy, Eds. T. Velisamy, V. R. Venugopal, 119.

Shulyak D. et al., 2008, Astron. Astrophys., **487**, 689.

Sigg D., 2006, Class. Quant. Grav., **23**, S51.

Sirothia et al., 2009, ASP Conf. Ser., **407**.

Smart W. M., 1956, Spherical Astronomy, Cambridge University Press, Cambridge.

Smith F. G., 1952, Nature, **170**, 1065.

Smith M. A. et al., 2004, Astrophys. J., **600**, 972.

Smith M. A., Robinson R. D., 1999, Astrophys. J., **517**, 866.

Soltau D., 2009, Private communication.

Southworth G. C., 1945, J. Franklin Inst., **239**, 285.

Srinivasan R., 2008, Private communication.

Stankov A., Handler G., 2005, Astron. Astrophys. Suppl., **158**, 193.

Stauffer J. R. et al., 2003, Astron. J., **126**, 833.

Stee P., 2005, 21st Century Astrophysics, Eds. S. K. Saha, V. K. Rastogi, Anita Publications, New Delhi, 149.

Stee P., de Araújo F., 1994, Astron. Astrophys., **292**, 221.

Stee P., de Araújo F., Vakili F., Mourard D., Arnold I., Bonneau D., Morand F., Tallon-Bosc I., 1995, Astron. Astrophys., **300**, 219.

Stee P., Vakili F., Bonneau D., Mourard D., 1998, Astron. Astrophys., **332**, 268.

Stéphan M., 1874, Compt. Rend. Acad. Sci. Paris, **78**, 1008.

Sterzik M. F., Tokovinin A. A., 2002, Astron. Astrophys., **384**, 1030.

Stokes G. G., 1852, Trans. Camb. Phil. Soc., **9**, 399.

Straizys V., Kuriliene G., 1981, Astrophys. Space Sci., **80**, 353.

Strehl K., 1902, Zeitschrift für Instrumentenkunde, **22**, 213.

Sudarshan E. C. G., 1963, Phys. Rev. Lett., **10**, 277.

Swain M. et al., 2003, Astrophys. J., **596**, L163.

Swarup G., Ananthakrishnan S., Kapahi V. K., Rao A. P., Subrahmanya C. R., Kulkarni V. K., 1991, Curr. Sci., **60**, 95.

Swinbank A. M. et al., 2010, Nature, **464**, 733.

Tacconi L. J. et al., 2008, Astrophys. J., **680**, 246.

Tai R. Z., 2000, Rev. Sci. Instrum., **71**, 1256.

Tango W. J., Booth A. J., 2000, Mon. Not. R. Astron. Soc., **318**, 387.

Tango W. J., Davis J., 2002, Mon. Not. R. Astron. Soc., **333**, 642.

Tango W. J., Twiss R. Q., 1980, Prog. Opt., **XVII**, 239.

Tanner A. A. et al., 2002, Astrophys. J., **575**, 671.

Tanner A. A., Ghez A. M., Morris M. R., Christou J. C., 2005, Astrophys. J., **624**, 724.

Tatarski V. I., 1961, Wave Propagation in a Turbulent Medium, McGraw-Hill Books, New York.

Tatarski V. I., 1993, J. Opt. Soc. Am. A., **56**, 1380.

Tatebe K., Chandler A. A., Wishnow E. H., Hale D. D. S., Townes C. H., 2007, Astrophys. J., **670**, L21.

Tatebe K., Wishnow E. H., Ryan C. S., Hale D. D. S., Griffith R. L., Townes C. H., 2008, Astrophys. J., **689**, 1289.

Tatulli E. et al., 2007, Astron. Astrophys., **464**, 29.

Tatulli E., Malbet F., Ménard F., Gil C., Testi L., Natta A., Kraus S., Stee P., Robbe-Dubois S., 2008, Astron. Astrophys., **489**, 1151.

Taylor G. L., 1921, in Turbulence, Eds. S. K. Friedlander, L. Topper, 1961, Wiley-Interscience, New York, 1.

ten Brummelaar T. A., 1999, in Principles of Long Baseline Stellar Interferometry, Ed. P. R. Lawson, JPL Publications, Pasadena, 87.

ten Brummelaar T. A. et al., 2005, Astroph. J., **628**, 453.

Thiébaut E., 2008, SPIE, **7013**, 43.

Thom C., Granès P., Vakili F., 1986, Astron. Astrophys., **165**, L13.

Thompson L. A., Gardner C. S., 1988, Nature, **328**, 229.

Thomson R. A., 1994, Synthesis Imaging in Radio Astronomy, Eds. R. A. Perley, F. R., Schwab, A. H. Bridle, Astronomical Society of the Pacific, San Francisco, 11.

Thompson R. A., Moran J. M., Swenson Jr. G. W., 2001, Interferometry and Synthesis in Radio Astronomy, Wiley, New York.

Thorne K. S., 1994, in Black Holes and Time Warps: Winstein's Outrageous Legacy, W. W. Norton & Co., NY.

Thureau N. D., Monnier J. D., Traub W. A., Millan-Gabet R., Pedretti E., Berger J.-P., Garcia M. R., Schloerb F. P., Tannirkulam A. -K., 2009, Mon. Not. R. Astron. Soc., **398**, 1309.

Torres G., Winn J. N., Holman M. J., 2008, Astrophys. J., **677**, 1324.

Townes C. H., 1984, J. Astrophys. Astron., **5**, 111.

Townes C. H., 1999, in Principles of Long Baseline Stellar Interferometry, Ed. P. R. Lawson, JPL Publications, Pasadena, 59.

Townes C. H., Bester M., Danchi W., Hale D., Monnier J., Lipman E., Everett A., Tuthill P., Johnson M., Walters D., 1998, SPIE, **3350**, 908.

Townes C. H., Wishnow E. H., Hale D. S., Walp B., 2009, Astrophys. J., **697**, L127.

Townsend R. H. D., Owocki S. P., Howarth I. D., 2004, Mon. Not. R. Astron. Soc., **2004**, 189.

Traub W. A., 1986, Appl. Opt., **25**, 528.

Traub W. A., 1999, in Principles of Long Baseline Stellar Interferometry, Ed. P. R. Lawson, JPL Publications, Pasadena, 31.

Traub W. A. et al., 2000, SPIE, **4006**, 715.

Traub W. A. et al., 2003, SPIE, **4838**, 45.

Tristram K. R. W. et al., 2007, Astron. Astrophys., **474**, 837.

Troxel S. E., Welsh B. M., Roggemann M. C., 1994, J. Opt. Soc. Am. A., **11**, 2100.

Tuthill P., Haniff C., Baldwin J., 1997, Mon. Not. R. Astron. Soc., **285**, 529.

Tuthill P. G., Monnier J. D., Danchi W. C., 2001, Nature, **409**, 1012.

Twiss R. W., Carter A. W. L., Little A. G., 1960, Observatory, **80**, 153.

Twiss R. W., Carter A. W. L., Little A. G., 1962, Aust. J. Phys., **15**, 378.

Tycner C. et al., 2003, Astron. J., **125**, 3378.

Tycner C., Hajian A. R., Armstrong J. T., Benson J. A., Gilbreath G. C., Hutter D. J., Lester J. B., Mozurkewich D., Pauls T. A., 2004, Astron. J., **127**, 1194.

Tycner C. et al., 2006, Astron. J., **131**, 2710.

Tyson R. K., 1991, Principles of Adaptive Optics, Academic Press, New York.

Tyson R. K., 2000, 'Introduction' in Adaptive optics engineering handbook, Ed. R. K. Tyson, Dekkar, NY, 1.

Udry S. et al., 2007, Astron. Astrophys., **469**, L43.

Unwin S., 2005, ASP Conf. Ser., **338**, 37.

Unwin S. et al., 2008, Pub. Astron. Soc. Pac., **120**, 38.

Vakili F., Mourard D., Bonneau D., Morand F., Stee P., 1997, Astron. Astrophys., **323**, 183.

Vakili F., Mourard D., Stee P., Bonneau D., Berio P., Chesneau O., Thureau N., Morand F., Labeyrie A., Tallon-Bosc A., 1998, Astron. Astrophys., **335**, 261.

Valley G. C., 1980, Appl. Opt., **19**, 574.

van Albada G. B., 1956, Contr. Bosscha Obs. No. 3.

van Albada G. B., 1962, J. Obs. Marseille, **45**, 1.

van Albada G. B., 1971, Astron. Astrophys., **11**, 317.

van Belle G. et al., 1999, Astron. J., **117**, 521.

van Belle G. T. et al., 2006, Astrophys. J., **637**, 494.

van Belle G. T., 2008, Olbin communication.

van Boekel R., Kervella P., Schöller M., Herbst T., Brandner W., de Koter A., Waters L. B. F. M., Hillier D. J., Paresce F., Lenzen R., Lagrange A. -M., 2003, Astron. Astrophys., **410**, L37.

van Brug H. et al., SPIE, 2003, **4838**, 425.

van Cittert P. H., 1934, Physica, **1**, 201.

Van de Hulst H. C., 1957, Light Scattering by Small Particles, Wiley, NY.

Verhost T., van Aarle E., Acke B., 2007, Astron. Astrophys., **470**, L21.

Vernin J., Roddier F., 1973, J. Opt. Soc. Am., **63**, 3.

von Zeipel H., 1924, Mon. Not. R. Astron. Soc., **84**, 665.

Walkup J. F., Goodman J. W., 1973, J. Opt. Soc. Am., **63**, 399.

Wallace J. K. et al., 1998, SPIE, **3350**, 864.

Wang L. et al., 2002, Astrophys. J., **579**, 590.

Wang Z., Chakrabarty D., Kaplan D. L., 2006, Nature, **440**, 772.

Ward M. J., Blanco P. R., Wilson A. S., Nishida M., 1991, Astrophys. J., **382**, 115.

Warner B., 2008, in Classical Novae, Eds. M. F. Bode, A. Evans, 16.

Weekes T. C., 2003, Very High Energy Gamma-Ray Astronomy, Institute of Physics Publishing, Bristol.

Wehinger P. A., 2004, Private communication.

Weigelt G., Baier G., 1985, Astron. Astrophys. **150**, L18.

Weigelt G., Balega Y., Blöcker T., Fleischer A. J., Osterbart R., Winters J. M., 1998, Astron. Astrophys., **333**, L51.

Weigelt G., Balega Y. Y., Blöcker T., Hofmann K. -H., Men'shchikov A. B., Winters J. M., 2002, Astron. Astrophys. **392**, 131.

Weigelt G., Balega Y., Preibisch T., Schertl D., Schöller M., Zinnecker H., 1999, astro-ph/9906233.

Weigelt G. et al., 2004, Astron. Astrophys., **425**, 77.

Weigelt G. et al., 2007, Astron. Astrophys., **464**, 87.

Weiler K. W., 1973, Astron. Astrophys., **26**, 403.

Weiner J. et al., 2006, Astrophys. J., **636**, 1067.

Wernecke S. J., D'Addario L. R., 1976, IEEE Trans. Comput., **C26**, 351.

Wieringa M., 1991, Radio Interferometry: Theory, Techniques, and Applications, IAU Colloquium 131, **19**, 192.

Wilkin F. P., Akeson R. L., 2003, Astrophys. Space Sci., **286**, 145.

Willson L. A., 2000, Ann. Rev. Astron. Astrophys., **38**, 573.

Wilson R. E., Liou J. -C., 1993, Astrophys. J., **413**, 670.

Wilson R. W., Dhillon V. S., Haniff C. A., 1997, Mon. Not. R. Astron. Soc., **291**, 819.

Wisniewski J. P. et al., 2007, Astrophys. J., **671**, 2040.

Wishnow E. H., Townes C. H., Walp B., Lockwood S., 2010, Astrophys. J., **712**, L135.

Wittkowski M., Boboltz D. A., Ohnaka K., Driebe T., Scholz M., 2007, Astron. Astrophys., **470**, 191.

Wittkowski M., Aufdenberg J., Kervella P., 2004, Astron. Astrophys., **413**, 711.

Wolf E., 1960, Proc. Phys. Soc., **76**, 424.

Wolter U., Robrade J., Schmitt J. H. M. M., Ness J. U., 2008, Astron. Astrophys., **478**, 11.

Wolszczan A., Frail D. A., 1992, Nature, **355**, 145.

Wood K., Bjorkman K. S., Bjorkman J. E., 1997, Astrophys. J., **477**, 926.

Wood P. R., Nulsen P. E. J., Gillingham P. R., Bessel M. S., Dopita M. A., McCowage C., 1989, Astrophys. J., **339**, 1073.

Worden S. P., Lynds C. R., Harvey J. W., 1976, J. Opt. Soc. Am., **66**, 1243.

Wyatt M. C., Dent W. R. F., Greaves J. S., 2003, Mon. Not. R. Astron. Soc., **342**, 876.

Xiang L., Dallacasa D., Cassaro P., Jiang D., Reynolds C., 2005, Astron. Astrophys., **434**, 123.

Xiang L., Reynolds C., Strom R., Dallacasa D., 2008, Astron. Astrophys., **454**, 729.

Yang L., McNulty I., Gluskin E., 1994, Rev. Sci. Instrum., **66**, 2281.

Yasuda M., Shimizu F., 1996, Phys. Rev. Lett., **77**, 3090.

Young J. S., Baldwin J. E., Boysen R. C., Haniff C. A., Lawson P. R., Mackay C. D., Pearson D., Rogers J., St. Jacques D., Warner P. J., Wilson D. M. A., Wilson R. W., 2000, Mon. Not. R. Astron. Soc., **315**, 635.

Young J. S., 2004, UK National Astronomy Meeting, 29 March-2 April 2004, Milton Keynes.

Yusef-Zadeh F., Roberts D. A., Biretta J., 1998, Astrophys. J., **499**, L159.

Zavala R. T., Hummel C. A., Boboltz D. A., Ojha R., Shaffer D. B., Tycner C., Richards M. T., Hutter D. J., 2010, Astrophys. J., **715**, L44.

Zernike F., 1938, Physica, **5**, 785.

Zhao et al., 2008, Astrophys. J., **684**, L95.

Ziad A. et al., 2004, Appl. Opt., **43**, 2316.

Zijlstra A. A., 1995, in Circumstellar Matter 1994, Eds. G.D. Watt, P. M. Williams, Kluwer, The Netherlands, 309.

Zuckerman B., Song I., 2004, Ann. Rev. Astron. Astrophys., **42**, 685.

Index